Lecture Notes in Physics

Springer-Verlag Berlin Heidelberg GmbH

The Editorial Policy for Proceedings

The series Lecture Notes in Physics reports new developments in physical research and teaching – quickly, informally, and at a high level. The proceedings to be considered for publication in this series should be limited to only a few areas of research, and these should be closely related to each other. The contributions should be of a high standard and should avoid lengthy redraftings of papers already published or about to be published elsewhere. As a whole, the proceedings should aim for a balanced presentation of the theme of the conference including a description of the techniques used and enough motivation for a broad readership. It should not be assumed that the published proceedings must reflect the conference in its entirety. (A listing or abstracts of papers presented at the meeting but not included in the proceedings could be added as an appendix.)
When applying for publication in the series Lecture Notes in Physics the volume's editor(s) should submit sufficient material to enable the series editors and their referees to make a fairly accurate evaluation (e.g. a complete list of speakers and titles of papers to be presented and abstracts). If, based on this information, the proceedings are (tentatively) accepted, the volume's editor(s), whose name(s) will appear on the title pages, should select the papers suitable for publication and have them refereed (as for a journal) when appropriate. As a rule discussions will not be accepted. The series editors and Springer-Verlag will normally not interfere with the detailed editing except in fairly obvious cases or on technical matters.
Final acceptance is expressed by the series editor in charge, in consultation with Springer-Verlag only after receiving the complete manuscript. It might help to send a copy of the authors' manuscripts in advance to the editor in charge to discuss possible revisions with him. As a general rule, the series editor will confirm his tentative acceptance if the final manuscript corresponds to the original concept discussed, if the quality of the contribution meets the requirements of the series, and if the final size of the manuscript does not greatly exceed the number of pages originally agreed upon. The manuscript should be forwarded to Springer-Verlag shortly after the meeting. In cases of extreme delay (more than six months after the conference) the series editors will check once more the timeliness of the papers. Therefore, the volume's editor(s) should establish strict deadlines, or collect the articles during the conference and have them revised on the spot. If a delay is unavoidable, one should encourage the authors to update their contributions if appropriate. The editors of proceedings are strongly advised to inform contributors about these points at an early stage.
The final manuscript should contain a table of contents and an informative introduction accessible also to readers not particularly familiar with the topic of the conference. The contributions should be in English. The volume's editor(s) should check the contributions for the correct use of language. At Springer-Verlag only the prefaces will be checked by a copy-editor for language and style. Grave linguistic or technical shortcomings may lead to the rejection of contributions by the series editors. A conference report should not exceed a total of 500 pages. Keeping the size within this bound should be achieved by a stricter selection of articles and not by imposing an upper limit to the length of the individual papers. Editors receive jointly 30 complimentary copies of their book. They are entitled to purchase further copies of their book at a reduced rate. As a rule no reprints of individual contributions can be supplied. No royalty is paid on Lecture Notes in Physics volumes. Commitment to publish is made by letter of interest rather than by signing a formal contract. Springer-Verlag secures the copyright for each volume.

The Production Process

The books are hardbound, and the publisher will select quality paper appropriate to the needs of the author(s). Publication time is about ten weeks. More than twenty years of experience guarantee authors the best possible service. To reach the goal of rapid publication at a low price the technique of photographic reproduction from a camera-ready manuscript was chosen. This process shifts the main responsibility for the technical quality considerably from the publisher to the authors. We therefore urge all authors and editors of proceedings to observe very carefully the essentials for the preparation of camera-ready manuscripts, which we will supply on request. This applies especially to the quality of figures and halftones submitted for publication. In addition, it might be useful to look at some of the volumes already published. As a special service, we offer free of charge LaTeX and TeX macro packages to format the text according to Springer-Verlag's quality requirements. We strongly recommend that you make use of this offer, since the result will be a book of considerably improved technical quality. To avoid mistakes and time-consuming correspondence during the production period the conference editors should request special instructions from the publisher well before the beginning of the conference. Manuscripts not meeting the technical standard of the series will have to be returned for improvement.

For further information please contact Springer-Verlag, Physics Editorial Department II, Tiergartenstrasse 17, D-69121 Heidelberg, Germany

Jochen Greiner Hilmar W. Duerbeck
Roald E. Gershberg (Eds.)

Flares and Flashes

Proceedings of IAU Colloquium No. 151
Held in Sonneberg, Germany,
5–9 December 1994

Springer

Editors

Jochen Greiner
Max-Planck-Institut für Extraterrestrische Physik
D-85740 Garching, Germany

Hilmar W. Duerbeck
Astronomisches Institut der Universität Münster
Wilhelm-Klemm-Strasse 10
D-48149 Münster, Germany

Roald E. Gershberg
Crimean Astrophysical Observatory
Nauchny, 19–17, Crimea, 334413, Ukraine

Scientific Organising Committee:
H.W. Duerbeck (Germany), G.J. Fishman (USA), R.E. Gershberg (Ukraine, chair), J. Greiner (Germany), M. Kato (Japan), A. King (UK), J.-P. Lasota (France), M. Rodonò (Italy), J.I. Smak (Poland), M.K. Tsvetkov (Bulgaria)

Local Organising Committee:
H.-J. Bräuer, B. Fuhrmann, P. Kroll, R. Luthardt (chair), S. Rößiger

Conference Secretaries:
A. Wicklein (Sonneberg), B. Ott (Münster)

Library of Congress Cataloging-in-Publication Data

IAU Colloquium (151 : 1994 : Sonneberg, Thuringia, Germany)
Flares and flashes : proceedings of the IAU Colloquium 151, held in Sonneberg, Germany, 5-9 December 1994 / Jochen Greiner, Hilmar W. Duerbeck, Roald Gershberg, eds.
p. cm. -- (Lecture notes in physics ; vol. 454)
Includes bibliographical references and index.
ISBN 978-3-662-14008-6 ISBN 978-3-540-49420-1 (eBook)
DOI 10.1007/978-3-540-49420-1
1. Solar flares--Congresses. 2. Stars--Congresses. 3. Cool stars--Congresses. 4. Astrophysics--Congresses. I. Greiner, Jochen, 1959- . II. Duerbeck, Hilmar W., 1948- . III. Gershberg, Roald, 1933- . IV. Title. V. Series: Lecture notes in physics ; 454.
QB526.F6I28 1994
523.7'5--dc20 95-21404
CIP

ISBN 978-3-662-14008-6

Originally published by Springer-Verlag Berlin Heidelberg New York in 1995
Softcover reprint of the hardcover 1st edition 1995

Typesetting: Camera-ready by the editors
SPIN: 10493051 55/3142-543210 - Printed on acid-free paper

Preface

The 151st Colloquium of the International Astronomical Union (IAU) was held in Sonneberg, Germany, from the 5th to the 9th December, 1994. With this Colloquium, the IAU has honoured G.A. Richter and W. Wenzel, two well-known Sonneberg astronomers. Both had their 65th birthdays in autumn 1994. This volume is dedicated to them for their outstanding contributions in the field of variable star research.

At the time of the Colloquium, Sonneberg Observatory was going to be closed within three weeks, and we originally had the hope that hosting a high-ranked international conference in Sonneberg would help in the fight for survival against the Thuringian Ministry for Science and Culture (as the main responsible authority). At the moment, this hope seems to have been fulfilled only partially.

This Colloquium was unique in trying to combine many different fields of current research under the aspect of rapid variability of astrophysical sources from radio up to gamma-ray wavelengths. This experiment proved to be very successful. Many discussions between the different groups and on the use of data collected in dedicated projects for different research projects took place. The principal focus of the Colloquium was the presentation and discussion of observational results, as well as their theoretical implications. The printed questions and answers and the edited roundtable discussion mirror only very imperfectly the lively discussions.

In these proceedings we attempted to group together similar issues sorted according to wavelength. However, many papers could have fit equally well in more than one section. The order in which the papers appear in these Proceedings is close, but not identical, to the order in which these talks were presented at the Colloquium. In addition to the papers printed in this volume we had the following additional presentations: by C. Kouveliotou on the *Temporal Properties of Gamma-Ray Bursts*, by F. Mahmoud on a *Renewed Brightness Decrease of the Flare Star V774 Her*, by F. Mirabel on *Relativistic Ejections from Gamma-Ray Transients*, by W.S. Paciesas on the *Monitoring of High-Energy Transients with BATSE*, by G.R. Ricker on *HETE*, and by R. Sunyaev on *Galactic Center Monitoring with Granat*.

Many persons and institutions helped to make the workshop a success.

This Colloquium would not have been possible without the dedication and vigor of the three most important members of the Local Organizing Committee, Angelika Wicklein and Bernd Fuhrmann from Sonneberg Observatory, and Beatrix Ott from the Astronomical Institute in Münster. We would like to thank them and the other staff members of Sonneberg Observatory very much for doing a very demanding job with great competence and providing at the same time a friendly atmosphere for the Colloquium.

We were fortunate to have received support for this Colloquium from a number of sources. We are very grateful to the Deutsche Forschungsgemeinschaft, which not only supported the Colloquium with a substantial grant, but in addition supplied travel grants for 19 astronomers from Eastern Europe. We thank the International Astronomical Union for its spiritual and financial support, which resulted in 18 additional grants for travel support. We also thank the International Science Foundation (New York), which financed the stay of three participants from Russia, Ukraine, and Usbekistan.

We thank the City authorities of Sonneberg for their permission to use the Assembly Hall of the Rathaus for our Conference. Professor Kippenhahn, who, in the 1940s, as a teenager gained his first exposure to astronomy at Sonneberg Observatory, gave a popular evening lecture on *The Life of the Stars* for the people of the city. In the free time, excursions to the cities of Coburg (visit to the Veste, organ concert in the St. Moriz church) and Lauscha (visit to the Glass Museum and a glassblower's workshop), and to the Sonneberg Toy Museum and Sonneberg Observatory took place.

We thank our colleagues A. Bruch, P.B. Byrne, D.H. Hartmann, R. Pallavicini, J. van Paradijs, and W.C. Seitter who served, on short notice, as 'referees' for some of the contributed papers.

Finally, we thank the Max-Planck-Institut für Extraterrestrische Physik and the Astronomisches Institut der Universität Münster for the excellent logistical support they supplied for the Organizers of this Colloquium.

The IAU Colloquium 151 presented the ideal opportunity to honour G.A. Richter and W. Wenzel, to whom this volume is gratefully dedicated.

München/Münster/Crimea
March 1995

J. Greiner
H.W. Duerbeck
R.E. Gershberg

Contents

Introductory Overviews

Flares in Late-type Stars: Radio and Optical

Flares in Late-type Stars: UV and X-ray

Theory of Flares

Flares in T Tauri Stars

Cataclysmic Variables: Eruptions and Flickering

High-Energy Transients

Gamma-ray Bursts

All–Sky Monitoring

Flares and Flashes: Past and Future

Appendix

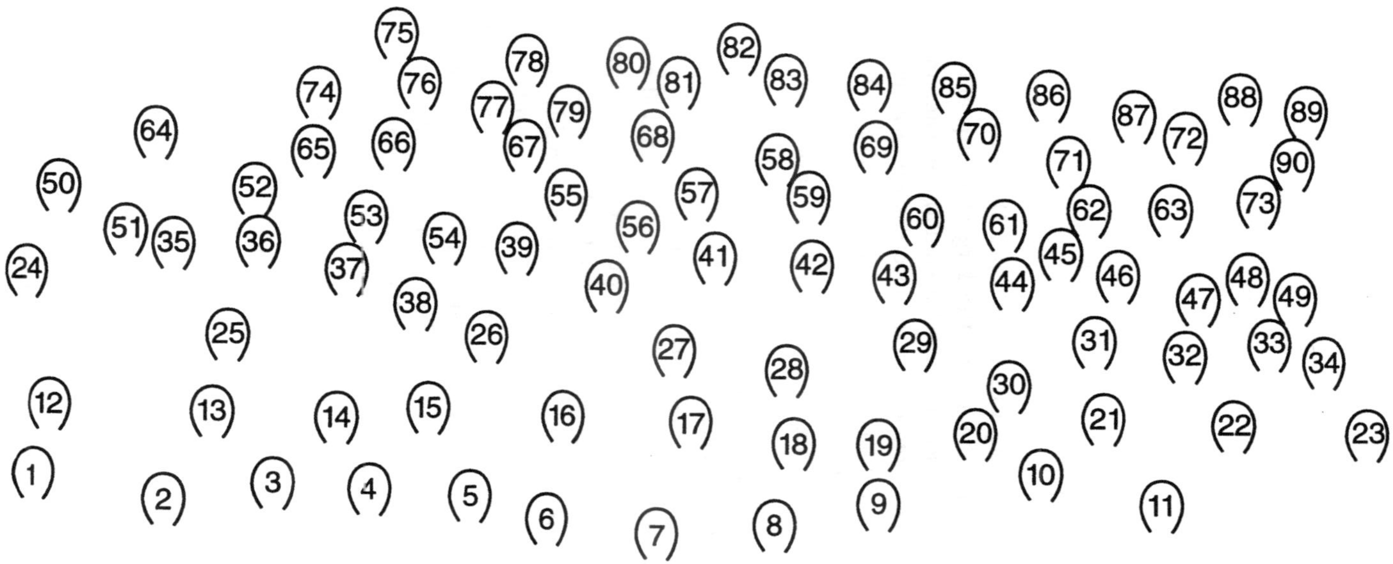

1. Welsh
2. Neizvestny
3. Luthardt
4. Fuhrmann
5. Ott
6. Jäger
7. Abada-Simon
8. Seitter
9. Richter
10. Shakhovskaya
11. Özkan
12. Bräuer
13. Notni
14. Hojaev
15. Yudin
16. Kurtanidze
17. Czepura
18. Luthardt-Menchenkova
19. Markina
20. Karetnikov
21. Pustilnik
22. Zhilayev
23. Verlyuk
24. Zinnecker
25. Benz
26. Glukhov
27. Semkov
28. Konstantinova-Antova
29. Roizman
30. Katsova
31. Shapiro
32. Livshits
33. Winterberg
34. Vogt
35. van Paradijs
36. Dravins
37. Trigilio
38. Ventura
39. Pallavicini
40. Franciosini
41. Minarini
42. Stegert
43. Gershberg
44. Beskin
45. Andronov
46. Byrne
47. Sanwal
48. Mirzoyan
49. Tsvetkova
50. van der Klis
51. Kouveliotou
52. Kroll
53. Pagano
54. Umana
55. Bruch
56. Bowyer
57. Ball
58. Ottmann
59. Stepanov
60. Pan
61. Pustylnik
62. Sokolov
63. Tsvetkov
64. Gullbring
65. Hazen
66. McHenry
67. Hildebrandt
68. Hudec
69. Preibisch
70. Reimer
71. Wicklein
72. Schwarz
73. Wang
74. Larsson
75. Ikhsanov
76. Castro-Tirado
77. Samus
78. Beskrovnaya
79. Torricelli
80. Kato
81. Casperson
82. Lapidus
83. Brandner
84. Paul
85. Aigner
86. Duerbeck
87. Poveda
88. la Dous
89. Greiner
90. Rößiger

Opening Remarks

J. Greiner

Dear participants of IAU Colloquium 151,

On behalf of the Scientific Organizing Committee and the Local Organizing Committee I welcome you in Sonneberg. For the largest group among us with English as non-native language I have prepared a similar sentence in Russian: От имени научного и локального организационного комитета я сердечно приветствую вас – из бывшего Советского Союза – в Соннеберге, в бывшей Германской Демократитческой Республике.

In organising this colloquium, we violated every possible basic rule: If you want to attract people then

1. you must give them a fashionable place to come to
2. you must chose a time with nice weather
3. you must chose a location which is conveniently to reach.

The consequences for us were terrible – I had to answer about 300 e-mails during the last 6 weeks, and the Local Organizing Committee has had a similar load with FAXes. The consequences for you are also terrible: if you had already a quick look into the latest version of the programme, you probably did not found much similarities to the version you got with the third announcement. But despite all these circumstances we are 110 participants from 28 countries.

The reason for choosing this bad combination of time and location is twofold:

1. We are honouring with this Colloquium G.A. Richter and W. Wenzel, two well known Sonneberg astronomers. Both had their 65th birthdays a few weeks ago.
2. Sonneberg Observatory is going to be closed within three weeks, and we originally had the hope that hosting a high-ranked international conference in Sonneberg would help in the fight against the Thuringian ministry (as the main responsible authority).

We have specialists with us from many different fields of astrophysical research and working in all wavelengths from radio to GeV. Though every specialist group is probably undercritical in number (except the GRB "mafia"), we hope that the following 5 days will trigger extensive discussions between the various groups and across the wavelengths bands. With this I want to close my opening remarks. I wish you a pleasant stay in Sonneberg (regardless of the weather) and a lot of new insights, exciting ideas and suggestions for your future research.

Opening Remarks

R.E. Gershberg

Dear Colleagues:

On behalf of the Scientific Organizing Committee I should like to welcome all of you, to thank you for being here and to wish you a successful activity: interesting lectures, intriguing posters, exciting talks, pleasant meetings with old friends and new acquaintances.

I shall not attempt to tell you about the importance of stellar flares and flashes: since you are here, you know this excellently. Instead, I should like to propose an unorthodox but attractive idea.

As you know, if somebody takes photographs of distant galaxies, sometimes he or she finds a supernova explosion. In photographs of nearby galaxies one can detect more frequent nova explosions as well. However, this result is not the only unique possible view of the world. For instance, a frog sees only moving bodies and thus lives in a strange world that essentially differs from ours. Continuing this thought, let me imagine that we have an equipment to detect only time derivatives of radiation, independent on flux levels and on flux variations due to quantum statistics. In such a case the picture of the stellar world would change drastically: we should see not only supernova and nova explosions on a static background but a boiling ocean of stellar flares. They would occur due to the most numerous red dwarfs and other stars with solar-type activity, due to young T Tau stars, to old cataclysmic variables, and many others. I think that just as pictures of surrealistic painters say something about the inner state of the human soul, the proposed stellar picture from this hypothetical detector would manifest the essence of stars. So, let us study the stellar flares, the stellar souls.

Thank you.

Laudation

With this Colloquium 151, the IAU is honouring the work of the two Sonneberg astronomers G. Richter and W. Wenzel. Both have spent their entire scientific life in discovering and investigating variable stars of basically all classes. Due to the impact of their work over the last 4 decades, IAU Commission 27 was enthusiastic about our proposal for this Colloquium.

However, part of their work is also relevant to other branches in astronomy, mainly X-ray astronomy. Unfortunately, this is widely ignored outside the variable star community. Who of you, in the X-ray community, knows the following sources by their optical names: V818 Sco - the brightest soft X-ray source in the sky (Sco X-1), or V1357 Cyg (Cyg X-1), or V725 Tau (A0535+26)? There are over 70 X-ray sources which have been either discovered as variable stars before their X-ray discovery or extensively studied thereafter at Sonneberg Observatory. The most prominent example is HZ Her = Her X-1 which was discovered by Cuno Hoffmeister, the founder and first director of Sonneberg Observatory, in 1936 and then was extensively studied here. Only in 1972, Liller identified the *Uhuru* source Her X-1 with the variable HZ Her. Further examples are AM Her, TT Ari, MV Lyr, GK Per, V616 Mon, X Per etc. This is another reason for getting considerable support from IAU Commission 42 and 44.

Already this short listing of sources makes clear that we are meeting here at a historical place in honouring G.A. Richter and W. Wenzel. About 30% of all known variable stars have been discovered at Sonneberg Observatory, most of these by C. Hoffmeister. It was Hoffmeister's activity and the open-minded, scientific atmosphere at Sonneberg in the early fifties which formed two young astronomers - and not only these two - for their career to follow.

1 Wolfgang Wenzel

W. Wenzel worked at Sonneberg Observatory in 1948 for one year before he graduated at the University of Jena with his diploma in astronomy in 1954. In the same year he joined the scientific staff of the Observatory, thus having had his 40th anniversary of working for the Observatory earlier this year. During these early times he introduced modern photoelectric photometry in Sonneberg and began the systematic photoelectric monitoring of variable stars.

After two years of observing campaigns at Heidelberg Observatory between 1957–1959 he (together with W. Götz) used the Sonneberg Schmidt telescope with an objective prism to determine the spectral classes of hundreds of variables. In 1961 he earned his doctoral degree on the properties of irregular and low-luminosity variables.

He strongly supported simultaneous photoelectric and spectroscopic observations of variables, in particular after the installation of the 2m Schmidt telescope in Tautenburg. Due to his early success, in 1967 he officially became director of the department "Variable stars" of Sonneberg Observatory which was part of the Central Institute for Astrophysics in Potsdam at that time. However, this year also saw his last opportunity for a business travel to a non-socialist country, and also the political pressure on the founder and director of Sonneberg Observatory, Cuno Hoffmeister, grew more and more severe.

In the mid seventies W. Wenzel initiated first experiments on infrared observations of T Tauri stars. Despite several years of intensive instrument developments this research branch was stopped before routine operations abroad could start. Many of you probably know that W. Wenzel acted as vice president of commission 27 of the IAU between 1973–1976. However, when he was nominated for president in 1976, the directorate of the Central Institute for Astrophysics Potsdam forced him to decline. This political decision clearly marks a break in his international reputation in that it marks the beginning of avoiding abroad business travels at all.

Nevertheless, he remained an active astronomer all the time. He initiated one of the first multi-wavelength campaigns, today a nearly normal business for many of us. Without e-mail or FAX they succeeded in simultaneous optical observations at Sonneberg Observatory and several other optical observatories and EXOSAT X-ray observations of TT Ari. In general, long-term photometry and statistical investigations of eruptive variables became the main topic of W. Wenzel in the eigthies.

In addition, he developed a strong interest in all new phenomena which could be targets for research at Sonneberg. One of the many examples are gamma-ray bursts (GRB). When I came to Sonneberg in February 1985 (nearly 10 years ago) to start searching for optical counterparts of GRBs, W. Wenzel had already investigated 5 GRB error boxes on about 3000 photographic plates. His enthusiasm and open minded character then started my personal collaboration with him. Recalling the many hours we spent together discussing scientific problems, I never will forget his ability of ironically questioning basic assumptions ("Oh, you know that there is a black hole in the Nova Muscae 1991 system?") and to stimulate new developments.

On the political side things didn't develop as smoothly. Because of his phantasy in finding ways to permanently by-pass official orders from the Central Astrophysical Institute in Potsdam he made enemies everywhere. It is one of the tragedies in his career that he always had to live with a security director which had the same authority and who even at the Observatory torpedoed many plans for enhanced independence from Potsdam, for more flexibility and improved

Gerold A. Richter

Wolfgang Wenzel

boundary conditions for scientific research. As a consequence of his strict character not to support any attempts to constrain the research conditions, W. Wenzel was forced to resign as a leader of the "Variable star" department in 1986. Officially he had to announce this fact as due to health problems. Given the fact that the responsible people are, at least partly, still in office, it is certainly time for rehabilitation.

Summarizing, W. Wenzel made a major contribution in the classification of variable stars, investigated the temporal behaviour of a huge number of variables and could clarify many of the interesting and widely known cases such as HZ Her, FG Sge, BO Cep etc. On a more local scale, he coordinated and headed a good deal of the scientific work at Sonneberg Observatory for more than 20 years. He also headed the bibliographical catalog of variable stars which now is a major contribution to the CDS Strasbourg. And last but not least, his willingness to educate and collaborate with amateurs has made Sonneberg Observatory also to a Mecca for (East-)German amateurs and guaranteed a lively use of the world's second largest plate collection by them.

2 Gerold Richter

G.A. Richter studied physics at Leipzig and Jena University and started his astronomical career at Sonneberg Observatory in early 1957. He also spent some observing time in Heidelberg before he made the Sonneberg 40 cm astrograph his instrument. Over decades he photographically monitored the northern sky in the long-term project of the Sonneberg field patrol, deep exposures of 80 well defined fields on the sky with nowadays 400–1000 plates per field. This was the observational basis for his discoveries of several hundred new variable stars, and the investigation of the lightcurves of further hundreds of known variables for statistical purposes. This statistical analysis of the various classes of variables, such as the frequency of variables of various classes in the solar neighbourhood and in our Galaxy, the question of population membership as well as the duration of variability states, allowed him to study the general structure of our Milky Way. With this result he earned his doctoral degree in 1966.

In the following years he continued this kind of investigation with the search for and spectroscopic investigation of UV excess objects at high galactic latitudes. For this, he several times went for observations to Zelenchukskaya and Tautenburg. In addition, G.A. Richter started to put emphasis also on extragalactic objects. He optically identified and investigated point-like radio sources. His investigations of the spiral structure of M31 led to the discovery of new OB associations and isolated OB stars.

In the eighties he concentrated on the study of many eruptive binary systems, in particular dwarf novae and recurrent novae. The main target of research was the determination of the cycle length of the outbursts. The statistical reduction of the huge observational material led to a widely used relation between outburst amplitude and cycle length, and served as an empirical basis for the understanding of the physics in cataclysmic variables.

Certainly all of you know the helium flash star FG Sge which was discovered as variable star in Sonneberg by Hoffmeister. But is was G.A. Richter who recognized the exceptional importance of this star for the theory of stellar evolution.

Summarizing, G.A. Richter did a huge work in investigating several 100 variable stars of all known classes. These were the basis for the far-reaching conclusions he could draw out of the statistical investigations of more than 1000 variable stars.

3 The future

One of the consequences of the German unification process in astronomy was the expiration of the contracts of all scientific personnel in December 1991. As the result of the evaluation process by predominantly West-German authorities, Tautenburg Observatory (which is about 100 km from Sonneberg) became the Thuringian State Observatory while Sonneberg was given a three year time limit for preparing either a new boost or the well ordered closure. To make it not too easy for Sonneberg Observatory, it was freed of all but two scientists at that time, among the fired people also G.A. Richter and W. Wenzel, who both were assessed to be old enough to retire. There are many voices in the air which claim that much of this never would have happened if W. Wenzel would still have been director in 1989/1990.

Fortunately, we could get funding for two small projects within a few months which allowed them to resume scientific work in the course of 1992. Due to mainly bureaucratic constraints the topics of these projects were devoted to correlative research to the ROSAT and GRO missions. It was extremely remarkable, how enthusiastic both started working on more or less completely new fields. This is especially honourable, because both were forced to observe more or less passively how politicians and frustrated former collaborators started damaging the reputation of the Observatory.

Things became worse with time. At present - three weeks before time is up - politics seems to have won against science. The deterioration of intelligence is not confined to Thuringia, however. Prominent politicians in our federal government officially claim that "experts are making proposals, but politicians decide" (Schwaetzer). There is much truth in this sentence. According to this realization politics obviously consists of getting the right things written down, putting them aside, and acting according to the own clientele. In the present case, it seems to be even worse, developing as in the script of a usual western: *Give them a fair trial and hang them.* The verdict is established, also the perpetrator, only the corpse is still missing.

Who then is astonished that some of the persons affected regard all this as a calculated gambit. As a consequence you will not have the possibility to meet W. Wenzel. He is not in town this week.

I think that it will be according to the intention of G. Richter and W. Wenzel if I do not only close with best wishes for themselves but with best wishes for the Observatory to which they devoted their lifes. So, let us hope together – and much more we cannot do anymore – that with the new Yhuringian government which formed last week and an unbiased new minister for science, research and culture also new decisions will be possible which eventually maintain scientific research at Sonneberg Observatory and which may allow G. Richter and W. Wenzel a dignified continuation of their scientific life.

December 5, 1994
Jochen Greiner

INTRODUCTORY OVERVIEWS

Solar Flares

Eric R. Priest

Department of Mathematical and Computational Sciences
The University, St Andrews, KY16 9SS, UK

Abstract: There are many types of solar flare, but the classic type is a two-ribbon flare with three phases – a preflare phase, a rise phase and a main phase. The properties of these phases are described, together with some recent observational advances in understanding the conditions for solar flares. Such flares are thought to be caused by an eruptive MHD instability which drives reconnection and therefore energy conversion. A review is given of our current understanding of the nature of this instability and the resulting reconnection process, including a recent attempt to describe its three-dimensional nature.

1 Description of typical large flare

Perhaps the most interesting flare star in the Universe is the Sun. However, the subject of solar flares is an enormous one, which I cannot hope to cover adequately in such a short review. It has been studied at many previous conferences (e.g. Sturrock 1980, Kundu et al. 1989, Schmieder & Priest 1991, Svestka et al. 1992, Uchida et al. 1991) and has been covered in several books (e.g. Svestka 1976, Priest 1981, Tandberg-Hanssen & Emslie 1988, Somov 1992). My aim here is simply to give a simple description of a typical large flare and to attempt to answer the two key questions that are addressed by MHD, namely: how is the energy converted in such a flare and how does the eruptive instability that drives it occur?

A large flare has three phases:
(i) the *preflare phase*, which lasts for about half an hour and during which there is a slight rise in soft X-ray emission; at the same time, an active-region prominence in a complex sunspot group starts to rise slowly;
(ii) the *rise phase*, for between 5 min and 1 hr, when there is a steep rise in intensity of soft X-ray and Hα emission; here the prominence undergoes a sudden rapid eruption and two Hα ribbons form in the chromosphere; at the same time, there are impulsive high-energy effects in the form of hard X-ray spikes, impulsive EUV and microwave bursts and type III and type II radio bursts;
(iii) the *main phase*, for an hour or up to 1 or 2 days, during which the intensity declines slowly; the Hα ribbons move apart and are joined by a rising arcade of cool loops which an enormous amount of downflowing plasma (more than would

fill the entire corona); the velocity of separation and rise is at first very rapid (up to 50 km s^{-1}) but it declines to much slower values later in the event (0.5 km s^{-1}); the density and temperature are 10^{17} m^{-3} and 40×10^6 K early on, falling to 10^{16} m^{-3} and 5×10^6 K later.

It may be noted that the typical energy in a flare is a few times 10^{25} J (10^{32} ergs), and flares only occur in complex active regions: complexity seems to be a necessary pre-requisite for a flare, probably because it helps the triggering and release of stored energy (Priest 1992). However, large quiescent prominences far from active regions can also erupt and have associated soft X-ray emission (and a coronal mass ejection): here the process is magnetically very similar, but the magnetic field is much weaker and so usually the high-energy aspects are absent.

Several new features of flares have been discovered by space satellite observations from Skylab (1973 – 1974), Solar Maximum Mission (1980 – 1989) and the Japanese *YOHKOH* mission (1990 – present):
(a) The density in the corona increases by one or two orders of magnitude by the following mechanism: fast particles are accelerated and heat is generated at the flare site up in the corona and they then propagate down to the chromosphere; there they heat the plasma which expands up into the loop by a process known as *evaporation*; subsequently, the loop plasma cools by conduction and radiation and drains back down to the feet of the magnetic fieldlines (e.g. Antiochos & Sturrock 1978, Peres 1989, Canfield et al. 1991);
(b) the high-temperature flare in the main phase consists of an arcade of hot X-ray loops that are located above the cool Hα loops (e.g. Pallavacini et al. 1977);
(c) the hard X-rays have been imaged and are found to come often from the footpoints in the impulsive phase (e.g. Hoyng et al. 1981);
(d) often a pre-existing coronal streamer overlying the flare erupts as a bubble, starting in the preflare phase; it is known as a *coronal mass ejection* (e.g. Hildner 1977);
(e) γ-ray lines and continua imply that the ions are accelerated to typically 100s of MeV (sometimes several GeV) and the electrons to 100s of keV within a few seconds (Chupp 1990, Forrest & Chupp 1983);
(f) narrow-band radio spikes imply a fragmentation of the acceleration site with a timescale of 0.01 sec (Benz 1994).

Another necessary condition for a flare (in addition to complexity) is shear up in the corona, because of the associated storage of free magnetic energy in excess of potential. One indication of this is the presence of a prominence and another is the presence of photospheric shear as revealed in vector magnetograms (e.g. Hagyard 1990, Canfield et al. 1991).

2 Basic magnetic reconnection process

2.1 Two-dimensional reconnection

The key process whereby magnetic energy is converted into other forms at the core of a flare is believed to be magnetic reconnection, whose two-dimensional MHD aspects are now fairly well understood (Priest 1990, Scholer 1991).

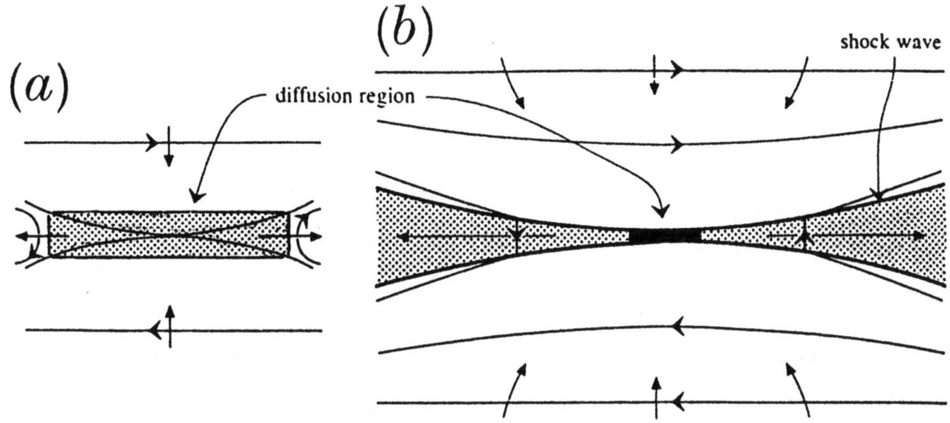

Fig. 1. The classical energy conversion mechanisms of (a) Sweet-Parker and (b) Petschek

Classical Regimes. The Sweet-Parker regime (Fig. 1) consists of a simple diffusion layer of length $2L_e$ between oppositely directed magnetic fields. The speed (v_i) with which magnetic energy is carried in and converted is called the reconnection rate and is written in dimensionless form in terms of the Alfvén speed (v_A) as

$$M_i = \frac{1}{R_{mi}^{1/2}}, \tag{1}$$

where $M = v/v_A$ and $R_m = L_e v_A/\eta$ is the magnetic Reynolds number. Since R_m is typically 10^8 or more, this gives an extremely slow reconnection rate, much too slow to explain flare energy release.

Petschek (1964), however, was able to set up a sufficiently rapid mechanism by making the diffusion region very small and having four slow-mode shock waves propagating from its ends and standing in the flow. Most of the energy release occurs at the shocks as magnetic energy is carried slowly in from large ("external") distances at a speed v_e and is converted at the shocks into the heat and kinetic energy of two hot fast streams of plasma. The maximum reconnnection rate (i.e. the dimensionless value of v_e) is now

$$M_e = \frac{\pi}{8 \log R_{me}}, \tag{2}$$

which is typically between 0.01 and 0.1.

New Generation of Fast Regimes. Now a new generation of fast reconnection regimes has been set up. First, there is an *Almost-Uniform Family* (Priest & Forbes 1986, Jardine & Priest 1990) which, like Petschek's mechanism, possesses weakly curving inflow magnetic fields (Fig. 2). The key point here is that a variety of different boundary conditions may be imposed on the inflow boundary and these produce a family of different regimes with different properties such as

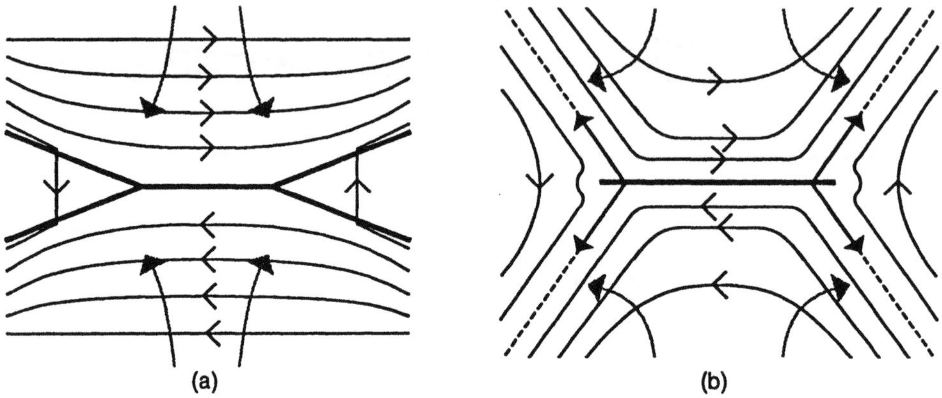

Fig. 2. Examples of members of the (a) Almost-Uniform and (b) Non Uniform regimes

sheet lengths and reconnection rates. Petschek's mechanism is one particular regime in this family.

Secondly, a *Non-Uniform Family* of regimes has been set up by Priest & Lee (1990), in which the inflow fields are now highly curved and there are jets of plasma expelled from the central current sheet along the separatrix field lines.

These two families have been used by Priest & Forbes (1992) to understand and explain results from some numerical experiments on reconnection (Biskamp 1986) and have also been carefully reproduced in other numerical experiments (Yan et al. 1992, 1993).

Reconnection in a Flare. In a large solar flare, the overall picture (Fig. 3) is that during the preflare phase a flux tube (containing a prominence) and an overlying arcade rises slowly because of some kind of eruptive instability. Then at the flare onset the rapid eruption is initiated by the beginning of reconnection in the stretched out field lines. During the main phase the reconnection continues and creates the hot flare loops and ribbons as the field closes down: the loops rise and the ribbons separate as the reconnection point rises.

This process of reconnection has been studied numerically by Forbes & Priest (1983), Forbes & Malherbe (1991) who start with open field lines which are line-tied at the base and then they watch them close down. Slow shocks develop, as in the Almost-Uniform family, and a new feature is that the stream of plasma flowing down from the reconnection site is slowed down by a fast shock where it meets a magnetic obstacle in the shape of closed loops attached to the solar surface.

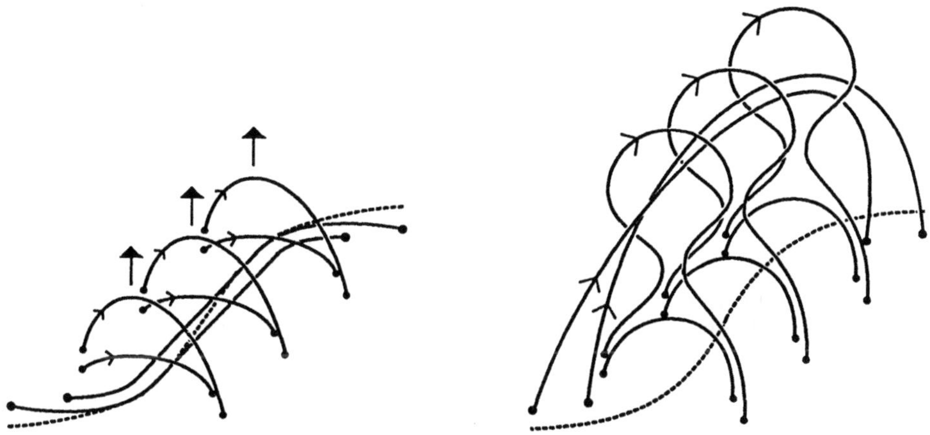

Fig. 3. The overall scenario for a large flare

Our present understanding for what is happening during the main phase (Fig. 4) is that the slow shock splits into a conduction front and an isothermal subshock. Plasma is evaporated up from the Hα ribbons to fill the hot loops with a marked cusp at their summits. The cool loops are loops which have cooled down from a previously hot state. A startling discovery from *YOHKOH* (Uchida 1993) is the presence in many flares of the cusp-shaped hot loops which had previously been suggested on theoretical grounds (e.g. Priest 1982).

2.2 Magnetic reconnection in three dimensions

Introduction. In two dimensions the global skeleton of a complex field consists of *separatrix curves* (Fig. 5) which separate the plane into topologically distinct regions in the sense that all the field lines in one region start at a particular source and end at a particular sink. The separatrices intersect in an X-type neutral point, where reconnection can take place as flux is transferred from one region to another.

In three dimensions the skeleton consists of *separatrix surfaces*, which divide the volume into topologically distinct regions (Fig. 5). They intersect in a curve known as a *separator*, which is a special field line that joins two null points.

Demoulin et al. (1993) have used the observed normal field component in the photosphere to calculate the overlying three-dimensional (potential or force-free) coronal magnetic field. They then determine the location of the separatrix surfaces and the curves in which they intersect the chromosphere (shown as dark solid curves in Fig. 6). What they find is that the resulting flare ribbons and

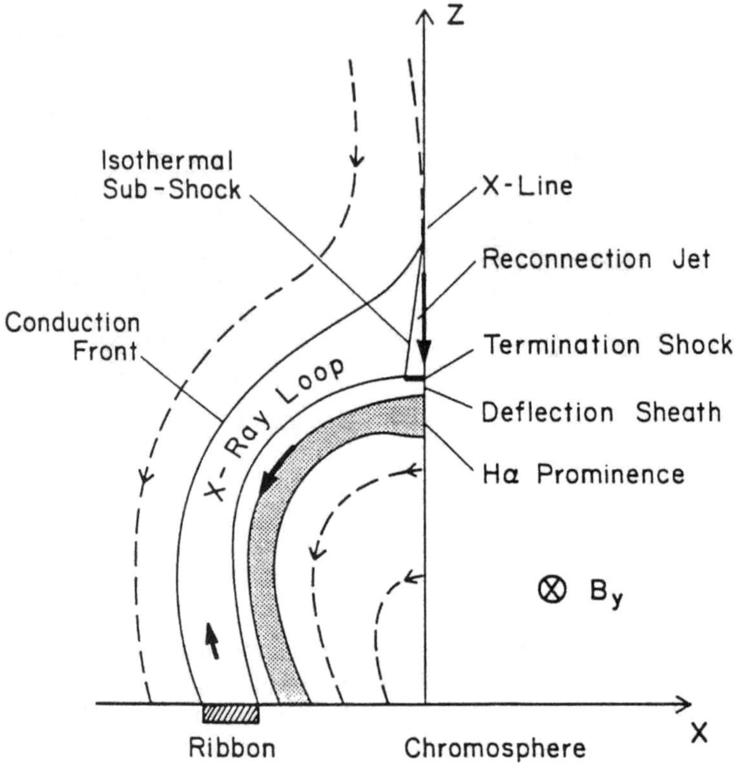

Fig. 4. Schematic of creation of flare loops and Hα ribbons

kernels are always located on or near the separatrices, which is clear confirmation of the role of reconnection at the separatrices.

It turns out that many flares posses no true null points (Demoulin et al. 1994) and no separatrices (Schindler et al. 1988), but nevertheless a theory has been developed for reconnection in such configurations at *quasi-separatrices* (Priest & Demoulin 1995).

Theory of 3D Reconnection (Priest & Titov 1995). The question then arises: how does reconnection occur at a null point in such a 3D configuration? This question is only just beginning to be tackled (Schindler et al. 1988, Lau & Finn, 1990). The simplest null point in three dimensions has field components $(B_x, B_y, B_z) = (x, y, -2z)$, so that $\nabla \cdot \mathbf{B} = 0$. There are two families of field lines through the null point at the origin (Fig. 7). The *null spine* along the z-axis is isolated, with neighbouring field lines forming two bundles around the z-axis which splay out as they approach the xy-plane. Also the *null fan* is a surface (the xy-plane) consisting of field lines which spread out from the origin.

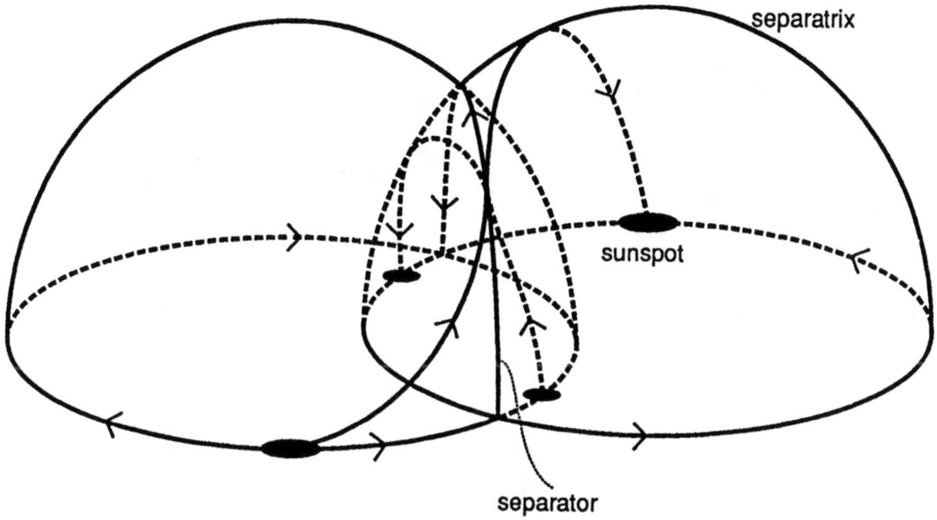

Fig. 5. The separatrix topology above four sunspots

Priest & Titov (1995) have studied the kinematic aspects of steady reconnection by solving the equations

$$\nabla \times \mathbf{E} = 0 \quad , \quad \mathbf{E} + \mathbf{v} \times \mathbf{B} = 0$$

subject to appropriate boundary conditions. They have discovered two distinct types of reconnection. In *spine reconnection* (Fig. 8) continuous footpoint motions are imposed on a surface, such as a cylinder, which encircles the spine and crosses the fan. Two flux surfaces come in, touch at the null and then reconnect at the null, unfurling from the spine like a bubble. Singular flows are driven at the spine. In *fan reconnection*, on the other hand, continuous footpoint motions are imposed on the top and bottom of the cylinder crossing the spine. Then singular swirling flows are driven at the fan.

3 Cause of eruptive behaviour

The other major question for MHD theory is: why does the prominence erupt? Why does the flare start? In theory, you may imagine this could be answered by solving the MHD equations numerically for evolution of the coronal field through a series of force-free equilibria in response to footpoint motions. Many such attempts have been made in two dimensions and a few in three dimensions, but it has proved remarkably difficult to produce an eruption.

One promising solution has been proposed by Priest & Forbes (1990), Demoulin & Priest (1988), Forbes & Isenberg (1991). They suggest that converging

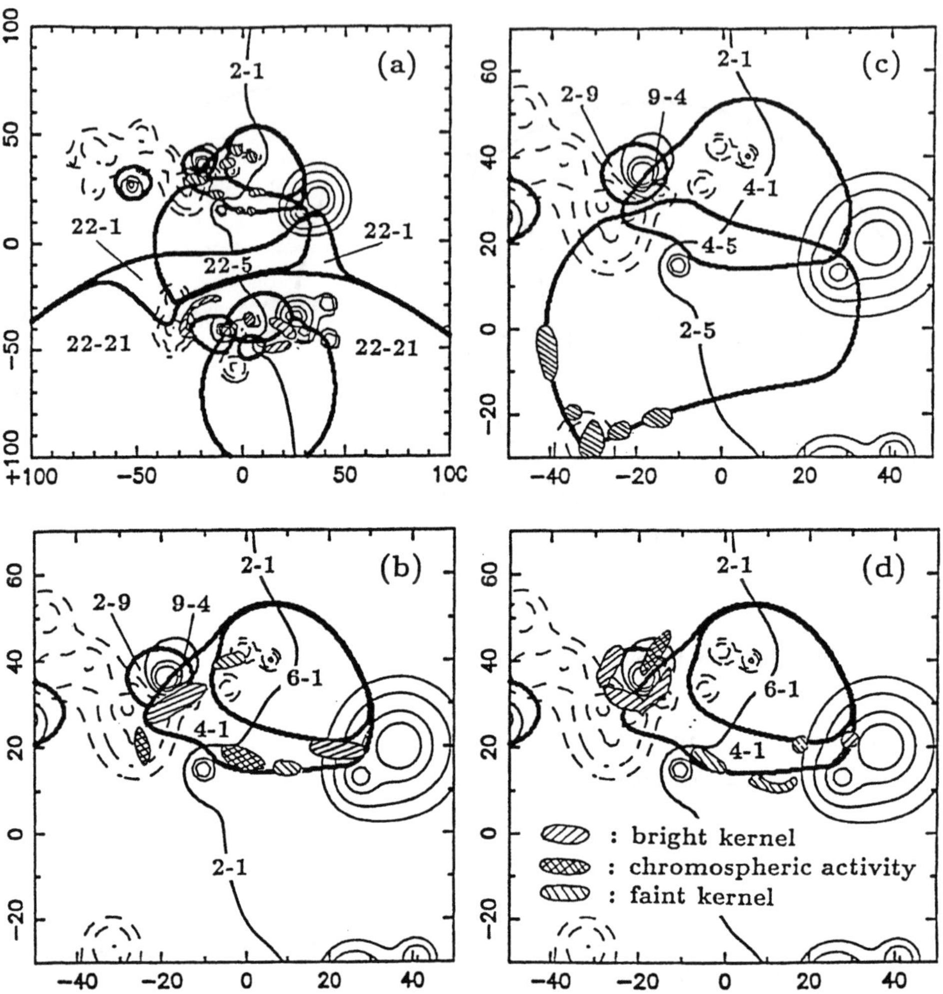

Fig. 6. Trace of the separatrices and flare features on the chromospheric plane showing (a) the general view and (b)–(d) a close-up for three flares (from Demoulin et al. 1993)

flow builds up the energy and flux in the magnetic field, so that, in a plane perpendicular to the arcade axis, a magnetic island is formed (Fig. 9), where a prominence can form. As the magnetic energy increases so the height of the prominence rises slowly. Eventually, however, a critical catastrophe point is reached, beyond which there is no neighbouring equilibrium. The imbalance in forces is upward and the prominence erupts, driving the formation of a current sheet and reconnection below the rising prominence. A simple model for this idea has been complemented by a detailed numerical experiment which elegantly explains many features of the observations.

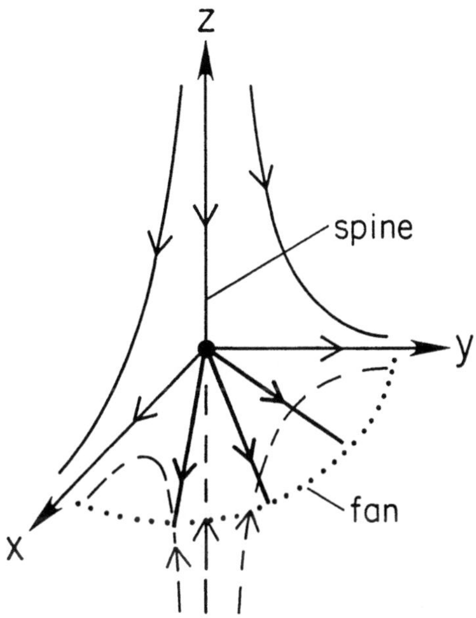

Fig. 7. The structure of a 3D null point

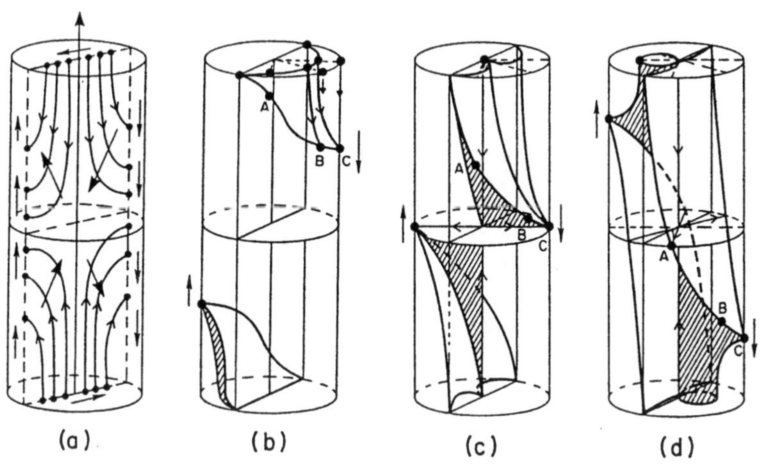

Fig. 8. Motion of field lines and flux surfaces in spine reconnection

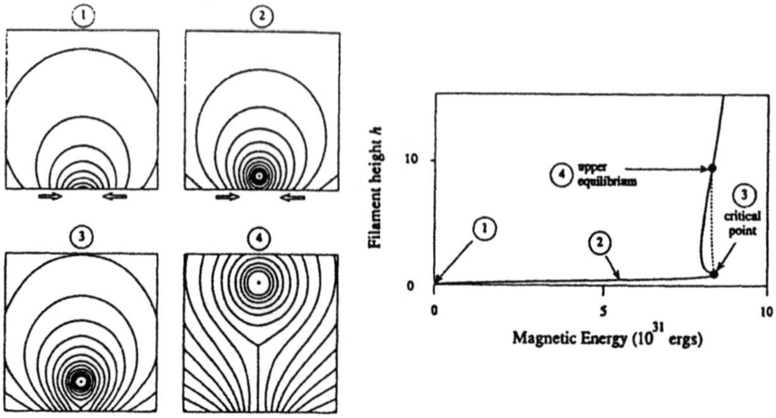

Fig. 9. A catastrophe model for prominence eruption

4 Conclusion

It is now well established that the energy release in a flare is by magnetic reconnection and it is likely that this is driven by an eruptive catastrophe. In future we expect progress from: three-dimensional understanding of reconnection; plasma theorists as they study the consequences of the MHD environment for microscopic processes such as particle acceleration; and finally comparisons with energy release processes under different parameter regimes in different parts of the universe.

References

Antiochos S.K., Sturrock P.A., 1978, ApJ 220, 1137
Benz A.O., 1994, Space Sci. Rev. 68, 135
Biskamp D., 1986, Phys. Fluids 29, 1520
Canfield R.C., Zarro D.M., Wulser J.P., Dennis B.R., 1991a, ApJ 367, 671
Canfield R.C., Beaujardiere J.F., Leka K.D., 1991b, Phil. Trans. Roy. Soc. Lond. A 336, 381
Chupp E.L., 1990, Science, 250, 229
Demoulin P., van Driel-Gesztelyi L., Schmieder B., Henoux J.C., Csepura G., Hagyard M.J., 1993, A&A 271, 292
Demoulin P., Henoux J.C., Mandrini C.H., 1994, A&A 285, 1023
Forbes T.G., Isenberg P.A., 1991, ApJ 373, 294
Forbes T.G., Priest E.R., 1983, Solar Phys. 84, 169
Forbes T.G., Malherbe J.M., 1991, Solar Phys. 135, 361
Forrest D.J., Chupp E.L., 1983, Nature 305, 291
Hagyard M.J., 1990, Mem. Soc. Astron. Ital. 61, 337
Hildner E., 1977, in Study of Travelling Interplanetary Phenomena, eds. M. Shea, D. Smart, S. Wu, D. Reidel, Dordrecht, p. 3
Hoyng P., Brown J.C., van Beek H.F., 1976, Solar Phys. 48, 197
Jardine M., Priest E.R., 1990, J. Plasma Phys. 43, 141
Kundu M., Woodgate B., Schmahl E., 1989, Energetic Phenomena on the Sun, Kluwer, Dordrecht

Lau Y.T., Finn J.M., 1990, ApJ 350, 672
Pallavacini R., Serio S., Vaiana G.S., 1977, ApJ 216, 108
Peres G., 1989, Solar Phys. 121, 289
Petschek H.E., 1964, AAS-NASA Symp. on Phys. of Solar Flares, NASA SP-50, 425
Priest E.R., 1981, Solar Flare MHD, Gordon and Breach, New York
Priest E.R., 1982, Solar Magnetohydrodynamics, Reidel, Dordrecht
Priest E.R., 1992, Proc IAU Coll. 133 on Eruptive Solar Flares, eds. Z. Svestka, B. Jackson, M. Machado, p. 15
Priest E.R., Forbes T.G., 1986, J. Geophys. Res. 91, 5579
Priest E.R., Forbes T.G., 1990, Solar Phys. 126, 319
Priest E.R., Lee L.C., 1990, J. Plasma Phys. 44, 337
Priest E.R., Forbes T.G., 1992, J. Geophys. Res. 97, 1521
Priest E.R., Demoulin P., 1995, submitted
Priest E.R., Titov V.S., 1995, submitted
Schmieder B., Priest E.R., 1991, Dynamics of Solar Flares, Obs. de Paris, Paris
Scholer M., 1991, Geophys. Astrophys. Fluid Dyn. 62, 51
Somov B., 1992, Physical Processes in Solar Flares, Kluwer, Dordrecht
Sturrock P., 1980, Solar Flares, Colo. Ass. Univ. Press
Svestka Z., 1976, Solar Flares, D. Reidel, Dordrecht
Svestka Z., Jackson B.V., Machado M.E., 1992, Eruptive Solar Flares, Springer, Berlin
Tandberg-Hanssen E., Emslie G.A., 1988, The Physics of Solar Flares, Cambridge University Press, Cambridge
Uchida Y., Canfield R.C., Watanabe T., Hiei E., 1991, Flare Physics in Solar Activity Maximum 22, Springer, Berlin
Uchida Y., 1993, in Physics of Solar and Stellar Coronae, eds. J.K. Linsky, S. Serio, Kluwer, p. 97
Yan M., Lee L.C., Priest E.R., 1992, J. Geophys. Res. 97, 8277
Yan M., Lee L.C., Priest E.R., 1993, J. Geophys. Res. 98, 7593

J. van Paradijs: What determines the direction of magnetic flux transport across a separator?

E.R. Priest: If the sources of magnetic flux (such as sunspots or photospheric magnetic fragments) move, this flux will be driven across the separatrix surfaces from one regime to another, so the direction of flux transport will depend on which sources move and the direction in which they move. Also it is possible to twist up or shear the field in each region and then for it to lose equilibrium and drive flux across the separatrices from the region where most energy is stored.

P.B. Byrne: What drives the converging flow at the footpoints of pre-flare loops?

E.R. Priest: The photospheric motions of footpoints are essentially due to convecture flows and general active region evolution. It has been observed by Sara Martin that under prominences photospheric magnetic fragments approach the prominence and cancel, thereby increasing the flux in the overlaying prominence field.

Flares: the Solar-Stellar Connection

P. Brendan Byrne

Armagh Observatory, Armagh BT61 9DG, N. Ireland

1 Introduction

The systematic study of stellar flares in now more than 30 years old. Throughout that time the solar flare paradigm has been used, highly successfully, to understand various aspects of the stellar phenomenon. In this introductory overview I will attempt to justify the use of the solar model in understanding the stellar equivalent. But I will also point out differences between the two which may require extension of the basic solar model and speculate briefly on what these extensions might be.

There are many excellent previous reviews of stellar flares in the literature (Haisch 1989, Byrne 1983, 1992), to which the reader is referred, along with the other reviews in this volume, for further reading.

1.1 In which stars do we find stellar flares?

The "classical" flare stars are the UV Cet stars, Main Sequence (MS) stars of spectral type later than ~K5Ve (the 'e' suffix indicates the Balmer Hα line in emission). Their characteristic property is rapid rotation, which, in combination with a deep convective zone, generates efficient dynamo action. As in the Sun, the resulting magnetic fields rise through the convection zone and at the photosphere they are locally intensified by surface convection into active regions, within which flares occur in stressed loop structures.

A second class of flare stars are the RS CVn stars. These are close detached binaries with one evolved component which is forced into co-rotation with the orbital period by tidal interaction. The resulting rapid rotation, in combination with a deepening convective zone, results in dynamo-generated magnetic fields which produce flares in the same general manner as in the UV Cet stars.

Flares are also found in a number of other kinds of objects which will not be discussed here, not for lack of interest, but rather lack of space. For instance, FK Com stars are isolated, extremely rapidly rotating field giants whose origin is a matter of vigorous debate. T Tau stars are pre-MS and in the final stages of contraction to the MS but the understanding of whose activity is complicated

Table 1. Representative parameters for dMe and RS CVn stars

Star	Type	Sp. Type	$R/R_\odot$	P(d)	d(pc)
V711 Tau	RS CVn	G5V/K1IVe	1.3/3.9	2.838	36
II Peg	RS CVn	K2IV/?	2.8	6.724	29
AU Mic	UV Cet	M2Ve	0.6	4.865	9.4
BY Dra	UV Cet	M0Ve/M0Ve	1.3/?	3.827	15.6

by massive winds and/or disks (see Gahm, these proceedings). Flares probably also occur on the low-mass secondaries of Cataclysmic Variables where rapid rotation is again tidally induced.

For the purpose of the present review we will discuss only the UV Cet and RS CVn stars. Representative parameters for members of these two classes will be found in Table 1.

1.2 What is a stellar flare?

At optical wavelengths a stellar flare appears as an impulsive heating resulting in a monotonic rise in light output, followed by a quasi-exponential cooling. Rise times of dMe flares can be as fast as several seconds or as slow as several tens of minutes, while they decay in times between 1 min and several hours (Fig. 1). In general, total energy and light curve complexity is correlated with time scale. Slower flares are usually more energetic and have more complicated light curves.

2 Comparison between dMe and solar flares

2.1 Optical

Optical spectroscopy of fast stellar flares shows that the rise and initial decay is dominated by broad-band continuum and high-excitation emission in such lines as He I and He II. The slower part of the decay shows strong emission in chromospheric lines, mainly H I Balmer and Ca II H&K (Bopp & Moffett 1973). This duality of time scales is also characteristic of solar flares. Continuum emission in solar flares was considered extremely rare until recently (Neidig 1989) but is now associated with the initial heating phase in large solar flares. The heating which gives rise to the observed continuum solar flares is believed to arise in the region of the temperature minimum between photosphere and chromosphere (Aboudarham & Henoux 1986). Typically the chromospheric flare, seen for example in Hα, evolves on a much longer time scale then the continuum.

Stellar flare light is very blue compared to the background photosphere. Broadband colours of flare light, after subtraction of the quiescent photospheric contribution, are far from blackbody but various models indicate plasma temperatures $\geq 10^4$ K. Lines of He I and even He II suggest that even higher temperatures are present. Furthermore, H I lines are broadened in their cores and may show extended and asymmetric wings (Doyle et al. 1988, Houdebine et al. 1993a,b). Core broadening is attributed, as in the Sun, to Stark effects resulting from heating denser layers to chromospheric temperatures. The asymmetric

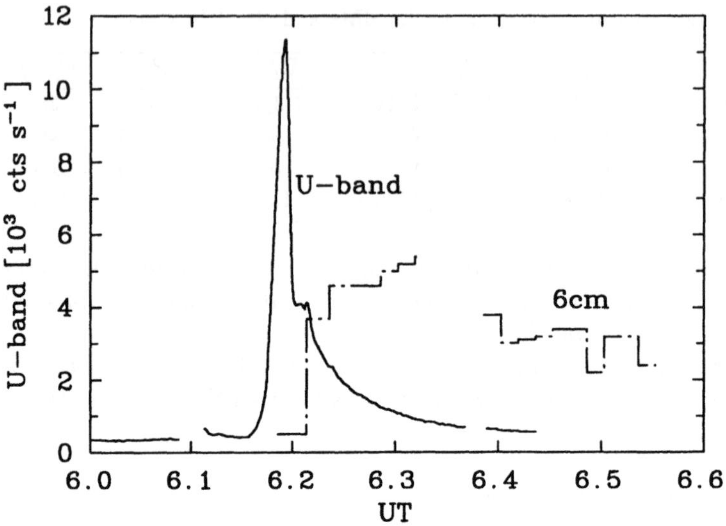

Fig. 1. A typical dMe flare U-band light curve (*continuous line*) (J.G. Doyle, private communication). Note the initially rapid decay followed by a quasi-exponential, and much slower, decay. The 6 cm microwave light curve of the same event is also shown (*broken line*).

wings cannot be explained in this way and gas motions must be assumed. Derivation of unique velocities from optically thick lines, however, must be treated with caution.

2.2 Ultraviolet

All flare stars exhibit quiescent ultraviolet emission line spectra having strong lines characteristic of the solar chromosphere (4,500 K$\leq T_e \leq$10,000 K) and low-to-mid-transition region (10,000 K $\leq T_e \leq$250,000 K) (see Byrne, these proceedinsg). Extrapolating the measured line fluxes at Earth to the star and dividing by the stellar surface area, yields a lower limit to the mean surface flux in each line. Surface fluxes calculated in this way are much higher than the same values for the Sun, even in quiescence.

During a flare all of the lines seen in the quiescent spectrum are enhanced by factors of up to 10. Generally the enhancement factor is greater the higher the excitation of the line, up to the temperature of formation of the C IV resonance lines ($\log T_e \sim 5$). Such behaviour is similar to that seen in solar flares in the same lines (Cohen et al. 1978, Feldman & Doschek 1978) and suggests a generally similar atmospheric structure in stellar and solar flare transition regions.

Stellar UV flares often exhibit strong continua (see e.g. Byrne & McKay 1989, 1990) as in the optical. The integrated energy of such flares may equal that emitted in transition region lines.

2.3 X-rays

dMe flares have mostly been observed in relatively soft X-rays (0.2–2 keV) by satellites such as *Einstein, EXOSAT* and, most recently, *ROSAT* and *ASCA*. Where some degree of spectral information is available, spectra may be fitted by 1- or 2-component thermal bremsstrahlung spectra. Maximum derived temperatures of the flaring plasma are similar to solar values and they are seen to cool as the flare progresses. Thus harder X-rays peak earlier than softer.

When simultaneous optical photometry is available, the optical broadband continuum flare is seen to have a faster evolution than the soft X-rays while the variation of the softest X-rays may be even slower than the chromospheric Balmer line component of the light evolution (see e.g. Doyle et al. 1988). This is in general agreement with the schematic behaviour of solar flares (see e.g. Simnett 1986, Fig. 1).

2.4 Energetics

dMe flares, while resembling their solar counterparts in both their time and wavelength evolution, differ fundamentally in their integrated energies. Fig. 2 shows a *cumulative frequency diagram* for optical U-band flares on a typical UV Cet flare star. This diagram plots the integrated U-band energy of flares against the frequency of a flare of that energy or greater. It will be seen that the largest flares have energies of $\geq 10^{34}$ erg in optical U-band alone. Furthermore, flares of energy equivalent to the *bolometric* energy (i.e. energy emitted in all wavebands) of the very largest solar flares, $\sim 10^{32}$ erg, occur at a rate $\sim$1–2 d^{-1}. Table 2 gives some details of the most energetic dMe broadband optical flares yet observed.

In general the total energy budget of flaring decreases towards stars of lower mass. However, the luminosity of these low-mass stars also decreases. As a result in some of these active latest type flare stars, the fraction of their bolometric luminosity emitted as flare visible light rises to ~1%, indicating a very high efficiency of convective energy to magnetic energy.

Table 2. Most energetic stellar flares observed in the optical U-band from dMe and RS CVn stars

Star	Type	Sp. Type	Date	ΔU_{max}	t_{rise}	$\log E_U$	Ref.
V1005 Ori	UV Cet	M0.5Ve	2 Oct '81	3.25	40s	$\geq$ 34.1*	Byrne et al. 1984
YY Gem	UV Cet	M0.5Ve	31 Oct '71	0.71	370s	34.1	Moffett 1974
AD Leo	UV Cet	M4.5Ve	17 Feb '74	–	–	34.0	Ichimura & Shimizu 1978
II Peg	RS CVn	K2IVe/?	17 Aug '89	0.72	43min	$\geq$ 35.3*	Doyle et al. 1992
V711 Tau	RS CVn	G5V/K1IVe	14 Dec '89	0.61	138min	34.9	Zhang et al. 1992

*This flare was still in progress when the observation was terminated by dawn.

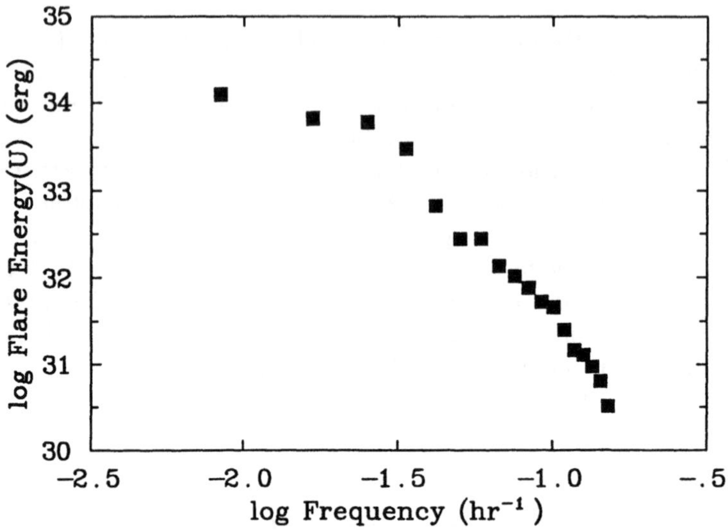

Fig. 2. Cumulative frequency diagram for U-band flares on the dMe flare star, YY Gem, prepared using data from Moffett (1974).

3 Comparison between RS CVn and solar flares

3.1 Optical

Until recently it was believed that RS CVn stars did not flare in broadband optical light. Observations made during the past 5 years, however, have shown this to be false. A small number of very large optical flares have been recorded (Table 2) from a small number of RS CVn stars (Doyle et al. 1992, Zhang et al. 1992). However, the frequency of optical flares is a matter of current debate. Two recent optical studies of the RS CVn star II Peg, appear to be in serious disagreement (Mathioudakis et al. 1992, Byrne et al. 1994). The former claims a rate of optical flaring in agreement with the rate of ultraviolet flares seen with the *IUE* satellite (see below), while the other failed to record any flares in a total observing time about 5 times the mean interval between ultraviolet flares. More recent results (Byrne et al. in prep.) appear to support the latter view, i.e. that broadband optical flares are rare events and are probably only associated with the most energetic flares.

Doyle et al. (1989) have speculated that, in view of the larger atmospheric scale height of the sub-giant active components of the RS CVn stars (compared to dMe's), particle beams, thought to be responsible for heating the chromospheres and photospheres of flares, may be absorbed too high in the atmosphere to yield optical continuum. The exception would be the very most energetic flares, which may produce particles of sufficient energy to penetrate to the upper photosphere/temperature minimum region where white light flare radiation is thought to arise (Aboudarham & Henoux 1986).

Spectroscopically, as with the dMe stars, chromospheric emission lines such as Balmer H I and Ca II H&K are seen strongly in emission in flare light. They are likewise broadened and show asymmetries, especially to the red. The interpretation of these effects is essentially the same as in the case of the dMe stars.

3.2 Ultraviolet

RS CVn stars are much better observed in the ultraviolet than is the case for dMe stars. This is because of the higher intrinsic luminosity of RS CVn stars, aiding the detection of their quiescent emissions, and the relatively greater energy and longer time scale of their flares.

The quiescent spectra of RS CVn stars exhibit the same chromospheric and transition region emission lines as in the case of the dMe stars. These emissions are enhanced during flares in a qualitatively similar way to dMe flares.

Because of their higher luminosity, even in quiescence, RS CVn stars provide sufficient signal-to-noise to be observed at high spectral resolution in the chromospheric Mg II *h&k* resonance lines. This has resulted in the detection of gross line broadening and asymmetries, similar to those seen in optical Balmer lines and usually interpreted as due to systematic mass motions within flares (Doyle et al. 1989). However, because these lines are optically thick, the derivation of velocities from such observations is again highly uncertain.

3.3 X-rays

As is the case in the ultraviolet, the higher luminosity of RS CVn stars has made it easier to observe them in X-rays, both in quiescence and while flaring. It has even been possible to carry out low resolution spectroscopy on some objects.

The general character of RS CVn flares in X-rays is similar to that of the dMe stars. However, as in other wavebands, they are longer in duration and of higher peak luminosity. It is also found that peak temperatures, measured by thermal bremsstrahlung fits to X-ray energy distributions, are considerably higher. This latter result is reinforced by the observation of line emission from very highly ionised species (e.g. Fe XXV) (Doyle et al. 1992).

3.4 Energetics

The most striking difference between solar and RS CVn flares lies in their energy. While dMe flares show total time-integrated energies up to 10^{2-3} times those of large solar flares, RS CVn flare energies are greater by a further factor 10^{2-3}. This has led some authors to consider the possibility that the solar analogue may no longer be applicable in unmodified form. Doyle et al. (1989) have considered the possibility that energy may be extracted from the orbital motions of the binary via tensioning of interconnecting loops between the components of the binary.

References

Aboudarham J., Henoux J.C. 1986, A&A 156, 73
Bopp B.W., Moffett T.J., 1973, ApJ 185, 329
Byrne P.B., 1983, in Activity in Red-Dwarf Stars, Proc. IAU Coll. No. 71, eds. P.B. Byrne & M. Rodonò, Reidel, Dordrecht, p. 157
Byrne P.B., 1989, in Solar and Stellar Flares, eds. B.M. Haisch & M. Rodonò, Solar Phys. 121, 61
Byrne P.B., 1992, in Physics of Solar and Stellar Coronae, eds. J.L. Linsky & S. Serio, Kluwer, Dordrecht, p. 489
Byrne P.B., McKay D., 1989, A&A 223, 241
Byrne P.B., McKay D., 1990, A&A 227, 490
Byrne P.B., Lanzafame A.C., Sarro L.M., Ryans R. 1994, MNRAS 270, 427
Byrne P.B., Doyle J.G., Butler C.J.., 1984, MNRAS 206, 907
Cohen L., Feldman U., Doschek G.A., 1978, ApJS 37, 393
Doyle J.G., Butler, C.J., Byrne P.B., van den Oord G.H.J., 1988, A&A 193, 229
Doyle J.G., Byrne P.B., van den Oord G.H.J., 1989, A&A 224,153
Doyle J.G., et al., 1992, A&AS 96, 351
Feldman U., Doschek G.A., 1978, ApJS 37, 443
Haisch B.M., 1989, in Solar and Stellar Flares, eds. B.M. Haisch & M. Rodonò, Solar Phys. 121, 3
Houdebine E.R., Foing B.H., Doyle J.G., Rodonò M. 1993a, A&A 274, 245
Houdebine E.R., Foing B.H., Doyle J.G., Rodonò M. 1993b, A&A 278, 109
Ichimura K., Shimizu Y., 1978, Tokyo Astr. Obs. Bull. 2nd Ser. 255, 2929
Moffett T.J., 1974, ApJS 29, 1
Mathioudakis M., Doyle J.G., Avgoloupis S., Mavridis L.N., Seiradakis J.H., 1992, MNRAS 255, 48
Neidig D., 1989, in Solar and Stellar Flares, eds. B.M. Haisch & M. Rodonò, Solar Phys. 121, 261
Simnett G.M., 1986, in Flares: Solar and Stellar, ed. P.M. Gondhalekar, Rutherford Appleton Labs.
Zhang R-X., Zhai D-S., Zhang X-B., Zhang J-T., Li Q-S., 1990, IBVS 3456

I. Pustylnik: Can you summarize in one ore two sentences whether the observed statistics of flare activity favours more the idea that flares are enhanced by the binary nature of the object or by the rate of axial rotation?

P.B. Byrne: There is no evidence of any difference in the energetics or the frequency of flares between dMe stars in binary systems or in single stars. In RS CVn systems all sources are binaries so the question is nor relevant.

H. Zinnecker: You did not discuss stellar flares with large changes in optical brightness ($\Delta m > 1$). May I ask you to comment on those stellar flares with large amplitudes?

P.B. Byrne: The flares I am discussing ARE large in term of energy. Estimating the energy of flares from the ΔU amplitude alone gives a false impression. This is because the background U brightness changes enormously from the earliest flare stars (=K5 Ve) to the latest (=M6 Ve). Thus, in magnitude terms the same flare on a later type star will have a much greater ΔU.

FLARES IN LATE-TYPE STARS:

RADIO AND OPTICAL

Radio Emission from Flares in Single Late-type Stars

Arnold O. Benz

Institute of Astronomy, ETH-Zentrum, CH–8092 Zürich, Switzerland

Abstract: Radio observations provide the most direct information on non-thermal electrons in stellar flares and in the coronae of late-type stars. Radio emissions of single main-sequence F, G, and of many K stars have recently been discovered, in addition to the well-known dwarf M stars. Their long-duration radio flares with low circular polarization, slow variations and broad bandwidth can be attributed to gyrosynchrotron emission of mildly relativistic electrons. The same holds for the low-level ('quiescent') radio emission. On the other hand, highly polarized radio flares of M stars have been interpreted by coherent emissions from loss-cone instabilities of magnetically trapped electrons. These conjectures are consistent with recent VLBI observations. The identification of the radio emission process allows to estimate the high-energy component of the flare and compare it to the total flare energy. The weakly polarized radio emission may serve as a proxy for hard X-ray signatures of relativistic electrons. The fraction of primary energy released into energetic electrons then appears to be large and similar to solar flares.

1 Introduction

Magnetic energy builds up by turbulent plasma motions in the interior of stars (where this phenomenon is called 'dynamo') and is transported into the corona by flux emergence and by footpoint motions of coronal magnetic fields. How is this magnetic energy and its associated currents reduced and released? Stellar flares in the coronae of protostars, interacting binary systems and rapidly rotating main-sequence stars present us with the challenge to develop models capable of interpreting the catastrophic annihilation of magnetic fields. The opportunity of stellar flares is to make the release process observable under a wide variety of physical conditions.

The Sun, a slowly rotating G2V star, is a convenient reference point. Solar hard X-rays (HXR) produced presumably by electron bremsstrahlung suggest that 20–40% of the energy in impulsive (regular) flares shows up initially in non-thermal electrons with energies in excess of 20 keV (Duijveman et al. 1982, Strong et al. 1984). However, the estimate of energy partition is model dependent and still a matter of research. In reality it may depend on flare conditions and flare type. The non-thermal electrons are distributed in energy approximately according to a power-law reaching into the GeV range (e.g. Vilmer 1994, Petrosian et al. 1994). In stellar flares, where no HXR have yet been observed, radio emissions provide the most direct probes of non-thermal electrons. Observations in recent years have given ample evidence that stellar radio flare emissions originate from two groups of processes:

1. *Gyrosynchrotron* emission of mildly relativistic electrons. It is characterized by a broad bandwidth, low polarization and relatively slow variations in time.
2. *Coherent* emissions by instabilities of non-thermal electrons, appearing often, but not always, in a narrow frequency band with fast variations and high circular polarization.

The solar emissions demonstrate clearly that these general characteristics must not be blindly used, and the classification of observations is not straightforward. However, the distinction is essential for quantitative models since coherent emissions can be many orders of magnitude more efficient (e.g. Melrose 1986, Kuijpers 1989, Benz 1993).

It is beyond the scope of this brief review to describe stellar flare radio emissions in general. I shall focus my attention on single, late-type main-sequence stars that have a convective outer layer and a corona, and display activity similar to the Sun, although in some cases at a much higher level.

2 Radio emissions from F and G stars

Güdel et al. (1993) and Güdel (1992) have established a close linear relation between the quiescent *thermal* soft X-ray emission and gyrosynchrotron radio emission of single, main-sequence M and K stars, respectively. The luminosities are related as $L_x \approx 10^{15.5} L_r$, where L_x is in erg/s and L_r in erg/(s Hz) at 3.5–6 cm wavelength. A similar ratio has been found in binary systems of the RS CVn type (Drake et al. 1989; Fox et al. 1994), Algols, BY Dra and pre-main-sequence T Tau stars (Güdel & Benz 1993). Benz & Güdel (1994) have measured about the same ratio also in solar flares, although with a significantly larger scatter. They have taken this to indicate that quiescent stellar coronae are heated by flare-like acceleration processes.

The radio/X-ray relation has recently inspired the search for radio emitting stars in regions of the Hertzsprung-Russell diagram where no non-thermal radio star has yet been known. Using the *ROSAT* survey as the guide, Güdel et al. (1994, 1995a) have discovered quiescent radio emissions from five G stars and four F stars. To avoid errors, the stars have to be checked for cooler companions using the available spectrographic and astrometric information. Although in some cases the possibility of confusion with a low-mass companion remains, it has become unlikely to explain all the observations.

A remarkable example of these newly discovered radio sources is EK Dra (HD 129333), a G0V star and member of the Pleiades Moving Group (probable age 50–100 Myr). The quiescent radio emission is modulated by the 2.7 d optical rotation period, thus confirming for the first time a G-star origin of radio waves despite the known, distant M-star companion (Güdel et al. 1995b).

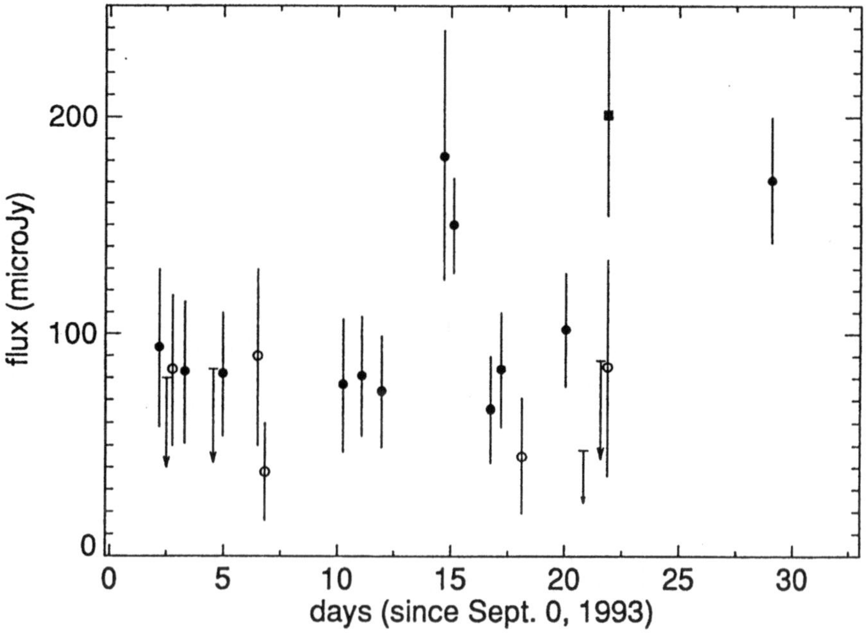

Fig. 1. Radio flux density of EK Dra observed in September–October 1993 by the VLA. Circles: 3.6 cm; squares: 6 cm; full symbols: $\gtrsim 2.5\sigma$; open symbols: $\lesssim 2.5\sigma$. Upper limits (arrows) refer to 2σ at 3.6 cm (adapted from Güdel et al. 1995b).

The time line displayed in Fig. 1 shows occasional flares on top of the quiescent background. The radio flares have surprisingly long durations of 10–180 min and no detectable circular polarization (< 13%).

Figure 2 displays an X-ray time line of EK Dra (not simultaneous with Fig. 1). It shows a variety of time scales ranging from about 1.5 hours to gradual changes over days. Its modulation with a period of 2.7 d suggests again a G-star origin. The flares, like events A, and possibly B and C, appear predominantly in the hard channel indicating high temperature. Events like A may occur at about the same rate as the radio flares apparent in Fig. 1.

The observation of weakly polarized flares on EK Dra are reminiscent of the radio emission observed from the G0V star χ^1 Ori. The weakly polarized (< 30%) radiation reached a peak within 20 minutes and slowly decreased on a time scale of about one hour (Gary & Linsky 1981, Linsky & Gary 1983). As astrometric observations and infrared speckle interferometry have revealed an M4V companion star, the G-star origin has later been questioned (Linsky &

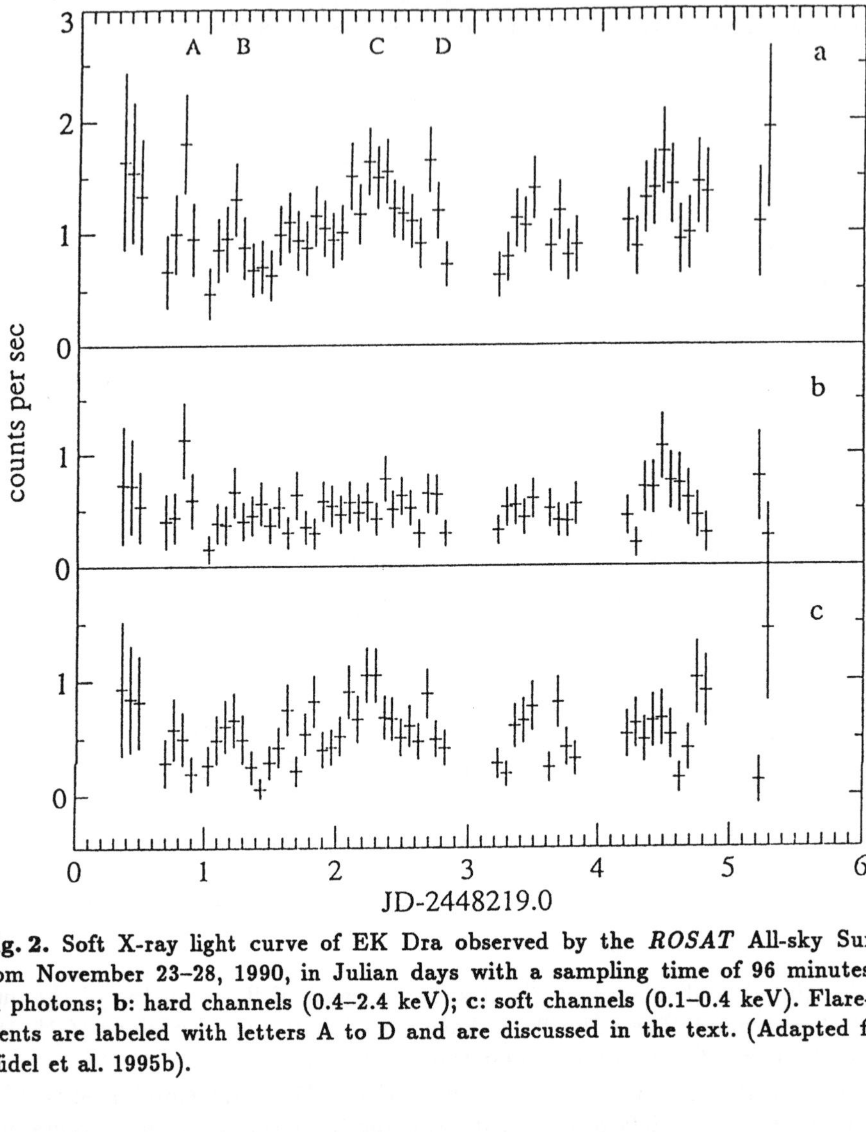

Fig. 2. Soft X-ray light curve of EK Dra observed by the *ROSAT* All-sky Survey from November 23–28, 1990, in Julian days with a sampling time of 96 minutes. **a:** all photons; **b:** hard channels (0.4–2.4 keV); **c:** soft channels (0.1–0.4 keV). Flare-like events are labeled with letters A to D and are discussed in the text. (Adapted from Güdel et al. 1995b).

Gary 1983, Pallavicini et al. 1985), but not disproved. A flare possibly similar to large solar flares was detected on the F5 IV-V star Procyon, giving further evidence of ubiquitous coronal activity along the cool main sequence (Drake et al. 1993).

A flare similar in polarization (upper limit 16%) and duration (about two hours), but an order of magnitude more luminous has recently been observed with high time resolution (Fig. 3). The flare radio luminosity has reached $2 \cdot 10^{16}$ erg $Hz^{-1}s^{-1}$. The binarity of the G8V star is unknown. The similarities to the EK Dra flares and their difference to typical dMe-star flares suggest that the flare has occurred on the G star.

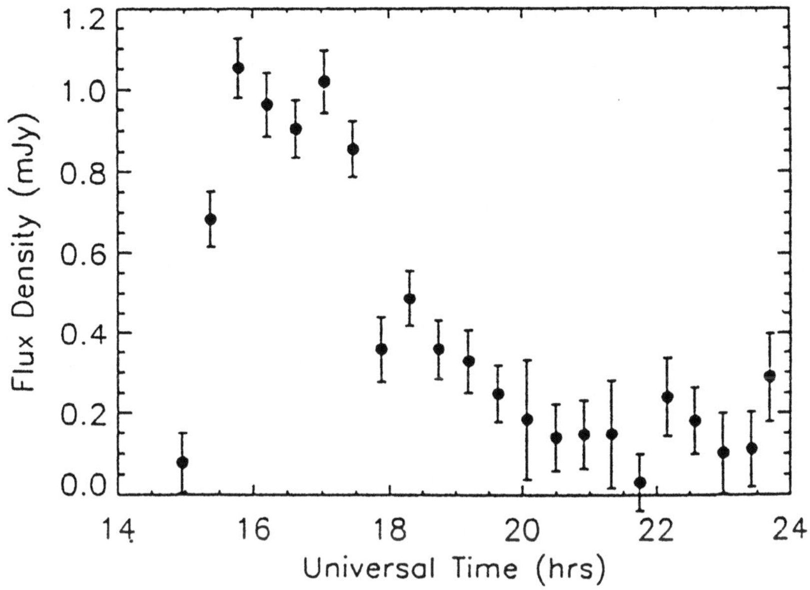

Fig. 3. Radio flare of the G8V star HII 1136 in the Pleiades observed on 1994 May 21 at 3.6 cm wavelength at the VLA (Lim & White 1995).

Not many radio flares have yet been reported from main-sequence K stars. Best known is the near or zero-age main-sequence K1 star AB Dor. Long-duration flares at 3.6 cm have been found with circular polarization of 13% and less (Slee et al. 1986). On the other hand, its radio flares at 6 cm wavelength lasting less than 1 hour have been reported to be highly circularly polarized (Lim et al. 1994). Furthermore, a prominent, weakly polarized ($< 14\%$) flare has been registered in 3.6 cm observations of HD 147365, an F3IV-V star with a K type companion. The radio flare coincided spatially with the K star (Güdel et al. 1995a).

The vast literature on radio flares of dwarf M stars has been reviewed by Bastian (1990) and Lang (1994). The reports on highly polarized flare emissions dominate (e.g. Gary et al. 1982, Lang et al. 1983, and many more). The degree of circular polarization in these flares is often, but not always, consistent with 100%. Only few weakly polarized flares on dMe stars have been observed (e.g. Kundu et al. 1988, Güdel et al. 1993).

From the few observed radio flares of main-sequence G and K stars, it appears that low-polarization events dominate in G stars, coexist with high-polarization events in K stars, and are a minority in M stars. Whether these observations constitute a general trend has yet to be established. In particular, the frequency dependence of polarization should be investigated in the various stellar types.

3 VLBI and emission process

The brightness temperature of a radio source can be used to test the proposed emission mechanism. Very long baseline interferometry (VLBI) of stellar flares can measure the radio diameter, d, or set an upper limit to it. The brightness temperature, defined as

$$T_b = \frac{c^2}{\pi k_B \nu^2} \left(\frac{d}{D}\right)^2 S \ , \tag{1}$$

is then calculated from the observed frequency ν, the source distance D, and from the flux density S of one mode (e.g. left or right circular).

Gyrosynchrotron emission is an incoherent process, in which each particle radiates independently. Thus its brightness temperature is limited by the mean kinetic energy, $\langle\epsilon\rangle$, of the radiating electrons

$$T_{syn} \leq \frac{\langle\epsilon\rangle}{k_B} \quad [\mathrm{K}] \ . \tag{2}$$

Equality is reached for optically thick emission. The synchrotron process encounters an absolute limit when the radiating electrons are cooled by inverse Compton scattering on their own emission. This 'inverse Compton catastrophe' sets in at about 10^{12} K (Kellermann & Pauliny-Toth 1969). Assuming equipartition of the energy densities between the radiating particles and the magnetic field and using the magnetic field estimate from the spectral peak caused by synchrotron self-absorption, Readhead (1994) has derived an upper limit on the synchrotron brightness temperature of a few times 10^{11} K (depending on the spectral index).

Since coherent emission is caused by non-thermal, but not necessarily relativistic particles, the inverse Compton effect can usually be neglected for this type of emission. Consequently, the catastrophe and equipartition upper limits do not apply.

Most VLBI studies of flares have concentrated on bright RS CVn type binaries. A weakly polarized flare on the RS CVn binary HR 1099 (V711 Tau) has been reported by Trigilio et al. (1993). The initial brightness temperature of the flare at 6 cm was $4.60(\pm1.16)\cdot10^9$ K. It decreased slowly and continuously to $1.72(\pm0.67)\cdot10^9$ K after 2.5 hours. These values are consistent with the interpretation by gyrosynchrotron emission of mildly relativistic electrons. Using Eq. (2), their initial mean energy is derived to be ≥ 400 keV.

Coming back to single main-sequence stars, a highly polarized flare with a very structured time profile has been observed in VLBI at 18 cm on the single dMe star EQ Peg by Benz et al. (1995). The flux densities measured by VLBI baselines do not deviate from the values observed by the VLA. It indicates that the radio source has not been resolved and must be smaller than 0.5 mas (or $5\cdot10^{10}$ cm). A brightness temperature $> 4\cdot10^{10}$ K results, which is below the limit of the equipartition value. The resolution was not sufficient to discard synchrotron emission, but puts the extreme requirement, that the mean electron energy would have to exceed 4 MeV. This exceeds the solar values by two orders of magnitude and is not a likely interpretation.

4 Radio/X-ray ratio of G-star flares

The detection of weakly polarized flares in single main-sequence G and K stars opens the prospect of estimating the flare energy released into energetic electrons. Direct energy calculations from synchrotron emissions are plagued by many free model parameters, such as trapping time, aspect angle, anisotropy, to name just a few. Instead I suggest a heuristic approach based on a comparison with solar flares.

The flux density of a typical radio flare of EK Dra is about 100 μJy (Fig. 1). This may be compared to the X-ray flux of the largest flare in a comparable time period, flare A in Fig. 2. It peaks at about 1 ct/s above background, corresponding to $7 \cdot 10^{29}$ erg/s at the distance of EK Dra if the same standard *ROSAT* (PSPC) conversion factor is adopted.

The resulting ratio of the two luminosities, $L_x/L_r \approx 10^{15.8}$ Hz. It is close to the value $10^{15.5}$Hz derived by Benz & Güdel (1994) for solar flares (impulsive and gradual). The observed X-ray flare on EK Dra produced an energy of $5 \cdot 10^{33}$ erg in thermal X-rays, about two orders of magnitude more than the largest solar flare; and the non-thermal radio flux was found at a proportionally higher level. It strongly suggests that the energy partition between thermal and non-thermal components is similar in solar and stellar flares.

5 Conclusions

F, G, and K main-sequence single stars have been discovered in radio emission in considerable numbers. The few radio flares that have yet been observed seem to be different from the typical flares on dMe stars:

- They tend to be weakly polarized and seem to be caused by synchrotron emission.
- They last longer, but are less frequent.
- Since their emission process is probably incoherent and thus less efficient, more non-thermal energy is needed to produce the same flux. Thus they may represent larger flares.

These trends need to be confirmed by more observations.

Multiwavelength flare observations of dMe stars have frequently resulted in the disappointing statement that there was no radio enhancement during an optical or X-ray flare, or that no optical or X-ray counterpart was observed during a radio flare (e.g. Karpen et al. 1977, Kahler et al. 1982). There is a strong suspicion that in all such cases the radio telescope was not sufficiently sensitive to detect the weakly polarized synchrotron emission or, respectively, the radio emission was strongly polarized, indicating highly efficient coherent emission by a possibly minor flare.

The radio emission of weakly polarized flares on a young G-dwarf seems to follow the same radio/X-ray relation for peak values as solar flares. This suggests that the stellar flares are comparable in acceleration efficiency for energetic electrons. It further suggests that the weakly polarized stellar radio flare emission is produced in a similar way. Emission models adapted from solar flares thus provide quantitative information on these electrons. The weakly polarized radio emission may serve as a proxy of hard X-ray bremsstrahlung of relativistic electrons used successfully to study the energetics of solar flares.
Acknowledgements: I thank Manuel Güdel for helpful discussions. This work has been supported by the Swiss National Science Foundation (grant 20-040336.94).

References

Bastian T.S., 1990, Solar Phys. 130, 265
Benz A.O., 1993, Plasma Astrophysics, Kluwer, Dordrecht, Holland
Benz A.O., Alef W., Güdel M., 1995, A&A, in press
Benz A.O., Güdel M., 1994, A&A 285, 621
Drake S.A., Simon T., Linsky J.L., 1989, ApJS 71, 905
Drake S.A., Simon T., Brown A., 1993, ApJ 406, 247
Duijveman A., Hoyng P., Machado M., 1982, Solar Phys. 81, 137
Fox D.C. et al., 1994, A&A 284, 91
Gary D.F., Linsky J.L., 1981, ApJ 250, 284
Gary D.F., Linsky J.L., Dulk G.A., 1982, ApJ 263, L79
Güdel M., 1992, A&A 264, L31
Güdel M., Benz A.O., 1993, ApJ 405, L63
Güdel M., Schmitt J.H.M.M., Bookbinder J.A., Fleming T., 1993, ApJ 415, 236
Güdel M., Schmitt J.H.M.M., Benz A.O., 1994, Science 265, 933
Güdel M., Schmitt J.H.M.M., Benz A.O., 1995a, A&A 293, L49
Güdel M., Schmitt J.H.M.M., Benz A.O., Elias N.M., 1995b, A&A, in press
Kahler S. et al., 1982, ApJ 252, 239
Karpen J. et al., 1977, ApJ 216, 479
Kellermann I.I., Pauliny-Toth I.I.K., 1969, ApJ 155, L71
Kuijpers J., 1989, Solar Phys. 121, 163
Kundu M.R., Pallavicini R., White S.M., Jackson P.D., 1988, A&A 195, 159
Lang K.R., 1994, ApJSS 90, 753
Lang K.R., Bookbinder J., Golub L., Davis M.M., 1983, ApJ 272, L15
Lim J., White S.M., Nelson G.J., Benz A.O., 1994, ApJ 430, 332
Lim J., White S.M., 1995, ApJ, in press
Linsky J.L., Gary D., 1983, ApJ 274, 776
Melrose D.B., 1986, Instabilities in space and laboratory plasmas, Cambridge Univ. Press, Cambridge
Pallavicini R., Willson R.F., Lang K.R., 1985, A&A 149, 95
Petrosian V., McTiernan J.M., Marschhäuser H., 1994, ApJ 434, 747
Readhead A.C., 1994, ApJ 426, 51
Slee O.B. et al., 1986, Proc. ASA 6, 312
Strong K.T. et al., 1984, Solar Phys. 91, 321
Trigilio C., Umana G., Migenes V., 1993, MNRAS 260, 903
Vilmer N., 1994, ApJS 90, 611

M. Shapiro: Dr. Benz has told us that coronal heating is related to the acceleration of electrons in active stars. Is it not likely that energetic protons also contribute significantly to the heating?

A.O. Benz: This is certainly possible. Some acceleration processes, such as shock acceleration or some current instabilities, put more energy into protons that electrons. Such models have not received much support from solar observations. Nevertheless, protons remain the big unknowns in both solar and stellar flares.

R.E. Gershberg: Did you estimate a probable contribution of the total stellar population radiation at microwaves in the galactic and cosmological background?

A.O. Benz: The quiescent flare star radio emission constitutes a background of about 10^{-5} of the cosmological background at 6 cm. In addition, it usually decreases with decreasing wavelength, contrary to the cosmological radiation that has a maximum around 1 mm.

I. Pustylnik: You have shown the beautiful pictures of radio isophotes for the two binaries EK Dra and HR 963. Do you see these objects as a single radio source with isophotes concentrated more to one or the other star, or as a binary radio source?

A.O. Benz: In the case of HR 963, the radio source is consistent with the optical position of the F7 IV star within the positional accuracies. There is no indication of plasma physical interactions with the G7 V companion at a distance of 1.8×10^{15} cm. EK Dra is probable a single star. In any case, the rotational modulation of the radio emission is consistet with a G0 V star.

High-Resolution Dynamic Spectrum of a Spectacular Radio Burst from AD Leonis

Meil Abada-Simon[1,2], *Alain Lecacheux*[2], *Monique Aubier*[2,3], *Jay A. Bookbinder*[4]

[1] Sterrekundig Instituut, 3508 TA Utrecht, The Netherlands
[2] ARPEGES - URA CNRS 1757, Observatoire de Paris, 92190 Meudon, France
[3] Université Paris 6, France
[4] Smithsonian Astrophysical Observatory, Cambridge, MA 02138, USA

1 Introduction

AD Leonis is a very active, single dMe flare star. The similarities between this type of star and the Sun has led to study their radio radiation, which originates from their corona. The high brightness temperatures and other characteristics of most dMe radio bursts can be attributed to a non-thermal, coherent mechanism: plasma radiation or a cyclotron maser instability (CMI) are both plausible explanations. Even for the strongest burst of AD Leo which reached 940 mJy at 21 cm, it was not possible to discriminate between these two mechanisms (Bastian et al. 1990).

Here we present an intense burst from AD Leo, exhibiting strong spikes for which the CMI seems to be the only reasonable explanation. In Sect. 2 we describe the observations, and in Sect. 3 we give an interpretation for this event.

2 Observations

We used the 305 m diameter Arecibo radio telescope to observe simultaneously the star ("ON") and another direction in the sky ("OFF") at 1.4 GHz, in both senses of circular polarization, to discriminate between flares of stellar origin and artificial radiofrequency interference. An acousto-optical spectrograph (AOS) recorded every 20 ms four spectra of 50 MHz bandwidth, and the frequency resolution was ~1 MHz. The observational technique used during that campaign is described in detail in Abada-Simon et al. (1994).

The event, starting at 4:12:55 UT on 13 February 1993, is the most interesting of the dozen bursts detected from AD Leo during the campaign mentioned above. This event is 100% right circularly polarized (RCP), and exhibits a variety of characteristics in frequency and time.

Fig. 1 shows the dynamic spectrum (1 s integration time) of the burst's most interesting part, but the total duration of the event is $\sim$ 7 min. The emission frequency increases and decreases with time at several instants; these positive and negative drifts give: $|df/dt| \simeq 1 - 5$ MHz.s^{-1}; they form "arches" which

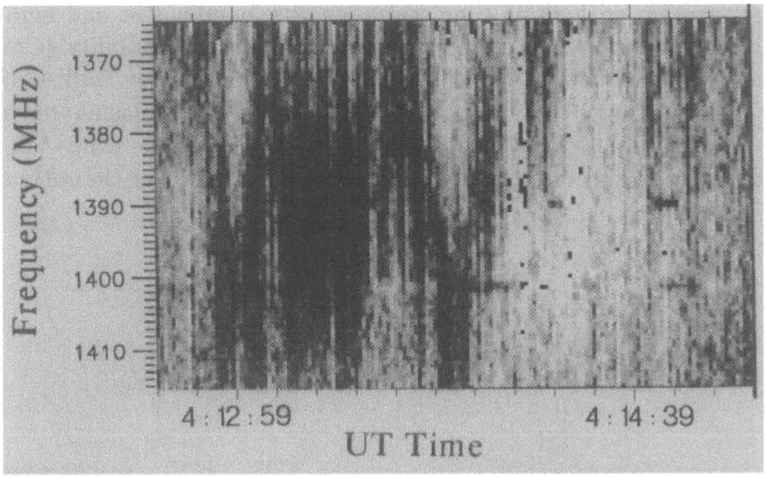

Fig. 1. RCP dynamic spectrum of 2.5 min observation of AD Leo on 13 February 1993. Each pixel represents 1 s of time and one channel of 0.5 MHz bandwidth. The frequency ranges from 1365 MHz to 1415 MHz. Time is in UT. Intensity is represented in grey levels. White-to-black represents an increase of $\sim$ 70 mJy.

spread over $\sim$ 10-20 s. Unfortunately, because of the limited observing bandwidth (50 MHz), one cannot see to which frequencies the emission actually extends. In addition, its instantaneous bandwidth varies with time from $\sim$ a few MHz to 50 MHz or more. Oscillations on a quasi-period of $\sim 5-10$ s are also present, which are more obvious at frequencies $\leq$ 1400 MHz.

Fig. 2a shows the time variations of four 10 MHz integrated bands over 5 s taken out of the burst's most intense parts, and plotted with 20 ms time resolution: one can clearly see the numerous spikes present in the burst, their flux density reaches up to 350 mJy from a quasi-zero level in 20 ms, which suggests that, maybe, a higher flux density could have been measured with a shorter time resolution. Fig. 2b shows a full resolution dynamic spectrum of four seconds taken out of the burst of AD Leo: the black vertical rectangles surrounded by two white ones show again that a high flux density is reached in no more than 20 ms from a zero level (in white). Fig. 2b also shows that the instantaneous bandwidth of each spike is Δf $\simeq$ 10-20 MHz, which corresponds to a relative bandwidth of $\Delta f/f \approx$ 1%. Finally, some of the spikes exhibit a frequency drift: $df/dt \simeq 400\,\mathrm{MHz\,s^{-1}}$.

3 Interpretation

The shortest time variations ($\Delta t = 20$ ms) infer a source size $\leq$ 6000 km. With flux densities reaching up to $\sim$ 350 mJy, we deduce a brightness temperature $T_b \geq 10^{15}$ K. A plausible mechanism explaining such a high brightness temperature, together with short time scales, narrow bandwidths and high degrees of circular polarization is the electron–cyclotron maser instability (CMI). This process has been reviewed by Wu & Lee (1979), Melrose & Dulk (1982), and recently favored (again) by Melrose (1994), rather than plasma radiation, to explain the flare star radio bursts. Similar spikes to those from AD Leo are observed from the auroral regions of magnetized planets and the solar corona.

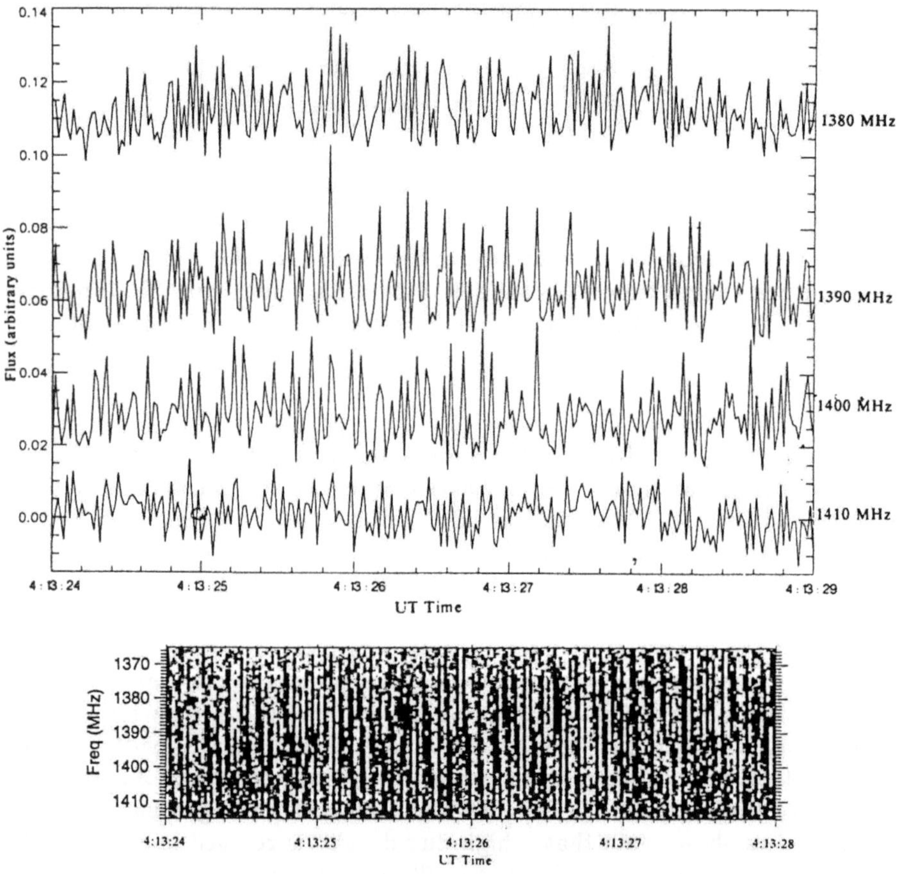

Fig. 2. *Top:* Ttime variations in RCP in AD Leo over 5 s for four 10 MHz integrated bands centered at 1380, 1390, 1400 and 1410 MHz, respectively, from top to bottom. The time resolution is 20 ms. The vertical axis is in arbitrary units, but the flux density is indicated by the r.m.s error bar: 1 $\sigma \simeq$ 39 mJy, which corresponds to 0.005 in arbitrary units. The most intense spike, which is best visible in the band centered at $\sim$ 1390 MHz, occurred at 4:13:25.84 UT and reached $\sim$ 350 mJy. *Bottom:* RCP dynamic spectrum of AD Leo on 13 February 1993 over 4 s. Each pixel represents 20 ms of time and one channel of 0.5 MHz bandwidth. White to black represents an increase of $\sim$ 400 mJy.

In the case of cyclotron maser emission, our observations at 1400 MHz lead to a magnetic field B $\simeq$ 500 G for an emission at the fundamental frequency ($s = 1$). The electron number density in the source is not known. Thus, an emission on the X-mode or on the O-mode, at $s = 1$, is possible, depending on the value of the ratio $\frac{\omega_p}{\Omega_e}$ (see Benz 1993).

In our application of this mechanism we will use the theory developed by Aschwanden & Benz (1988a,b): an electron-cyclotron masing source operating in a stationary loss-cone can explain the spiky intense radio emission, as one of us developed in the solar case (Barrow et al. 1994). Detailed calculations will be presented in a future paper.

Although the burst in AD Leo presents similar features to those observed from the Sun, one should keep in mind one of the most striking differences: AD Leo's spikes are $\sim 10^4$ times stronger than their solar analogues.

References

Abada-Simon M., Lecacheux A., Louarn P., Dulk G.A., Belkora L., Bookbinder J.A., Rosolen C., 1994, A&A 288, 219

Aschwanden M.J., Benz A.O., 1988a, ApJ 332, 447

Aschwanden M.J., Benz A.O., 1988b, ApJ 332, 466

Barrow C., Zarka P., Aubier M., 1994, AA 286, 597

Bastian T.S., Bookbinder J.A., Dulk G.A., Davis M., 1990, ApJ 353, 265

Benz A., 1993, Plasma Astrophysics, Kluwer, Dordrecht

Melrose D.B., 1994, Proceedings ASA 10, (3) 254

Melrose D.B., Dulk G.A., 1982, ApJ 259, 844

Wu C.S., Lee L.C., 1979, ApJ 230, 621

L. Pustil'nik: You only use the Melrose model on the cyclotron maser emission. Another process with polarized burst generation by plasma waves on hybrid-frequency generation (Stepanov and Zaitsev models) is possible, and gives the same result for burst emission. Why did you not analyse these possibilities? Do you have any observational reasons for this?

M. Abada-Simon: It is the spikes that I attribute to electron-cyclotron maser emission, using not only Melrose's recent publication in favour of this mechanism, but also by analogy with spikes having similar characteristics to those observed from the magnetized planets and the sun. A solar burst similar to that observed in AD Leo was studied by Aschwanden & Benz (1988): they also explain several of the observed features in the frame of the cyclotron maser occuring in certain conditions.

I am not aware of another mechanism which produces very intense spikes with strong circular polarization and narrow bandwidth in the same context (same frequency, etc.). However, I shall consider your suggestion.

VLBI Observations of Flares in RS CVn Binary Systems

Corrado Trigilio[1], *Grazia Umana*[1], *Victor Migenes*[2]

[1] Istituto di Radioastronomia del C.N.R., C.P. 169 Noto, Italy
[2] C.S.I.R.O., ANU Campus, GPO Box 3023, Canberra 2601, Australia

1 Introduction

RS Cvn systems have been studied very extensively at radio wavelengths and a large fraction of them have been found to be strong radio sources, with 6 cm radio luminosities in the range between 10^{15} and 10^{18} erg s^{-1}. The radio flux density is highly variable and it usually shows two different regimes: active periods, characterized by a continuous strong flaring which can last for several days, and quiescent periods, during which the flux density goes down to a few mJy. These characteristics indicate the non-thermal nature of the radio emission which is driven by magnetic activity of these systems, seen in other spectral regions. In this scenario, the radio flux arises from the interaction between the magnetic field of one or both components with mildly relativistic particles, i.e. gyrosynchrotron emission.

Due to the erratic nature of radio flares in RS CVn binaries, most VLBI observations have been performed during quiescent periods. Only the 1983 UX Ari observation (Mutel et al. 1985), carried out during an intense outburst, allowed the first, and still the unique, hybrid map of a RS CVn binary, which revealed a core-halo morphology of the radio source.

It is still not clear if the flare events are localized in very compact regions of the corona of one component or if the entire binary system is involved. VLBI measurements in both active and quiescent periods are extremely important to determine the characteristic dimensions and the structure of the radio source.

2 Single dish monitoring program

In 1991 we have started a 6 cm, single dish monitoring of active binary systems, using the 32 m telescope at the Noto VLBI station of the Istituto di Radioastronomia of C.N.R. The main aim of these systematic total power observations is to activate ad-hoc VLBI observations once a flaring period is detected. Four binary systems have been selected as targets for the monitoring program: HR 1099, UX Ari, HR 5110 and Algol. Those system are the only ones for which information of flaring events with flux densities higher then 250 mJy are available.

One of the most interesting results of this program comes from the monitoring of HR 1099 (V711 Tau). This is a non-eclipsing RS CVn binary system consisting of a K1 subgiant and a G5 main sequence star, with an orbital period of 2.8 days and a separation between the two components of about 11.5 $R_\odot$.

The single dish measurements show that there are long active periods during which strong radio flares occur one after the other and the flux density never reaches its quiescent value. On the other hand, the quiescent phase can last for several weeks.

3 VLBI observations

3.1 The 1991 July 26 observation

In July 1991 HR 1099 experienced a series of strong radio flares, one of which reached a flux density of $\approx$ 800 mJy. Just after the main flare event, which occurred on July 25, an ad hoc VLBI observation run was carried out (Trigilio et al. 1993), giving for the first time the opportunity to measure the brightness temperature trend during a flare decay in a RS CVn binary. The observations were made, on July 26, at 6 cm in standard VLBI Mark II mode, with a 2 MHz bandwidth. Three telescopes (Noto, Jodrell Bank and Cambridge) were involved. From the single dish measurements performed just before and after the VLBI run, it was clear that we were following the decay of the flare; the flux went from $\approx$660 mJy to $\approx$360 mJy in about five hours. Due to the source variability we split the 3 hours observing time into 6 different, 30 min scans. We have separately fitted the data from each scan to a two-dimensional Gaussian brightness distribution and derived the source angular dimension and the corresponding linear size, assuming a distance of 35 pc. The angular size of the radio source was found to be about 4 mas, which corresponds to a linear size almost twice as large as the overall binary size. The dimensions and the observed flux density allowed us to determine the brightness temperature, using the relation:

$$T_{\rm B} = 1.97 \times 10^6 \text{ Flux (mJy)} \frac{\lambda^2(\text{cm})}{\theta^2_{\rm mas}} \tag{1}$$

Because the source size did not change during our observation, apart for a possible expansion between the first and the second scan, we can exclude any contribution of structural changes of the radio source to the flare decay. We can, therefore, conclude that the observed flux density variation is due entirely to a decay of emitting particles energy.

3.2 The 1992 April 29 and 30 observations

The April 29 and 30 VLBI observations were performed at the frequency of 5 GHz in MK II recording mode, with a bandwidth of 2 MHz. Three European antennas started the observations, joined by five VLBA antennas, with two hours of simultaneous observations; the VLBA ended the observations at 02 UT of 30

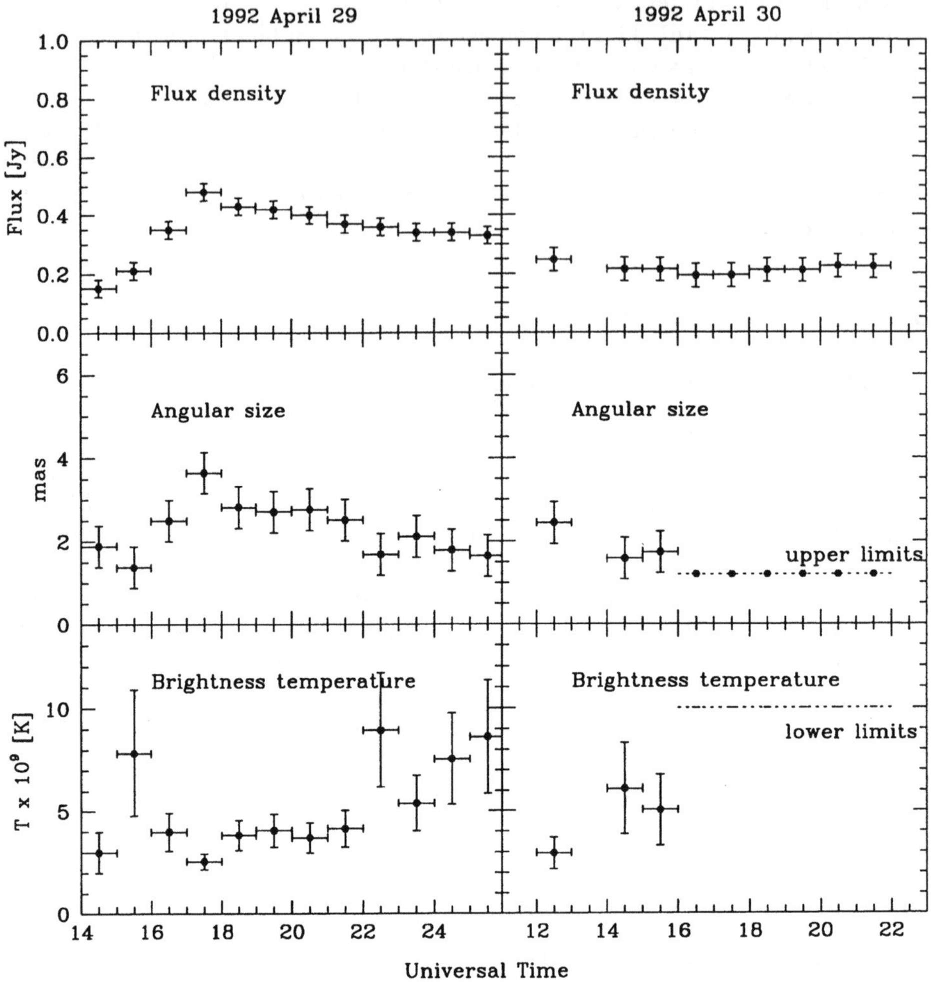

Fig. 1. Radio light curve, source size and brightness temperature evolution for the 29 and 30 April 1992 VLBI observations.

April (for the first day) and at 22 UT of 30 April (for the second day). The visibility function was analyzed as for the July 91 observations. The flux density radio light curve, together with the angular size and brightness temperature evolution are shown in Fig 1.

On April 29, we observed a complete flare: the source reached a maximum flux density of about 500 mJy in about 3 hours, then a very slowly decay, lasting more than 8 hours, occurred. The source was very compact at the beginning, being comparable with the size of the K star; an expansion followed which involved the whole binary system. During the decay, the size of the source was almost constant, showing a decrease at the end. The brightness temperature was high at the beginning of the flare, decreased during the expansion, and at the end, it increased drastically.

The decay of this flare is much longer than the one in July 1991. A possible interpretation can be given. The brightness temperature beaviour indicates that an energy release occurred at the beginning, when the expansion started, then loss mechanisms operated in the corona, as the decay seems to indicate. In the second part of the observations a new ejection of energetic particles occurred. The long flux decay seems to be sustained by a new flare.

During the April 30 observations, the activity level was still high, showing, however, a different trend. The flux density was almost constant (about 250 mJy) during the whole observation, lasting more than 10 hours. The source size decreased, beeing comparable with the entire binary system at the beginning of the observations. The source was not spatially resolved in the second part, so only an upper limit of 1.2 mas can be estimated. The brightness temperature was increasing, showing a lower limit of 10^{10} K at the end.

These data seem to indicate that in the first part the decay of a previous flare has been observed, with a source as big as the whole system (halo) and a low brightness temperature. Then multiple flares, or a continuous particle ejection occurred in a very compact layer (core).

4 Conclusions

The single dish monitoring program gives the possibility to organize VLBI observations during flares. In fact, due to the erratic nature of the radio emission in RS CVn binary systems, it is impossible to foresee when an active period will start. VLBI observations give the possibility to follow the size and the brightness temperature evolution during these events, and this provides inportant constraints to any attempt to model stellar coronae.

From the three VLBI observations of HR 1099 a qualitative picture of the radio flare dynamic can be drawn. A flare first develops in a compact region (comparable with the stellar size), then the source expands, involving the whole system. The relativistic electrons lose energy via radiative decay, collisions and other possible mechanisms, and the typical decay time is about 5-6 hours. A flare is often a non unique event, and sometimes new flares occur during the decay, providing energetic electrons. Longer decay times can be observed in this case.

References

Mutel R.L., Lestrade J.F., Preston R.A., Phillips R.B., 1985, ApJ 289, 262

Trigilio C., Umana G., Migenes V., 1993, MNRAS 260, 903

Coherent Radio Bursts from RS CVn Binaries

Elena Franciosini[1], *Stephen M. White*[2]

[1] Department of Astronomy and Space Science, University of Florence, Largo E. Fermi 5, I-50125 Florence, Italy
[2] Department of Astronomy, University of Maryland, College Park, MD 20742, USA

We present VLA observations of RS CVn binaries which indicate that the low–frequency polarization inversion usually ascribed to the quiescent radio emission of these systems is in fact due to the presence, much of the time, of a coherent burst emission. Fig. 1 shows two examples of this coherent emission in the behaviour of the light curves of HR 1099 in left and right circular polarization at 1.4 GHz. The emission is highly circularly polarized, with one polarized component steady or slowly varying in a way similar to the higher frequency emission (Fig. 1a), while the other shows rapid fluctuations on timescales of ~10 s and a slower modulation on longer timescales. The similarity of the time profiles in two 50 MHz–wide sidebands 30 MHz apart indicates that the fluctuating component is broadband; moreover, the sense of polarization at 1.4 GHz is opposite to that of the quiescent emission at higher frequencies (attributed to gyrosynchrotron emission), which is exactly the behaviour attributed to the 1.4 GHz quiescent emission of these systems by Mutel et al. (1987).

The rapid fluctuations and high polarization are characteristic of a coherent emission process. We argue that a combination of two sources of emission is generally present at 1.4 GHz: a very highly (possibly 100%) polarized coherent component, and a weakly-polarized, steady gyrosynchrotron component. Since the polarization at higher frequencies should be the x-mode of (optically thin) gyrosynchrotron emission, we infer that the low frequency polarization probably represents the o-mode. The observed characteristics of the emission then suggest that the most likely mechanism for the highly-polarized component is plasma emission at the fundamental of the plasma frequency. Such plasma emission is relatively rare on the Sun at 1.4 GHz, but could occur in the hotter coronae of RS CVn systems because the higher temperature reduces the free–free opacity which is thought to absorb 1.4 GHz plasma emission on the Sun.

These observations suggest a different interpretation for the reversal in the sense of polarization at low frequencies previously attributed to the quiescent radio emission of RS CVn binary systems, which has proven very difficult to explain in terms of plausible gyrosynchrotron models (see White & Franciosini, 1995). We propose that a coherent o-mode emission is frequently present also during low-activity periods, when it may be too weak to be distinguished from the gyrosynchrotron component. The gyrosynchrotron component is probably

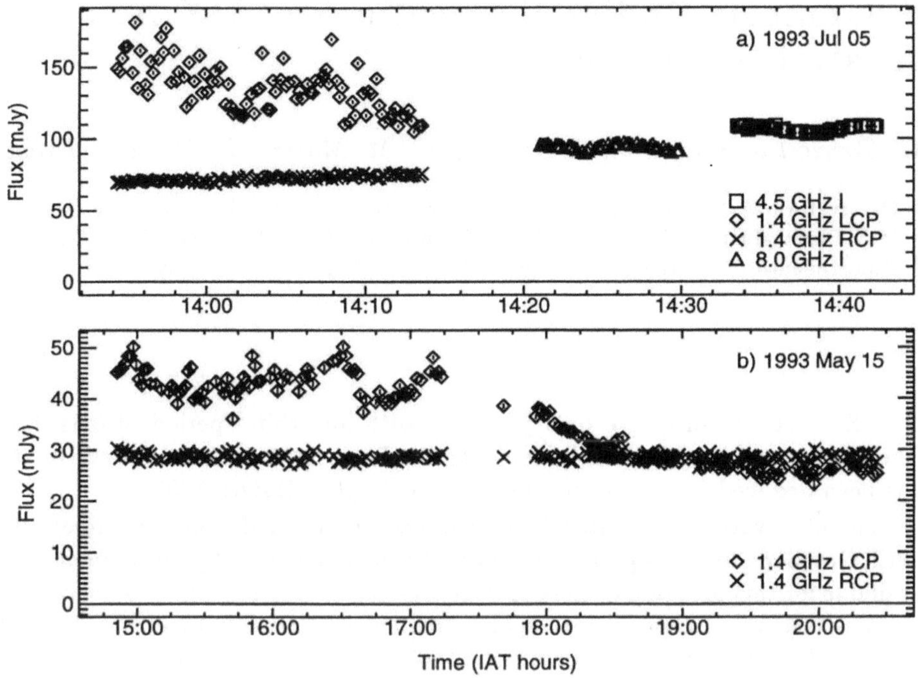

Fig. 1. The time profiles of right (R) and left (L) circular polarization at 1.4GHz during a VLA observation of HR 1099 on 1993 July 5 (upper) and 1993 May 15 (lower). In the first case the total intensity (I) in subsequent scans at 4.5GHz and 8.1GHz is also plotted. Error bars of $\pm\sigma$ are plotted on all points.

weakly polarized, so that the polarization of the integrated emission is dominated by the coherent component. We have some data suggesting that the true polarization of the quiescent emission at 1.4 GHz is in the sense of the x-mode as at higher frequencies. An example is shown in Fig. 1b, where the steady emission observed after the decay of the rapidly-varying coherent component is clearly polarized in the opposite sense.

The long duration of the observed coherent emission (20 min – several hours) together with the rapid variability (~10 s) suggest that there must be a continuous supply of energetic electrons to the source, which requires a continuous acceleration mechanism and therefore also a continuous energy release in the corona of these stars. Such emission may therefore be evidence for steady, rather than sporadic, sites of energy release in the coronae of active stars.

References

Mutel R.L., Morris D.H., Doiron D.J., Lestrade J.-F., 1987, AJ 93, 1220
White S.M., Franciosini E., 1995, ApJ (in press)

Systematic Radio-Observations of UX Arietis: Analysis of Its Variability

G. Torricelli-Ciamponi[1], *J. Neidhöfer*[2], *M. Massi*[1], *F. Chiuderi-Drago*[3]

[1] Osservatorio Astrofisico di Arcetri, Largo E. Fermi 5, 50125 Firenze, Italy
[2] Max-Planck-Institut für Radioastronomie, Auf dem Hügel 69, 53121 Bonn, Germany
[3] Dipartimento di Astronomia, Largo E. Fermi 5, 50125 Firenze, Italy

UX Arietis is an active binary system with an orbital period of 6.44 days. The presence of long lasting large spots on the surface of the more active star has been deduced from optical observations (Vogt & Hatzes 1991).

The observations presented here have been performed from December 1992 until October 1994 using the Effelsberg 100 m telescope. The instrument was available for this program in the gaps between previously scheduled observations. For this reason our observations span a frequency interval between 1.4 GHz (21 cm) and 43 GHz (7 mm), depending on the scheduled receiver at each observing time. These data, which are an extension of those published by Neidhöfer et al. (1993), are shown in Fig. 1, plotted in terms of the known orbital period (6.44 days). The search for different periodicities (by using the Phase Dispersion Minimization method) has shown the existence of two main other periods: at 156 days and at 24 days (Massi et al., in preparation).

The presence of a minimum, discernible in Fig. 1 around phase 0.4 of the orbital period, becomes more evident if only radio fluxes above 50 mJy are selected, averaged in phase and plotted. In contrast to this, radio fluxes below 50 mJy appear to be evenly distributed over the orbital period. This minimum could be interpreted in terms of a geometrical effect strictly linked to the rotation of the system. A source of enhanced radio emission localized near the surface of the active star, e.g. above a star spot, would be a good candidate for the explanation of this periodicity. Alternatively, a change in the orientation of the magnetic field could also give a flux decrease, appearing as a shadowing of the source.

Since the strong emission *only* suffers from the geometrical shadowing induced by the system rotation, it follows that the high intensity emission is more localized (and organized, if one thinks in terms of magnetic field) than the low intensity emission. We deduce that the emitting system is composed by a compact region where the strong initial phase of the flare takes place and by an extended halo, too large to be obscured, where the flare then develops and fades. This picture is in complete agreement with the conclusion derived by Mutel et al. (1985) from VLBI observations and with the model proposed by Franciosini & Chiuderi-Drago (1995). Since the shadowing in the emission does not last very long, $< 1/10$ of the orbital period, the compact region must have a typical size

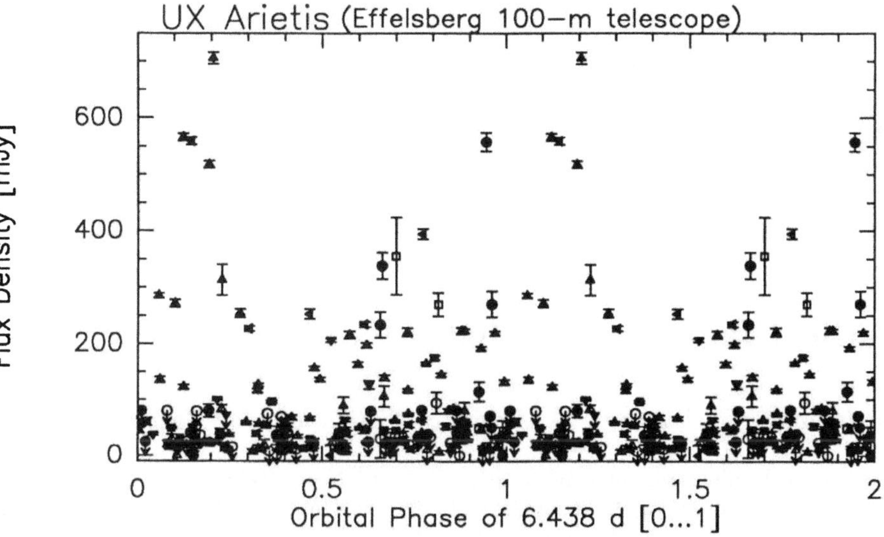

not much smaller than the stellar disk.

Some flare spectra can also be derived from the available data. A first analysis shows that:

a) A long-lasting ejection of accelerated particles is necessary to explain the time scale of the rising phase of some flares which can last as long as two days.

b) The flares observed in the rising phase always show a range of frequencies in which the spectrum slope is positive ($F \sim \nu^{\alpha}$, $\alpha > 0$) but not as steep as in the self-absorbed limit (i.e. it is $\alpha < 2.5$). A possible interpretation in terms of accelerated particles diffusing along the magnetic field lines in wider and wider regions of space is in progress (Torricelli et al. in preparation).

c) The flare decaying phase can be explained as due to the temporal evolution of a population of non-thermal electrons undergoing radiative and collisional losses (Chiuderi-Drago & Franciosini 1993, Franciosini & Chiuderi-Drago 1995).

References

Chiuderi-Drago F., Franciosini E., 1994, A&A 410, 301
Franciosini E., Chiuderi-Drago F., 1995, A&A, in press
Mutel R.L., Lestrade J.F., Preston R.A., Phillips R.B., 1985, ApJ 289, 262
Neidhöfer J., Massi M., Chiuderi-Drago F., 1993, A&A 278, L51
Vogt S.S., Hatzes A.P., 1991, Proc. IAU Coll. 130, 297

Radio Flares in HR 1099 (V711 Tau)

G. Umana[1], *C. Trigilio*[1], *G. Leto*[1], *S. Catalano*[3], *M. Rodonò*[23]

[1] Istituto di Radioastronomia del C.N.R., C.P. 169 Noto, Italy
[2] Istituto di Astronomia, Università di Catania, Viale A. Doria 6, 95125 Catania, Italy
[3] Osservatorio Astrofisico di Catania, Viale A. Doria 6, 95125 Catania, Italy

We present the results of the first three years of a single dish radio monitoring of HR 1099 we are carrying out using the the 32 m radiotelescope at the Noto VLBI station of the Istituto di Radioastronomia of the Italian Consiglio Nazionale delle Ricerche (C.N.R.).

These data, together with other observations available from literature, constitute a database which shed light on the radio emission from this system.

HR 1099 tends to undergo extended periods of strong activity, interrupted by periods of low flux density level, as can be seen from the radio behaviour on a timescale of several months. The switching between flaring and quiescent periods is not random and, at the moment, it is not possible to associate a characteristic duration to each phase.

It is therefore not possible to derive an unique "luminosity function" that can be used to predict the probability to observe the system at a certain luminosity level.

We observed several major radio flares, during which the flux density reached 600 – 800 mJy. There is evidence of periods during which the probability to catch these major radio event is particularly high.

The comparison between the long-term trend of the V magnitude of HR 1099 with its radio behaviour reveals a possible correlation ($R = 0.66$).

The mean V magnitude in the period 1976 – 1993 and of the radio emission flux density, at least at 3.6 and 6 cm, indicates that the strongest radio flares have an high probability to occur close to the average brightness minima, i.e. at the times of maximum spot coverage.

More details can be found in Umana et al. 1995, A&A (in press).

Sunspot Proper Motion and Flare Frequency

G. Csepura[1], *L. Győri*[1], *A.A. Galal*[2]

[1] Heliophysical Observatory of the Hungarian Acad. Sci., H-4010 Debrecen, P.O. Box 30, Hungary
[2] National Research Institute of Astronomy and Geophysics, Helwan, Cairo, Egypt

Flare activity of solar active regions is generally believed to depend on a sheared configuration of magnetic fields (Hagyard et al. 1984). There are cases when the shear necessary for a flare can be attributed to the emergence of a new flux in the spot group (Wang 1992). But, perhaps, a newly born active region can also influence the magnetic field configuration in a nearby active region (Poleto et al. 1993, Gesztelyi et al. 1993). In this paper we are interested primarily in the influence of a newly emerging spot group on a nearby one.

The three neighbouring active regions NOAA AR 6412(B-C), 6413(A) and 6415(D) have been studied between 13-22 December 1990. White-light pictures for studying sunspot proper motion and area evolution were taken at Gyula Observing Station (Hungary), Debrecen Heliophysical Observatory (Hungary) and Helwan Observatory (Egypt). Times and positions of the flares were taken from the Solar Geophysical Data (No. 558, part 1, February 1991).

Interesting events have been found on December 16 and 18, 1990.

1990 December 16: This was the day when the emergence of the newly born region D began (Fig. 1). Up to this day the area of the active region B-C continuously increased but afterwards it began to decrease (Fig. 1.IV). From this day on, the spots B1 and B2 in the active region B-C started a relative revolution around each other and in the same active region, there was a sudden increase in the relative revolution of spots C1 and C2 (Fig. 1.I, 1.III). Up to December 16, the occurrence of flares in the active region B-C was scattered over the whole region but on this day they were concentrated between B and C and increased, while, contrary to this, the daily flare frequency in these active regions was found to have a local minimum (Fig. 1.I, 1.II).

1990 December 18: On this day, group D reached its maximal area, and at the same time, the total umbral area had its maximal value also (Fig. 1.IV). It is interesting that up to this time the umbral areas of the spots had practically not changed (except group D) but on this day there was a slight increase in their values and after this day the previous values were restored. The positions of the flares after this day is shifted to group D (Fig. 1.I, 1.II). As was found on December 16, the rate of the relative revolution of spots C1 and C2 increased on this day, too (Fig. 1.I, 1.III). The above events can probably be explained by the fact that this was the day when group D started to decline.

Considering the above coincidences, we suggest that the emergence or the fast decline of an active region can exert an influence on another active region in its vicinity.

Acknowledgement: This work was supported by the Hungarian Foundation for Scientific Research under grant No. OTKA 007422.

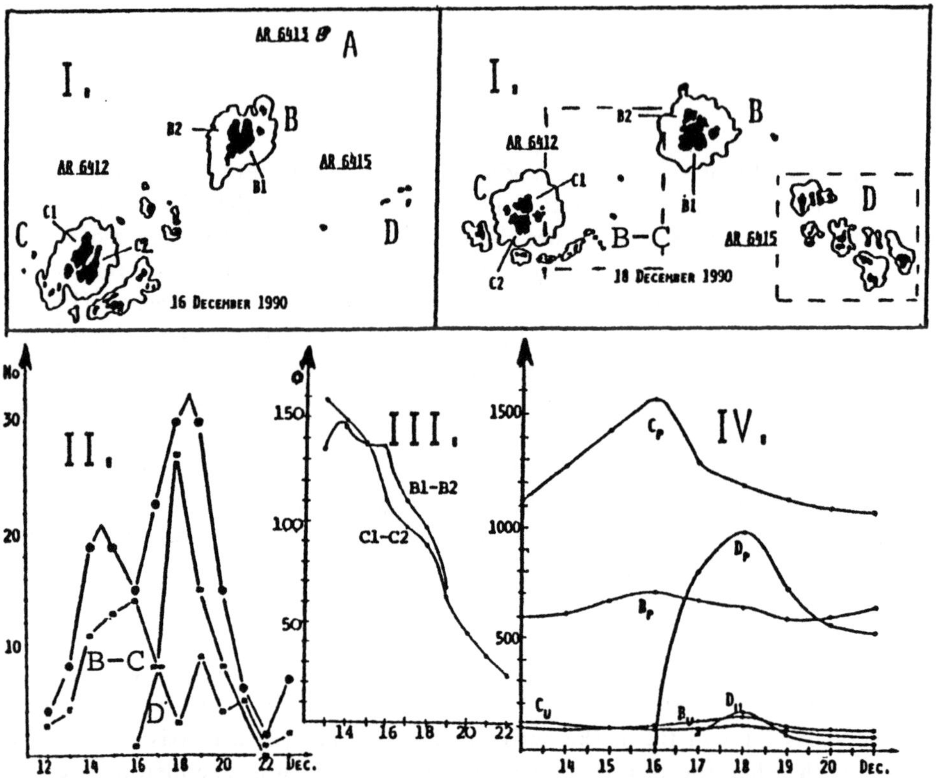

Fig. 1. I. Development of the sunspot groups. II. Unmarked curve shows the daily numbers of flares. The other two curves show the daily number of flares in the two flare centra (B-C and D). III. The rate of the daily relative revolution of spots B1 and B2 and spots C1 and C2. IV. Development of the total umbral and penumbral area of groups B, C and D, expressed in millionths of the Sun's visible hemisphere.

References

Hagyard M.J., Smith J. Jr., Teuber D., West E. A., 1984, Solar Phys. 91, 115
Wang H., 1992, Solar Phys. 140, 85
Poletto G., Gry G.A., Machado M.E., 1993, Solar Phys. 144, 113
van Driel-Gesztelyi L., Csepura L., Nagy I., Gerlei O., Schmieder B., Rayrole J., Demoulin P., 1993, Solar Phys. 145, 77

Optical Flares: Observations and Interpretations

L. V. Mirzoyan

Byurakan Astrophysical Observatory, Armenia

Abstract: Observational results on flare stars in the Galaxy and their interpretation, based on the observational approach to early stages of stellar evolution at Byurakan, are considered. The evolutionary path of red flare stars is determined. Stellar flare activity is explained as an internal phenomenon of stars. The significance of photographic observations of stellar flares in star clusters and associations is emphasized.

1 Introduction

The discovery and study of red dwarf flare stars in the solar vicinity and in star clusters and associations had a fundamental significance for the physics and evolution of red dwarf stars. Flare stars were discovered in the 1950s in the solar vicinity and named UV Cet type variables. The quiescent radiation of these stars is interrupted by randomly distributed flares. UV Cet stars have very low luminosities and are seen in the vicinity of the Sun. The total number of known UV Cet stars is at present about 150 (Shakhovskaya 1995).

Ambartsumian (1954) noted that UV Cet stars at minimum light are similar to normal red dwarfs, but during flares exhibit some peculiarities of the T Tau stars, the characteristic population of stellar associations. This suggests a common nature of UV Cet and T Tau stars, an idea which was strengthened by the discovery of flare stars of higher luminosity in the Orion association (Haro & Morgan 1953). The existence of luminous young flare stars, also in the Monoceros and Taurus Dark Cloud associations, indicates a bridge between the UV Cet and the T Tau stars. Haro (1957) suggested that the stage of flare activity is an evolutionary stage of red dwarfs which follows the T Tau stage. Ambartsumian (1970) found that this path of evolution is the only possible one for all red dwarf stars.

Photoelectric observations of UV Cet stars allow us to study flare events in detail. Photographic observations of stellar flares in clusters and associations, in spite of the low time resolution, turned out to be essential for the study of red dwarf stars in large numbers. We here consider the results of optical observations of stellar flares and their interpretation from the point of view of physics and evolution of red dwarf stars. It is based mainly on investigations carried out at the Byurakan Astrophysical Observatory.

2 Photographic observations of flares in star clusters and associations

After Ambartsumian (1969) had shown that all low luminosity stars in the Pleiades must be flare stars, numerous photographic observations of stellar flares in regions of star clusters and associations were made with wide-field telescopes at the Tonantzintla, Asiago and Byurakan, and subsequently also at the Abastumani, Konkoly and Rozhen observatories. Photography with wide-field telescopes allows to detect all sufficiently powerful stellar flares which occur during the exposure time in a given field.

The search for stellar flares by photographic observations is usually carried out by the method of stellar chains. The exposure time for each image in a stellar chain is 5-15 minutes, depending on the spectral band of observation. The analysis of photoelectric observations of UV Cet flares by Moffett (1974) shows that out of 297 flares registered in the U-band, only 21, and of 342 flares in the B-band, none could have been detected by photographic observations with an exposure time of 5 minutes for each image due to short duration and/or low amplitude of the flares.

Photographic flare observations permit the detection of flares with amplitudes larger than $0\overset{m}{.}5$. The efficiency of photographic observations of flares in regions of star clusters and associations is due to the wide field of the telescope, which monitors the large number of flare stars concentrated in these areas. The probability of observing a flare in a given flare star in the Pleiades with a 1-m Schmidt telescope may be as large as 1/1000 per hour. Because of the large number of flare stars in this region, a flare is detected usually every 1-2 hours on photographic plates (Mirzoyan 1981).

3 Light curves and amplitudes of stellar flares

Because of the low time resolution, photographic light curves of flares are distorted and can give only a general impression (Mirzoyan 1980). The amplitude of a flare increases towards shorter wavelengths. As a consequence, the effectivity of flare observations for fairly bright stars is largest in the U-band. For stars near or below the magnitude limit of photographic observations in the UV, it can be higher in the B-band. Results of photographic flare observations at Byurakan Observatory in regions of star clusters and associations (Table 1) show that they were actually more effective in the B- than in the U-band.

Table 1. Effectivity of photographic observations by the method of stellar chains in the Pleiades region, carried out with the 1m-Schmidt telescope of the Byurakan Observatory in 1967 – 1976 (Mirzoyan 1981).

Spectral band	B	U
Duration of observations [h]	224	202
Number of detected flares	127	85
Mean number of flares/hour	0.57	0.42

4 Colours of stellar flares

Photographic colour indices $B-V$ and $U-B$ of stellar flares by the chain method were obtained in 1979-1981 (Mirzoyan et al. 1981, 1983) using three wide-field telescopes: the 1m and 0.53m Schmidt cameras of the Byurakan Observatory (in U and B, respectively) and the 0.70m Maksutov telescope of the Abastumani Observatory (in V). The exposure time for an image in the chain was 5 minutes. These observations were synchronized with the accuracy of a few seconds. They were carried out in the Orion association (total duration about 35 hours) and in the Pleiades cluster (104 hours). 13 flares in Orion and 38 flares in the Pleiades were detected. Their photographic $B-V$ and $U-B$ indices are in general agreement with those of UV Cet flares (Moffett 1974). The mean $U-B$ index for 45 flares observed near maximum is -0.64 (5 min), as compared to -0.88 for 153 flares of Moffett's UV Cet stars (1 second time resolution). In both cases, the colours of flare emissions became generally redder after maximum.

5 Spectra of flare stars

Originally it was thought that flare (flash) stars in associations and the UV Cet stars also show other differences: the absence of emission lines in the flare star spectra, and their connection with diffuse matter (Herbig 1958, 1968). The former found a natural explanation, the absence of emission lines in flare star spectra was caused by the low spectral resolutions of the objective prism surveys. Spectra of Pleiades flare stars taken with the 6 m-telescope of the Special Astrophysical Observatory showed intense H and Ca II emission (Mirzoyan et al. 1990), similar to of UV Cet spectra (Pettersen & Hawley 1989).

Spectral changes in the flare emission of UV Cet stars were studied by several authors, the most detailed studies being made by Bopp & Moffett (1973) and Moffett & Bopp (1976). Using synchronized spectral and photoelectric observations of UV Cet flares with high time resolution, they found spectral changes for flares on different time scales and confirmed Joy & Humason's (1949) result that during the sharp increase of flare emission the continuum radiation plays the dominant role.

6 Frequencies of photographically detected flares

The mean frequency of flares is different for different flare stars. For a given star, it depends on the spectral band and on the flare amplitude, and usually increases with shorter wavelengths and smaller amplitudes. The mean flare frequency of photographic flares is very low compared to that of photoelectric flares of UV Cet stars. A study of the mean flare frequencies of Pleiades flare stars (the duration of photographic observations exceeds 3000 hours) shows that there is a very large range of frequencies: from one flare per some ten hours to one flare per some ten thousand hours (Mirzoyan 1981).

Ambartsumian (1978) developed a statistical method for determining the distribution function of flare frequencies, and applied it to the Pleiades cluster. The number of flare stars with high flare frequencies is larger in the Pleiades than in Orion (Parsamian 1980), the number of flare stars with lower flare frequencies is larger in Orion. Ambartsumian (1980) showed that the distribution of mean flare frequencies depends on the luminosities of flare stars. This result has been confirmed for UV Cet stars (Mirzoyan 1981, Mirzoyan & Hambarian 1988).

7 Energy spectra of stellar flares

Changes of evolutionary nature are observed in the energy spectra of flare stars. Gershberg (1972) and Krasnobabtsev & Gershberg (1975) showed a correlation between energy spectra and luminosities of flare stars, for increasing stellar luminosity the contribution of frequent, less powerful flares decreases. This correlation was confirmed by Lacy et al. (1976) for UV Cet stars. Korotin & Krasnobabtsev (1986) showed that energy spectra form a natural sequence: energy spectra of UV Cet stars are the continuation of those of flare stars in clusters and associations towards lower energies. This is evidence in favour of a similar physical nature for flare stars of all types.

8 A possible interpretation of flares. Fast and slow flares

Knowledge about detailed properties of flares is needed for the study of their physical nature, although the many photographic flare light curves have also proven to be very useful. Haro (1964) noted that all photographically detected flares in clusters and associations can be divided into two groups according to flare rise time. For most flares, the rise time is less than 20 – 30 minutes. These are the 'fast' flares. In the second group of 'slow' flares, the rise time is longer than 20 – 30 minutes. Haro's division supports Ambartsumian's (1954, 1971) idea about the intrastellar origin of flares: all differences between fast and slow flares (rise time, colour of flare emission, ratio of numbers of flares in the two groups) can be explained by the difference in depth of the layer where the flares originate (corresponding energies are liberated): in the atmosphere of the star or above it. The fast or slow behaviour of a flare does not depend on the physical properties of the flare star on which it occurs. Almost all flare stars showing 'slow' flares have also shown fast flares (Mirzoyan 1981).

In the case of photoelectric flares of UV Cet stars (Moffett 1974) it was found that flare rise times are distributed continuously (Mirzoyan & Melikian 1986). The correlation between mean flare rise time and luminosity of the flare star is in line with the interpretation of 'fast' and 'slow' flares and also in agreement with the fact that the percentage of 'slow' flares increases towards younger systems, i.e. towards flare stars of higher luminosity.

9 On the evolution of red dwarf stars

Photographic observations of flare stars in the Galaxy allow us to outline the evolutionary path of red dwarf stars (Mirzoyan 1990, 1993). The main results were presented by Haro & Chavira (1965), Haro (1976), and Ambartsumian and Mirzoyan (1970, 1975). Here we give some, partially new, observational results, which confirm the evolutionary concept of red dwarf stars.

1. T Tau and flare stars have physical and dynamical similarities: they coexist in stellar associations (Haro & Chavira 1965, Rosino 1969); characteristic features of T Tau stars appear distinctly in flare stars, usually during flares (Ambartsumian 1954); some T Tau stars show flares (Haro & Chavira 1965, Rosino 1969); multiple stars of the trapezium type are known to consist of combinations of T Tau and flare stars (Hambarian 1988). These results indicate a physical and evolutionary connection between T Tau and flare stars. However, T Tau stars belong to the characteristic population of stellar associations, they present the first stage of evolution of red dwarfs after formation.

The coexistence of young flare stars with T Tau stars in stellar associations, and the existence of comparatively old flare stars in star clusters and the general field where T Tau stars are absent, show that flare stars represent the next stage of evolution of red dwarfs after the T Tau stars (Haro 1976, Ambartsumian &Mirzoyan 1970). The question is: is the evolutionary stage of flare stars obligatory for all red dwarf stars?

A positive answer was given by Ambartsumian (1969), who showed that all low luminosity stars in the Pleiades should be flare stars, and concluded that all red dwarf stars pass through this stage of evolution. The discovery of T Tau stars with flare activity showed that both evolutionary stages are overlapping (Ambartsumian 1970). Independent confirmation of the evolutionary status of flare stars was obtained by Mirzoyan et al. (1990): of 34 red dwarf stars studied by Stauffer (1980), 27 turned out to be known flare stars from the catalogue of Haro et al. (1982), and one may suppose that the remaining ones are as yet unknown flare stars.

The high percentage of double and multiple stars among flare stars is noteworthy. Statistics by Gershberg (1978) showed that more than half of the known UV Cet stars are members of multiple systems. Evans (1977) gives a value of about 85% double stars. Rodono (1979) has shown that double stars occur among flare stars more frequently than among other stars.

2. Flare stars in star clusters and associations and UV Cet stars in the solar vicinity are objects of the same physical class by their observed properties (light curves, colours, spectra, energy spectra, physical nature, etc.) (Mirzoyan 1993). In this context, the space distribution of flare stars of different luminosities in the Galaxy, and the origin of the UV Cet stars of the solar vicinity, provide decisive arguments, which have been studied in detail by Arakelian (1969) and Garibjanian (1976). They conclude that the space density of these stars decreases with the distance from the Sun, in agreement with Ambartsumian's (1957) suggestion that UV Cet stars form a cluster, in which they originated, around the Sun. However, Arakelian's and Garibjanian's conclusion is in contradiction with the large variety of ages and kinematical characteristics of the UV Cet stars (Gershberg 1978, Mirzoyan 1981). It appears now that UV Cet stars of the lowest luminosities are distributed uniformly in the galactic disk (Mirzoyan et al. 1988), after the disintegration of their 'maternal' systems.

The observational data indicate that all flare stars originate in star clusters and associations. Young and middle aged flare stars of comparatively high luminosity are observed in star clusters and associations where they originated. Their flare activity can be considered as a criterion for their membership in young systems (Mirzoyan et al. 1993). Old flare stars of low luminosity kept their flare activity because of their slow evolution (small masses). They are mainly observed in the general galactic field, after disintegration of their 'maternal' systems. Kunkel (1975) has shown that the duration of flare activity stage for stars of absolute magnitude $M = 15$ amounts to some billion years. The majority of flare stars of middle ages are observed in star clusters, and the flare activity of comparatively high luminosity stars can be considered as a criterion for their cluster membership (Mirzoyan et al. 1993).

The observed space distribution of flare stars of different luminosities in the Galaxy is thus a natural consequence of their different ages and the increase of duration of the flare activity stage of evolution towards flare stars of lower luminosities (Mirzoyan 1993).

Haro (1962, 1964, 1968) and Haro & Chavira (1965) have shown that in every stellar system there is a boundary spectral class (luminosity) which divides all stars of the system into two groups: flare and non-flare stars. The limit shifts towards later spectral classes (lower luminosities) when older systems are considered. The comparison of flare stars in young (Orion association) and comparatively old (Pleiades cluster) systems shows also that flare stars of the same luminosity are more active in the younger system.

3. The total number of red dwarf stars in the Galaxy is about 10^{10} (Mirzoyan 1993). The overwhelming majority is found in the general field and only a small part in clusters and associations.

4. The physical nature of flare events is not completely understood. Even the flares observed in the Sun, which differ from stellar flares in scale and in nature (see, e.g. Mirzoyan 1981 and Lang 1990) have no generally accepted interpretation (Mirzoyan 1993). However, the discovery and study of the 'fast' and 'slow' flares, their similarity with the processes on the T Tau stars and related objects seem to offer a perspective.

10 Conclusion

A number of observatories take part in photographic observations of flare stars in star clusters and associations (Mirzoyan 1981, 1993). Among them, the most active ones were Tonantzintla and Byurakan. Although the photographic method of observations is evidently very coarse when compared with the photoelectric method, it is the most effective one for collecting a large amount of material. Flares detected by photographic observations are generally more energetic than those registered by the photoelectric method. Photographic observations are successfully used for the search of flare stars in star clusters and associations. Without these, the overwhelming majority of results on the evolution of red flare stars would not exist. On the basis of all these observations, the probable evolutionary path of red dwarf stars is as follows (Mirzoyan 1981):

T Tau stars → Flare stars → red dwarf stars of constant radiation

The possibility to observe all three stages together in the nearest associations is due to the favourable circumstance that evolution progresses more rapidly in stars of higher mass (luminosity). This is the first successful attempt to find a complete evolutionary path of stars, on the basis of observations (see, e.g. Ambartsumian 1947, Ambartsumian & Mirzoyan 1982, Mirzoyan 1984), without any theoretical assumption. After the discovery of stellar associations, which for the first time indicated that star formation in the Galaxy continues to the present time (Ambartsumian 1947, Ambartsumian & Mirzoyan 1982), the concept of red dwarf evolution (Haro 1976, Ambartsumian & Mirzoyan 1970) can be considered the most important achievement of the observational approach to early stages of stellar evolution.

Although the fundamental concept (Haro 1957) about the evolutionary status of flare stars seems to be beyond doubt, further studies are needed. It is necessary to organize synchronous observations of young and more luminous flare stars in clusters and associations and old and less luminous UV Cet stars in different spectral regions (optical, far UV, radio, X-ray) using ground based and space observatories as noted by Rodonò (1990). This approach may bring a solution concerning the physical nature of flare events, and maybe of stellar instability in general (Mirzoyan 1991). Let us hope that such cooperation will be realized before the beginning of the XXIth century.

References

Ambartsumian V.A., 1947, Stellar Evolution and Astrophysics, Acad. Sci. Armenia, Yerevan

Ambartsumian V.A., 1954, Comm. Byurakan Obs. 13, 3

Ambartsumian V.A., 1957, in Non-Stable Stars, M.A. Arakelian (ed.), Acad. Sci. Armenia, Yerevan, p. 9

Ambartsumian V.A., 1969, in Stars, Nebulae, Galaxies, V.V. Sobolev (ed.), Acad. Sci. Armenia, Yerevan, p. 283

Ambartsumian V.A., 1970, Astrofiz. 6, 31

Ambartsumian V.A., 1971, Astrofiz. 7, 557

Ambartsumian V.A., 1978, Astrofiz. 14, 367

Ambartsumian V.A., 1980, in Flare Stars, Fuors and Herbig-Haro Objects, L.V. Mirzoyan (ed.), Acad. Sci. Armenia, Yerevan, p. 85

Ambartsumian V.A., Mirzoyan L.V., 1970, in IAU Coll. No. 15, New Directions and New Frontiers in Variable Star Research, Veröff. Bamberg 9, No. 100, p. 98

Ambartsumian V.A., Mirzoyan L.V., 1975, in IAU Symp. 67, Variable Stars and Stellar Evolution, V. Sherwood and L. Plaut (eds.), Reidel, Dordrecht, p. 3

Ambartsumian V.A., Mirzoyan L.V., 1982, ApSS 84, 317

Arakelian M.A., 1969, in IAU Coll. 4, Non-Periodic Phenomena in Variable Stars, L. Detre (ed.), Academic Press, Budapest, p. 161

Bopp B.W., Moffett T.J., 1973, ApJ 185, 239

Evans D.S., 1977, in Flare Stars, L.V. Mirzoyan (ed.), Acad. Sci. Armenia, Yerevan, p. 40

Garibjanian A.T., 1976, Comm. Byurakan Obs. 49, 63

Gershberg R.E., 1972, ApSS 19, 75

Gershberg R.E., 1978, Flare Stars of Small Masses, Nauka, Moscow

Hambarian V.V., 1988, Astrofiz. 28, 149

Haro G., 1957, in IAU Symp. No. 3, Non-Stable Stars, G.H. Herbig (ed.), Cambridge University Press, Cambridge, p. 26

Haro G., 1962, in Stellar Evolution, J. Sahade (ed.), Astr. Obs. Univ. La Plata, p. 37
Haro G., 1964, in IAU-URSI Symp. No. 20, The Galaxy and the Magellanic Clouds, F.J. Kerr and A.W. Rodgers (eds.), Australian Acad. Sci., Canberra, p. 30
Haro G., 1968, in Stars and Stellar Systems, Vol. 7, B.M. Middlehurst and L.H. Aller (eds.), Chicago Univ. Press, Chicago, p. 141
Haro G., 1976, Bol. Inst Tonantzintla 2, 3
Haro G., Chavira, E. 1965, Vistas Astr. 8, 89
Haro G., Morgan W., 1953, ApJ 118, 16
Haro G., Chavira E., Gonzalez G., 1982, Bol. Inst. Tonantzintla 3, No. 1
Herbig G.H., 1958, in Stellar Populations, D.J.K. O'Connell (ed.), North Holland, Amsterdam, p. 127
Herbig G.H., 1968, ApJ 135, 736
Joy A.H., Humason M.L., 1949, PASP 64, 133
Korotin S.A, Krasnobabtsev V.I., 1986, in Flare Stars and Related Objects, L.V. Mirzoyan (ed.), Acad. Sci. Armenia, Yerevan, p. 147
Krasnobabtsev V.I., Gershberg R.E., 1975, Izv. Krym. Astrofiz. Obs. 53, 154
Kunkel W., 1975, in IAU Symp. 67, Variable Stars and Stellar Evolution, V.Sherwood and L. Plaut (eds.), Reidel, Dordrecht, p. 15
Lacy C.H., Moffett T.J., Evans D.S., 1976, ApJS 30, 85
Lang K.R., 1990, in IAU Symp. 137, Flare Stars in Clusters, Associations, and the Solar Vicinity, L.V. Mirzoyan et al. (eds.), Kluwer, Dordrecht, p. 125
Mirzoyan L.V., 1980, in Flare Stars, Fuors and Herbig-Haro Objects, L.V. Mirzoyan (ed.), Acad. Sci. Armenia, Yerevan
Mirzoyan L.V., 1981, Stellar Instability and Evolution, Acad. Sci. Armenia, Yerevan
Mirzoyan L.V., 1984, Vistas Astr. 27, 77
Mirzoyan L.V., 1990, in IAU Symp. 137, Flare Stars in Clusters, Associations, and the Solar Vicinity, L.V. Mirzoyan et al. (eds.), Kluwer, Dordrecht, p. 1
Mirzoyan L.V., 1991, Early Stages of Stellar Evolution, Acad. Sci. Armenia, Yerevan
Mirzoyan L.V.. 1993, Astrofiz. 36, 277
Mirzoyan L.V., Hambarian V.V., 1988, Astrofiz. 28, 375
Mirzoyan L.V., Melikian N.D., 1986, in Flare Stars and Related Objects, L.V. Mirzoyan (ed.), Acad. Sci. Armenia, Yerevan, p. 153
Mirzoyan L.V., Chavushian H.S., Natsvlishvili R.Sh., Ohanian G.B., Hambarian V.V., Garibjanian A.T., 1981, Astrofiz. 17, 197
Mirzoyan L.V., Chavushian H.S., Melikian N.D., Natsvlishvili R.Sh., Ohanian G.B., Hambarian, V.V., Brutian G.H. 1983, Astrofiz. 19, 725
Mirzoyan L.V., Hambarian V.V., Garibjanian A.T., Mirzoyan A.L., 1988, Astrofiz. 29, 44
Mirzoyan L.V., Hambarian V.V., Garibjanian A.T., Mirzoyan A.L., 1990, Astrofiz. 33, 5
Mirzoyan L.V., Hambarian V.V., Mirzoyan A.L., 1993, Astrofiz. 37, No. 3
Moffett T.J., 1974, ApJS 29, 1
Moffett T.J., Bopp B.W., 1976, ApJS 31, 61
Parsamian, E.S. 1980, Astrofiz. 16, 677
Pettersen B.R., Hawley S.L., 1989, A&A 217, 187
Rodono M., 1979, Flare Active Binary Systems, Catania Obs. preprint
Rodono M., 1990, in IAU Symp. 137, Flare Stars in Clusters, Associations, and the Solar Vicinity, L.V. Mirzoyan et al. (eds.), Kluwer, Dordrecht, p. 371
Rosino L., 1969, in Low Luminosity Stars, S.S. Kumar (ed.), Gordon & Breach, New York, p. 181
Shakhovskaya N.L., 1995, this volume, p. 61
Stauffer J.R., 1980, AJ 85, 1341

The Red Dwarf Star Population in the Galaxy

L. V. Mirzoyan

Byurakan Astrophysical Observatory, Armenia

Red dwarf stars in the Galaxy occur in three forms: as flare stars, T Tau stars and red dwarf stars of constant brightness. Haro (1957) suggested that all these stars present regular evolutionary stages of red dwarf stars.

The space distribution of UV Cet flare stars in the solar vicinity indicates that these low luminosity stars belong to the disk population of the Galaxy. Therefore one can suppose that all red dwarf stars have a space distribution which is similar to that of the flare stars (Mirzoyan et al. 1988a). Only a small part of them is found in star clusters and associations. The T Tau stars are an exception: because of their very short lifetime, all of them are found in stellar associations.

The space distribution of red dwarf stars is determined by the fact that all of them are formed in star clusters and associations and are finally lost from their stellar systems and merge into the general galactic field (Mirzoyan 1995).

The total number of flare stars can be estimated under the assumption that the mean density of these stars in the general galactic field is equal to that in the solar vicinity. The density of UV Cet stars in a sphere of radius 20 pc around the Sun is about $4 \cdot 10^{-3}$ pc^{-3} (Gershberg 1978), but this is a lower limit. Taking 10^{-2} pc^{-3}, and 250 pc for the thickness of the galactic disk, the total number of flare stars in the Galaxy is $5.6 \cdot 10^{8}$ (Mirzoyan et al. 1988b).

The total number of red flare stars in the Galaxy can be estimated from the percentage of flare stars among all red dwarfs. Using the Catalogue of Nearby Stars (Gliese 1969, Gliese & Jahreiss 1979), and the catalogue of UV Cet stars (Gershberg 1978), this percentage is estimated for the red dwarf stars of spectral classes K and M (Mirzoyan et al. 1989). Table 1 shows the distribution of this percentage for red dwarfs in absolute magnitude bins (M_{pg}), where N is the total number of red dwarfs and N_{UV} the number of flare stars among them.

A regular increase of the percentage towards fainter magnitudes is indicated. It can be supposed that the bright red dwarf stars in the galactic field have already terminated their flare activity, while the stars of lower luminosities have kept their ability to produce flares until today.

Table 1 shows that the total number of red dwarf stars in the solar neighborhood is larger than that of flare stars by about one order of magnitude. Therefore the total number of red dwarf stars in the Galaxy, including the total number of

Table 1. Percentage of UV Cet type flare stars among the near red dwarf stars of spectral classes K and M (Mirzoyan et al. 1989)

M_{pg}	N	N_{UV}	N_{UV}/N (%)
7.5 – 8.5	50	1	2
8.5 – 9.5	100	3	3
9.5 – 10.5	131	7	5
10.5 – 11.5	67	10	15
> 11.5	113	47	42
Total	461	68	15

stars which are found today in star clusters and associations, is $\approx 10^{10}$ (Mirzoyan et al. 1988b). This is in agreement with the estimate of the total number of stars, which originated in stellar associations during the lifetime of the Galaxy (10^{10} years), supposing that the star formation process had a constant intensity.

Ambartsumian (1950) showed that the total number of stars that originated in associations in the Galaxy for 10^{10} years is about 10^9 in OB-associations, and more than 10^{10} in T-associations. Since all stars that originated in T-associations and the overwhelming majority of those that originated in OB-associations are red dwarf stars, the total number of red dwarf stars is at least 10^{10}. We do not take into account the red dwarf stars of spectral classes F and G, which must be mainly in the general galactic field. This does not change our order-of-magnitude estimate.

The masses of red dwarf stars are from a few $M_\odot$ for the T Tau stars to some hundredths for the flare stars of lowest luminosities. When taking $0.1 - 1.0\ M_\odot$ as the mean mass of red dwarf stars, their total mass in the Galaxy is of the order of $10^9 - 10^{10}\ M_\odot$, which does not contradict Oort's (1958) estimate.

Thus, the space distribution of red dwarf stars in the Galaxy corresponds to their ages: the stars are found in 'maternal' systems if they are young enough, and in the general galactic field, if they are older.

References

Ambartsumian V.A., 1950, Izv. Acad. Sci. USSR Ser. Phys. 14, 15

Gershberg R.E., 1978, Flare Stars of Small Masses, Nauka, Moscow

Gliese W., 1969, Veröff. Astr. Rechen-Inst. Heidelberg, No. 22

Gliese W., Jahreiss H., 1979, A&AS 38, 423

Haro G., 1957, in IAU Symp. No. 3, Non-Stable Stars, G.H. Herbig (ed.), Cambridge University Press, Cambridge, p. 26

Mirzoyan L.V., 1995, this volume, p. 47

Mirzoyan L.V., Hambarian V.V., Garibjanian A.T., Mirzoyan A.L., 1988a, Astrofiz. 29, 44

Mirzoyan L.V., Hambarian V.V., Garibjanian A.T., Mirzoyan A.L., 1988b, Astrofiz. 29, 531

Mirzoyan L.V., Hambarian V.V., Garibjanian A.T., Mirzoyan A.L., 1989, Astrofiz. 31, 259

Oort J.H., 1958, in Stellar Populations, D.J.K. O'Connell, S.J. (ed.), North Holland, Amsterdam, p. 25

Kinematics and Ages of UV Ceti Stars

*A. Poveda, C. Allen, M.A. Herrera**

Instituto de Astronomia, Universidad Nacional Autonoma de Mexico, Mexico

Abstract: The kinematic properties of 93 UV Ceti stars of the solar neighborhood are studied, based on a list of flares within 25 pc of the Sun ($\pi \geq 0\farcs04$). With updated values for their distances, proper motions and radial velocities (Gliese & Jahreiss 1991) space velocity dispersions are calculated for these stars. It is found that the total velocity dispersion of the flare stars (σ=30$\pm$3 km s^{-1}) is similar to that of the F5 V stars from the same catalogue, for which the conventionally estimated mean age is about $3 \cdot 10^9$ years. A number of flare stars are identified as members of the Hyades, Sirius or Pleiades groups. The velocity dispersions found for the nearby flare stars, as well as their scale height and the membership of some of them to young kinematic groups, indicate that they belong to the young disk population. A small number (7) of UV Ceti stars have kinematics corresponding to the thick disk or halo population. Their long-lived chromospheric activity is interpreted as due to coalescence of old contact binaries. The question of the age of Proxima Centauri is examined in the context of our results, and found to be compatible with the ages of α Centauri A and B.

1 Introduction

It is well known that pre-main sequence late type stars in young clusters and associations exhibit eruptive behavior similar to that of the classic UV Cet stars in the solar vicinity, as was found for the first time by Haro & Morgan (1953) in Orion, and by Haro extensively in Orion, NGC 2264, the Pleiades, the Hyades, etc. (Haro 1968, and references therein). Because the physical and environmental characteristics of such groups are clearly very different from those of the solar vicinity, Haro called "flash stars" the flare-like stars found in clusters and associations.

From the work by Haro and Chavira in Tonantzintla, Rosino in Asiago, and Ambartsumian and his collaborators in Byurakan (see Haro 1968, Gershberg 1971 and Gurzadyan 1980 for extensive references), a number of statistical regularities emerged both among the flash stars and also among the UV Cet stars in

* Also at Programa Universitario de Investigacion y Desarrollo Espacial (PUIDE) UNAM

the solar vicinity. One of the regularities of particular interest to us here involves the correlation between the age of a cluster (or association) and the spectral type of the earliest flares in it: the older the cluster, the later its flare stars, ranging from K0 – K2 for the flash stars in Orion to M3 – M5 in the Hyades, the latter being similar to the spectra of the UV Cet stars in the solar vicinity. This similarity led Haro to an estimate of about 10^9 years for the ages of the oldest UV Cet stars.

Boesgard & Hagen (1974) pointed out an inconsistency between the ages found for components A and B of the α Cen system and Haro & Chavira's mean age for UV Cet stars. Specifically, by means of the intensities of the H and K emission lines, the rotation periods, and the intensity of the lithium lines, Boesgard & Hagen estimated an age of about $3 - 4 \cdot 10^9$ years for α Cen A and B, while for Proxima, a well-known flare star and a convincing member of the system, an expected age of only $4 \cdot 10^8$ years followed from Haro & Chavira's scheme. The question of the age of the flare stars was further discussed by Kunkel (1975) and others, who pointed out evidence indicating that faint UV Cet stars flare throughout the life of the Galaxy.

With the motivation provided by the preceding comments, we study in the present paper the kinematic properties of the nearby UV Cet stars, and we compare them with those of other types of stars in the solar vicinity, whose nuclear ages are well-known. We note that the velocity dispersion of stars in the galactic disk is a monotonically increasing function of age (Wielen 1977) and therefore can be used, as is done here, to date a group of stars. To this end, we use modern data on nearby UV Cet stars, mostly from Gliese & Jahreiss' (1991) catalogue of stars closer than 25 pc.

2 Kinematic properties of nearby UV Cet stars

A working list of stars nearer than 25 pc and exhibiting reasonably certain flare activity has been compiled. The basis for this list is the catalogue provided by Pettersen (1991), which contains all known flare stars with a trigonometric parallax $\pi \geq 0\farcs04$ and a time-resolved record (usually a photometric light curve in optical or X-ray wavelengths) of one or more flares. The lists of Gurzadyan (1980), Kunkel (1975) and Rodono (1978) and the notes to the Gliese (1969) catalogue were also consulted, but we only added stars from these lists to ours when the star was nearer than 25 pc and there seemed to be no doubt about its flare activity. Information on distances, proper motions and radial velocities for these systems was checked and updated using the data provided by Gliese & Jahreiss (1991). In such a way, a list of 86 systems containing a total of 108 stars with reliably reported flare activity was obtained. However, only 74 of these systems, with a total of 93 stars, have sufficient information to allow the calculation of their space velocities.

A small number of flare stars exhibit very discordant kinematics (high space velocities), clearly implying that they do not belong to the normal disk population. The most extreme example is Groombridge 1830B, whose visual companion

Table 1. Velocity dispersions of nearby UV Ceti and main sequence F stars

	σ_U (km/s)	σ_V (km/s)	σ_W (km/s)	σ (km/s)
All 86 flare stars	27.4	12.5	14.7	33.5
43 brightest flare stars	26.7	13.8	14.8	33.5
43 faintest flare stars	27.8	11.6	13.7	33.1
33 flare stars with $M_V \leq 13$, $d \leq 13$'pc	22.7	10.0	10.4	26.9
56 F4+F5+F6 stars	25.7	12.5	15.1	32.3

is spectroscopically classified as a subdwarf; this system has a heliocentric space velocity of 304 km s^{-1}. Other examples, although less extreme, are Gl 412B, Gl 424 and Gl 781, photometrically classified as metal-poor, and Gl 166C, Gl 630.1 and Gl 866.

Velocity dispersions were calculated by the method of percentiles for the 86 flare stars with reliable space velocities, after eliminating the 7 discordant ones (see Table 1). This study shows that there are no significant differences in the velocity dispersions of the faint flare stars compared to the brighter ones. A subgroup of the working list, namely systems closer than 13 pc and with visual magnitudes brighter than $M_V = 13$, which should be less affected by incompleteness and kinematical bias than the entire sample, yields a total velocity dispersion of $\sigma = 27\pm3$ km s^{-1}. This value is somewhat smaller than the velocity dispersion for the whole list, namely $\sigma=34\pm3$ km s^{-1}, pointing to the presence of small kinematical bias in favor of the fainter and remote flare stars. Therefore, we may adopt $\sigma = 30\pm3$ km s^{-1} as a reliable estimate of the velocity dispersion of the UV Cet stars. A comparison of the velocity dispersion of those flares that are members of double and multiple systems, with those that are single, shows no significant difference.

3 Discussion and conclusions

The kinematic properties of the bulk of the UV Cet stars most closely resemble those of main sequence stars of types F4 – F6 in the solar vicinity. For these stars, the estimated mean nuclear ages are about $3 \cdot 10^9$ years (Iben 1967). Assuming that there has been on the average a constant rate of star formation in the galactic disk over the last $8 \cdot 10^9$ years, we conclude that the age of individual UV Ceti stars may span the whole interval from 0 to $6 \cdot 10^9$ years, or even from 0 to $8 \cdot 10^9$ years, when we consider the longer timescales obtained by Maeder & Maynet (1988) for models with convective overshooting. In other words, we may find normal UV Cet stars as old as the Sun (or somewhat older), a fact which would resolve the inconsistency pointed out by Boesgard & Hagen (1974) for the α Cen system.

There remains a small group of apparently single, high velocity – and hence presumably very old – flare stars whose persistent activity remains unexplained.

One conjecture worth exploring involves the occasional coalescence of the components of old contact binaries, which would produce in most cases a fully convective single star with high rotational velocity; such a star would become chromospherically active, with Hα emission, flare activity, etc., but with kinematic properties that would advertise its membership to an old population. Because of the loss of angular momentum through flare activity and stellar winds the fate of these short-period binaries is to coalesce into a rejuvenated rapidly rotating single star. In fact, van't Veer & Maceroni (1989) have found the time-scale for coalescence of short period ($P < 4\,\mathrm{d}$) late type main sequence binaries to be $\approx 7.5 \cdot 10^8$ years. Some of the high velocity stars of our list may have this origin.

The supercluster membership of a significant fraction of the flare stars in the solar vicinity (Hyades: 10, Pleiades-Perseus-Taurus: 4; Sirius: 4; Wolf 630: 3) suggests that UV Cet stars may indeed be identified with the oldest members of the flash star population once belonging to galactic clusters that have long ago been disrupted, and whose members have populated the general galactic field.

Acknowledgements: This work is part of a more comprehensive investigation submitted to A&A. We wish to acknowledge support of the University Space Program (PUIDE). Thanks are due to James L. Green of World Data Center, and to Wayne H. Warren, Jr., of the Astronomical Data Center for providing the CD-ROM of selected astronomical catalogues, and to G. Cordero, C. Lavalley and A. Nigoche for computing assistance. This research has made use of the Simbad database, operated at CDS, Strasbourg, France.

References

Boesgard A.M., Hagen W., 1974, ApJ 189, 85
Gershberg, R.E. 1971, Flares of red dwarf stars, Crimean Astroph. Obs. Translated by D.J. Mullan, Armagh Obs. North Ireland
Gliese W. 1969, Veröffentl. Astron. Rechen-Inst. Heidelberg, 22
Gliese W., Jahreiss H., 1991, in L.E. Brotzman, S.E. Gessner: Selected Astronomical Catalogs, Volume I, National Space Science Center, NASA GSFC
Gurzadyan G.A., 1980, in Flare Stars, Y. Terzian (ed.), Pergamon, Oxford
Haro G., 1968, in Stars and Stellar Systems, Vol. 7, Nebulae and Interstellar Matter, B.M. Middlehurst & L.H. Aller (eds.), University of Chicago Press, Chicago, p. 141
Haro G., Chavira E., 1966, in Vistas in Astronomy, Vol. 8, A. Beer (ed.), Pergamon, Oxford, p. 89
Haro G., Morgan W.W., 1953, ApJ 118, 16
Iben I., Jr. 1967, ARAA 5, 571
Kunkel W.E., 1975, in Variable Stars and Stellar Evolution, V.E. Sherwood & L. Plaut (eds.), Reidel, Dordrecht, p. 15
Maeder A., Meynet G., 1988, A&AS 76, 411
Pettersen B.R., 1991, Mem. Soc. Astr. Ital., 62, 217
Rodonó M., 1978, A&A 66, 175
van't Veer F., Maceroni C., 1989, A&A 220, 128
Wielen R., 1977, A&A 60, 263

Flare Activity Among Nearby Stars

N.I. Shakhovskaya

Crimean Astrophysical Observatory, Nauchny, 334413 Crimea, Ukraine

I selected the common objects from our Data Base on 234 UV Cet Flare Stars and Related Objects and the Third Catalogue of Nearby Stars by Gliese & Jahreiss (CNS-3). It is found that manifestation of flare activity (FA) (flares in any spectral region or photometric variability caused by spots) is observed on 143 flare stars (FS) out of the 3803 CNS-3 stars within 25 pc of the Sun with absolute magnitudes M_V from 5.2 (Sp G2 V) to 18.8 (Sp dM5e – dM6e).

Magnitudes and spectra of flare stars: For each M_V-interval, Table 1 gives the "mean" spectral class (Sp), the number of FS (N_f) and the portion of FS among other red dwarf stars from CNS-3 (N_f/N). Absolutely fainter stars include a higher portion of known FS. However, observational selection effects can cause this dependence, because the probability is higher to register flares over the background radiation of a faint star. The observed portion of FS among all red dwarfs from CNS-3, which are fainter than M_v=5, is about 5 %.

M_V	Sp	N_f	N_f/N
5 : 6.99	G2	18	0.03
7 : 8.99	K5	28	0.05
9 :10.99	M0	36	0.05
11 :12.99	M3	25	0.03
13 :14.99	M4	21	0.12
15 :16.99	M5	11	0.18
17 :18.99	M6	4	0.30

In the quiet spectrum of most FS from our Data Base H I or Ca II emission is present, but in 10 % of FS these features are absent, and their quiet spectra differ in no way from spectra of normal red dwarfs.

Binary systems among flare stars. 28 from 143 FS are components of close visual binary systems, and flare activity was observed in the total light of both components. In addition, on 13 visual binaries the flare activity was observed on both components separately. Also, on 7 visual binaries the flare activity was observed only on the bright component, and on 4 only on the faint component. Morever, on 4 stellar systems the flares were observed on the third (C) component. Only 77 objects among the FS, or 53 % are not known as visual binaries. The appropriate portion of single stars in the same M_v range and Sp from CNS-3 is 78 %. Thus, among FS the portion of visual binary systems is higher then among normal stars, and in these systems the flare activity can be observed on the bright as well as on the faint components.

Mean density of flare stars in the solar neighbourhood Following Arakelian (1968) I have arranged the above 143 FS in order of their increasing dis-

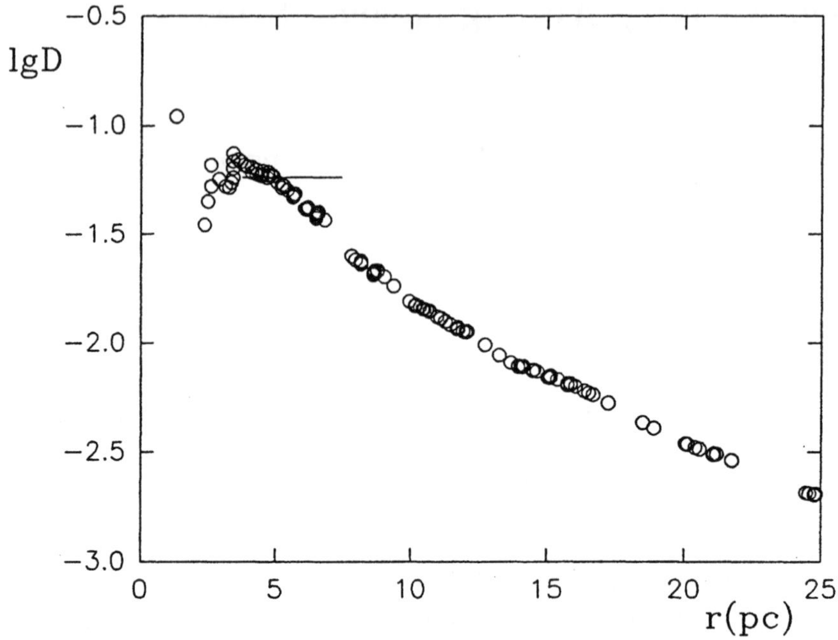

Fig. 1. Determination of the mean density of FS in the solar neighbourhood.

tance from the Sun and have estimated their mean density from the formula $D(r) = 3k/4\pi r_k^3$, (r is the distance from the Sun to the kth star).

The $D(r)$ values based on these data are given in Fig. 1. The first point corresponds to the flare star Proxima Cen, which is the nearest star to the Sun. The next group of two dozen points form a plateau, and beginning from 5 pc $D(r)$ decreases monotonically. It is suspected that beyond this distance more and more low luminosity stars remain undetected. The plateau allows to estimate the mean density of flare stars to be ~ 0.056 star pc^{-3}. Earlier, with only 67 objects, Gershberg & Shakhovskaya (1976) using the some method have estimated the mean density of flare stars to be ~ 0.045 star pc^{-3}. It is possible to think that the real number of FS is considerably higher because FA is found on even distant active stars. Thus, the new value $D = 0.056$ star pc^{-3} should be considered as a lower limit of the mean density of FS in the solar neighbourhood.

The total density of all stars in the solar neighbourhood is $0.111 - 0.118$ star pc^{-3} according to Arakelian (1968). Therefore it is possible to suggest that at least half of the stars in the solar neighbourhood are objects of the UV Cet type.

Acknowledgement: I am grateful to the International Science Foundation which funded my participation at IAU Coll. 151.

References

Arakelian M.A., 1968, Astrofiz. 4, 617
Gershberg R.E., Shakhovskaya N.I., 1976, ApSS 44, 463

Young and Old Subsystems of Flare Stars

L.V. Mirzoyan, V.V. Hambarian

Byurakan Astrophysical Observatory, Armenia

Photographic observations with wide-field telescopes of flare stars in the Orion association and the Pleiades cluster have been carried out since 40 years. About 500 flare stars were found in each of these systems. The ages of these stellar systems differ by one or two orders of magnitude (Mirzoyan 1991). This fact allows us to compare them, in order to show that they fit into the concept, suggested first by Haro (1957), that the evolutionary status of flare stars is a stage of red dwarf star evolution which follows that of the T Tau stars (Haro 1976, Ambartsumian & Mirzoyan 1970).

The basic difference between the Orion and Pleiades subsystems of flare stars has been known for a long time: the coexistence of flare stars and T Tau stars in the Orion association, and the absence of the latter group in the Pleiades. Some of the T Tau stars show flare activity (Haro 1964). This fact and the existence of multiple systems of trapezium type, which are dynamically unstable, in the Orion association, are arguments in favour of an evolutionary connection between these two types of stars.

Another noteworthy difference between the Orion and Pleiades subsystems is the luminosity of the brightest flare stars, which is 3^m brighter in Orion. For flare stars of the same luminosity, the flare activity is higher in Orion.

The distribution of the mean flare frequencies is also different for the Orion and Pleiades subsystems, which can be explained by the difference of their luminosity functions. The reason is that the mean flare frequency increases towards lower luminosities.

All abovementioned differences between the flare star subsystems of different ages are in remarkable accordance with the evolutionary status of flare stars. For example, the absence of T Tau stars in the middle-age Pleiades cluster is naturally explained by the young age of T Tau stars. They belong to the characteristic population of stellar associations (Ambartsumian 1947) and are transformed later into flare stars (Mirzoyan 1981). The higher luminosity of the brightest flaring stars in the younger subsystem is a consequence of the fact that the rate of evolution increases with mass (or luminosity).

It should be added that the subsystem of flare stars in Orion is much larger than that of the Pleiades. However, the estimated total number of flare stars in the Orion association is only two times larger than in the Pleiades cluster. It seems that this difference is a consequence of observational selection caused by the noticeable difference in distance. The total number of flare stars in Orion must be larger than that in the Pleiades by at least one order of magnitude.

References

Ambartsumian V.A., 1947, Stellar Evolution and Astrophysics, Acad. Sci. of Armenia, Yerevan

Ambartsumian V.A., Mirzoyan L.V., 1970, in IAU Coll. No. 15, New Directions and New Frontiers in Variable Star Research, Veröff. Bamberg 9, No. 100, p. 98

Haro G., 1957, in IAU Symp. No. 3, Non-Stable Stars, G.H. Herbig (ed.), Cambridge University Press, Cambridge, p. 26

Haro G., 1964, in IAU-URSI Symp. No. 20, The Galaxy and the Magellanic Clouds, F.J. Kerr and A.W. Rodgers (eds.), Australian Acad. Sci., Canberra, p. 30

Haro G., 1976, Bol. Inst. Tonantzintla 2, 3

Mirzoyan L.V., 1981, Stellar Instability and Evolution, Acad. Sci. of Armenia, Yerevan

Mirzoyan L.V., 1991, Early Stages of Stellar Evolution, Acad. Sci. of Armenia, Yerevan

Flare Activity and Cluster Membership Probability of Flare Stars

L.V. Mirzoyan, V.V. Hambarian, A.L. Mirzoyan

Byurakan Astrophysical Observatory, Armenia

Mirzoyan (1976) showed that the concentration of flare stars around the center of the Pleiades cluster (Alcyone) was the same, irrespective of their proper motions. At that time, however, proper motions of only a few flare stars were known. The result was confirmed on the basis of more extensive observational material (Chavushian 1979, Mirzoyan 1983). Photographic observations of stellar flares in the general galactic field during 181 hours yielded the detection of only a single flare, i.e. the percentage of flare stars in the general galactic field is about 10% of the total number of flare stars detected in the regions of stellar clusters and associations (Chavushian 1979, Mirzoyan et al. 1988).

This result shows that flare activity can be considered as a definitive cluster membership criterion, which appears to be a better one than the cluster membership probability, which is based on proper motions. To prove this, cluster membership probabilities (Stauffer et al. 1991) for 408 Pleiades cluster flare stars from the catalogue by Haro et al. (1982), are used (Table 1).

Table 1. The distribution of the cluster membership probabilities P for Pleiades flare stars according to Stauffer et al. (1991)

P	Number of stars
0.0 – 0.2	193
0.2 – 0.4	14
0.4 – 0.5	17
0.5 – 0.6	13
0.6 – 0.8	56
0.8 – 1.0	135
Total	408

All flare stars of Table 1 were divided into two groups: those with $P > 0.5$, being probable members of the Pleiades cluster, and those with $P < 0.5$, probable non-members. A plot of their surface distributions around the centre of the Pleiades shows the absence of essential differences between both distributions, nor are any differences seen between the distributions of surface stellar density $d(r)$ around the centre of the Pleiades, when weights P and $(1 - P)$ are taken (Mirzoyan et al. 1993). Hence, the new cluster membership probabilities of flare stars based on proper motions confirm the conclusion that the overwhelming majority of flare stars is connected with the Pleiades cluster.

On the other hand, the inspection of the new cluster membership probabilities indicates that the concentration of flare stars around the center of the Pleiades is stronger for flare stars of comparatively high luminosities: almost all flare stars of comparatively high luminosities belong to the cluster. Such a regularity is also observed for the flare stars of the Orion association (Mirzoyan et al. 1993). Therefore it can be supposed that all flare stars of comparatively high luminosities are members of star clusters and associations, i.e. they are formed in these stellar systems and at present, the further they are from their centres, the lower are their luminosities (smaller masses).

The detection of flare activity, especially for stars with small flare frequencies, is very difficult. For practical purposes, it is convenient to use another criterion which can be determined more easily. The presence of strong emission lines in the spectra of flare stars is such a criterion.

Almost all spectroscopically studied flare stars of the Pleiades region have rather intense emission lines of H and Ca II (e.g. Stauffer 1980, Mirzoyan et al. 1990). In their spectral study of the Hyades and Pleiades red dwarfs, Kraft & Greenstein (1969) found that the existence of strong H and Ca II emission can be considered as a powerful criterion for cluster membership.

Prosser at al. (1991) showed that there is a substantial difference in $H\alpha$ emission line strength between flare and non-flare stars in the Pleiades. A few exceptions from this rule can be explained by the fact that all low luminosity stars are flare stars, but not all of them are yet known as flare stars (Ambartsumian 1969). Hence, on the basis of existing observational data on flare stars, the flare activity can be considered as a powerful criterion for their membership in nearby clusters and associations.

References

Ambartsumian V.A., 1969, in Stars, Nebulae, Galaxies, V.V. Sobolev (ed.), Acad. Sci. of Armenia, Yerevan, p. 283

Chavushian H.S., 1979, A Study of Flare Stars in the Region of the Pleiades Aggregate, Byurakan Astrophys. Obs.

Haro G., Chavira E., Gonzalez G., 1982, Bol. Inst. Tonantzintla 3, No. 1

Kraft R.P., Greenstein J.L., 1969, in Low Luminosity Stars, S.S. Kumar (ed.), Gordon and Breach, New York, p. 65

Mirzoyan L.V., 1976, in Stars and Galaxies from Observational Points of View, E.K. Kharadse (ed.), Georgian Acad. Sci., Tbilisi, p. 121

Mirzoyan L.V., 1983, Astrofiz. 18, 588

Mirzoyan L.V., Hambarian V.V., Garibjanian A.T., Mirzoyan A.L., 1988, Astrofiz. 29, 531

Mirzoyan L.V., Hambarian V.V., Mirzoyan A.L., 1993, Astrofiz. 36, No. 3

Mirzoyan L.V., Hambarian V.V., Garibjanian A.T., Mirzoyan A.L., 1990, Astrofiz. 33, 5

Prosser C.F., Stauffer J.R., Kraft R.P., 1991, AJ 101, 1361

Stauffer J.R., 1980, AJ 85, 1341

Stauffer J.R., Klemola A., Prosser C., Probst R., 1991, AJ 101, 980

New EUV Selected Flare Stars

Bill Ball, Gordon Bromage

Centre for Astrophysics, University of Central Lancashire, Preston PR1 2HE, UK

1 Introduction

The ROSAT Wide Field Camera all-sky EUV survey of 1990-91 identified over 200 active late-type stars (Pounds et al. 1993). Most of these were not known to be active prior to the survey and follow-up optical spectroscopy has been successful in selecting the most active dMe stars. These stars are characterized by their Balmer line emission which is direct evidence for an active chromosphere. We present photometric monitoring data for 4 new dMe stars, including the discovery of flaring activity on all 4 stars.

2 Observations

The observations were made at the South African Astronomical Observatory at Sutherland near Capetown in October 1994, using the 0.5m and 1.9m telescopes for photometry and spectrometry, respectively. The 0.5m telescope is equipped with an automated, pulse-counting photometer. Flare monitoring of all the stars was done in the U-band with either 5 or 10 s integration times. All the stars monitored were between magnitude $V = 11$ and 13. A total of 15 flares were detected, including at least one on each star (Table 1). Comparison stars were measured in standard *UBVRI* photometric bands for each star.

There are some unique aspects of this observing run. This is the first time that candidate stars for flare monitoring have been chosen systematically from their EUV/bolometric flux ratio. Several flares, including the largest, were seen on previously unmonitored dMe stars and the $4\overset{m}{.}0$ flare was one of the largest optical flares to be observed since the giant $4\overset{m}{.}5$ flare in 1985 on AD Leo (Hawley & Pettersen 1991).

On the last but one night of simultaneous observing a large 4^{m} flare was observed on the EUV selected star RE J0241-53N, with a time resolution of 10 sec (Fig. 1). This star is the northerly, fainter component of a binary dMe pair with a separation of 25″. The flare started at 00:20 UT on 23rd., rose $\approx 4\overset{m}{.}0$ in about 15 min. and had not returned to quiescence by the end of the observations

Table 1. Summary of flares observed in October 1994

Star	Monitoring time (hours)	Largest Flare ($\Delta\, U$)	No. flares seen
RE J0241-53N	36	4.0	5
RE J0604-34	9	2.5	5
RE J0625-60	10	1.5	3
RE J2055-17B	21	0.3	1

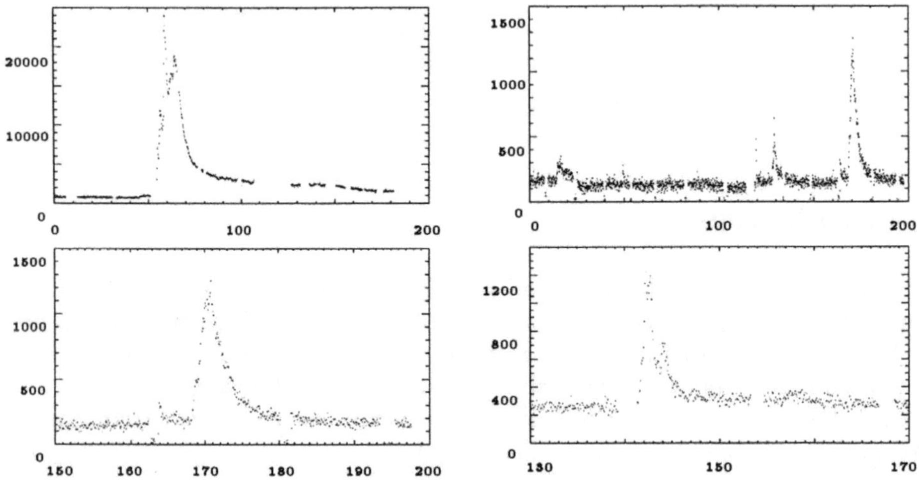

Fig. 1. Light curves of counts vs. time in min from start of monitoring on RE J0241-53N (top left), RE J0604-34 and RE J0625-60 (bottom right).

at 03:00. The breaks in the curves are due to sky readings. It can be seen that the number of counts started to increase just prior to the main flare onset. This star also flared on October 24 and 25, rising $0\overset{\mathrm{m}}{.}5$ and $1\overset{\mathrm{m}}{.}2$, respectively, and twice on October 27, rising $0\overset{\mathrm{m}}{.}7$ each time.

The star RE J0604-34 produced a sequence of events on October 31 (Figs. 1b, c). Two small short-lived flares of $\approx 0\overset{\mathrm{m}}{.}8$ occurred at 22:15 and 22:45, followed at 23:58 by a 20 sec $1\overset{\mathrm{m}}{.}4$ and one of $1\overset{\mathrm{m}}{.}7$ at 24:07. The largest flare on RE J0604-34 ($2\overset{\mathrm{m}}{.}5$) occurred at 24:48 and again there was some preflare activity. The integration time was 5 s for this star.

The star RE J0625-60 was seen to flare at 22:19, 23:44 and 24:44 on October 30 (Fig. 1d), also with 5 s integration time. The largest was a $\approx 1\overset{\mathrm{m}}{.}5$ event lasting $\approx$ 30 min.

Finally, a small $0\overset{\mathrm{m}}{.}3$ flare was seen on RE2055-17B at 18:32 on October 31 using 5 secs integration time.

References

Hawley S.L., Pettersen B.R., 1991, ApJ 378, 725
Pounds K.A., et al., 1993, MNRAS 260, 77

Flare Stars in a Field Near the North Galactic Pole: ROSAT and Optical Observations

G.A. Richter[1], *H.-J. Bräuer*[1], *J. Greiner*[2]

[1] Sternwarte Sonneberg, 96515 Sonneberg, Germany
[2] Max-Planck-Institut für Extraterrestrische Physik, 85740 Garching, Germany

In a field of 100 square degrees around 26 Com, 238 X-ray sources have been discovered by *ROSAT* during the All-Sky-Survey observations. These sources were identified using objective prism spectrograms taken with the Hamburg Schmidt telescope on Calar Alto (courtesy N. Bade) and including the positional accuracy of typically 30″ and the X-ray to optical luminosity ratio. In an ongoing investigation of these X-ray sources on more than 400 plates of the Sonneberg astrographs 400/1600 mm and 400/2000 mm, a total of 5 of the M type counterparts were found to be flare stars. In Table 1 we give the *ROSAT* name (column 1), the new designation as a Sonneberg variable (2), the position of the optical counterpart (3), the magnitude range of the newly discovered flare stars (4), the *ROSAT* PSPC countrate during the All-Sky-Survey (5), the distance D between X-ray and optical position (6). All 5 objects had not been known to be flaring. The real amplitudes must be larger than those given in column 3 because the (unknown) duration of the flares is usually much shorter than the exposure time of the plates. The figures give APM finding charts of these new flare stars.

The following are remarks on the brightness behaviour of the individual objects (in parentheses: exposure time of the plate in min.; estimated magnitudes):

RX J1235.5+1954: Eclipsing star with flaring activity. Preferred magnitude level about 15^{m}. Flares: 243 8817.660 (60; $14\overset{\mathrm{m}}{.}6$), 244 2513.449 (140; $14\overset{\mathrm{m}}{.}6$), 6121.582 (30; $14\overset{\mathrm{m}}{.}5$), 6881.447 (20; $14\overset{\mathrm{m}}{.}6$). Minima near $15\overset{\mathrm{m}}{.}5$: 243 8106.464, 8142.581, 244 5077.426, 7262.415, 7566.592.

RX J1253.6+2247: Eclipsing star with flaring activity. Preferred magnitude about 16^{m}. Flares: 243 7822.417 (40; $15\overset{\mathrm{m}}{.}7$), 244 5056.400 (30; $15\overset{\mathrm{m}}{.}5$), 8683.507 (25; 15^{m}). Observed minima near $16\overset{\mathrm{m}}{.}5$: 243 8083.523, .566, 9616.380, 244 4367.370, .384, 4636.657, 7566.592, 7672.411.

RX J1256.8+2329: Eastern component of two stars of similar magnitude and colour. Observed flares: 243 7668.641 (60; $13\overset{\mathrm{m}}{.}5$), 8088.536 (60; $14\overset{\mathrm{m}}{.}5$), 9205.483 (40; 14^{m}), 9592.342 (60; 14^{m}), 9945.500 (40; 14^{m}), 244 0981.514 (40; 14^{m}), 2107.533 (20; 15^{m}), 4663.524 (40; $14\overset{\mathrm{m}}{.}5$), 6552.376 (20; 15^{m}).

RX J1258.9+2112: Eclipsing star with flares. Preferred magnitude level about 17^{m}. Flares: 243 7790.418 (60; $15\overset{\mathrm{m}}{.}5$), 8851.470 (60; 16^{m}), 9537.451 (60; 16^{m}), 244 5815.404 (18; 16^{m}), 5820.389 (18; $15\overset{\mathrm{m}}{.}5$). Observed minima (about $17\overset{\mathrm{m}}{.}5$): 243 7765.468, 8084.545, 9535.592, 9618.504.

RX J1300.5+2255: Numerous flares: 243 7764.405 (60; $16^{\rm m}.5$), 7783.440 (24; $15^{\rm m}$), 7790.459 (60; $15^{\rm m}.5$), 7822.445 (40; $16^{\rm m}.5$), 7823.455 (60; $16^{\rm m}.5$), 9611.507 (60; $16^{\rm m}.5$), 244 2513.449 (100; $16^{\rm m}.5$), 4342.372 (20; $16^{\rm m}.5$), 5384.628 (60; $16^{\rm m}$), 6534.403 (18; $15^{\rm m}.5$), 7613.457 (14; $16^{\rm m}$), 8357.474 (22; $16^{\rm m}$).

Table 1. Newly discovered flare stars in the Coma field

ROSAT Name	Variable Star designation	Position of flare star (2000.0)	Magnitude range (pg)	X-ray intensity (cts/sec)	D
RX J1235.5+1954	S10935	$12^{\rm h}35^{\rm m}32^{\rm s}.9$ $+19°54'43''$	14.5–15.5	0.018	17″
RX J1253.6+2247	S10936	$12^{\rm h}53^{\rm m}36^{\rm s}.2$ $+22°47'35''$	15.0–16.5	0.075	2″
RX J1256.8+2329	S10937	$12^{\rm h}56^{\rm m}52^{\rm s}.2$ $+23°29'50''$	13.0–15.5	0.071	5″
RX J1258.9+2112	S10938	$12^{\rm h}59^{\rm m}00^{\rm s}.3$ $+21°12'56''$	15.5–17.5	0.029	15″
RX J1300.5+2255	S10939	$13^{\rm h}00^{\rm m}33^{\rm s}.2$ $+22°55'41''$	15.0–17.0	0.010	27″

At the same time there are another 6 known flare stars in the field covered by our investigation. For two of these, finding charts are available, and a check of the coordinates given in the GCVS revealed these to be wrong by $\sim 8'$ (CT Com, Haro 1968) and $\sim 10'$ (CV Com, Argue & Kenworthy 1968), respectively. Using the new positions, both sources are not detected during the *ROSAT* All-Sky-Survey. The other four sources (GCVS positions) have also not been detected in X-rays by *ROSAT*. However, since no finding charts are available for these stars, we have to assume that the coordinates given in the discovery papers (references see GCVS) are correct. The upper limit for the X-ray intensity of all the sources is 0.01 cts/s in the 0.1–2.4 keV range. These flare stars are listed in Table 2. The magnitude range and the coordinates for the four stars without available finding charts were taken from the GCVS, the latter have been precessed to equinox 2000.

Table 2. Flare stars not detected in the *ROSAT* Survey

Star	coordinates (2000.0) R.A.	DEC.	Magnitude range (pg)	Comment on coordinate
HO Com	$12^{\rm h}\,19^{\rm m}32^{\rm s}$	$+\,25°13'.3$	16.2 – 21.0	from GCVS
CT Com	$12^{\rm h}\,23^{\rm m}49^{\rm s}.0$	$+\,24°07'56''$	12.5 – 16.0	newly determined
HQ Com	$12^{\rm h}\,25^{\rm m}12^{\rm s}$	$+\,23°36'.4$	14.7 – 19.6	from GCVS
CV Com	$12^{\rm h}\,24^{\rm m}26^{\rm s}.6$	$+\,25°45'07''$	14.3 – 16.8	newly determined
HS Com	$12^{\rm h}\,26^{\rm m}24^{\rm s}$	$+\,25°54'.5$	15.1 – 21.0	from GCVS
HV Com	$12^{\rm h}\,31^{\rm m}53^{\rm s}$	$+\,25°09'.4$	15.3 – 16.6	from GCVS

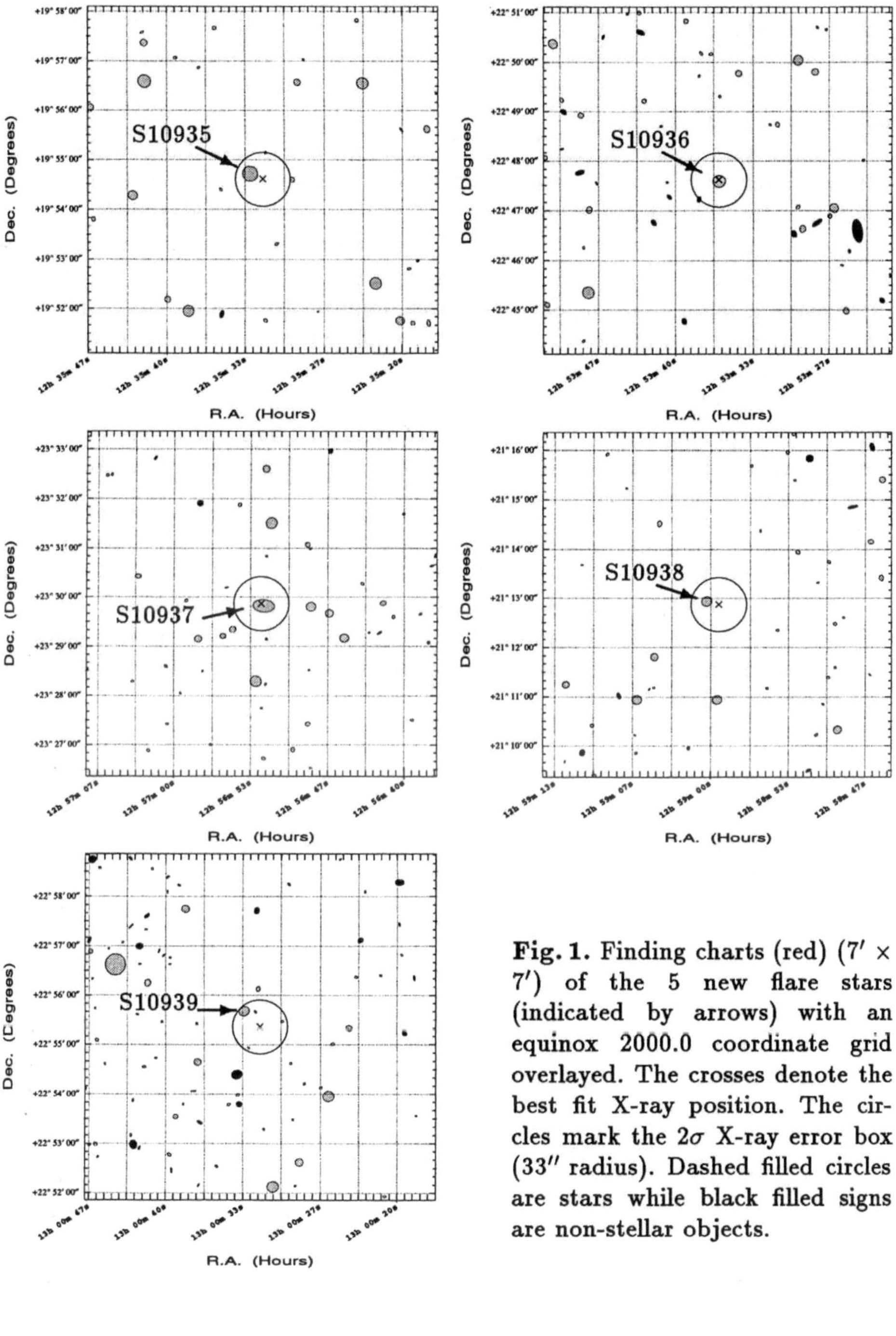

Fig. 1. Finding charts (red) (7′ × 7′) of the 5 new flare stars (indicated by arrows) with an equinox 2000.0 coordinate grid overlayed. The crosses denote the best fit X-ray position. The circles mark the 2σ X-ray error box (33″ radius). Dashed filled circles are stars while black filled signs are non-stellar objects.

The flare stars of Table 2 have been found according to the classical method for searching for eruptions (Haro 1968) by taking multiple exposures. In principle, this method allows to calculate the discovery probability of flare stars according to Poisson statistics (Ambartsumian 1970). In practice, the determination of this very low probability is rather laborious and gives reasonable results only with large samples.

Though statistics is small, the above reported discovery of 5 new flare stars approximately doubles the number of known flare stars in this high galactic latitude field. The fact that none of the X-ray selected flare stars was optically discovered prior to *ROSAT*, is a further indication of a very small discovery probability of faint flare stars. This implies a high frequency and, because of the low luminosity, a large space density. However, quantitative conclusions cannot be drawn without investigations of additional fields (at different galactic latitudes).

It is clear, however, that the use of a complete X-ray selected sample of sources represents an effective way of identifying optically faint, active dwarf stars, (and in particular if a large number of archival plates can be used). Further identification work of the *ROSAT* All-Sky-Survey sources promises many more flare stars to be discovered.

Acknowledgement: This work has been supported by funds of the German Bundesministerium für Forschung und Technologie under grant 05 2SO524 (GAR) and FKZ 50 OR 9201 (JG). We are grateful to R. McMahon for installing the software used for the APM finding charts. The *ROSAT* project is supported by the German Bundesministerium für Forschung und Technologie (BMFT/DARA) and the Max-Planck-Society.

References

Ambartsumian V., 1970, Astrofizika 6, 31
Argue A.N., Kenworthy C.M., 1968, Observatory 88, 109
Haro G., 1968, Stars and Stellar Systems 7, 141

Some Results of the Coordinated Observations of the Red Dwarf Flare Star EV Lac in 1986–1994

R.E. Gershberg

Crimean Astrophysical Observatory, Nauchny, Crimea 334413, Ukraine

In 1986–94 a series of coordinated monitorings for the flare star EV Lac has been performed. The observations were aimed to collect homogeneous data on the star both in its active and quiet states, to study its flares and starspots. These photometric patrols were accompanied by monitorings of EV Lac of different other kinds: polarimetric, spectral, radio, UV and high energy studies. In total, these monitorings covered a range from decametre radio wavelengths through IR, optical and UV windows up to very high energy γ-rays. The list of participants of these studies is very large. Up to date the results of 1986, 1987, 1990 and 1991 are published (Gershberg et al. 1991a,b, 1993, Alekseev et al. 1994), the data of 1989 are submitted (Berdyugin et al. 1995), the data of 1992 (Abdul-Aziz et al. 1995) will be submitted in the near future. In 1993, photometric observations were accompanied by monitoring observations of *EUVE* and *IUE*, in 1994, VHE γ-ray observations were included. The publication of data collected in 1993 and 1994 is in preparation. In this communication I shall concentrate on several new findings related to the EV Lac flares only.

1. Colour-Colour diagrams: Among more than two hundred flares detected during these runs with the 1.25 m AZT-11 telescope of the Crimean Astrophysical Observatory and the photometer-polarimeter by Piirola (1984), several flares were registered in the total *UBVRI* bands. From these flares we selected those for which the pure flare radiation colour indices were determined with $\sigma < 0\overset{m}{.}25$ at brightness maximum. In a second step, we rejected double peak flares, since their light curves may be superpositions of independent events. As a result, we obtained 7 flares with amplitudes ΔU from $3\overset{m}{.}7$ to $2\overset{m}{.}0$ and considered their pure flare radiation colour indices at brightness maxima, at $1-2$ minutes before and $5-8$ minutes after maxima.

To analyze their pure flare radiation we have constructed theoretical colour-colour diagrams covering the whole *UBVRI* range. Their construction was described earlier, and the last version of such a diagram is presented in Fig. 1 (Abdul-Aziz et al. 1995). They include theoretical colours of black body radiation, emission of hydrogen plasma, both optically thin and thick in Balmer continuum, and radiation emitted by the upper layers of red dwarf stars heated by fast particles.

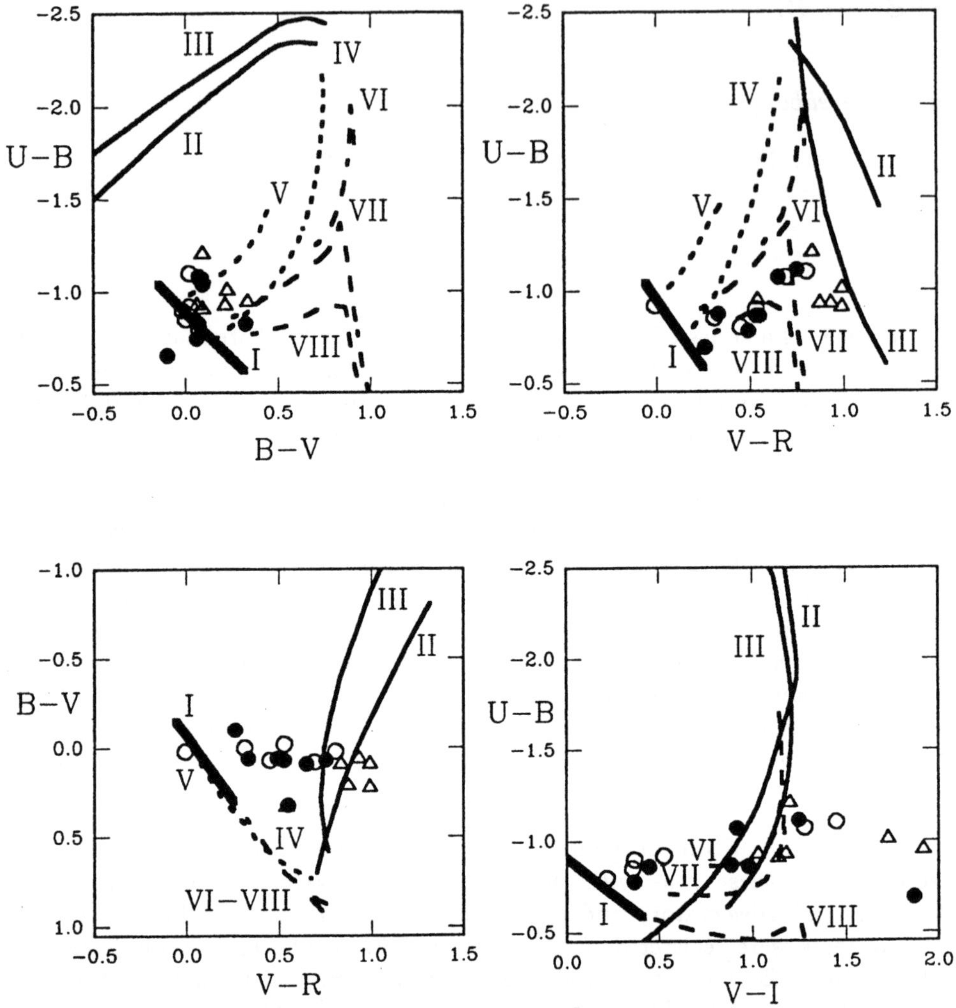

Fig. 1. Colour-colour diagrams of the pure flare radiation of powerful flares on EV Lac. Open circles – premaximum flare radiation, full circles – maximum flare radiation, and triangles – flare decay phase radiation. I – black body radiation for temperatures ranging from 6 000 to 20 000 K; II and III – emission of hydrogen plasmas, optically thin in the Balmer continuum (Ba_c), with $T = 10\,000$ K and $n = 10^{12}$ and 10^{14} cm^{-3}, respectively; IV and V – emission of optically thick in the Ba hydrogen plasmas, optically thick in the Ba_c, with $T = 10\,000$ and $15\,000$ K, respectively; VI, VII and VIII – radiation emitted by upper atmospheric layers heated by fast particles of different energies.

Fig. 1 shows that at flare maxima (full circles) the flare radiation in the *UBV* bands is rather close to that of a blackbody. However, in the *R* and *I* bands we find significant deviations from such a radiation model. Examining the $(U-B,\ V-R)$ diagrams, one finds a shift of the flare maximum radiation from the black body line toward the region occupied by radiation emitted by the upper layers of a stellar photosphere heated by fast particles. In the $(B-V,\ V-R)$ diagrams we find even larger deviations from the black body line, and the flare maximum radiation reaches the region corresponding to optically thin hydrogen plasma emission.

Multicolour measurements at premaximum phases (open circles) show that at these phases the colours of the pure flare radiation were rather close to the respective colours at subsequent flare maxima.

At the decay phase (triangles) in the $(U-B,\ B-V)$ and $(U-B,\ V-R)$ diagrams we find shifts of the pure flare radiation from the black body line towards the regions of radiation coming from hydrogen plasma and/or of heated upper atmospheric layers. However, in the $(B-V,\ V-R)$ diagram we see once again shifts toward the region of optically thin hydrogen plasma only.

In the $(U-B,\ V-I)$ diagram, the pure flare radiation is spread over the whole region from black body to hydrogen plasma or heated upper atmospheric layers, and in some cases even across the regions corresponding to any of the above considered models.

Thus, Fig. 1 permits to conclude that none of the four models considered above can alone represent the flare emission over the whole *UBVRI* range at any flare phase. The observations are better represented by a combination of two computed models, for instance by the superposition of short-lived black body radiation and a slowly developing emission of hydrogen plasma. However, in general, these models may be regarded only as starting blocks for a real model of flares.

2. Polarimetry: In 1989 and 1991, during the polarimetric monitoring of the star with the 2.6 m Shajn reflector of the Crimean Astrophysical Observatory, about ten EV Lac flares were detected. The most powerful of these was detected on September 15 at 01:29 UT with $\Delta U = 3\overset{m}{.}7$. The data analysis leads to the conclusion that the polarization of the star during this powerful flare was less than 2 % at 10 s time resolution and less than 1 % at 50 s time resolution. Detailed discussion on all earlier polarimetric data on the UV Cet type flares permit to conclude that the published data on discoveries of flare polarization or on significantly smaller upper limits of polarization are questionable. These conclusions are described in detail by Alekseev et al. (1994).

3. Spectroscopy: During the spectroscopic monitoring of EV Lac with the CCD spectrograph of the Shajn reflector in 1992 we (1) confirmed a kinematic heterogeneity of flare matter responsible for the hydrogen emission, (2) measured a veiling of intensity jumps in the TiO molecular band heads by flare continuum emission, and (3) detected several noticeable emission lines near Hβ. These lines were registered already earlier and have been attributed to He I and Mg I emission. However, we suggest that they are Fe II (42) emission, well known in solar flare spectra and in T Tau type star spectra. These results are in preparation (Abdul-Aziz et al. 1995).

4. Simultaneous radio monitoring: The EV Lac flare on September 10, 1993, at UT 22:32 was detected in Crimea in the U band, and was monitored simultaneously by M. Kidger at the Carlos Sanchez telescope on the Canary Islands in the K band. Data analysis shows that this flare with $\Delta U = 3\overset{m}{.}1$ did not vary noticeably in the K band. Thus, the dip in the K band detected during the AD Leo flare on 1984 March 28, at 03:22 UT (Rodonò 1986), is not a universal feature of UV Cet type flares.

5. Simultaneous Radio monitoring: During the optical monitoring of EV Lac in 1992, a decametre radio wavelength patrol has also been carried out. As is known, decametric emission from a flare of a red dwarf star has been detected only once before: on 1963 November 3, Slee et al. (1963) detected a flare on V 371 Ori at frequencies of 19.7, 410 and 1410 MHz. However, such events were not recorded later by any other observers.

The synphase grating UTR-2 of the Radio Astronomical Institute of the Ukrainian Academy of Sciences (Braude et al. 1978), the largest decametric radio telescope in the world, was used. Radio brightness monitoring of the star was made by a beam supplied to two-channel radiometers at 25 MHz and 20 MHz. Such a two-frequency receiver permits the discrimination of broadcasting interference and allows to estimate a bandwidth and spectral index of the detected signal. Simultaneously, a third, similar radiometer was used to monitor an area offset by 1° in declination from EV Lac. Local impulse interference would be expected to occur in all beams.

Radio monitoring of EV Lac was made on September 1 – 6, between 21 to 26 UT each night. During these 30 hours, more than thirty bursts were detected. About half of them were identified with terrestrial interference, the rest were individual bursts or compact groups of bursts with typical durations of 1 – 3 s and 10 s, respectively. Those bursts which could not be identified with terrestrial interferences, show a general correlation with the optical activity of the star. Indeed, if one considers the temporal intervals from 21 to 26 UT during which the radio monitoring was carried out, a ranking of nights according to the stellar optical activity from higher to lower levels is the following: September 1, 4, 3 and 2. On the other hand, the number of radio bursts detected during these nights are 9, 5, 2 and 1. During the same nights the number of radio bursts attributed to terrestrial interference are 7, 5, 3 and 5. In spite of small statistics, it is clear that the second distribution is noticeably more even than the first one. This general correlation permits to suggest that radio bursts coincident in time with optical flares are most probably of stellar origin. Groups of bursts were detected only during the night September 1/2, when the optical activity level was high. During the other nights, with moderate and low optical activity of the star, only weak individual bursts with a duration of about 1 s were detected. These two types of bursts exhibit different signal intensities as well. If one supposes that the radio source size is equal to the stellar area with $R_* = 0.35\,R_\odot$, the fluxes detected correspond to brightness temperatures from $1 \cdot 10^{18}$ to $4 \cdot 10^{17}$ K. These values are 2 orders of magnitudes lower than the corresponding value for the burst of V 371 Ori at 19.7 MHz (Slee et al. 1963) and not more than $3-4$ orders of magnitude higher than the strongest type III solar bursts (Melrose 1989). It is appropriate to mention here that if one considers optical flares with a mean occurrence interval of about a dozen hours both on EV Lac and on the Sun, the total optical energy of the stellar events is also 3–4 orders of magnitude higher than the respective solar flares (Gershberg 1989).

Thus, the probability seems to be high that we detected decametric radio emission from the flare star EV Lac. However, remaining doubts lead to some caution in the conclusions and require continuation of this observational program. This study is described in detail by Abranin et al. (1994).

References

Abdul-Aziz H., Abranin E.P., Alekseev I.Yu., Avgoloupis S., Bazelyan L.L., Beskin G.M., Brazhenko A.I., Chalenko N.N., Cutispoto G., Fuensalida J.J., Gershberg R.E., Kidger M.R., Leto G., Malkov Yu.F., Mavridis L.N., Pagano I., Panferova I.P., Pustil'nik L.A., Rodonò M., Seiradakis J.H., Sergeev S.G., Spencer R., Shakhovskaya N.I., Shakhovskoy D.N., 1995, (in prep.)

Abranin E.P., Alekseev I.Yu., Bazelyan L.L., Brazhenko A.I., Gershberg R.E., Shakhovskaya N.I., 1994, Kinematics and physics of celestial bodies 10, N 4, 70

Alekseev I.Yu., Gershberg R.E., Ilyin I.V., Shakhovskaya N.I., Shakhovskoy N.M., Avgoloupis S., Mavridis L.N., Seiradakis J.H., Kidger M.R., Panferova I.P., Pustil'nik L.A., 1994, A&A 288, 502

Berdyugin A.V., Gershberg R.E., Ilyin I.V., Malanushenko V.P., Shakhovskaya N.I., Shakhovskoy N.M., Pagano I., Panferova I.P., Pustil'nik L.A., 1995, Izvestiya Krymsk. Astrophys. Obs. 89, (in press)

Braude S.Ya., Megn A.V., Ryabov B.P., Sharykin N.K., Zhouck I.N., 1978, ApSS 54, 3

Gershberg R.E., 1989, Mem. S.A. It. 60, 263

Gershberg R.E., Grinin V.P., Ilyin I.V., Nesterov N.S., Shakhovskaya N.I., Getov R.G., Ivanova M.S., Panov K.P., Tsvetkov M.K., Tsvetkova A.G., Leto G., 1991a, Astron. Zh. 68, 548

Gershberg R.E., Ilyin I.V., Shakhovskaya N.I., 1991b, Astron. Zh. 68, 959

Gershberg R.E., Ilyin I.V., Rostopchina A.N., Shakhovskaya N.I., Garbuzov G.A., Pettersen B.R., Korhonen T., Avgoloupis S., Mavridis L.N., Seiradakis J.H., Konstantinova-Antova R.K., Antov A.P., Melkonyan A.S., Panferova I.P., Pustil'nik L.A., Herouni P.M., Oskanyan A.V., 1993, Astron. Zh. 70, 984

Melrose D.B., 1989, Solar Phys. 120, 369

Rodonò M., 1986, in Flare stars and related objects, L.V.Mirzoyan (ed.), Publ. House Armenian Acad. Sci., Yerevan, p. 19

Slee O.B., Solomon L.H., Patston G.E., 1963, Nature 199, 991

A Possible Discovery of a Flaring 10^{12} eV Gamma-Ray Source near the Red Dwarf EV Lac

I.Yu. Alekseev, N.N. Chalenko, V.P. Fomin, R.E. Gershberg, O.R. Kalekin, Yu.I. Neshpor, A.A. Stepanian

Crimean Astrophysical Observatory, Nauchny, Crimea 334413, Ukraine

During the 1994 coordinated observations of the red dwarf flare star EV Lac, the star was monitored in the very high energy (VHE) γ-ray range around 10^{12} eV with the Crimean ground-based γ-ray telescope GT-48. This telescope consists of two identical optical systems (Vladimirsky et al. 1994) which were directed in parallel on EV Lac.

The detection principle of the VHE γ-rays is based on the Čerenkov radiation emitted by relativistic electrons and positrons. The latter are generated in the interaction of the γ-rays with nuclei in the Earth's atmosphere that leads to an appearance of a shower of charged particles and γ-quanta. The duration of the Čherenkov radiation flash is very short, just about a few nanoseconds. The angular size of the shower is $\sim 1°$. To detect such flashes we use an optical system with large area mirrors and a set of 37 photomultipliers (PMs) in the focal plane. Using the information from these PMs which are spaced hexagonally and correspond to a field of view of 2°.6 on the sky, we can obtain the image of an optical flash. The electronic device permits us to detect nanosecond flashes (40 ns exposure time and 12 μs readout dead-time).

Unfortunately, the same kind of flashes are generated by VHE charged particles of cosmic rays. Using the differences in angular distributions of flash light permits us to reject the majority of this charged cosmic-ray particle background (Hillas 1985). The average number of flashes initiated by the cosmic-ray background is equal to 0.5 per second.

During several nights we carried out simultaneous optical (with the 1.25m reflector AZT-11) and VHE γ-ray monitoring of EV Lac. The most statistically reliable result in γ-rays was obtained just before the optical flare on August 31 at UT 19:40 (Fig. 1). The data from GT-48 were analysed to reject events due to charged cosmic-ray particles. It remains an excess in the number of selected flashes. The probability of 21 selected events occuring over one minute to be due to a random fluctuation is equal to $7 \cdot 10^{-9}$ according to Poisson statistics. We thus conclude that a VHE gamma-ray burst was detected with a duration of about 1 min.

The effective area for detecting Čerenkov flashes in the Earth atmosphere, initiated by γ-quantum, is about $4 \cdot 10^8$ cm^2 (e.g. Stepanian 1994). Therefore, the detection of 21 flashes correspond to a fluence of $0.5 \cdot 10^{-7}$ γ-rays (of 10^{12} eV

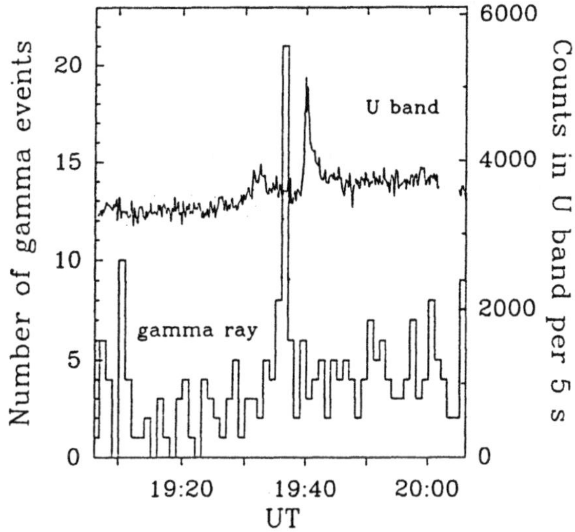

Fig. 1. EV Lac light curve in the U band around the flare on 1994 August 31 at UT 19:40, and the histogram of the 10^{12} eV γ-ray events.

energy) per cm^2 from the burst. With a mean energy of $\sim$ 1.6 ergs per gamma-ray, placing the burst source at the distance of EV Lac of 5 pc, and assuming isotropic gamma-ray emission, we obtain a total burst energy of $\sim 2 \cdot 10^{32}$ erg.

Using the imaging capability of the camera for Čerenkov flashes we have determined the position of the γ-ray source. This position does not coincide with that of the flare star EV Lac: its 2000.0 coordinates are RA = $22^h43^m45^s \pm 68^s$ and Dec = $+44°37' \pm 12'$, and thus is 0.°6 distant from EV Lac. We call this VHE γ-ray source Lac γ–1. Our attempts to identify Lac γ–1 with a known object has so far been unsuccessful. It would be very desirable to extend the survey by Seiradakis et al. (1993) to include the error box of Lac γ–1, and also to look for candidates in the ROSAT data and other short-wavelength surveys.

The upper limit of the VHE γ-ray flux from the EV Lac flare mentioned above is estimated to be 3 photons. The total optical energy release of the flare was $8 \cdot 10^{30}$ erg, and thus $E_\gamma / E_{\rm opt} < 4$. This limit is too weak to give any essential physical constraint on stellar flare theory.

References

Hillas A.M., 1985, in Proc. 19th Internat. Cosmic Ray Conf. 3, F.C. Jones, J. Adams, G.M. Mason (eds.), p. 445

Seiradakis J.H., Avgoloupis S., Mavridis L.N., Varvoglis P., Fürst E., 1993, Proc. 1st Panhellenic Astron. Meeting, Athens, Sept. 21-23, 1992, p. 261

Stepanian A.A., 1994, Izv. Krymsk. Astrofiz. Obs. 91, 56

Vladimirsky B.M., Zyskin Yu.L., Kornienko A.P., Neshpor Yu.I., Stepanian A.A., Fomin V.P., Shitov V.G., 1994, Izv. Krymsk. Astrofiz. Obs. 91, 74

Short-Time Flare Events of EV Lac

B.E. Zhilyaev, I.A. Verlyuk

Main Astronomical Observatory, National Academy of Sciences of Ukraine,
252650 Kiev, Ukraine

1 Observations

Moffett (1972) pointed out the existence of short-time outbursts in flare stars as long ago as 1972. Gershberg & Petrov (1986) had registered a flare in EV Lac with a duration of $2^{\mathrm{s}}.4$ and an amplitude of $3^{\mathrm{m}}.1$ in 1984 with *ASTRON*. In recent years, a few reports appeared about observations of short-time outbursts in EV Lac with a duration in the range from $0^{\mathrm{s}}.1$ to several seconds. These results were summarized briefly by Zhilyaev (1993). Our observations confirm the existence of short-time flare activity on EV Lac.

The observations reported here were obtained at the 60 cm telescope at the Peak Terskol high-altitude Observatory (3 100 m above sea level) with a high-speed two-channel photometer (Zhilyaev et al. 1992) in B as well as simultaneously in both, U and V bands, with a time resolution down to $0^{\mathrm{s}}.01$. A new prefiltering technique (Zhilyaev 1995) has been used for the detection of small-scale activity of the flare star EV Lac.

2 Results

1. The time span of the detected flare events lies in the subsecond range down to $0^{\mathrm{s}}.1$ (Zhilyaev et al. 1990). Fig. 1 represents an extreme short-term flare in U and V bands simultaneously, with a duration of about $0^{\mathrm{s}}.3$ and with amplitudes of $\Delta V \approx 0^{\mathrm{m}}.8$, $\Delta U \approx 3^{\mathrm{m}}$.
2. The above-mentioned new prefiltering technique for high time resolution data allows to detect an event amplitude comparable with that caused by counting statistics. An example of such a small-scale flare with $1^{\mathrm{s}}.5$ duration and $0^{\mathrm{m}}.4$ amplitude can be seen in Fig. 1 of Zhilyaev (1995).
3. A new kind of flares was detected in EV Lac. A few events show a flare in the U band accompanied by a dip in the V band at the same time. Fig. 2 represents such an event with a duration of about 10^{s} and amplitudes in U and V of $-1^{\mathrm{m}}.2$ and $+0^{\mathrm{m}}.05$, respectively.

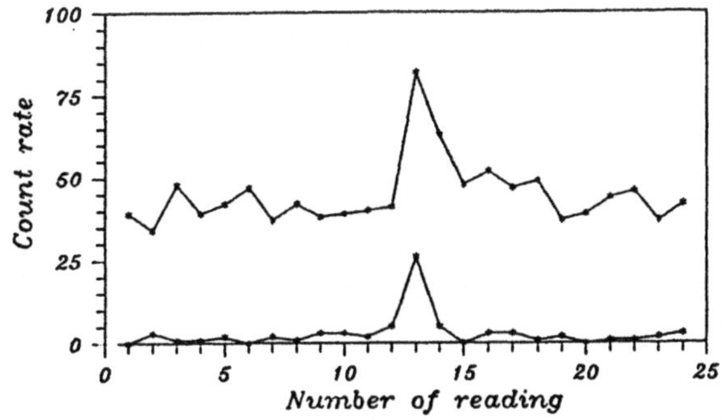

Fig. 1. EV Lac on 1990 Aug. 26, 00:04 UT. Time resolution: $0^{\mathrm{s}}.05$. V band (top), U band (bottom).

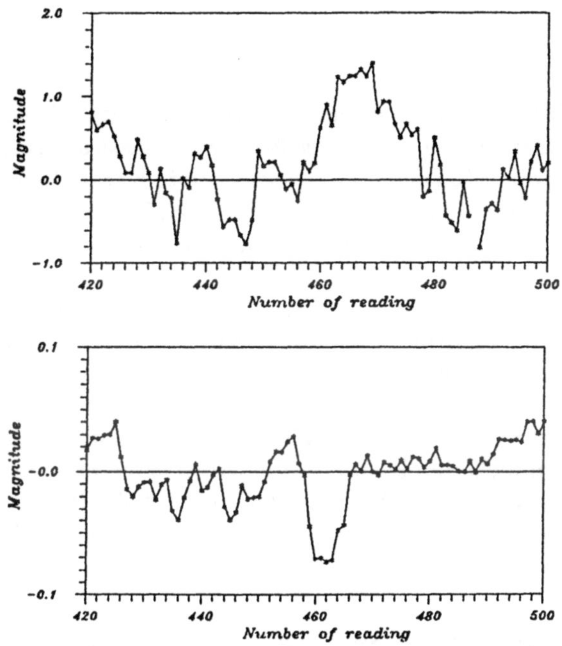

Fig. 2. EV Lac on 1990 Aug. 27, 00:45 UT. Time resolution: $0^{\mathrm{s}}.05$. U band (top), V band (bottom). Prefiltering technique was used.

References

Gershberg R.E., Petrov P.P., 1986, Flare Stars and Related Objects, Publishing House of Acad. of Sci., Yerevan, p. 38

Moffett Y.J., 1972, Nature 240, 41

Zhilyaev B.E., Romaniuk Ya.O., Svyatogorov O.A., 1990, IAU Symp. 137, 35

Zhilyaev B.E., Romaniuk Ya.O., Svyatogorov O.A., 1992, AZh 69, 895

Zhilyaev B.E., 1993, Kinematics and Physics of Celestial Bodies, 10, 32

Zhilyaev B.E., 1995, this volume, p. 133

Observations of High-Frequency Activity of EV Lacertae

B.E. Zhilyaev, I.A. Verlyuk

Main Astronomical Observatory, National Academy of Sciences of Ukraine, 252650 Kiev, Ukraine

1 Introduction

The flare star EV Lac shows a wide variety of activity phenomena on a wide range of time scales. These are the well-known "classical" solar-like flares, the short-time flare events (Gershberg & Petrov 1986, Karapetian & Zalinian 1991, Tovmassian & Zalinian 1988, Tsvetkov et al. 1986, Zhilyaev 1994), and the long-duration variations caused by starspots and rotation (Petterson 1980). However, the "classical" flares are not such a frequent phenomenon as one gets used to think. As noted by Roizman & Kabitchev (1985), a considerable fraction of flares shows a complicated multipeak structure with sudden onset and decay. Their interpretation along the lines of classical models favored for solar-like flares is very problematic.

Here, we report the detection of high-frequency chaotic variability (flickering) in EV Lac. Optical flickering is observed in cataclysmic binaries, and in X-rays it is observed in low-mass X-ray binaries. As a rule, flickering is an accretion phenomenon. Strange as it may seem, EV Lac as a visual binary has the analogous property.

2 A test for detection and estimation of chaotic variability

Denote the sampling values of the mean count rate and of the variance as $\bar{n}$ and s^2. The variance has the physical meaning of the power of fluctuations. It consists of three components caused by Poisson noise, scintillations and intrinsic stellar variability:

$$s^2 = s_{\mathrm{p}}^2 + s_{\mathrm{sc}}^2 + s_*^2 \tag{1}$$

The first term is equal to the mean count rate of star and background

$$s_{\mathrm{p}}^2 = \bar{n} = \bar{n}_* + \bar{n}_{\mathrm{b}} \tag{2}$$

The second term evaluates a contribution of atmospheric scintillations to the total variance. It may be defined as

$$s_{\mathrm{sc}}^2 = \frac{1}{T}\int_0^T \left[f(t)n_*(t)\right]^2\, dt = \epsilon_{\mathrm{sc}}\bar{n}_*^2 \tag{3}$$

where $f(t)$ is the light wave envelope in the turbulent atmosphere. The quantity ϵ defined as

$$\epsilon = \frac{s^2 - \bar{n}}{\bar{n}_*^2} = \frac{s_{sc}^2 + s_*^2}{\bar{n}_*^2} = \epsilon_{sc} + \epsilon_* \tag{4}$$

specifies the power of scintillations in the operating frequency range $1/T \div 1/2\Delta t$, where Δt is the sampling time. The quantity ϵ_{sc} depends on atmospheric conditions as well as on the aperture of the telescope. It is independent of intensity $\bar{n}_*$ (Tatarsky 1967).

Using the standard procedure (Kendall & Stuart 1946), the error variance for ϵ can be found

$$D[\epsilon] = \frac{1}{\bar{n}^4}\left\{D\left[s^2\right] + \left(\frac{2s^2}{\bar{n}} - 1\right)^2 D[\bar{n}]\right\} \tag{5}$$

where D is the variance operator. For variances of $\bar{n}$ and s^2 we have

$$D[\bar{n}] = \frac{s^2}{N}, \qquad D\left[s^2\right] = \frac{2s^4}{N-1} \tag{6}$$

where N is the length of the measurement series.

The actual value of ϵ_{sc} caused by scintillations can be determined from measurements of a reference star assuming $\epsilon_* \equiv 0$. To detect atmospheric scintillations, one can assume the null hypothesis: $s^2 = \bar{n}$. In view of (5) a detection threshold can be set

$$\epsilon_{th} = \alpha\sqrt{D[\epsilon]} = \frac{\alpha}{\bar{n}}\sqrt{\frac{2}{N}}\ . \tag{7}$$

For a time series with normal density, $\alpha = 1.96$ corresponds to a 95% confidence level. The quantity ϵ specifies an accuracy limit of stellar photometry in the operating frequency range at given atmospheric conditions. The above quoted expressions open the way for the detection of intrinsic chaotic fluctuations of stars masked by atmospheric noise.

3 Observations and results

High-frequency chaotic variabiliy was found in EV Lac in the B band. Observations were obtained with a high-speed two star photometer with $0\overset{s}{.}05$ integration time at the 60 cm telescope at the Mount Terskol Observing Station in North Caucasus (3 100 m above sea level). Monitoring was carried out on 1991 October 13, and on 1993 November 6, for about half an hour. Simultaneous measurements of a reference star were made in order to control and estimate atmospheric noise.

Our observations reveal that intrinsic flickering takes place in EV Lac. The results are presented as a power vs. time curve in Fig. 1. Each dot is the result of a $1\overset{s}{.}0$ estimate of ϵ. In addition, the curve has been smoothed by a moving-average filter for the purpose of noise cancelling. Flickering in EV Lac shows a nonstationary quasiperiodical character on a time scale of 140 sec. The maximum power of the variable component is equal to about $5 \cdot 10^{-3}$ in the frequency band of 10 Hz. At minimum it drops below the detection threshold equal to about

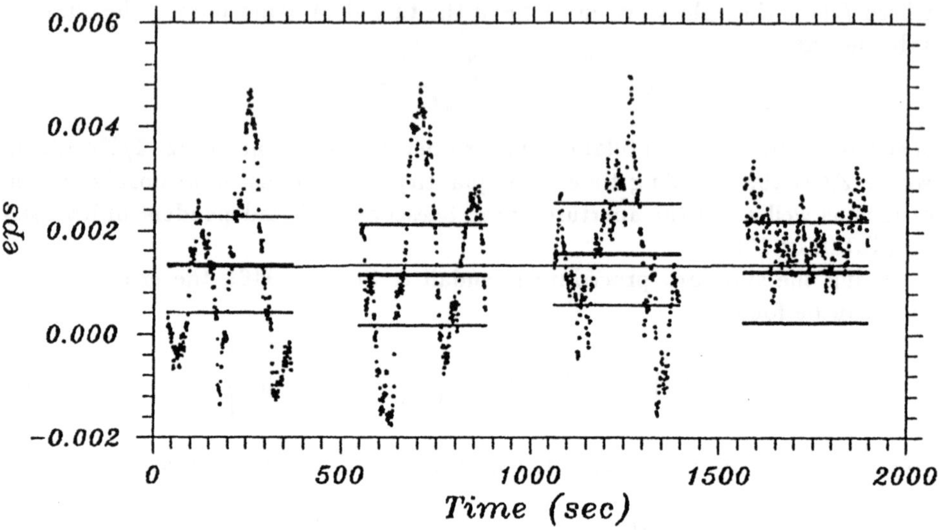

Fig. 1. EV Lac on 1991 Oct. 13. Variations in the power of fluctuations in the B band. Each dot is a result of a 1.0 sec estimate of ϵ. The mean values and the error limits of ϵ are presented.

$4 \cdot 10^{-4}$. The mean power of the variable component remains constant during the observational run and equals $(9 \pm 2) \cdot 10^{-4}$.

The power of atmospheric scintiallations derived from the reference star data remains constant to an accuracy of 3% and equals $(322 \pm 9) \cdot 10^{-6}$. EV Lac demonstrates the same behaviour of the power curve in the second observing run on 1993 November 6. Both runs were performed at fair atmospheric conditions with a low level of atmospheric scintillations.

Acknowledgements: The research described in this publication was made possible in part by Grant No. UCBOOO from the International Science Foundation.

References

Gershberg R.E., Petrov P.P., 1986, in Flare Stars and Related Objects, Publishing House of the Acad. of Sci., Yerevan, p. 38
Karapetian A., Zalinian V., 1991, IBVS 3550
Kendall A., Stuart A., 1946, The Advanced Theory of Statistics, Vol. 1, Charles Griffin & Comp., London
Petterson R.B., 1980, AJ 85, 871
Roizman G.S., Kabichev G.I., 1985, AZh 62, 1095
Tatarsky V.I., 1967, Propagation of Waves in Turbulent Atmosphere, Moscow, Nauka
Tovmassian H, Zalinian V., 1988, Astrofiz. 28, 131
Tsvetkov M., Antov A., Tsvetkova K., 1986, Comm. Konkoly Obs. No. 86, 435
Zhilyaev B.E., 1994, Kinematics and Physics of Celestial Bodies 10, 32
Zhilyaev B.E., Verlyuk I.A., 1995, this volume, p. 80

Spectral and Photometrical Investigation of EV Lac in Different States with High Time Resolution

G.M. Beskin, S.N. Mitronova, I.P. Panferova

Special Astrophysical Observatory, Nizhnij Arkhyz, Karachaevo-Circassia, 357147 Russia

Spectroscopic and photometric monitoring of the flare star EV Lac was carried out with the 6-m telescope of the Special Astrophysical Observatory on the following 3 nights: 1993 September 10 (18:54–20:16 UT), 11 (17:46–20:03 and 21:25–22:16 UT) and 12 (19:46–20:08 and 21:23–22:21 UT).

We used the observation method described in detail by Berdyugin et al. (1994) and Alekseev et al. (1994). Observations were carried out with the 1024 channel TV scanner (Drabek et al. 1986) installed on the spectrograph SP-124 at the Nasmyth focus which covered a wavelength range 3480–5530Å with a spectral resolution 2Å per channel. Exposure times were 3 – 5 minutes, the signal/noise ratio was 20 at 3500Å and 40 at 5500Å and dead time between exposures was 10 – 15 sec. 92 spectra were obtained in total.

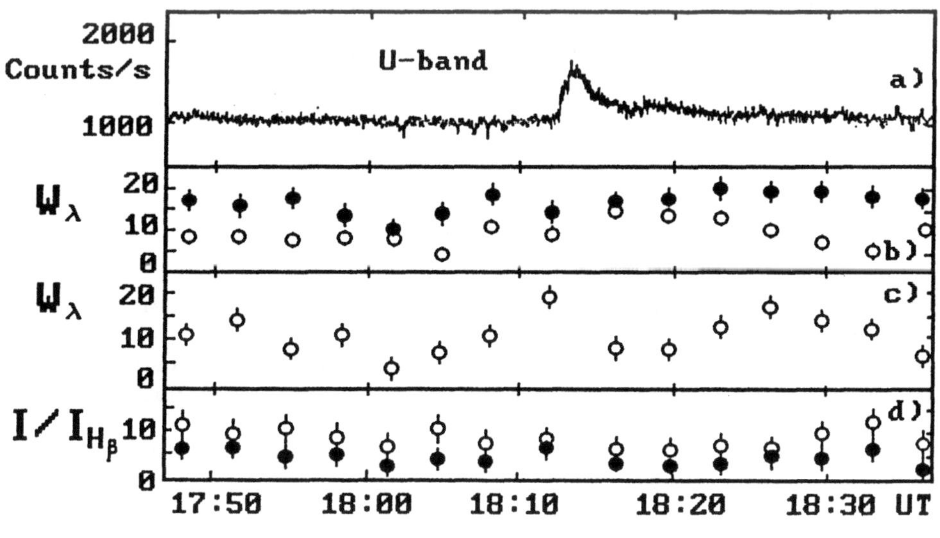

Fig. 1. Monitoring of EV Lac on the 6-m telescope SAO RAS 11/12.09.1993. (a) Light curve. Eequivalent widths of lines: (b) • – $H\gamma$, ○ – $H\beta$; (c) ○ – $H\delta$. (d) The Balmer decrements: • – $I_{H\delta}/I_{H\beta}$, ○ – $I_{H\gamma}/I_{H\beta}$.

All spectroscopic observations were simultaneous with U band photometry made using the *MANIA* (Multichannel Analysis of Nanosecond Intensity Alterations) complex. It registers the arrival time of each photon with an accuracy of 20 ns. The time resolution is 10^{-7} s (Beskin et al. 1982, Zhuravkov et al. 1992). 19 flares with amplitudes $\Delta U > 0.1$ and 4 flares with $\Delta U > 0.5$ were detected. The light curve and spectroscopic variations during a flare of $\Delta U = 0.6$ on 1993 September 11 at 18:12 UT are shown in Fig. 1. An increase of equivalent widths of the lines Hβ, Hγ and Hδ before and during the photometric flare was detected (Fig. 1b,c). No such emission line variations were recorded during a fainter and shorter flare with an amplitude $\Delta U > 0.4$ on 1993 September 11 at 19:50 UT. 3 segments of photometric data (Fig. 1a) were analyzed with the y_2- and d_2-function formalism (Shvartsman 1977, Plokhotnichenko 1983) to search for stochastic variability of EV Lac's brightness. The results of this statistical analysis (see Table 1) showed that there is no fast variability on time scales of 1 μs - 10 sec with a confidence of 99%.

The value of the intensity jump near the limit of the TiO band at 4955Å and the equivalent widths of photospheric Ca I and MgH lines were practically constant during the whole observation. Sporadic variations of the Balmer decrement of up to a factor 2, outside of flares, were observed on 1993 September 12.

Table 1. Upper limits for the relative power of the brightness variable component (in %) (filling factor 0.2, confidence probability $\geq$ 99%).

Phase of brightness	Time scales (s)				
	$10^{-6}-10^{-5}$	$10^{-5}-10^{-3}$	$10^{-3}-10^{-2}$	$10^{-2}-1$	$1-10$
1. Top of flash	55	34	14	6.3	2.2
2. Tail of flash	39	21	9	4.6	1.7
3. Quiet	37	23	9	4.5	1.5

References

Alekseev I.Yu., Gershberg R.E., Ilyin I.V. et al., 1994, A&A 288, 502

Berdyugin A.V., Gershberg R.E., Ilyin I.V. et al., 1994, Izv. Krymsk. Astrofiz. Obs. 89 (in press)

Beskin G.M., Neizvestny S.I., Pimonov A.A. et al., 1982, in Proc. IAU Coll. 67, Instrumentation for astronomy with large optical telescopes, C.M. Humphries (ed), Reidel, Dordrecht, p. 181

Drabek S.B., Kopylov I.M., Somov N.N., Somova T.A., 1986, Astrofiz. Issled. (Izvestiya Spets. Astrofiz. Obs.) 22, 64

Plokhotnichenko V.L., 1983, Soobshcheniya Spets. Astrofiz. Obs. 38, 29

Shvartsman V.F., 1977, Soobshcheniya Spets. Astrofiz. Obs. 19, 5

Zhuravkov A.V., Pimonov A.A., Plokhotnichenko V.L., 1992, SAO RAS Prepr. 91

Photoelectric Observations of AD Leo: 1989–1994

R. Konstantinova-Antova, A.P. Antov

Institute of Astronomy, Bulgarian Academy of Sciences, 72 Tsarigradsko shose, BG-1784 Sofia, Bulgaria

1 Observations

An investigation of the activity of the flare star AD Leo was made using U-band patrol observations. 60 cm computer-controlled telescopes with identical single channel photon-counting photometers at the National Astronomical Observatory at Rozhen and at the Belogradchik Observatory were used. The integration time was 1 sec. Differential photometry was carried out, the AD Leo measurements were made relative to BD+20°2475. The data was reduced with the program package APR (Kirov et al. 1991).

Two simultaneous observing runs were made. The results of the first have already been published (Antov et al. 1991). The other was part of the observational campaign in May, 1991 with the *ROSAT* satellite. In Bulgaria, the observations were carried out at Rozhen in the U-band and at Belogradchik in the B-band. Four flares were detected by both.

2 Results

37 flares with amplitudes more than 4σ were detected during $71^{\mathrm{h}}05^{\mathrm{m}}17^{\mathrm{s}}$ total monitoring time. Approximately 16% are so called fast flares, with a duration <1 min. If flares with amplitude $< 4\sigma$ are added, the fast flares are ~23%. Usually the fast flares had an amplitude $\leq 0\overset{\mathrm{m}}{.}5$ (except one flare, detected on 29.3.1993 with an amplitude $0\overset{\mathrm{m}}{.}72$).

The reality of so-called single spikes on AD Leo, with a duration 1–2 sec was not confirmed during the period of the observations. Several such events were detected at Rozhen (altitude 1750 m) but none were detected at Belogradchik (altitude 630 m). One possible explanation is that the spikes were due to cosmic rays or artificial sources.

A rough investigation of the flare activity of AD Leo was carried out. The rate of the flare activity (the number of flares hr^{-1}) was calculated for each year of the period (Table 1) and the result is shown in Fig. 1.

Table 1. Rate of the flare activity of AD Leo

Year	No. of nights	Effective monitoring time	No. of flares	Flares hr^{-1}
1989	2	2^h 51^m 45^s	2	0.95±0.47
1990	15	25^h 47^m 00^s	18	0.70±0.16
1991	10	12^h 53^m 12^s	7	0.58±0.20
1992	10	14^h 59^m 30^s	5	0.33±0.15
1993	4	3^h 25^m 08^s	1	0.29±0.28
1994	7	11^h 08^m 42^s	4	0.36±0.18

It appears that the rate of flare activity changes from year to year in a systematic way. This fact could be explained by the existence of a change in the chromospheric activity on a long time scale.

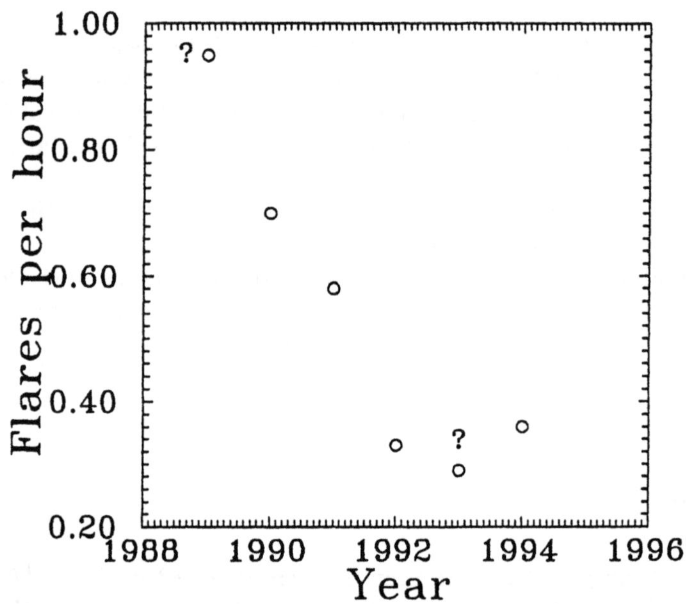

Fig. 1. Flare activity rate of AD Leo.

Acknowledgements: R.K.-A. is thankful to the German Science Foundation for the financial support to attend the Colloquium.

References

Antov A., Genkov V., Konstantinova-Antova R., Kirov N., 1991, IBVS 3577
Kirov N.K., Antov A.P., Genkov V.V., 1991, Compt. Rend. Bulg. Acad. Sci. 44, 5

Multifrequency Analysis of a UV Ceti Flare on 1991 December 31

A.V. Stepanov[1], *E. Fürst*[2], *A. Krüger*[3], *J. Hildebrandt*[3], *H. Barwig*[4], *J.H.M.M. Schmitt*[5]

[1] Crimean Astrophysical Observatory, RT-22 Katzively, 334247 Crimea, Ukraine
[2] MPI für Radioastronomie, Auf dem Hügel 69, D-53121 Bonn, Germany
[3] Astrophysikalisches Institut, An der Sternwarte 16, D-14482 Potsdam, Germany
[4] Institut für Astronomie und Astrophysik der Universität München, Scheinerstr. 1, D-81679 München, Germany
[5] MPI für Extraterrestrische Physik, Giessenbachstrasse, D-85748 Garching, Germany

Abstract: An analysis of optical, radio and X-ray observations of the flare of December 31, 1991 on UV Ceti is presented. Radio spikes at 4750 GHz are interpreted in terms of both ECM and plasma emission. Resulting parameters of the flare plasma are discussed.

1 Observations

Optical data of UV Cet were obtained in *UBVRI* with the 0.8 m telescope of the Wendelstein Observatory. Two flares were recognized, near 17:00 and 18:25 UT. The second one was a strong white light flare visible in almost all colors. Radio observations were carried out with the Effelsberg 100 m radio telescope at 4750 MHz with a total bandwidth of 50 MHz splitted into 32 channels. The time constant was 0.125 s. The burst started about 5 min after the second optical flare and showed $\gtrsim 75\,\%$ LH circular polarization with a peak flux density of 250 mJy. The polarized radio data are integrated to 0.9 s for various bands each about 3.1 MHz wide and display several larger spikes with duration $\lesssim 0.125$ s. Soft X-ray emission (0.1–2.5 keV) was observed with *ROSAT* PSPC before and after the optical and radio flares only, because UV Cet was unfortunately occulted by the Earth. Nevertheless, the data associate suggestively the decay phase of the X-ray flare with the optical and radio flare observations.

2 Discussion

The brightness temperature of the radio spikes was estimated to be $3 \cdot 10^{12}$ K indicating coherent radiation. Usually two coherent radiation mechanisms are considered for the spikes: Electron Cyclotron Maser (ECM) and plasma emission. Two problems arise if we assume ECM emission. The first one is that the electron gyrofrequency to plasma frequency ratio should be $\nu_B/\nu_{\rm p} > 1$, but the corresponding Alfvén velocity ($v_{\rm A} \approx 10^4$ km/s) seems to be too large for the stellar corona. The second problem is the escape of the ECM emission from the corona, because the emission should be absorbed at the gyroresonance layers $\nu = s\nu_B$ ($s = 2, 3$). To resolve this problem Robinson (1984) suggested an escape window for the o-mode at $\nu = 2\nu_B$ near the direction perpendicular to the magnetic field. However, our calculations of optical depths for gyroabsorption show that there are no "perpendicular" windows for the o-mode at $s = 2, 3$. Instead, quite large "parallel" windows for both the x- and o-mode exist, enabling the escape of ECM emission if some wave isotropization mechanism works. Supposing ECM emission, one can derive the magnetic field $B \approx 850\,\mathrm{G}$ and the plasma density $n \lesssim 7 \cdot 10^{10}\,\mathrm{cm}^{-3}$ in the source region.

It is not excluded that highly polarized spikes are due to o-mode radiation from plasma emission at the fundamental $\nu \approx \nu_p$. Thus, the radio observation corresponds to $n \approx 3 \cdot 10^{11}\,\mathrm{cm}^{-3}$, and from the condition for the x-mode cut-off, we estimate $B \gtrsim 200\,\mathrm{G}$. Genkin et al. (1990) have shown that emission at $\nu \approx \nu_p$ can be very irregular due to scattering of plasma waves on density irregularities induced by plasma waves themselves and thermodiffusion. The correlation time $t_c = 2\,v/(\nu_{ei} v_g)$ (for $t > t_c$ no spikes exist!) depends on the energetic electron velocity v, the group velocity of plasma waves v_g and the electron-ion collisional frequency ν_{ei} and is compared with the spike duration observed ($t_i \lesssim 0.13\,\mathrm{s}$).

From the multifrequency observations of the UV Cet flare on 1991 December 31 we find the following parameters for the radio source: $B \approx 200$–$300\,\mathrm{G}$, $n \approx 10^{11}\,\mathrm{cm}^{-3}$, $T \approx 10^7\,\mathrm{K}$, and the source size $D \approx 10^4\,\mathrm{km}$, which do not contradict known diagnostics of flare plasmas on red dwarfs. The temperature of the "cool" plasma deduced from optical data is $T \approx 1.6 \cdot 10^4\,\mathrm{K}$ and the linear scale $L \approx 2 \cdot 10^3\,\mathrm{km}$. The time delay of about 5 min between the optical and radio flare can be understood either in terms of magnetic trapping and storage of accelerated particles or in terms of evaporated plasma filling up the flare loop. The flare scenario can be described using both the Impulsive Explosion Model (De Jager et al. 1989) and the Advanced Circuit Model (Zaitsev & Stepanov 1992). A more detailed discussion will be published elsewhere (Stepanov et al. 1995).

References

De Jager C., Heise J., van Genderen et al., 1989, A&A 211, 157

Genkin L.G., Erukhimov L.M., Levin B.N., 1990, Solar Phys. 128, 423

Robinson P.A., 1984, ApJ 341, L99

Stepanov A.V., Fürst E., Krüger A., Hildebrandt J., Barwig H., Schmitt J.H.M.M., 1995, A&A (submitted)

Zaitsev V.V., Stepanov A.V., 1992, Solar Phys. 139, 343

A Large Amplitude Flaring dMe Star in the 1978 October 6B γ-ray Burst Error Box*

J. Greiner[1], *C. Motch*[2]

[1] Max-Planck-Institut für Extraterrestrische Physik, D-85740 Garching, Germany
[2] CNRS, UA 1280, Observatoire de Strasbourg, 11 rue de l'Université, F-67000 Strasbourg, France

1 Introduction

During a search for flaring optical counterparts to γ-ray burst sources (GRBs) on archival patrol plates, an optical outburst image was found in the 3200 arcmin^2 error box of the GRB which occurred on 1978 October 6 (GB 781006B = GBS 0008+13; Greiner et al. 1991) and was named S 10933 according to the convention of new Sonneberg variables. This outburst image appears on three simultaneously exposed plates taken on 1966 August 14/15 from 23.33 to 0.13 UT. The brightness estimates yielded $m_{pg} = 13\overset{m}{.}3 \pm 0\overset{m}{.}1$ and $m_v = 13\overset{m}{.}8 \pm 0\overset{m}{.}2$ with a high negative colour index of $C = -0\overset{m}{.}5 \pm 0\overset{m}{.}3$.

The coordinates of these images (measured on the Tessar plate Te_3 5154) are (equinox 1950.0) $\alpha = 0^h\ 10^m\ 07\overset{s}{.}6$, $\delta = +\ 12°\ 51'24''$ (1σ error of $\pm\ 2''$). Within the 3σ error circle of the position of the object a star of $m_B = 18$ is visible on the Palomar Observatory Sky Survey prints. Here, we report spectroscopic observations of this faint object.

2 Observations and Results

CCD B and R images of the field of the optical transient were obtained from 1991 November 10 to 12 with the 1.2 m telescope at Observatoire de Haute-Provence (OHP). Low and medium resolution spectra of the brightest optical object A were obtained on 1992 February 4, 7 and 8 with the CARELEC spectrograph attached at the OHP 1.93 m telescope. Further spectroscopic observation were obtained using the 3.5 m telescope at Calar Alto, equipped with the Cassegrain spectrograph, on 1992 Sep 28 – Oct 1. The slit was put parallel to the row of faint objects apparently aligned with the bright candidate. On 1992 Sep 30, three spectra with a grating of 60 Å/mm were acquired.

* Partly based on observations collected at the German-Spanish Astronomical Centre, Calar Alto, operated by the MPI für Astronomie, Heidelberg, jointly with the Spanish National Commission for Astronomy, and on observations obtained with the 1.2 m and 1.9 m telescopes of the Observatoire de Haute-Provence, CNRS, France.

Table 1. Other large amplitude flares

Star	Δm_B (mag)	References
CZ Cnc	≈ 9.5	Lovas 1977, Schaefer 1990, Greiner & Motch 1994
AF Psc	≈ 7.5	Bond 1976, Greenstein 1977
UV Cet	≈ 6.5	Jarrett & Gibson 1975
EV Lac	≈ 6.0	Roizman & Shevchenko 1982
AM Her (comp)	≈ 5.9	Shakhovskoy *et al.* 1993
S 10933	≥ 5.2	this work, Greiner & Motch 1995
AD Leo	≈ 4.1	Hawley & Pettersen 1991
G102–21	≈ 3.9	Pagano *et al.* 1995

The spectral type of the central object was estimated by comparing our flux calibrated spectra with spectra in electronic atlases (Jacoby et al. (1984) and Turnshek *et al.* (1985)). We conclude that the central star is a M3–M5e late type dwarf. The lack of Li absorption tells us that the star is older than $\approx 10^7$ yr (Rebolo 1991). Comparison with photometric standards in NGC 7790 (Christian *et al.* 1985) yielded $B = 18.50 \pm 0.20$.

Two additional objects lie within or slightly outside the 3σ error circle of the archival flare event. These are at least $2^{\rm m}$ fainter than the dMe star. They are aligned southeastwards from the central dMe star together with a third one which is already outside the 3σ position. Our long slit spectra of the three aligned objects show that the brightest notch (object D) located at the end of the apparent alignment is an M star, however without noticeable Balmer emission. No usable spectra could be obtained from notches B and C located in between the dMe star and object D. These objects exhibit flat continua without recognizable emission or absorption features.

3 Discussion

3.1 The flare star

Our observations leave no doubt that the bright dMe star is responsible for the large optical flare of 1966. The spectral type indicates an absolute magnitude M_V in the range of 10 to 14 and a distance of 40 to 250 pc. The apparent magnitudes during the flare and in quiescence translate into visual luminosities (blue band) of 4×10^{31} $(d/100\ \mathrm{pc})^2$ erg/s and 3.5×10^{29} $(d/100\ \mathrm{pc})^2$ erg/s, respectively.

Assuming that the flare lasted as long as (or longer than) the exposure time of the discovery plates (40 min), implies a peak amplitude of about $\Delta m_B \approx 5\overset{\rm m}{.}2$. Alternatively, if the event belongs to the class of fast flares with typical duration shorter than $\approx$ 10 minutes, the peak amplitude could have been as high as 7–$9^{\rm m}$. For comparison, other large amplitude flaring stars are listed in Table 1. Even a peak amplitude of $5\overset{\rm m}{.}2$ (corresponding to a flare energy release of 10^{35} $(d/100\ \mathrm{pc})^2$ erg in the optical bandpass) are rare for a flare star and we can conclude that the event witnessed in 1966 was certainly unusual.

3.2 The relation to GBS 0008+13

At the time of the discovery (Greiner *et al.* 1991) it was assumed that the recorded optical flash was a likely counterpart candidate for a GRB. Indeed, this optical flash is well positioned within a 3200 arcmin^2 error box of a GRB which occured on October 6, 1978 (GB 781006B = GBS 0008+13, Atteia *et al.* 1987). The rather extreme value for the flare amplitude might favour speculations whether or not this flare is indeed related to a GRB. In fact, the possible relation with stellar flares was among the first models proposed to explain GRBs. Nearby flare stars are isotropic in the sky and have a $-3/2$ slope in their $\log N - \log S$ distribution (for a sampling distance smaller than $\approx$300 pc) and thus meet one of the statistical GRB criteria. Recently, Liang and Li (1993) have shown that flare stars could account for as many as half of the BATSE triggers without violating the observed $\log N - \log S$ distribution of GRBs.

Estimating the a priori probability of discovering such a large amplitude flaring star in a rather large (3200 arcmin^2) γ-ray error box is difficult because of the various statistical biases in flare star catalogs and flare observational data bases. There are convincing evidences that flare activity preferentially occurs in rather young stars (e.g. Mirzoyan 1990). Most flaring M stars exhibit strong Balmer emission (Pettersen 1991) while kinematical studies of dMe stars show that they belong to the young disk population with a mean age of $\approx 2\times10^9$ yrs (Giampapa & Liebert 1986). However, old stars in close binaries may also exhibit strong flaring activity (Haisch et al. 1991).

Using the stellar population model of Robin & Crézé (1986) which explicitly handles the age parameter, we predict that $\approx$10 stars with $B \leq 18.5$ and younger than 10^9 yrs should be present inside the GRB error box, including two late M type dwarf stars. Therefore, in the absence of stronger constraints on the age of the Me dwarf and reliable statistics on flare amplitudes, we cannot prove that this particular flare star has caused the γ-ray burst GB 781006B, nor can we dismiss the flare star explanation for this particular GRB.

4 Summary

We have identified the 1966 optical flash to be caused by a B=$18\overset{\mathrm{m}}{.}5$ mag dMe flare star. We think that the relation between this flare star and the γ-ray burst source GBS 0008+13 is possible but unlikely. Even in this case of an unique archival object observed on synchronously exposed plates inside a GRB error box we are still left with an explanation other than a γ-ray burst optical counterpart.

Our identification confirms the feasibility of searching archival plate collections for real optical flashes of astrophysical origin. Unfortunately, the existence of an unknown number of such large amplitude flaring stars as the one reported here poses serious problems for wide-field GRB optical flash searches. More details can be found in Greiner & Motch (1995).

Acknowledgements: JG is partly supported by the Deutsche Agentur für Raumfahrtangelegenheiten (DARA) GmbH under contract number FKZ 50 OR 9201. CM acknowledges support from a CNRS-MPG cooperation contract.

References

Atteia J.-L., Barat C., Hurley K., Niel M., Vedrenne G., Evans W.D., Fenimore E.E., Klebesadel R.W., Laros J.G., Cline T., Desai U., Teegarden B., Estulin I.V., Zenchenko V.M., Kuznetsov A.V., Kurt V.G., 1987, ApJS 64, 305

Bond H.E., 1976, IBVS 1160

Christian C.A., Adams M., Barnes J.V., Butcher H., Hayes D.S., Mould J.R., Siegel M., 1985, PASP 97, 363

Giampapa M.S., Liebert J., 1986, ApJ 305, 784

Greenstein J.L., 1977, PASP 89, 304

Greiner J., Naumann Ch., Wenzel W., 1991, A&A 242, 425

Greiner J., Motch C., 1994, IBVS 4092

Greiner J., Motch C., 1995, A&A 294, 177

Haisch B., Strong K.T., Rodono M., 1991, ARA&A 29, 275

Hawley S.L., Pettersen B.R., 1991, ApJ 378, 725

Jacoby G.H., Hunter D.A., Christian C.A., 1984, ApJS 56, 257

Jarrett A.H., Gibson J.B., 1975, IBVS 979

Liang E.P., Li H., 1993, A&A 273, L53

Lovas M., 1977, IBVS 1345

Mirzoyan L.V., 1990, IAU Symp 137, L.V. Mirzoyan, B. Pettersen & M.K. Tsvetkov (eds.), Kluwer, Dordrecht, p. 1

Pettersen B.R., 1991, Mem. S.A. It. 62, 217

Pagano I., Ventura R., Rodonò M., Peres G., Micela G., 1995, this volume p. 95

Rebolo R., 1991, in IAU Symp 145, Evolution of Stars: The Photospheric Abundance Connection, G. Michaud & A. Tutukov (eds.), p. 85

Robin A., Crézé M., 1986, A&A 157, 71

Roizman G.Sh., Shevchenko V.S., 1982, Sov. Astron. Lett. 8, 85

Schaefer B.E., 1990, ApJ 353, L25

Shakhovskoy N.M., Alexeev I.Yu., Andronov I.L., Kolesnikov S.V., 1993, in Cataclysmic variables and related physics, O. Regev & G. Shaviv (eds.), Inst. Phys. Publ., Bristol, p. 237

Turnshek D.E., Turnshek D.A., Craine E.R., Boeshaar P.C., 1985, An Atlas of Digital Spectra of Cool Stars, Western Research Company, Tucson

H. Zinnecker: Have any of the large amplitude flare stars that you have listed been observed with ROSAT pointed observations, and if so, have any been detected?

J. Greiner: The objects CZ Cnc, AF Psc, and S 10933 are not covered by ROSAT pointed observations. The other flare stars of Tab. 1 are all observed by ROSAT and to my knowledge are also detected (quiescent emission).

D. Dravins: What accuracy in γ-ray burst positions will be required in order for flare stars not to contribute any significant background in searches for optical counterparts?

J. Greiner: With the present number density of known flare stars (outside of stellar clusters) one needs 1 arcmin^2 sized error boxes to push the flare star contamination below 0.1%. However, discoveries of quite a substantial number of new flare stars even in the galactic field (see e.g. Richter *et al.* 1995, these proceedings, p. 69), make this estimate an upper limit.

A Major Optical Flare on the Recently Discovered X-ray Active dMe Star G102-21

I. Pagano[1], *R. Ventura*[1], *M. Rodonò*[1], *G. Peres*[2], *G. Micela*[2]

[1] Osservatorio Astrofisico di Catania and Istituto di Astronomia dell'Università degli Studi, viale A. Doria 6, I-95125 Catania, Italy
[2] Istituto e Osservatorio Astronomico di Palermo, Palazzo dei Normanni, I-90134 Palermo, Italy

During *UBV* photoelectric measurements of G102-21 – a dM3 star recently discovered as a remarkably active X-ray source by Micela et al. (1995) – we have observed the flare shown in Fig. 1 (left panel). The observed flux increase at flare maximum and the energy output (cf. Table 1) make this event one of the largest amplitude and most energetic flares ever detected on UV Cet-type stars.

The ratios E_U/E_B (=8.9) and E_U/E_V (=3.9) are more than 7 and 2 times larger, respectively, than the mean values found for typical dMe flares by Lacy et al. (1976). Moreover, assuming the empirical relation between the absolute magnitude M_V and the mean energy dissipation during a flare (Gurzadyan 1980), the energy released in the U band exceeds the predicted one by a factor of $\sim 3 \cdot 10^3$. Clearly, we observed a rather peculiar and rare event that does not appear to be a typical flare on red-dwarf stars.

An aspect that deserves attention is the apparently abrupt descent to the quiescent level ($\sim 0\overset{m}{.}7\ \mathrm{min}^{-1}$). Only very few similar cases have been observed (Henry & Hall 1991, Cristaldi & Rodonò 1970, Foing et al. 1994). A possible interpretation is that we observed only the upper flaring region of a flare occuring close to or behind the stellar limb, so that the energy emission from low lying and densest flare sites, where the slow decay is produced, was not observed. Otherwise, the flare might have occurred just at or behind the star limb, so that, during the descent phase, the low-lying flare site eventually disappeared behind the limb due to the rotation of the star or was invisible all through the flare duration, respectively.

If the energy released during the course of the flare had been previously stored in a filament, as for solar two-ribbon flares, then a surface magnetic field of few thousands Gauss for a loop semi-length of 1 R_*, and a binary separation of $3 - 16 R_\odot$, are implied (see Pagano et al. 1995 for details).

Table 1. Characteristics of the flare shown in Fig. 1

	U	*B*	*V*
Δm max (mag)	7.29	3.88	3.03
Flare energy[a] ($\times 10^{34}$ erg)	8.9±3.0	1.0± 0.3	2.3±0.8
Equivalent duration[a] (hours)	79	3.4	1.5

[a]assuming d=21.3±3.6 pc (Gliese & Jahreiss 1991)

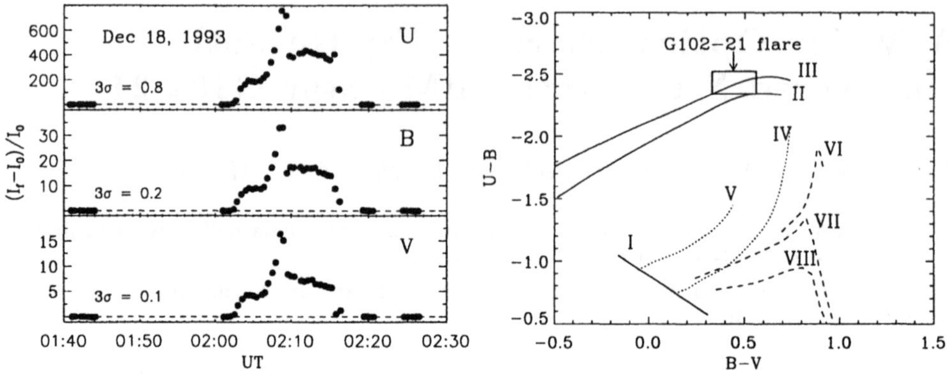

Fig. 1. *Left panel:* the light curves of the G102-21 flare. Note the quite different ordinate scale on the left subpanels. *Right panel:* two colour diagram for the flare energy emission. The labelled lines correspond to the following models for flare emission, (**I**) blackbody at 8,000-20,000 K; (**II-III**) hydrogen plasmas, optically thin in the Balmer continuum, with $T_e = 10,000$ K and electron density 10^{12} and 10^{14} cm^{-3}, respectively; (**IV-V**) an optically thick plasma at T_e=10,000 and 15,000 K, respectively; (**VI-VII-VIII**) upper layers of a dwarf star heated by proton beams with threshold proton energy 1, 2 and 5 MeV, respectively. The theoretical model grid is from Gershberg et al. (1991).

The rectangle in Fig. 1 (*right panel*) delimits the range of variability of $U-B$ and $B-V$ colours for the pure flare radiation during the event. The $U-B$ (-2.43 ± 0.09) is extremely negative when compared to typical $U-B$ flare colours on dMe stars, usually in the range from -0.5 to -1.5 (Gurzadyan 1980, Cristaldi & Rodonò 1975). Instead, the $B-V$ colour (0.45 ± 0.09) is quite typical for stellar flares. In the same Fig. 1, theoretical models from Gershberg et al. (1991) are also shown. The colours of the pure flare radiation fit quite well with the colours predicted by Minikulov (1987) for a hydrogen plasma at 10,000 K, optically thin in the Balmer continuum, with electron density of 10^{14} cm^{-3}.

Acknowledgements: Stellar activity research at Catania Observatory and University is funded by the Italian MURST-CRA, CNR-GNA and ASI.

References

Cristaldi S., Rodonò M., 1970, A&AS 2, 223
Cristaldi S., Rodonò M., 1975, in IAU Symp. 67, 75
Foing B.H. et al., 1994, A&A, 292, 543
Gershberg R.E. et al., 1991, SvA 35, 269
Gliese W., Jahreiss H., 1991, 3th Catalogue of Nearby Stars, ADC, GSFC
Gurzadyan G.A., 1980, Flare Stars, Pergamon, Oxford
Henry G.W., Hall D.S., 1991, ApJ 373, L9
Lacy C.H., Moffett T.J., Evans D.S., 1976, ApJS 30, 85
Micela G., Favata F., Pye J., Sciortino S., 1995, A&A, in press
Minikulov N.Kh., 1987, Izv. Krym. Astrofiz. Obs. 76, 43
Pagano I., Ventura R., Rodonò M., Peres G., Micela G., 1995, A&A, submitted

Flares Detected on Some Late Giants: Are They Real?

R. Konstantinova-Antova[1], *M.M. Ivanov*[2], *A.P. Antov*[1]

[1] Institute of Astronomy, Bulgarian Academy of Sciences, 72 Tsarigradsko shose, BG-1784 Sofia, Bulgaria
[2] Department of Astronomy, Sofia University, 5 James Bourchier Av., BG-1126 Sofia, Bulgaria

1 Introduction

Recently, some apparently single red giants with evidence of chromospheric activity were reported. In this work we observed those single late giants with reported flares, referenced by Pettersen (1989) and Schaefer (1989), as part of a program for investigation of the chromospheric activity in G and K giants.

According to Simon & Drake (1989), active single G and K giants may be first crossing stars that have just entered the phase of rapid spin-up as their convection zones advance upon their more rapidly rotation radiative cores. These stars are situated on the base of the red giant branch on the HR diagram. As a result they may possess dynamo driven activity at this stage of evolution.

2 Observations and Results

The aim of our research is to investigate suspected chromospherically active late giants and especially those with reported flares, using some typical indicators, accessible with our equipment, i.e. Ca II K+H, Hα and flares (Ulmschneider 1979).

For this purpose, spectroscopy and photometry were carried out for ν Oph (G9 III), τ CrB (K0 III) and V654 Her (K2 III). The 2 m telescope at the National Astronomical Observatory at Rozhen, equipped with a Coude spectrograph and CCD detector was used along with two 60 cm telescopes equipped with single-channel photon-counting photometers and a U-filter. The integration time was 1 sec.

High-resolution spectra were obtained for each star on more than one night with S/N>100 at Hα and S/N>50 at Ca II K+H. No evidence of chromospheric activity, such as emission cores in Ca II K+H were observed (Fig. 1). Hα was in absorption and seems not to be filled-in (Fig. 2). Hα spectra of OP And, an apparently single, chromospherically active K1 giant, were obtained with the same equipment. Its Hα absorption is partially filled-in and, according to Strassmeier et al. (1990), it also exhibits strong Ca II K+H emission.

V654 Her is too faint for us to obtain a spectrum in Ca II K+H, so we used those presented in Tsvetkov & Pettersen (1985). Their low-dispersion spectrum shows no emission cores.

$17^h52^m34^s$ total monitoring time was obtained for V654 Her and $13^h53^m44^s$ for τ CrB. No flares were detected. ν Oph was too bright to be observed with our photometric equipment.

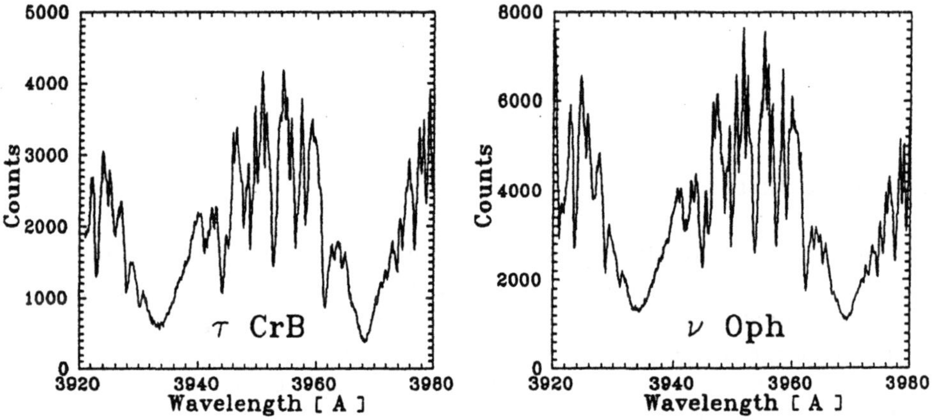

Fig. 1. Spectra of ν Oph and τ CrB in the region of Ca II K+H.

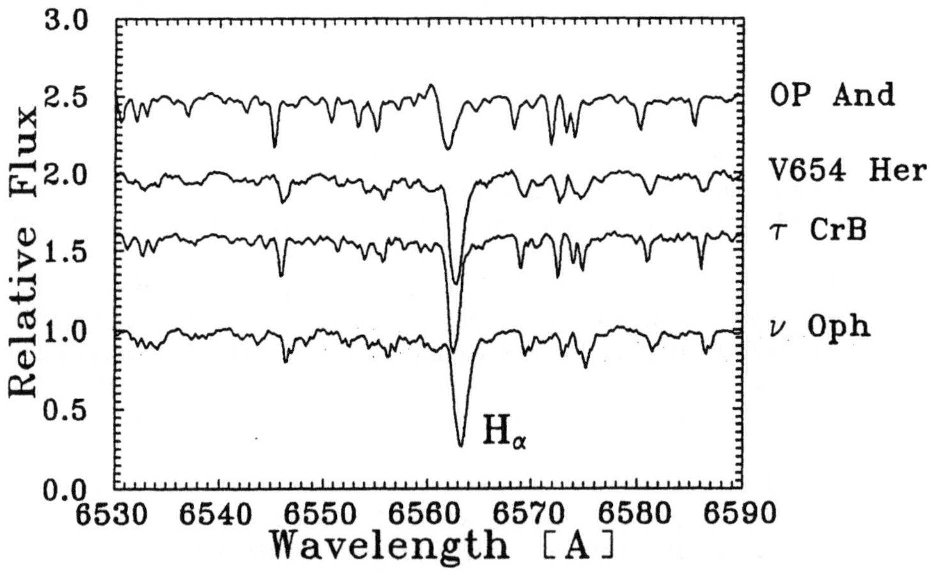

Fig. 2. Hα spectra of the stars investigated. OP And is also presented for comparison.

3 Discussion

Our work revealed a lack of typical chromospheric activity indicators in all three stars. In Fig. 3 we show a HR diagram derived using the evolutionary tracks of Schaller et al. (1992) for solar abundance and $M_{\rm bol}$ and $T_{\rm eff}$ from Allen (1973).

M_V and $B - V$ for ν Oph and τ CrB were taken from Wilson (1976). The M_V of τ CrB differs significantly from other K0 giants and so we took its M_V, obtained from DDO photometry by Brown et al. (1989). For V654 Her we used $B - V$ value obtained by us (a mean value from 8 nights of observation at the Belogradchik Observatory) and M_V was derived using standard models from Allen (1973). The stars are situated high above the base of the giant branch and thus are inconsistent with Simon & Drake's (1989) hypothesis.

These stars seem to be normal giants and cannot be considered as chromospherically active. The nature of the events reported in Pettersen and Schaefer and presented in Table 1 remains a puzzle. We point out that the so-called flares were detected only once, and no other have been observed since.

If we analyze the reported flares we come to the following conclusions:

1. The event, detected on τ CrB visually, may have been due to a satellite glint, meteor or other sky effect.
2. The same might be possible in a lesser degree for the flare on ν Oph.
3. The nature of the so-called flare, detected photometrically on V654 Her, is unclear. We consider the star as chromospherically inactive. But we don't know artificial causes which might simulate a flare. Since the event seems not to be due to chromospheric activity, its nature remains unknown.

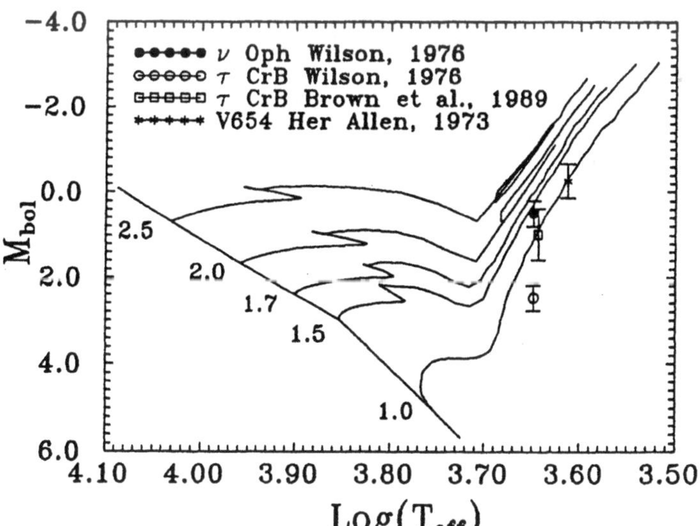

Fig. 3. Positions of the investigated stars in the HR diagram.

Table 1. The table is taken from Schaefer (1989), where: Δm is the amplitude of the event; t is the minimum e-folding rise or fall time; D is the total event duration; N is the number of independent measurements which confirm the existence of an anomaly; and q is the angular resolution of the detector.

Star	Sp. Type	m_V	Δm	t(sec)	D(sec)	N	q(")	Detector
ν Oph	G9 III	3.3	7	10	>80	10	300	visual
τ CrB	K0 III	4.7	1.7	<0.25	0.5	1	300	visual
V654 Her	K2 III	10.0	0.15	140	300	50	10	photometer

Acknowledgements: The authors are thankful to Drs. N. Tomov and R. Zamanov for their help in obtaining the spectroscopic material. R.K.-A. is grateful to the German Science Foundation for financial support to attend the Colloquium. She is also thankful to Dr. R. Pallavicini for useful discussions. This work is partially supported by the Bulgarian National Science Foundation under contract F-347/1993.

References

Allen C.W., 1973, Astrophysical Quantities, Athlone Press, London, p. 289ff.
Brown J.A., Sneden Ch., Lambert D.L., Dutchover Jr. E.., 1989, ApJS 71, 293
Pettersen B.R., 1989, Solar Phys. 121, 299
Schaefer B.E., 1989, ApJ 337, 927
Schaller G., Schaerer D., Meynert G., Maeder A., 1992, A&A 96, 269
Simon T., Drake S.A., 1989, ApJ 346, 303
Strassmeier K.G., Fekel F.C., Bopp B.W., Dempsey R.C., Henry G.W., 1990, ApJS 72, 191
Tsvetkov M.K., Pettersen B.R., 1985, A&A 150, 160
Ulmschneider P., 1979, SSR 24, 71
Wilson O.C., 1976, ApJ 205, 823

P.B. Byrne: The closest analogy to active giants are the RS CVn stars. There, optical flares are rare. So, is 14–18 hours monitoring sufficient to rule out optical flares?

R. Konstantinova-Antova: Of course, more observations are required, because the giants are more luminous than the dwarf flare stars. Also, we don't know whether the dynamo mechanism, suspected to operate on them is with a comparable efficiency as on the active dwarfs. But for the three giants I reported, there are no other evidences for chromospheric activity than the fact that they exhibited (in two of the cases under suspect) such flashes. In the light of our investigations (including spectral ones) I cannot say that they are due to chromospheric activity.

BD+47°819, an A-type Flare Star with Peculiar Spectroscopic Variation

Jun-Jie Wang

Beijing Astronomical Observatory, Chinese Academy of Sciences,
Beijing 100080, China

Stellar flares are one of the most challenging problems in astrophysics. Up to now, most of the detected flare stars are of late spectral type (e.g. UV Cet type stars etc.). Non-classical flare stars with early spectral type are very rare.

Recently, an A5-type flare star BD+47°819, near the open cluster α Persei, was discovered (Wang 1993). The flare U amplitude was $1^{\mathrm{m}}.5$ in ten minutes. The V magnitude of the star was $9^{\mathrm{m}}.25$ during the quiescent state.

On 1993 February 28 and March 1, spectroscopic observations of BD+47°819 were madewithn the 2.16 m telescope of Beijing Astronomical Observatory. A grating with a dispersion of 5nm/mm was used. Figs. 1 (a) and (b) show the observed spectra in the two nights. The exposure time for each spectrum was 15 minutes. Fig. 1 (a) shows a strange feature, an "emission line" at around 676 nm. Is this a real feature?

Clearly, the six Hα profiles in (a) were nearly of the same shape. There were no large variations. But in (b), the Hα profiles show relatively large variations, which present the tendency from broad to narrow in order. It seems that the stellar photospheric activity on March 1 was more violent than that on February 28.

Some phenomena could be explained by a binary model, the bright component being an A type star and the faint one a dMe star. Another interpretation would be in terms of an envelope model. Since both the binary model and the envelope model cannot fully explain the phenomena in BD+47°819, the question is: what kind of variable star is it? Certainly, further observations, both spectroscopical and photometrical, are needed. Also, achival plates should be checked.

Acknowledgements: I would like to acknowledge beneficial discussions with Dr. R. Mundt, Dr. M. Tsvetkov and Prof. J.J. Hu, as well as Dr. J. Greiner for his good suggestion. I also thank for support from CNSF.

References

Wang J.-J., 1993, IBVS 3836

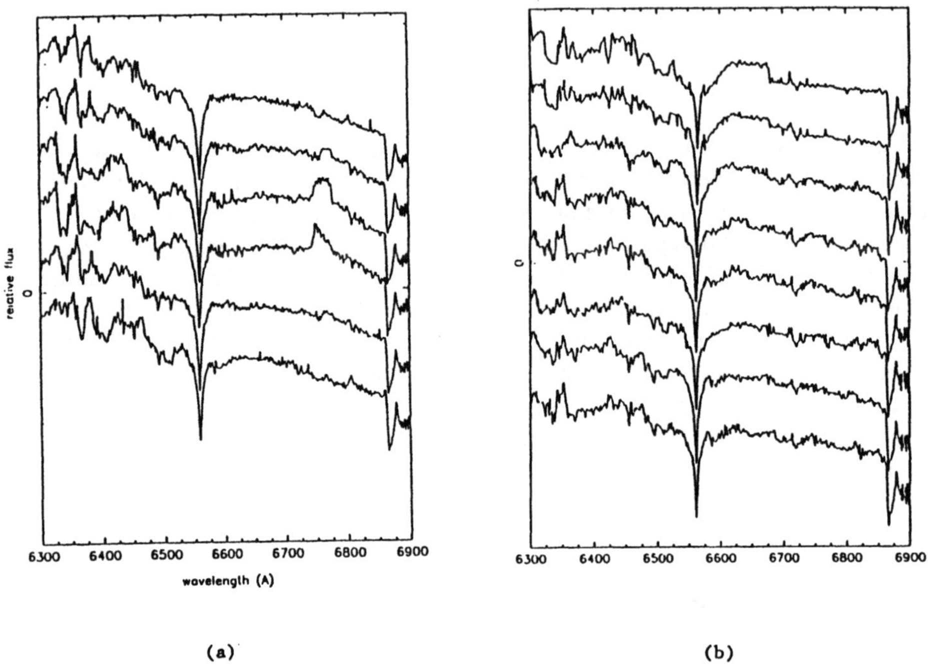

Fig. 1. Results of the spectroscopic observations of BD+47°819

The Long-Term Behaviour of BY Dra

K.P. Panov[1], *M.S. Ivanova*[1], *J.S.W. Stegert*[2]

[1] Institute of Astronomy, Bulgarian Academy of Sciences, Zarigradsko Shose 72, 1748 Sofia, Bulgaria
[2] Sternwarte der Universität Bonn, Auf dem Hügel 71, 53121 Bonn, Germany

BY Dra has been extensively monitored for flares and spot-modulation of brightness at the Bulgarian National Astronomical Observatory (Panov & Ivanova 1993). Its mean light level exhibited remarkable changes in the past, due to the emergence and disintegration of photospheric spots, and different cycles have been proposed for the long-term changes of the BY Dra activity. This star is still believed to be a good candidate for cyclic variations, presumably similar to the sunspot activity cycle. After a spectacular drop in brightness in 1965 – 1967, and a subsequent rise in the early 70s, the mean light remained approximately constant during 1977 – 1991. A brightening of BY Dra occurred in 1992.

In Fig. 1, the light curves are shown for 1991, 1992 and 1993 in the V-band and in $B-V$. Phases were calculated with the ephemeris of Chugainov (1966):

$$\mathrm{JD\ (min.\ light)} = 2438983.612 + 3.836 \cdot E$$

From Fig. 1, it is apparent that the mean visual light, which was about $V = 8.20$ in 1991, increased in 1992 by $0\overset{\mathrm{m}}{.}10$, and a further increase by another $0\overset{\mathrm{m}}{.}03$ occurred in 1993. The phase of minimum light in 1992 indicates that the position of the spot remained approximately the same in stellar longitude as in 1991. However, in 1993 the position of the spot changed by almost half a period.

In Fig. 2, the phase of minimum light is plotted versus JD for 1986 – 1992. The smooth variation indicates that the light modulation during that time was dominated by the same spot (or spot-group), which migrated either in stellar longitude, or in latitude. In the second case, the phase variations reflect the rotation on different stellar latitudes, i.e. differential rotation. This spot disintegrated in 1992, resulting in an increase of the mean light. The spot phase of 1993 indicates that probably a new spot emerged on the opposite side of the star.

In the 1992 light curve (Fig. 1), the deviation of the encircled point is probably due to the rapid shift of the light curve in ordinate during the observing season. It reflects a brightening of $0\overset{\mathrm{m}}{.}06$ from JD 2448780.5 to 2448835.4. The $B-V$ curves have an amplitude of about $0\overset{\mathrm{m}}{.}015$ in both 1992 and 1993, and the colour is reddest at light minimum. This is in agreement with a dark-spot model. It is interesting to note, that the $0\overset{\mathrm{m}}{.}13$ brightening of BY Dra resulted in a blue shift of only $0\overset{\mathrm{m}}{.}01$ in $B-V$.

In Fig. 3, the long-term behaviour of the BY Dra is shown in the V band and in $B-V$. The 1993, the mean visual light is about the same as in the early

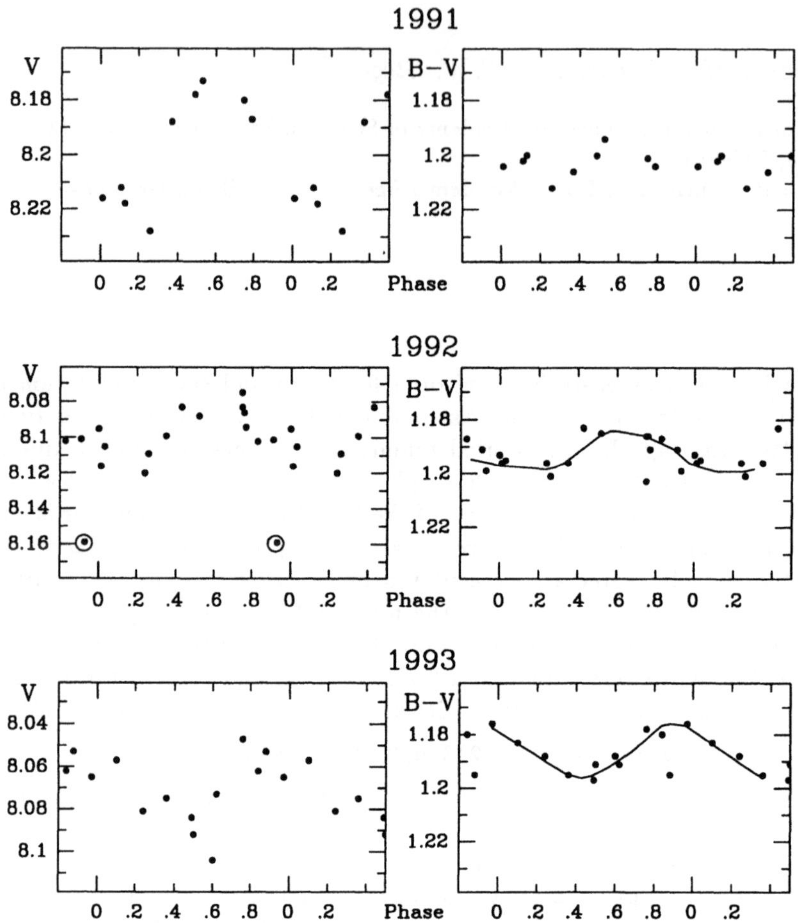

Fig. 1. Light curves of BY Dra in 1991, 1992 and 1993. Data from the Bulgarian National Observatory.

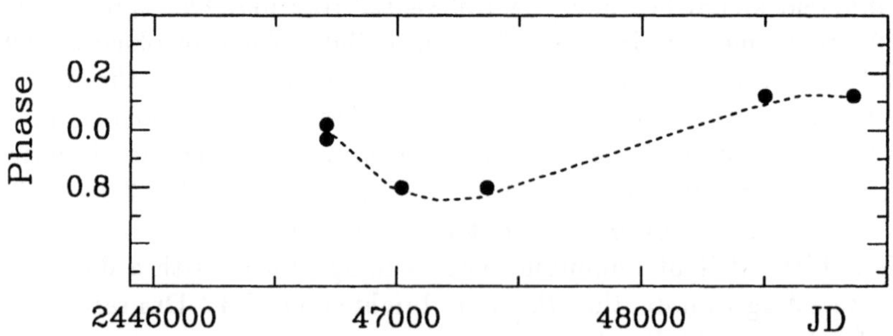

Fig. 2. Phase of minimum light of BY Dra versus JD for 1986 – 1991.

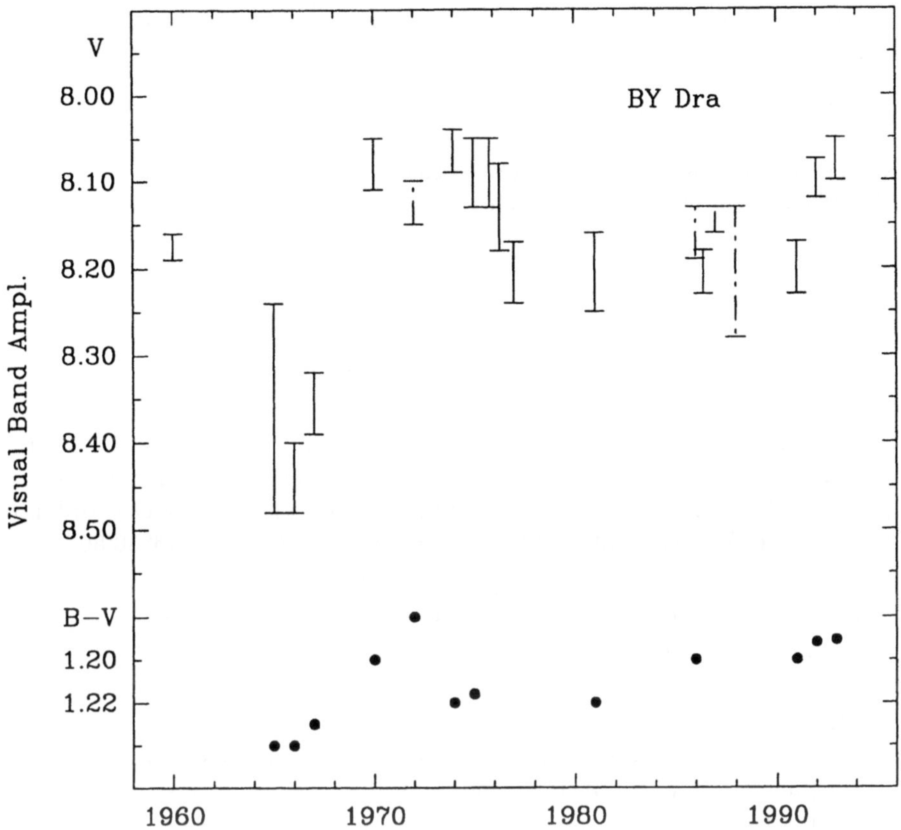

Fig. 3. The long-term mean light variations of BY Dra (from published data and the present work)

70s. The complicated character of the long-term variations does not confirm previously proposed activity cycles for BY Dra, and any suggestion for a cycle length seems as yet premature.

Acknowledgements: This work has been supported by the Bulgarian National Science Foundation with grant F-465 and by the Deutsche Forschungsgemeinschaft (Projekt 436 BUL 113).

References

Chugainov P.F., 1966, IBVS No. 122
Panov K.P., Ivanova M.S., 1993, ApSS 199, 265

CN Com: a Typical RS Canum Venaticorum Star

G.A. Richter[1], *J. Greiner*[2]

[1] Sternwarte Sonneberg, 96515 Sonneberg, Germany
[2] Max-Planck-Institut für Extraterrestrische Physik, 85740 Garching, Germany

CN Com = CSV 6907 was dicovered by Romano (1958). First we note that the GCVS (and Simbad) coordinates are wrong by ≈2′. Using the finding chart of Romano (1958) and the APM data (digitized POSS), the correct coordinates of CN Com are R.A.(2000.0) = $12^h19^m47\overset{s}{.}0$, DEC(2000.0) = $+16°30'50''$.

CN Com is only 8″ distant from the *ROSAT* X-ray source RX J1219.7+1630 detected during the All-Sky-Survey at a PSPC countrate of 0.022 cts/s. Due to the positional proximity and the absence of other optical objects brighter than 20^{rmm} within the about 30″ error circle, CN Com is very probably the optical counterpart of RX J1219.7+1630.

CN Com was investigated on some 600 photographic plates (Sonneberg astrographs 400/1600 mm and 400/2000 mm) of the fields 26 Com and 5 Com, covering the time interval from 1962 to 1994 (with only a very few plates from the years 1967-1974).

The elements are:

$$T(\text{min}) = 243\ 7668.521 + 0\overset{d}{.}7354422 \cdot E,$$

which deviate only negligibly from the dates given by Meinunger (1968).

Figure 1 gives the year by year light curves (from January to June, each). One can see migrating waves typical for RS Canum Venaticorum stars. Sometimes the amplitude of the primary minimum is very small (only $0\overset{m}{.}3$ in 1963, 1975, 1982), at other times large ($0\overset{m}{.}7$ in 1962, 1984, 1985, 1991). In some years the amplitude is unknown because there are no observations near phase 0 (1965, 1976-1980, 1983, 1986-1988, 1993/1994). In these cases, other plate collections should be checked. Probably it is not merely the primary that is eclipsed but rather a disk or a halo of varying luminosity and size.

Acknowledgement: This work has been supported by funds of the German Bundesministerium für Forschung und Technologie under grant 05 2S0524 (GAR), and FKZ 50 OR 9201 (JG). We are grateful to R. McMahon for installing the software used for the APM finding charts. The ROSAT project is supported by the German Bundesministerium für Forschung und Technologie (BMFT/DARA) and the Max Planck Society.

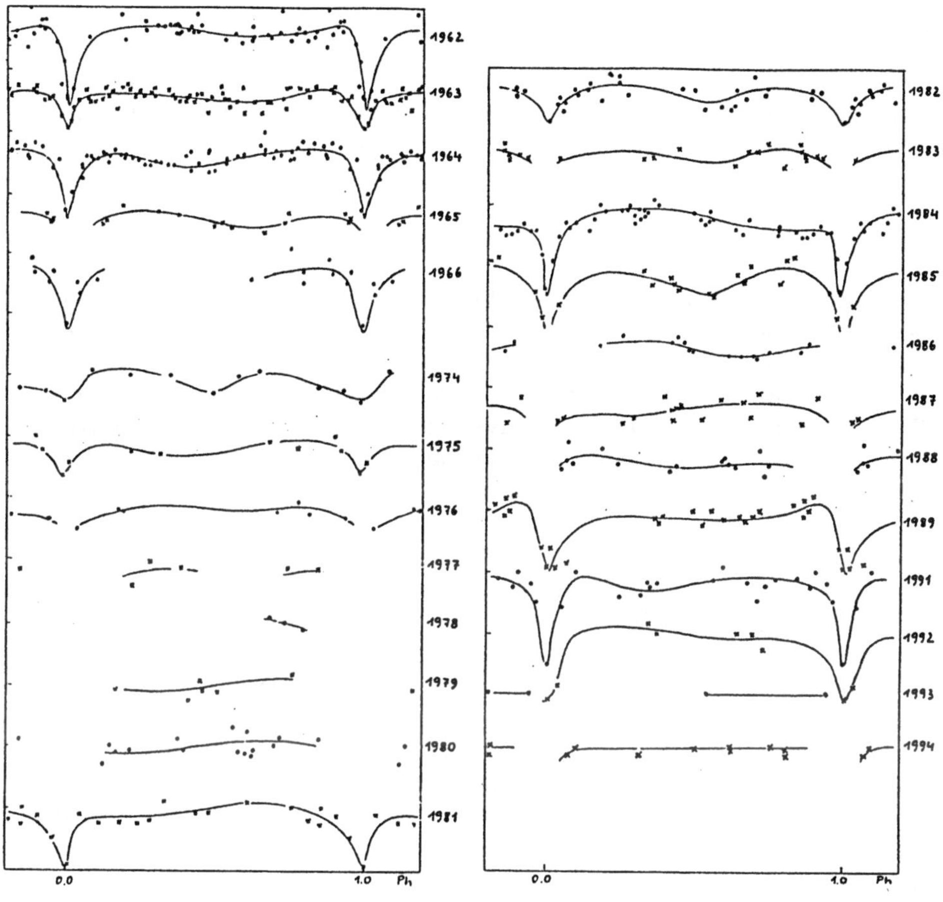

Fig. 1. Light curve of CN Com as deduced from the astrographic plates (field patrol) of Sonneberg Observatory. The migrating waves clearly classify CN Com as a typical RS CVn star.

References

Meinunger L., 1968, Veröff. Sternw. Sonneberg 7, 385
Romano G., 1958, Coelum 26, No. 108

A Long-Duration Pre-flare Dip in the Light Curve of V1054 Oph

R. Ventura[1], *I. Pagano*[1], *G. Peres*[2], *M. Rodonò*[1,3]

[1] Osservatorio Astrofisico di Catania, Città Universitaria, I-95125 Catania, Italy
[2] Istituto e Osservatorio Astronomico, Palazzo dei Normanni, I-90134 Palermo, Italy
[3] Istituto di Astronomia, Università degli Studi, viale A. Doria 6, I-95125 Catania, Italy

Photometric observations of V1054 Oph (Wolf 630 AB, $m_V = 9.69$) and of the comparison star HD 152678 ($m_V = 8.1$) were carried out in the *UBV* bands at Catania Astrophysical Observatory on 1994, June 15, using the 91 cm Cassegrain telescope equipped with an improved version of the twin-beam photometer URSULA (De Biase et al. 1988). We have detected a relatively low-intensity flare event preceded by a pre-flare dip of low amplitude and the exceptionally long duration of 36 min. The integration times were 10, 5 and 5 sec in the *U*, *B* and *V* bands, respectively. The variable and the comparison stars were observed simultaneously in the two independent channels of the twin-beam photometer. The sky background was subtracted from the photon count rates of V1054 Oph and its comparison star in each spectral band, and the flux of the variable was normalized to that of the comparison star.

Owing to the very small angular separation (0″.201) of the two brightest components of the system (A and B components) the measured flux was that of the combined total flux of the system. Since both components are classified as flare stars, the observed activity phenomena may have occurred in any of the two stars. Therefore we have considered the possibilities that either A or B had flared. In both cases we removed the flux contribution of the non-flaring star from the observed combined flux.

Fig. 1 shows the differential light curves in the *UBV* bands on 1994 June 15. The characteristics of the flare and the pre-flare dip are reported in Table 1.

Table 1. Magnitude variations at flare maximum and at dip minimum, and total energy in the three spectral bands involved in the two events. The corresponding flare values in the *V* band is not reported because the ΔV was below our sensitivity level.

	FLARE				DIP					
	ΔU (mag)	ΔB (mag)	E_U (erg)	E_B (erg)	ΔU (mag)	ΔB (mag)	ΔV (mag)	E_U (erg)	E_B (erg)	E_V (erg)
A	1.01	0.14			0.19	0.10	0.06			
			$1.93\ 10^{32}$	$5.68\ 10^{31}$				$1.97\ 10^{31}$	$2.75\ 10^{31}$	$7.08\ 10^{31}$
B	1.14	0.17			0.23	0.12	0.07			

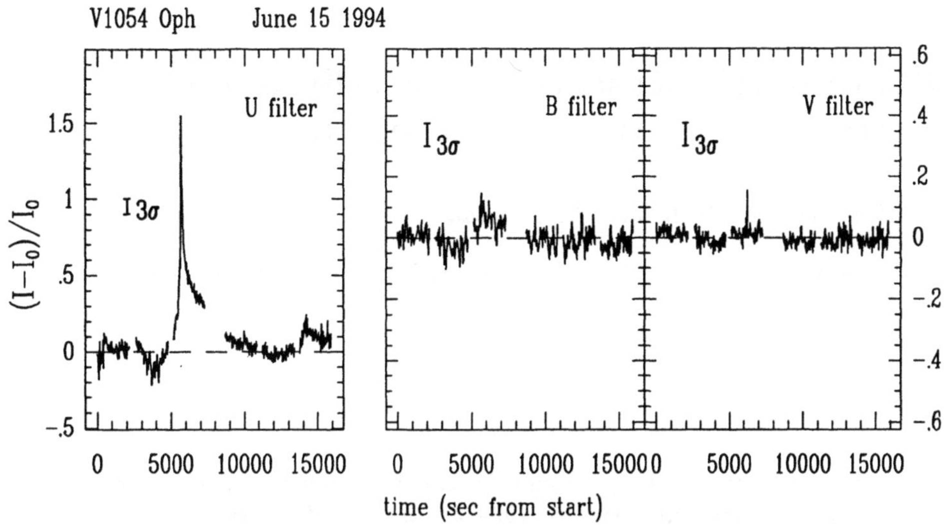

Fig. 1. The U, B and V light curves of the flare detected on V1054 Oph, on 1994 June 15 beginning at 22:59:54 UT. Time is reckoned in seconds from 21:00 UT. The ordinate axis scale is normalized to the quiescent star level, assuming that the flare had occurred on the A component. Note that the B and V plots are expanded by a factor of 2 with respect to the U plot.

The present observation of such a long duration pre-flare dip is hardly reconcilable with any theoretical model of dips so far proposed (Gurzadyan 1968, 1980; Mullan 1975; Grinin 1976; Giampapa et al. 1982) and poses strong constraints on them.

References

De Biase G.A., Paternò L., Fedel B., Santagati G., Ventura R., 1988, 2nd Workshop on Improvements to Photometry, NASA CP-10015, Borucki (ed.), p. 17

Giampapa M., Africano J.L., Klimke A., Parks J., Quigley R.J., Robinson R.D., Worden S.P., 1982, ApJ 252, L39

Grinin V.P., 1976, Izv. Krymsk. Astrofiz. Obs. 55, 179

Gurzadyan G.A., 1968, Astrofiz. 4, 154

Gurzadyan G.A., 1980, Flare Stars, Pergamon, Oxford

Mullan D.J., 1975, A&A 40, 41

Photometric Research of Pre-flare and Extra-flare Activity of Flare Stars

G. Roizman, L. Pustil'nik, D. Pundak

Observatory of the Science-Educational Center, Jordan Valley Academic College, Israel

Preliminary results of a photometric search for non-flare activity of the flare star EV Lac during June-August 1994 are presented. These observations are part of the long-term project entitled "Investigation of pre-flare processes on flare stars" being carried out at the Observatory of the Jordan Valley Regional College since July 1994. The project was motivated by reports of the existence of slow, low-amplitude non-periodic variability of both spectral lines and continuum outside/between flares (Rodono et al. 1979, Roizman et al. 1984).

1 Observations

We used a 14-inch Compustar telescope with an SBIG ST-6 CCD-photometer and a set of *RGB* filters. The $8' \times 6'$ field of view of the photometer allows the measurement of EV Lac and four comparison stars on each frame. We exposed the EV Lac region every 2 min and then recorded the image. In order to reach the necessary accuracy we carried out differential photometry. The magnitude differences were calculated without reduction to the standard *BVR* photometric system. As the analysis of the data shows, we achieve the expected Poisson accuracy in spite of difficult observational conditions (our Observatory is located 200 m below sea level (and is unique in this sense!). So, we were able to detect low-amplitude BY Dra variability of EV Lac. The amplitude is 2.3%, for an estimated standard deviation ~2%. During our observations the EV Lac ephemeris was: $JD_{min} = 2449544.627 + 4.375 \cdot E$.

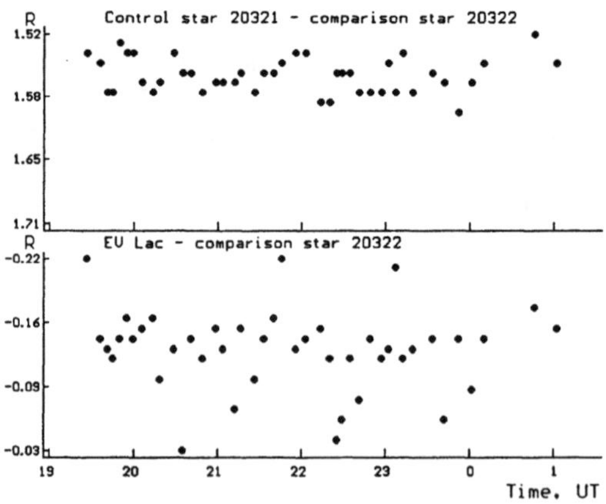

Fig. 1. Brightness differences of EV Lac and its two comparison stars (EV Lac-C1) and (C2 - C1) on 1994 August 13/14 in the red band.

2 Discussion

It is important to stress that all stars were measured simultaneously and that EV Lac is brighter than the check star C2, at least in the green and red bands. This means that the difference cannot be due to fast variations of the atmospheric extinction or to different numbers of photons. Thus it appears that there are real fluctuations of EV Lac's brightness on a time scale of about several minutes or less. This would not be so surprising if we were only dealing with the blue band where the quiescent flux (both hydrogen lines and continuum) may be disturbed by numerous small flares (microflaring). The observed scatter is maximum in the red (according to the F-test the significance level is < 0.01). Taking into account that the real flux from EV Lac increases from the blue to the red and so the Poisson error, as well as the flare contribution, decreases, we may conclude that the cause of the additional scatter increases toward long wavelength.

This unusual "inverse" dependence of the observed scatter on wavelength offers us a possible explanation. Sometimes before flares short dips below the quiescent level are observed – negative preflares (NPF) (Cristaldi et al. 1982). According to the model of Grinin (1983), an NPF is due to weak impulsive heating leading to ionization of metals, and thus an increase of the H^- opacity, aresulting in a temporary decrease of the flux. Grinin's calculations show that this effect occurs at wavelengths excluding the strong molecular bands of TiO, and that the amplitude of an NPF increases towards long wavelength, as in our data.

Thus, in the first approximation the assumption that the negative preflares are the main reason for the observed fluctuations of the brightness of EV Lac does not contradict our observational data. In fact we assume that the "usual microflaring" observing in the U and B bands causes the "negative microflaring" that may be observed in the green and near infrared bands.

Acknowledgements: The authors are grateful to S.V. Pustil'nik for her help in the data reduction. This work is supported by a grant from Ministry of Science of Israel.

References

Cristaldi S., Gershberg R., Rodono M., 1982, A&A 89, 123

Grinin V.P., 1983, in IAU Coll. 71, Activity in Red-Dwarf Stars, P.B. Byrne & M. Rodonò (eds.), Reidel, Dordrecht, p. 613

Mathioudakis. M., Doyle J., Rodono M., Gibson D., Byrne P., Avgoloupis S., Linsky J., Gary D., Mavridis L., Varvoglis P., 1991, A&A 244, 155

Rodono M., Pucillo M., Sedmak G., de Biase G., 1979, A&A 76, 242

Roizman G., Kabichev G.I., 1985, Soviet Astron. 29, 639

Stream Parameters in the Neighbourhood of the L_1 Point in W Serpentis-type Binaries

V.G. Karetnikov, E.V. Luthardt-Menchenkova, V.V. Nazarenko

Astronomical Observatory of the Odessa State University, Park Shevchenko, 270014 Odessa, Ukraine

The gas motion in close binary systems has been investigated by several authors using a hydrodynamic method (e.g. Prendergast & Taam 1974, Flannery 1975, Lubow & Shu 1975, 1976). All these researches have, besides important results, a common deficiency: in the analysis of the gas motion, the extent and velocity of the stream in the neighbourhood of the inner Lagrangian point L_1 was not calculated. They have been chosen on the basis of general assumptions (the same for all types of close binaries). A fundamental assumption was the small extent of the stream in the neighbourhood of L_1 perpendicular to the line connecting the centers of the stars. At the same time the extent and the velocity of the stream near L_1 exert a great influence on the further stream motion, on the formation of the circumstellar envelope and finally on the evolution of the close binary system. Therefore the first step in the investigation of the gas motion in a close binary must be the determination of the initial parameters of the stream in the neighbourhood of the point L_1.

A method of calculation of the parameters and the evolution of the stream near L_1 in binary systems was proposed by Nazarenko (1993). Using this method we calculated the physical conditions in the stream near point L_1 for three interacting binary systems of W Ser-type: β Lyr, V367 Cyg, and RY Sct. For β Lyr and V367 Cyg several versions of stream parameters were calculated, corresponding to different orbital elements obtained by different authors. For

Table 1. Main parameters of the investigated systems

N	Star	M_1 ($M_\odot$)	M_2 ($M_\odot$)	T_1 (K)	A ($R_\odot$)	P (days)	References
1	β Lyr	3.8	14.6	12000	61.0	12.914	Skul'ski (1975)
2	β Lyr	2.0	11.7	11000	55.0	12.914	Ziolkowski (1976)
3	V367 Cyg	2.3	3.6	12000	53.0	18.598	Menchenkova (1990)
4	V367 Cyg	2.5	4.4	8000	56.7	18.598	Pavlowski et al. (1992)
5	V367 Cyg	19	12	11600	93.0	18.598	Li & Leung (1987)
6	RY Sct	10	33	28000	75.0	11.125	Antokhina & Cherepashchuk (1988)

Table 2. Stream parameters in the neighbourhood of point L_1

Star	N_L (cm^{-3})	T_L (K)	R_S	R_R	$\dot{M}$ ($M_\odot$/yr)	V_L (km/s)	References
β Lyr	$6.4 \cdot 10^{14}$	46000	0.15	0.22	$6.4 \cdot 10^{-6}$	29.5	Ziolkowski (1976)
β Lyr	$2.9 \cdot 10^{14}$	32200	0.10	0.25	$1.3 \cdot 10^{-6}$	23.7	Skul'ski (1975)
V367 Cyg	$5.2 \cdot 10^{14}$	51780	0.19	0.34	$1.8 \cdot 10^{-5}$	28.5	Menchenkova (1990)
V367 Cyg	$3.8 \cdot 10^{14}$	23040	0.14	0.33	$4.0 \cdot 10^{-6}$	19.3	Pavlowski et al.(1992)
V367 Cyg	$2.7 \cdot 10^{14}$	36300	0.10	0.42	$1.0 \cdot 10^{-6}$	25.0	Li&Leung (1987)
RY Sct	$1.4 \cdot 10^{15}$	69100	0.15	0.27	$1.0 \cdot 10^{-5}$	32.3	Antokhina & Cherepashchuk (1988)

RY Sct, only one variant was calculated, because the orbital elements, obtained for this system by different authors, have similar values. The main parameters of the investigated systems are collected in Table 1, together with the references. M_1 and T_1 are the mass and the temperature of the mass losing star, M_2 is the mass of the accreting star, A is the distance between the centres of the stars, P is the orbital period.

The main results of our calculations are collected in Table 2. N_L is the concentration of matter in the point L_1, T_L is the temperature and V_L is the stream velocity (along the line connecting the centres of the stars) in the point L_1. We calculated the mass exchange rate $\dot{M}$ through the point L_1 on the basis of the velocity V_L and the concentration of matter in the neigbourhood of the point L_1 (using the correlation between the concentration in the stream N and the distance from the stream axis and taking into consideration the axial symmetry of the stream near point L_1). R_S is the stream radius (assuming that at the edge of the stream the concentration of matter has decreased by three orders), R_R is the radius of Roche lobe of the mass losing star. R_S and R_R are given in units of the distance between the centres of the stars.

The analysis of the results permits the following conclusions:

1. The extent and the velocity of the stream in the neighbourhood of the point L_1 depend mainly on the degree of filling the Roche lobe by the mass losing star and on the conditions in its atmosphere.
2. The stream has an axial symmetry in the neighbourhood of the point L_1 (the symmetry axis coincides with the line connecting the centres of the stars). The stream radii in the investigated systems are very large (0.10...0.19). They are only two times less than the radii of the mass losing stars.
3. The stream velocity along the line connecting the centres of the stars in the neighbourhood of the point L_1 amounts 20...30 km/s and is equal to the sound velocity in this point. This result confirms the supposition by Lubow & Shu (1975) about the stream velocity in the point L_1. In perpendicular direction the stream is in hydrostatical equilibrium (its velocity does not exceed a few hundred m/s). It also agrees well with the earlier assumption of Prendergast & Taam (1974).

4. The stream radius and the mass transfer rate are in good correlation with the parameter F ($F = U_L/K$, where U_L is the total gravitational potential in the point L_1, K is the specific kinetic energy of the ideal gas in the atmosphere of the mass losing component in the point L_1). For small values of F the stream radius and the mass transfer rate are large. For large values of F the stream radius and the mass transfer rate are small. For the investigated systems the rate of mass transfer reaches 10^{-5} $M_\odot$/yr.
5. For V367 Cyg the stream radii calculated on the basis of the system parameters, obtained by the different authors, differ considerably. The maximum value of the stream radius and the rate of mass transfer were obtained by using the parameters of the system from Menchenkova (1990). The stream radius and the rate of mass transfer calculated on the basis of the parameters from Pavlowski et al. (1992) and Li & Leung (1987) are smaller. These small values are in contradiction to the observational data about the existence of circumstellar matter with developed structure in this system. Therefore we draw the conclusion, that for V367 Cyg the parameters obtained by Menchenkova (1990) are more realistic: $M_1 = 2.3$ $M_\odot$, $M_2 = 3.6$ $M_\odot$, $T_1 = 12000$ K, $A = 53$ $R_\odot$.
6. For β Lyr the maximum values of the stream radius and the rate of mass transfer obtained for the system parameters calculated by Ziolkowski (1976) are: $M_1 = 2.0$ $M_\odot$, $M_2 = 11.7$ $M_\odot$, $T_1 = 11000$ K, $A = 55$ $R_\odot$.
7. For RY Sct the stream radius calculated on the basis of the different variants of the system parameters are equal.

References

Antokhina E.A., Cherepashchuk A.M., 1988, Pis'ma Astr. Zhurn. 14, 252
Flannery B.P., 1975, ApJ 201, 661
Li Y.F., Leung K.C., 1987, ApJ 313, 801
Lubow S.H., Shu F.H., 1975, ApJ 198, 383
Lubow S.H., Shu F.H., 1976, ApJ 207, L53
Menchenkova E.V., 1990, Astr. Zhurn. 67, 653
Nazarenko V.V., 1993, Astr. Zhurn. 70, 12
Pavlowski K., Schneider H., Akan M.C., 1992, A&A 258, 329
Prendergast K.H., Taam R.E., 1974, ApJ 189, 125
Skul'skij M.Yu., 1975, Astr. Zhurn. 52, 710
Ziolkowski J., 1976, ApJ 204, 512

Monitoring of Flare Stars at Nainital Observatory

B.B. Sanwal

Uttar Pradesh State Observatory, Manora Peak, Nainital, India

1 Observations and Reductions

Five UV Cet-type flare stars were photoelectrically monitored during the period 1970 to 1981 using either the Carl Zeiss Jena 1.04 m reflector or the 0.56 m reflector of the Uttar Pradesh State Observatory Nainital. Thermoelectrically cooled EMI 6094S and 1P21 photomultipliers were used with the 1.04 m and 0.56 m telescopes, respectively. The sky brightness was also recorded frequently for short intervals during continuous monitoring of the stars.

The light curves of the recorded flare events were represented by a graph showing the dependence on time of the quantity $i = I_f/I_0$, which represents the flow of excess radiation I_f, expressed in units of the flow I_0 from the star in the normal non-excited state. The fraction of the flow of the outburst is given in stellar magnitudes: $\Delta m = 2.5 \log (I_f + I_0)/I_0$. The energy released during a flare event was computed using the relation given by Cristaldi & Rodonò (1973):

$$E_c = 4\pi d^2 \times 10^{-0.4 m_c} Pi_c \times 60 \times P_c \text{ erg} \quad (1)$$

where d is the distance of the star, m_c is the apparent magnitude of the star in colour c, Π_c is the energy flux produced by a zero magnitude star outside the terrestrial atmosphere which is taken from Cristaldi & Rodonò (1973) and

$$P_c = \int \left(\frac{I_{f+o}}{I_0} - 1 \right) dt \quad (2)$$

$$\Pi_B = 6.3 \times 10^{-6} \text{ erg cm}^{-2} \text{ sec}^{-1}, \quad \Pi_U = 2.4 \times 10^{-6} \text{ erg cm}^{-2} \text{ sec}^{-1}$$

We estimated Δm, E_c, P_c, rise and decay times and energy released by the flare outbursts for most of the recorded events. Some of the statistics of the recorded events are given in Table 1. The results of the observed events have been published in the Information Bulletin on Variable Stars Nos. 557, 750, 758, 810, 848, 901, 932, 998, 1180, 1210, 1572, 2143 and IAU Coll. No. 15.

Name of star	UV Cet	YZ CMi	AD Leo	Gliese 867B
Visual magnitude	12.95	11.24	9.43	11.8
Spectral type	dM5.5e	dM4.5e	dM3.5e	dM4e
Total nights	5	36	46	1
Filter used	*B*	*B*	*B*	*U*
Total monitoring duration(hours)	11.6	103.8	107.6	3.1
Total flares recorded	17	20	20	2
Quiescent energy (10^{28} erg sec^{-1})	0.73	20.6	80.0	2.2
Energy range of flare events (10^{30} ergs)	1.2–0.07	22–0.8	70.6–1.0	1.17–1.08
Mean energy per flare (10^{30} ergs)	.26	8.2	14.0	1.12
Mean flare duration (min)	5.6	9.5	10.6	2.6
Mean rise time (min)	.5	.8	1.9	.38

2 Conclusions

The following conclusions support those drawn from statistical studies done by other authors:

1. The mean energy per flare increases with increasing intrinsic quiescent luminosity of the parent star.
2. Brighter flare stars have more energetic but less frequent flares.
3. The rise times become more rapid as the quiescent energy of the parent star decreases.
4. Most of the flare light curves are either combinations of spike and slow flares or successive overlapping flares, supporting the view that energy can be added at various times during the development and decay of a flare event.
5. Flares with greater energy release occasionally last longer.

Acknowledgements: The author is thankful to IAU and to the SOC of IAU Coll. 151 for providing the travel grant and the opportunity to participate.

References

Bhatt T.R., Sinvhal S.D., 1971, IBVS 557
Bhatt T.R., Sinvhal S.D., 1971, IAU Coll. 15, 124
Cristaldi S., Rodono M., 1973, A&A Suppl. 10, 47
Kapoor R.C., Sinvhal S.D., 1972, IBVS 750
Kapoor R.C., 1973, IBVS 758
Kapoor R.C., Sanwal B.B., Sinvhal S.D., 1973, IBVS 810
Kapoor R.C., 1973, IBVS 848
Kapoor R.C., Sinvhal S.D., 1974, IBVS 901
Sanwal B.B., 1974, IBVS 932
Sanwal B.B., 1975, IBVS 998
Sanwal B.B., 1976, IBVS 1180
Sanwal B.B., 1976, IBVS 1210
Sanwal B.B., 1979, IBVS 1572
Sanwal B.B., 1982, IBVS 2143

CCD Monitoring of Flare Stars in Stellar Aggregates

Omar M. Kurtanidze

Abastumani Astrophysical Observatory, 383762 Abastumani, Georgia

1 Introduction

After Haro's fundamental discovery of flare stars in stellar associations and young clusters, their importance was fully recognized. The flare star system provides one of the most important records of the stellar aggregate's history. This record can be used to establish the chronology of these systems and to test theories of star and aggregate formation.

Unfortunately, the observational material contains an unavoidable strong selection with respect to the statistics and the physical characteristics of flare stars in aggregates. This explains the difficulties which arise when one tries to compare physical and statistical characteristics of flare stars in aggregates and in the solar neighborhood, since the latter objects are studied almost exclusively by photoelectric methods. The flare frequency in aggregates is more than one order of magnitude smaller than that of flare stars in the solar neighbourhood. This is probably a selection effect, since the photographic observations in aggregates have been carried out with exposure times 5-10 min. Small amplitude flares cannot be recorded at all. Flare stars in aggregates are usually objects whose recorded flares have amplitudes $> 0\overset{m}{.}6 - 0\overset{m}{.}7$, and which last for at least 5 min. The long exposure masks the true amplitude of the flare.

CCD photometry allows precise monitoring of stellar flare radiation for many stars simultaneously at good temporal resolution. The photometric precision is as good as that with photoelectric photometry. Such observations will provide additional new constraints on stellar evolution.

2 The Telescope and CCD Camera

The Abastumani Observatory 70-cm meniscus telescope has a field of view of 4.8×4.8 square degrees. For spectral observations, it is equipped by 8, 4, 2 and 1° prisms with reciprocal linear dispersions 16.7, 66, 125 and 250 nm/mm at Hγ, respectively (Kiladze 1960).

The model ST-6 Professional CCD Imaging Camera system consists of a camera head, IBM PC XT-type CPU and software. The camera head contains the CCD chip, a two stage thermoelectric cooler and 16-bit A/D converter. It is now attached to the 70-cm meniscus telescope through a Newtonian focus photometer which is equipped with a focussing system and filter wheel. The pixel size corresponds to $2''\!.46$ and the readout noise is 30 e. The gain is $\approx$ 6 e per ADU. A typical twilight flat field frame has a best rms over the whole image of 2%. The CCD is from 350 nm to 1.2 μm, with a maximum efficiency at 740 nm (Kurtanidze et al. 1994).

3 The CCD monitoring programme

Flare stars in aggregates are normally identified photographically (Gurzadyan 1980, Mirzoyan 1981) by the multiple exposure method with exposures 5-10 min. Photographic flare star search in stellar aggregates has several disadvantages: (1) Flares of stars with amplitudes about 2^m and a duration of 1 min cannot be recorded by photographic observation with duration 5-10 min. (2) The limiting magnitude for photographic observations is relatively bright, while the CCD goes by $1.5 - 2^m$ deeper in a 5-10 times shorter exposure. For these reasons, we undertake at Abastumani a monitoring programme of flare stars in the Orion association by a CCD based photometer.

The only disadvantage of CCD relative to the astronomical emulsion is the small field of view – $10' \times 14'$. Therefore we decided to carry out the observations in the central part of this aggregate, where the Great Orion Nebula is located. As is well known, this is the center of the flare star system. The small view of CCD is compensated by the fainter limiting magnitude, and the observed number of stars should strongly increase, at least by a factor 5-10 (Jones et al. 1988). Furthermore, it was practically impossible to identify flare stars in this region by photography because of the strong nebular background radiation.

The C1 glass filter was used for preliminary observations. It has a peak transmission of 85% at 390 nm and a FHWM of about 80 nm, and is similar to the C filter of the Washington system. Its central wavelength is over 40 nm redwards of that of the Johnson U filter, leading to two significant advantages: it is less affected by reddening, and the detection of K-M stars is favored; required exposure times for late type stars in C1 are a factor 3 shorter than those in U.

We plan to carry out also CCD observations of this area through interference filters to suppress the nebular radiation. Observations through H+K and Hα filters for identification of emission line stars are also planned, using the 8° prism with resolution 0.3 nm at H+K and 2 nm at Hα.

Acknowledgements: The author is grateful to the SOC for financial support to attend.

References

Gurzadyan G.A. 1980, Flare Stars, Pergamon Press, Oxford
Jones B.F., Walker M.F. 1988, AJ 95, 6
Kiladze R.I. 1960, Bull. Abast. Aph. Obs. 30, 42
Kurtanidze O.M., Kvernadze T.M. 1994, in IAU Coll. No 148, p. 467
Mirzoyan L.V. 1981, Nestatsionarnost i evolutsia zvezd, Izdatelstvo Akademii Nauk Armianskoi SSR.

Automated Search for Flare Stars Continued

J. Winterberg[1], *M. Nolte*[1], *W. C. Seitter*[1], *H. W. Duerbeck*[1]
M. K. Tsvetkov[2], *K. P. Tsvetkova*[2]

[1] Astronomical Institute of Muenster (AIM), Germany
[2] Institute of Astronomy, Bulgarian Academy of Sciences, Sofia, Bulgaria

1 Introduction

To date we obtained 511 blue plates with multiple exposures of nine young clusters and associations to be used for studying the flare activity-age relation. First attempts showed already that automatic search increases the number of discovered flares by 50% compared to visual inspection (Aniol et al. 1989). Improved algorithms have now been implemented and applied. Automated object identification in existing catalogues (FSDB, GCVS, SIMBAD) is also available.

2 Reductions

All plates are digitized with the PDS 2020 GM^{plus} microdensitometers at AIM and reduced with the AIM software. Steps following the reconstruction of images include the transformation of plate to sky coordinates and matching of the multiple exposure chains with the stars on the deep reference plate. All complete chains (generally six exposures) are used to derive average light curves for all objects. For each magnitude interval 10–11, 11–12,...15–16 the mean deviation and standard error of an image from the overall mean value of the respective chains is determined. If any image in a chain differs by more than a preset multiple of the mean error it is considered a potential flare event and registered for visual inspection, together with the chains immediately preceding and following the suspected flare event, to obtain information on the total duration of the outburst. The candidates are filed and the data of all stars are stored for checks of long-term variability.

3 Results in the Orion Region

The stellar aggregate M42/M43 in Orion is the region best observed for flare stars (Tsvetkova et al. 1995). In our campaign from 1985 to 1990 we obtained 53 blue plates with a total of 59 hours observing time at the GPO astrograph at ESO/La Silla. 23 flares were observed, examples of two slow flares are are given in Figure 2, 5 flare were found on hitherto unknown flare stars.

4 Statistics

The detection rate on GPO astrograph plates, using automated search procedures, is 0.10 flares per hour and square degree. The mean rate of six large published investigations is 0.06±0.02 in the same units. This shows that astrographs, in spite of missing the bright ultraviolet contributions, together with automated search procedures are suitable tools for flare detection.

Another aspect of automated search of flare stars are the types of flares registered. A test between visual inspection of plates by an experienced observer showed that an extremely short bright event was only registered by eye. Otherwise, as might be expected, our automatic procedure found more slow flares and flares on bright stars than detected by eye. This is supported by the fact that on all plates reduced so far, 4 of our 23 detected flares in Orion had rising times >20 min in comparison with only 25 slow flare events registered in the region up to 1993 (Parsamian et al. 1993).

Some of the flares occurred in known In- and Ins-type variable stars. Their variations are visible in the long term light curves derived from all exposures (see also Aniol et al. 1989).

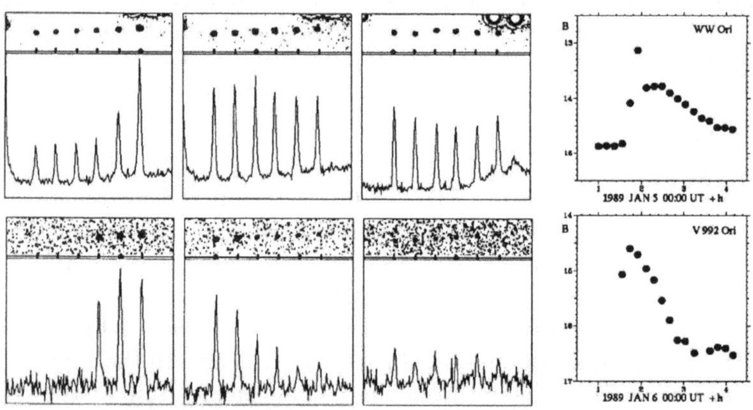

Fig. 1. Light curves of flares in WW Ori and V992 Ori. Isopleths from the plates (upper part left) and tracings (lower part left) are shown together with the reconstructed, calibrated light curves (right).

Acknowledgements: M.K.T. and K.P.T. thank the DFG and BNSF (project F-340/93) for support, J.W. acknowledges support from the IAU.

References

Aniol R. et al., 1989, IAU Symp. 137, p. 85
Parsamian E.S., Chavira E., Gonzalez G., 1993, Rev. Mex. Astron. Astrof. 25, 71
Tsvetkova K.P. et al., 1995, this volume, p. 121

Flare Stars Database: Present Status

K.P. Tsvetkova, M.K. Tsvetkov, K.Y. Stavrev

Institute of Astronomy, Bulgarian Academy of Sciences,
Tsarigradsko Shose 72, BG-1784 Sofia, Bulgaria

Abstract: An updated data collection for UV Cet type stars and their registered flare events in stellar aggregates and in the solar neighbourhood during more than 9580 hours observing time is described. The Flare Stars Database (FSDB) presently includes 3127 entries for 1622 stars.

1 Introduction

Among the stellar content of young star clusters and associations, stars with flare activity constitute a significant fraction. Up to February 1993, the number of known variable stars in Orion is 1380, 559 of them are UV Cet stars. This indicates that flare activity is a common characteristic in the early evolution of red dwarf stars. The distribution of flare stars in stellar aggregates and in the solar neighbourhood and the same nature of both types of stars, differing only in age and mass, led to the conclusion that UV Cet type stars in the solar neighbourhood have survived their original stellar system, which has disintegrated, so that these stars belong now to the general galactic field. Strange biases exist on these stars in the literature, obviously due to the lack of extensive flare star data collections. In their overview of low mass star formation in the Orion region, Brand & Wouterloot (1991) include only 116 of the 559 known flare stars. In order to avoid that significant numbers of low mass stars are neglected when processes of low mass star formation are considered, and in order to facilitate statistical investigations, we have collected all flare star data known to us.

2 Status of Flare Star Observations

Flare activity is a random time process, thus the parameters describing this process and the evaluation of the total number of flare stars in a given aggregate (those with registered flare events and those which are not yet discovered) must be based on large samples of observational material. The status of flare star observations from the beginning of systematic monitoring up to the present is shown in Table 1. Aggregates with monitoring of more than 100 hours are listed.

Table 1. Optical flare star monitoring observations up to December 1994

Region	No. of flare stars	No. of registered flares	Observing time (hours)
Orion M42/M43	563	826	1623
Pleiades	547	1635	3250
Taurus dark clouds	102	122	870
Cygnus NGC 7000	83	120	1168
Praesepe	59	146	680
Monoceros NGC 2264	42	43	105
Cygnus IC 1318	17	18	300
Coma open cluster	14	21	338
Scorpius-Ophiuchus	14	15	321
Alpha Persei cluster	6	6	128
Others	70	70	800
Solar neighbourhood	105	> 105	
Total	1622	>3127	>9583

3 The Flare Stars Database

In order to provide direct access to the original flare star data, we assembled catalogues, which are compilations of parameters drawn from many sources (e.g. for Cygnus, Praesepe), and we made computer-readable versions of already existing catalogues (Pleiades region: Haro et al. 1982; Orion region: Natsvlishvili 1989).

The original catalogues were checked with the help of the special programme package FLAREBASE (Tsvetkov et al. 1993). The aim was not only the preparation of master catalogues but also a critical evaluation to ensure the homogeneity of the data, which are needed for basic statistical statements, e.g. the number distribution of flare stars with stellar magnitude in quiescence or with outburst amplitude, time and energy parameters for the calculation of the total flare-up energy. FLAREBASE also helped to correct the errors in the original catalogues, to supplement the data, and to add new information.

After analysing the existing flare star data we adopted the following table structure for the FSDB (Table 2).

Each flare event is identified uniquely by its identifier, which is the primary key for searching and linking the information. The structure of the identifier is:

5 bytes	3 bytes	6 bytes	1 byte
Name of the star according to GCVS	Constellation	Date of the flare event	Consecutive Nos. of events occurring on the same date

For instance, `CU␣␣␣CNC7301302` identifies the second flare event on 30 January 1973 on the star CU Cnc.

Table 2. Structure of the FSDB

MAIN TABLE

↓

Flare event identifier
Equatorial coordinates (R.A., Decl.) of flare star for equinox J2000.0
Equatorial coordinates (R.A., Decl.) of flare star for equinox B1950.0
Galactic coordinates in the l(II), b(II) system
Julian date of the flare event
First registered flare event
Telescope used
Stellar magnitude at minimum in U or pg/POSS B bands
Stellar magnitudes at maximum in U and pg bands
Amplitudes in U and pg bands
Criterion - Dm > 5s
V magnitude, $B-V$ and $U-B$ indices at minimum
Spectral class
Aggregate membership
Pointers to tables SIMBAD and NOTES

↓↓↓↓

Table CROSS-IDENTIFICATION
Table REFERENCES
Table SIMBAD
Table NOTES

All preparatory work on the FSDB has been executed on the IBM/4381 main frame computer at the Computer Centre of Physics of the Bulgarian Academy of Sciences, using the facilities of the operational system VM/SP Rel. 4. The FSDB is a database not only for UV Cet type stars but also for their registered flare events, which are missing in other astronomical catalogues and data sets: the General Catalogue of Variable Stars (GCVS, Kholopov et al. 1982), the Set of Identifications, Measurements and Bibliography for Astronomical Data (SIMBAD-version June 1993), the Database of UV Cet Type Flare Stars and Related Objects (Gershberg et al. 1993). The information on stellar magnitude at minimum and maximum given in the GCVS is not sufficient for assigning flare event parameters. Very often, the lack of identification charts is the reason for some misidentifications and errors (Tsvetkova & Tsvetkov 1989). Approximately 20% of the UV Cet type stars included in the FSDB are not present in SIMBAD. For other flare stars, stellar magnitudes and coordinates are not given in SIMBAD. Very often only one reference is given, usually the Name-List of Variable Stars is quoted. SIMBAD is more complete for UV Cet type stars in the solar neighbourhood. We find the given pointer to SIMBAD in these cases very useful. Gershberg et al.'s (1993) database contains the data of 230 UV Ceti type stars mainly from the solar neighbourhood and some related objects.

4 Future Developments

The collection of data for all observed flare events in the solar neighbourhood and the application of various statistical methods to the data in the FSDB are the next steps. A serious obstacle for the FSDB is the lack of some necessary parameters of the flare event (beginning of event, duration etc.) in the literature of some stars. A further step will be the display of flare star identification charts, based on scans with a HP ScanJet IIcx. This is especially important for astrometric work, which now reaches much fainter magnitudes, but usually does not yet include flare stars (exceptions are the Pleiades flare stars and the work of Stauffer et al. 1991 and Prosser et al. 1991).

Acknowledgements: This project was supported by the Bulgarian National Science Foundation (projects F-340/93 and F-311/93), the Alexander von Humboldt Foundation, and the Deutsche Forschungsgemeinschaft. This research made use of the SIMBAD database, operated at CDS, Strasbourg, France.

References

Brand J., Wouterloot J.G.A., 1991, in Low Mass Star Formation in Southern Molecular Clouds, ESO Scientific Report No. 11, B. Reipurth (ed.), p. 1
Gershberg R.E., Shakhovskaya N.I., Katsova M.M., 1993,IBVS 3939
Haro G., Chavira E., Gonzales G., 1982, Bol. Inst. Tonantzintla, 3, No. 1, 3
Kholopov P.N. (ed.), 1985, General Catalogue of Variable Stars, Nauka, Moscow
Natsvlishvili R.Sh., 1991, Astrofiz. 31, 107
Prosser C.F., Stauffer J., Kraft R., 1991, AJ 101, 1361
Stauffer J., Klemola A., Prosser C.F., 1991, AJ 101, 980
Tsvetkov M., Chukova M., Tsvetkova K., 1994, in Proc. IAU Symp. 161, Astronomy from Wide-field Imaging, H.T. MacGillivray et al. (eds.), Kluwer, Dordrecht, p. 380
Tsvetkova K.P., Tsvetkov M.K., 1989, IBVS 3366

Flare Star Observations with a Single-Photon Counting Imaging Detector

Donald E. Casperson, William C. Priedhorsky, Miles H. Baron, Cheng Ho

Astrophysics and Radiation Measurements Group, Los Alamos National Laboratory, Los Alamos NM 87545, USA

Abstract: At Los Alamos National Laboratory we are developing a new imaging sensor which combines high spatial and high temporal resolution over a large area format, while maintaining single-photon counting sensitivity and sustaining a high count rate. The detector is called a microchannel plate with crossed delay line readout, or MCP/CDL. This detector is ideally suited to the observation of weak transient events, such as stellar flares from red dwarf flare stars in our Galaxy. At present we are initiating an experiment with the MCP/CDL detector which will utilize a 30-cm aperture f/7 telescope to characterize U-band, B-band, and V-band emission from such low-luminosity flare stars, and to search for weak optical transients associated with other astrophysical sources.

1 The MCP/CDL Detector

Over the past several years, new photon-counting microchannel plate imagers have been developed that allow the location and time-tagging of individual photons with high accuracy (Baron & Priedhorsky 1993). The newest version utilizes a crossed delay line for readout of individual photon events, and so it is designated as an MCP/CDL detector.

The current MCP/CDL design uses a transmissive S-20 visible photocathode deposited on the inside surface of the vacuum window of the detector. Photoelectrons emitted from this surface are accelerated across a narrow gap to the 40 millimeter active diameter microchannel plates in a z-stack (three plates in series) configuration, providing an electron gain of 10^7. Each incident photoelectron on the front surface of the stack generates a cloud of electrons out the back that passes through a pair of orthogonal helical windings – one in the X-direction and the other in the Y-direction. Readout of each photon is accomplished by precisely measuring, at the ends of each delay line, the time difference of arrival of the electron-cloud induced pulses.

The maximum photon counting rate available from the detector is presently one-half million per second, limited by pileup of events on the 100-ns long delay lines. Within a few months a new front-end electronics package will boost this rate to approximately five million photons per second. The demonstrated spatial resolution in our prototype detector is better than 20 microns, corresponding to a 2000×2000 pixel circular format. Background count rates from this uncooled detector are approximately 100 counts/second at room temperature, out of the $5 \cdot 10^5$ counts/second overall.

The output of the detector's electronics is in digital format. Photon events are stored as X and Y positions plus a time tag. This enables us to reconstruct images from photons integrated over long exposures, and to remove any effects of apparent image motion, and to correct for the non-uniformities associated with each detector. It also permits us to easily examine the time history of the photons arriving at any given pixel. Data are stored on multi-gigabyte disk pedestals during observations, and then transferred to mass storage devices for later analysis.

2 Application to Flare Star Observations

This detector is well suited for the observation and characterization of magnetically active red dwarf flare stars which, in addition to their quiescent output, randomly emit copious quantities of broadband electromagnetic radiation. Observations are typically made in the U-band or B-band to provide the greatest contrast between quiescent and flare output. These flare stars are relatively common in the Galaxy (Mirzoyan 1990), with an estimated density of around 0.01 pc^{-3}. Recent calculations suggest that the density of flare stars in the solar vicinity is several times higher than that value (Mirzoyan 1995, Shakhovskaya 1995). A typical value of U-band energy released in a flare is 10^{32} ergs. Previous calculations have shown a relationship between the rate of flaring and the energy released in the flare (Gershberg and Shakhovskaya 1983). That relationship is

$$R(> E_{U32}) = 10^{-2}(E_{U32})^{-0.75} \tag{1}$$

where R is the rate (hr^{-1}) of flares with energy greater than E_{U32}, and the quantity E_{U32} is defined as the U-band energy in units of 10^{32} ergs. From this relationship it is evident that the weaker the flare energy one can observe, the higher the rate at which those flares will be observed, and the best statistical results will be obtained from a detector which can image very weak flares from low luminosity flare stars. The MCP/CDL can do this, and it should provide an improved observational value for the density of flare stars in the solar vicinity.

Under moonless and cloudless observing conditions at viewing angles near the zenith, the background photon count rates in the U-band (O'Connell 1987) correspond to an average of a few tenths of a count per pixel per second with our MCP/CDL detector placed at the focal plane of our telescope. This is equivalent to a $U = 19$ star in each pixel. A flare event which generates photons at a rate that is statistically significant against this background, for example 20 photons over 20 seconds in a given pixel, will contribute to our survey.

The minimum detectable energy of a flare is a function of distance. In the U-band our MCP/CDL detector and 30-cm telescope should detect a flare of $3.5 \cdot 10^{30}$ ergs at a distance of 100 parsecs, and one of $3.5 \cdot 10^{28}$ ergs at 10 parsecs, both of which correspond to recording 20 photons. The expected flare rate seen by our detector is obtained by integrating out along the $1\overset{\circ}{.}1$ FoV of the telescope. Our U-band calculations show that, assuming the 20-photon observational threshold, event scaling according to equation (1), and a flare star density of 0.01 per cubic parsec, flares should be detected at a rate of several per hour, when viewing a general galactic field.

3 Current Status of the Experiment

At the present time we have one working version of the MCP/CDL detector. As a result of a degraded photocathode response due to some suspected high-voltage breakdowns over time, this unit is sufficiently insensitive that it cannot be used for routine flare star survey sessions. A few images, however, were obtained with this unit in the summer of 1994 using a 15 cm aperture refractor telescope and with an 18 cm aperture Questar telescope. These images have provided verification of background counting rates, as well as a data base with which to develop the algorithms necessary for image reconstruction. One example of a V-band image obtained in a July observation is shown in Fig. 1. This exposure was integrated over 400 seconds with the fixed refractor telescope. The resulting raw image shows streaks corresponding to the apparent motion of the stars. This motion has been removed in software, and the integrated stellar images are as shown in the figure. Some inherent non-uniformity in the present detector's response has not yet been removed from this image. The two remaining streaks in the image were generated from fixed hot spots in the detector that were spread out during image reconstruction.

At the present time we also have a small amount of data obtained in the U-band, taken with the 18 cm Questar telescope. It is sufficient to verify background photon counting rates for comparison to theoretical values. We do not yet have sufficient data to begin any search for flare events. That search will begin with our acquisition of a new detector and telescope in early 1995.

4 Observation Schedule

We are presently awaiting delivery of a new telescope that has been designed specifically for the flare-star project with this MCP/CDL detector. The modified Ritchey-Chretien optical configuration has been designed to provide a flat-field image over the 40-millimeter diameter of the MCP/CDL detector for the U, B, and V bands. Its aperture is 30 cm and the focal ratio f/7. These values have been selected in order to match the maximum photon counting rate capability of the MCP/CDL electronics, where it operates at highest efficiency. The angular resolution will be $2''$. This telescope is under construction, and due for delivery in January, 1995. We also expect delivery of a new MCP/CDL detector in March, 1995. Observations will begin soon thereafter, at a remote site in the Jemez mountains located thirty five miles from Los Alamos National Laboratory.

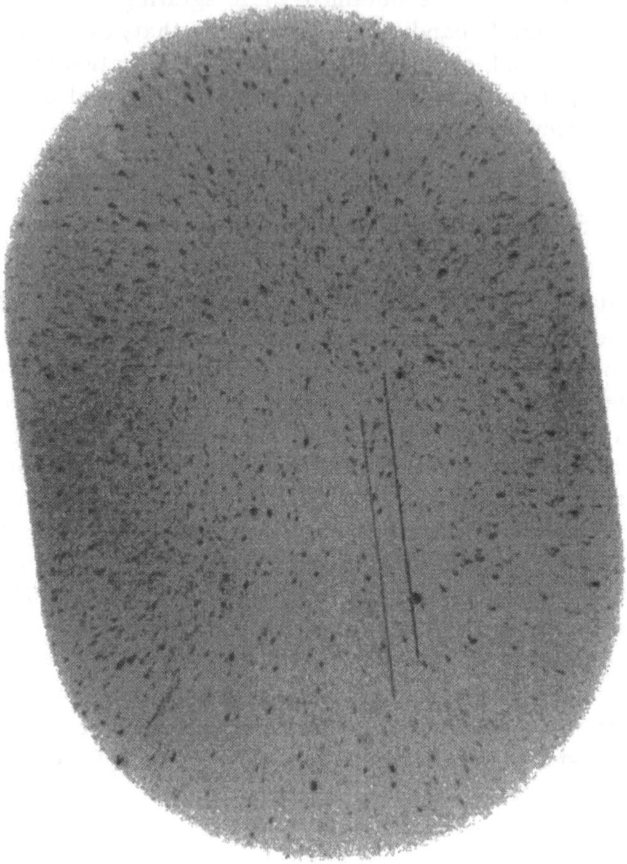

Fig. 1. *V*-band stellar image from the vicinity of the constellation Lacerta obtained with the MCP/CDL detector. Exposure time was 400 seconds. Apparent stellar motion has been removed in software leaving two detector hot-spot streaks (see text).

References

Baron M.H., Priedhorsky W.C., 1993, SPIE Proc. 2006, 188

Gershberg R.E., Shakhovskaya N.I., 1983, ApSS 95, 235

Mirzoyan L.V., 1990, in IAU Symp 137, Flare Stars in Star Clusters, Associations, and the Solar Vicinity, p. 5

Mirzoyan L.V., 1995, this volume p. 55

O'Connell R.W., 1987, AJ 94, 876

Shakhovskaya N.I., 1995, this volume, p. 61

Optical Observations on Milli-, Micro-, and Nanosecond Timescales

D. Dravins, L. Lindegren, E. Mezey

Lund Observatory, Box 43, S-22100 Lund, Sweden

Abstract: Instrumentation and observing methods are developed for optical high-speed astrophysics, aiming at exploring milli-, micro-, and nanosecond variability. Such rapid fluctuations can be expected from instabilities in accretion flows, and in the fine structure of photon emission. For the optical, we have constructed a dedicated instrument, whose first version was tested on La Palma to study atmospheric scintillation on very short timescales. A second version is now under development, using photon-counting avalanche photodiodes as detectors.

1 The challenge of high-speed astrophysics

A goal is to explore the possible *very* rapid variability in astrophysical objects. Some processes may occur over scales of only kilometers or less, and there is no immediate hope for their spatial imaging. Insights could instead be gained through studies of their small-scale instabilities, such as hydrodynamic oscillations or magneto-hydrodynamic flares. Events which may be observable in the time domain include:

- Atmospheric intensity scintillation of stars on the shortest timescales
- Plasma instabilities and the fine structure in accretion flows onto white dwarfs and neutron stars
- Small-scale [magneto-]hydrodynamic instabilities in accretion disks around compact objects
- Radial oscillations in white dwarfs ($\simeq$ 100-1000 ms), and non-radial ones in neutron stars ($\leq$ 100 μs)
- Optical emission from millisecond pulsars ($\leq$ 10 ms)
- Fine structure in the emission ('photon showers') from pulsars and other compact objects
- Photo-hydrodynamic turbulence ('photon bubbles') in extremely luminous stars
- Stimulated emission, e.g. synchrotron radiation, from magnetic objects ('cosmic free-electron laser')
- Non-equilibrium photon statistics (i.e. non-Bose-Einstein distributions) from certain sources

2 Optimum observing techniques

A number of criteria can be defined for optimizing an observing instrument, and we have designed such a unit, named *QVANTOS* ('Quantum-Optical Spectrometer'). Its first version was used on La Palma (Dravins et al. 1994; 1995). Design criteria included:

- *Data flow:* 1 ms resolution means 3.6 million points an hour, and 100 Mb in three nights. However, 1 μs gives 100 Gb, demanding real-time data analysis.
- *Faint sources:* To explore timescales shorter than typical intervals between successive photons, a statistical analysis of their arrival times is required.
- *The terrestrial atmosphere* causes rapid fluctuations of the source intensity, demanding an accurate calibration.
- *Efficient detectors:* New photon-counting detectors are emerging, e.g., silicon avalanche photodiodes are now being tested for a new instrument version.

Previous work shows that meeting all such requirements is non-trivial. The pioneering *MANIA* experiment (Beskin et al. 1982; 1994) has limitations in the maximum photon count rates that can be processed. Instruments in space avoid the atmosphere: the *High Speed Photometer* on the *Hubble Space Telescope* was a major effort (Bless 1982), but had limits on the amount of data that could be stored onboard. Our instrument also has limits in the sense that the full data stream is not saved, only certain statistical functions.

2.1 The role of very large telescopes

The new generation of very large optical telescopes will permit enormously more sensitive searches for rapid phenomena. On submillisecond timescales, data rates become very high and light curves are of little use. Measurements instead have to be of power spectra or other statistical functions, which increase with the light collected to a power of 2 or more, making the gains very much greater than in ordinary photometry or spectroscopy (Dravins 1994).

References

Beskin G.M., Neizvestnyi S.I., Pimonov A.A., Plakhotnichenko V.L., Shvartsman V.F., 1982, in C.M. Humphries, ed. *Instrumentation for Astronomy with Large Optical Telescopes*, IAU Coll. 67, Reidel, Dordrecht, p. 181

Beskin G., Neizvestny S., Plokhotnichenko V., Popova M., Zhuravkov A., Benvenuto O.G., Feinstein C., Méndez M., 1994, A&A 289, 141

Bless R.C., 1982, in D.N.B. Hall, ed. *The Space Telescope Observatory*, NASA CP-2244, p. 106

Dravins D., 1994, ESO Messenger No. 78, (in press)

Dravins D., Hagerbo H.O., Lindegren L., Mezey E., Nilsson B., 1994, in Instrumentation in Astronomy VIII, D.L. Crawford & E.R. Craine (eds.), SPIE 2198, p. 289

Dravins D., Lindegren L., Mezey E., Young A.T., 1995, Appl. Opt., (to be submitted)

MANIA Complex – Hardware and Software for the Investigation of Optical Variability on Time Scales 10^{-7}–10^{2} s

G.M. Beskin, S.I. Neizvestny, V.L. Plokhotnichenko, A.I. Zhuravkov

Special Astrophysical Observatory, Nizhnij Arkhyz, Karachaevo-Circassia, 357147, Russia

One of the basic astrophysical problems is the search for and investigation of brightness fluctuations of celestial objects on time scales comparable with the mean time between registered photons. To facilitate this a special mathematical formalism was developed (Shwartsman 1977, Plokhotnichenko 1983).

At the Special Astrophysical Observatory, photometric equipment has been developed which allows these ideas to be explored observationally and which also allows to study light curves of astrophysical objects using classical methods.

This equipment and these methods are used in the search for and study of variable objects with time resolution 10^{-7} s in the *MANIA* (Multichannel Analysis of Nanosecond Intensity Alterations) experiment. The hardware consists of a photon-counting photometer for synchronous detection in different colour bands, a specialized "time-code" converter (Quantochron 3-16), measuring photon arrival times with an accuracy of 20 ns and a PC AT 386 and DAT-cassette recorder. Our equipment allows the uninterrupted accumulation of 10^{8} photocounts with a rate up to 375 kHz without distortions in 2^{8} or 2^{16} parallel channels (Zhuravkov et al. 1994). The software permits the investigation of variability on time scales $10^{-7} - 10^{2}$ s.

Two different versions of this complex are installed on the 6-m telescope of the Special Astrophysical Observatory and on the 2.15-m telescope of the Leoncito Observatory in Argentina.

The following astrophysical results have been obtained recently:

- Observations of 40 objects – candidates for isolated black holes – have shown that their luminosity is constant on time scales of $10^{-6} - 10^{2}$ s. On the basis of these results we obtained the upper limit of the number of black holes – $5 \cdot 10^{-4}$ of the number of usual stars (Swartsman at al. 1989).
- A light curve of the Crab pulsar with a resolution of 3.3 μs was obtained (see Fig. 1). The main pulse width at 95% of maximum level is $\sim$ 200 μs. There are no modulations which exceed 10% on 5μs. (Swartsman at al. 1988a).
- The pulsar PSR 1937+21 (P = 1.6 ms) has a pulsating component fainter than $B = 26^{m}.2$. As a consequence, the lower limit of the distance to the pulsar was determined to be 5 kpc (Swartsman at al. 1988b).
- The temporal structure of 100 flashes of UV Cet type stars was investigated. On the basis of these data it was proven that the thermal hydrodynamic mechanism takes place in these objects (Swartsman et al. 1988c).

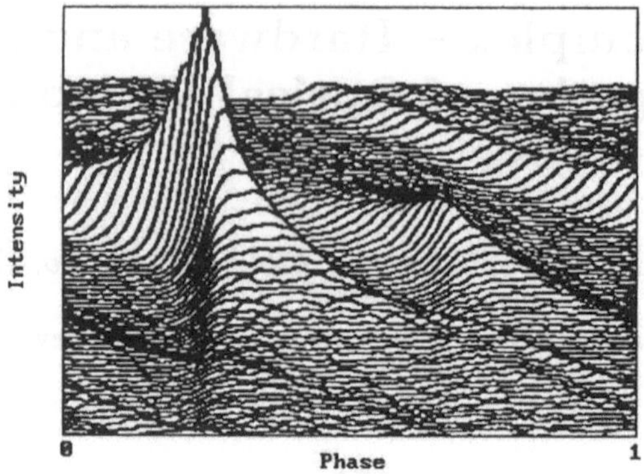

Fig. 1. Series of Crab pulsar light curves with close periods

- Non-thermal flashes were detected in observations of V616 Mon = A0620-00 and MXB 1735-44. Their duration is between 1 ms and 0.2 s, their rise time between 0.1 ms and 0.1 s. Fine structure on time scales of 5 – 10 ms was detected. Brightness temperatures exceed 10^8 – 10^9 K (Swartsman at al. 1989, Beskin et al. 1994).
- The variability of V518 Per = GRO J0422+32 = X-ray nova Per 1992 in the high optical state ($V < 15\overset{m}{.}5$) on time scales from 4 ms to 200 s has been detected (Bartolini et al. 1994). The shortest flashes have nonthermal origin since their brightness temperatures exceed 10^8 K.

References

Bartolini C, Guarnieri A., Piccioni A., Beskin G., Neizvestny S., 1994, ApJS 92, 455

Beskin G., Neizvestny S., Mitronova S., Plokhotnichenko V., Popova M., Benvenuto O., Feinstein C., Mendez M., 1994, A&A 289, 141

Plokhotnichenko V.L., 1983, Soobshch. Spets. Astrofiz. Obs. 38, 28

Shvartsman V.F., 1971, Astron. Zh. 48, 479 = Sov. Astron. J. 15, 377

Shvartsman V.F., Beskin G.M., Plokhotnichenko V.L., 1988a, in The Physics of Neutron Stars. Pulsars and Bursters, Leningrad, p. 178

Shvartsman V.F., Beskin G.M., Plokhotnichenko V.L., 1988b, in The Physics of Neutron Stars. Pulsars and Bursters, Leningrad, p. 184

Shvartsman V.F., Beskin G.M., Gershberg R.E., Plokhotnichenko V.L., Pustilnik, L.A., 1988c, Pis'ma v Astron. Zh. 14, 223

Shvartsman V.F., Beskin G.M., Mitronova S.N., 1989, Pis'ma v Astron. Zh. 15, 337

Zhuravkov A.V., Pimonov A.A., Plokhotnichenko V.L., 1994, Bull. Spec. Astrophys. Obs. 37, 159

A New Technique for Precise Stellar Photometry: Application to Small-Scale Activity of EV Lacertae

B.E. Zhilyaev

Main Astronomical Observatory, National Academy of Sciences of Ukraine, Kiev 252650, Ukraine

Abstract: A new prefiltering technique (PFT) for precise stellar photometry is described. It has been proved that the prefiltered and added high-time resolution data provide an increase in S/N ratio by factors $1.3 - 1.7$ in comparison with simple integration. The PFT allows to detect event amplitudes which are comparable to that caused by counting statistics. Adopting the PFT, our observations reveal small-scale flare events in the subsecond range in EV Lac.

1 Brief sketch of the PFT algorithm

For lack of space consider a specific application of the method. Let readings x_i be mutually independent random variables, i.e. $\mathrm{cov}(x_i, x_j) = s^2\delta_{ij}$. Let us denote initial values of the mean count rate and the variance as $\bar{x}$ and s^2. Let the variation coefficient be $V_0 = s/\bar{x}$. Then the S/N ratio in the light curve is $S/N = 1/V_0$. Consider in addition raw readings x_i in groups of three points, and readings $\tilde{x}_i$, smoothed by a moving-average three-point filter. Let us denote by M and D the expectation and the variance operators of the random variable. It is easy to show for raw and prefiltered data that

$$y_i = x_{i-1} + x_i + x_{i+1}, M[y] = 3\bar{x}, D[y] = 3s^2, V_3 = 0.577V_0$$

and

$$\tilde{y}_i = \tilde{x}_{i-1} + \tilde{x}_i + \tilde{x}_{i+1}, M[\tilde{y}] = 3\bar{x}, D[\tilde{y}] = 19/9s^2, \tilde{V}_3 = 0.484V_0 .$$

Applying the PFT, the gain in S/N is equal to 1.192. As shown in detail by Zhilyaev et al. (1994), in the common case of an arbitrary l-point moving-average filter and an m-point addition, the gain is equal to

$$\frac{V_m}{\tilde{V}_{l,m}} = \left(\frac{m}{\sum_{i=1}^{m}\sum_{j=1}^{m} r(i-j)}\right)^{1/2}$$

with

$$r(k) = \frac{l-k}{k}, \quad | k | \leq l-1 ,$$

where $r(k)$ is the autocorrelation function of prefiltered data. The gain can reach a value of $1.3 - 1.7$ for appropriate values of l and m. In this case, the radius of correlation of modified data is no more than $l - m + 1$ points.

2 Application of the PFT to EV Lacertae

High-time resolution observations were obtained with 0.01 s integration time in B-band with a high-speed photometer. The advantage of the PFT is shown in Fig. 1. The first panel shows a simple 5-point addition of readings, the second one is obtained by adoption of the PFT with $l = 11$, $m = 5$. The outburst seen in Fig. 1 is significant at the $> 4\sigma$ confidence level. Miscellaneous features of the flare-shaped event can be deduced from the PFT light curve. One can see a short-lived flare with an amplitude of $-0\overset{m}{.}4$, superimposed on an intensity drop of longer duration with an amplitude of $+2\overset{m}{.}2$. Significant intervals of the flare and of the drop are of the order of 1.5 and 3.5 s, respectively. The paper attempts to describe the principle of operation of the prefiltering technique, and this example is a typical demonstration. Note that a reference star showed no presence of flare events at such a high confidence level during an overall observing run of half an hour.

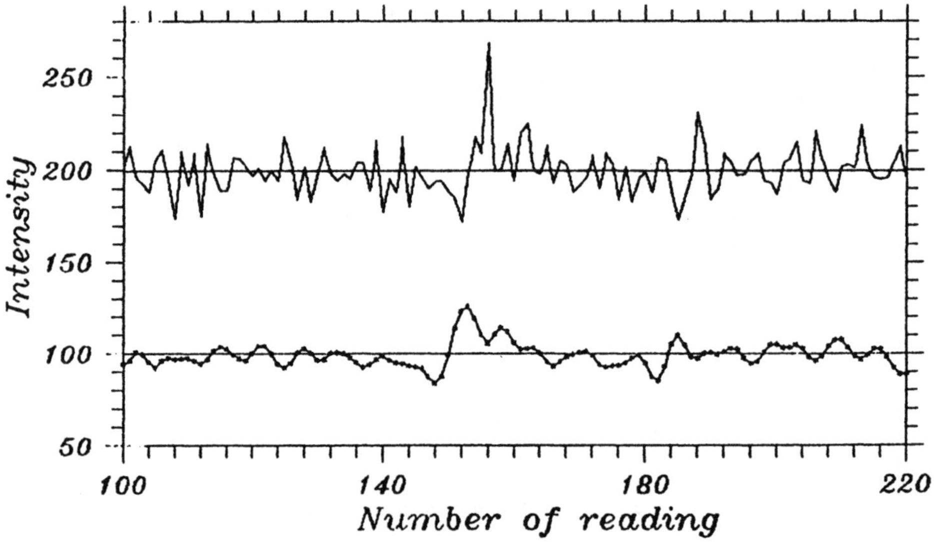

Fig. 1. EV Lac on 1992, Sept. 5, 22:05 UT, B-band. Time resolution is 0.05 s. Upper curve – raw data, lower curve – PFT light curve.

Acknowledgement: The research described in this publication was made possible in part by Grant No UCBOOO from the International Science Foundation.

References

Zhilyaev B.E., Romaniuk Ya.O., Svyatogorov O.A., 1994, Kinemat. Phys. Celestial Bodies 10, No. 6 (in press)

FLARES IN LATE-TYPE STARS:

UV AND X-RAY

Flare in Late-type Stars: UV

P. Brendan Byrne

Armagh Observatory, Armagh BT61 9DG, N. Ireland

1 Introduction

Early studies of stellar flares were made entirely in the optical regime. It was recognised that flares arose from the generation of hot plasma within the stellar chromosphere at whose temperature (indicated, for instance, by the presence of a strong, blue optical continuum) a substantial emission in the ultraviolet would be expected. It was not until the advent of space-borne instruments of adequate sensitivity, however, that direct confirmation of this prediction was forthcoming. In this review I examine some results of more than a decade of observation of stellar flares in the UV.

1.1 Sources of UV data

The major source of ultraviolet data on stellar flares has been the International Ultraviolet Explorer (*IUE*) satellite (Boggess et al. 1979). In order to understand the limitations of our current understanding in this area it is important to appreciate some of the characteristics of its instrumentation. *IUE*'s telescope is of 40 cm aperture and it is equipped with a spectrograph which can operate at two resolutions, i.e. $\Delta\lambda/\lambda$ ~350 (LORES) and ~17 000 (HIRES). Its detectors are optimised for operation in two wavebands, i.e. ~1150–1950 Å (SW) and ~1950–3200 Å (LW). *IUE*'s small aperture results in a limited sensitivity, a consequence of which is a modest time resolution when studying stellar flares (a long exposure time is needed to gain adequate signal-to-noise). *IUE*'s elliptical 24-hour quasi-geosynchronous orbit and its resulting interactive mode of operation make continuous monitoring feasible, a feature suiting flare star work.

The recent advent of the Hubble Space Telescope (*HST*) (Giacconi et al. 1982) with its 2.4 m aperture represents in some ways such a huge gain in light-gathering power that it might be expected to render *IUE* obselete. Its prime UV spectroscopic instrument is the Goddard High Resolution Spectrograph (GHRS) whose detectors span essentially the same spectral bands as *IUE*. This offers spectral resolutions of $\Delta\lambda/\lambda$ ~2 000, ~25 000, and ~80 000. However, the spectral ranges available at each of these resolutions are ~285 Å, ~30 Å, and ~8 Å respectively. Thus while *HST* offers a giant leap in sensitivity, it fails to offer the wide spectral coverage of *IUE*. The orbit of *HST* imposes a further restriction on its use. Its low orbit means that, for objects at low ecliptic latitude, the Earth occults many targets for up to half the observing time.

Thus most of the ultraviolet flare data available to date has come from *IUE*. Observations from *HST* are beginning to appear, however, and will, no doubt, increase in the near future.

2 Flares on dMe stars

2.1 dMe quiescent emission

Quiescent emission from dMe flare stars is characterised by strong line emission from neutral and ionised atomic species arising in temperatures characteristic of the solar chromosphere ($6 \cdot 10^3$ K $leqT_e \leq 10^4$ K) and transition region ($3 \cdot 10^4$ K $\leq T_e \leq 2 \cdot 10^5$ K). A weak continuum is sometimes present in the long wavelength portion of the LW spectra but is entirely absent in SW. An example of a LORES *IUE* spectrum of dMe flare star in quiesence will be found in Fig. 1 and the corresponding mean surface fluxes in Table 1.

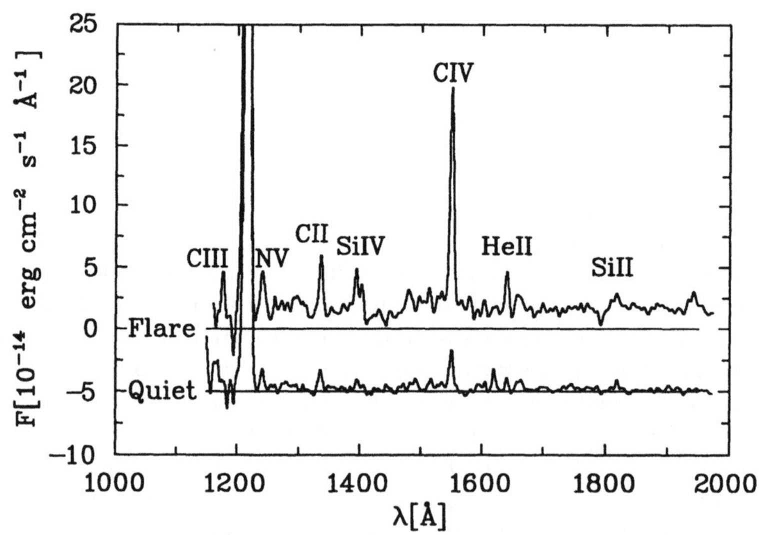

Fig. 1. An example *IUE* SW LORES spectrum of the dMe flare star, YZ CMi, in quiescence and while flaring (J.G. Doyle, *private communication*). Prominent emission lines are indicated. The quiescent spectrum has been displaced to aid visibility of the two spectra. The horizontal line indicates the zero level for each spectrum. Note the strong continuum in the flare spectrum, which is not detectible in quiescence.

Table 1. Typical mean emission line surface fluxes, in units of $10^4 \mathrm{erg\,cm^{-2}\,sec^{-1}}$, for dMe stars in quiescence and during flare, compared to the mean solar quiescent flux in these lines. Note that the flaring values are lower limits for the reasons discussed in Section 2.2. Figures are also given for continuum surface fluxes in *IUE*'s SW region (1150–1950 Å).

Star	V1005 Ori[1]		BY Dra[2]		YZ CMi[3]		HK Aqr[4]		Quiet
Line	$\log f_q$	$\log f_f$	$\log f_q$	$\log f_f$	$\log f_q$	$\log f_f$	$\log f_q$	$\log f_f$	Sun
C II	7	32	16	24	6	22	12	19	0.5
C IV	11	229	26	33	13	75	22	98	0.6
N V	—	—	8	14	6	18	4	11	0.1
cont.	—	1200	—	—	—	280	—	1200	

[1]Mathioudakis et al. 1991; [2]Butler et al. 1987; [3]Doyle in prep.; [4]Byrne & McKay 1990

2.2 dMe flaring emission

During flares all of the lines visible in the quiescent spectrum are seen to increase, some by factors of up to ~10 or greater. In general the degree of increase is dependent on the excitation of the relevant line, up to mid-transition region ($T_e \sim 10^5$ K). So in the *IUE* wavebands the greatest flux enhancements are seen in the resonance doublets of C IV (λ1548/51 Å) and N V (λ1238/42 Å). An example flare spectrum is shown in Fig. 1 and some representative integrated flare line surface fluxes at source are given in Table 1. It should be noted, however, that, since areas of individual flares are unknown, it has been assumed in Table 1 that they cover an entire hemisphere. Obviously, if a flare covers a fraction $\frac{1}{n}$ of a hemisphere, the fluxes in Table 1 should be increased by the factor n. It will be seen from Fig. 1 that a strong continuum is also present. Such continua are recorded frequently in the more energetic dMe flares, but by no means universally. Lack of continuum detection in some flares may be a matter of *IUE*'s limited sensitivity, however. The origin of dMe flare continua has been discussed by Phillips et al. (1992).

Flaring emission in dMe's is usually only detected in a single spectrum. This is a result of the relatively long exposure times needed by *IUE* to detect flare stars with adequate signal-to-noise compared with the typical duration of a flare as gauged from other wavebands. In practise, exposure times in LW LORES need be longer than ~10 min and in SW LORES longer than ~20 min for flare detection. At the s/n levels attainable by *IUE* the effect of the flare on the spectrum is usually undetectable in subsequent spectra. Thus flare parameters derived from such observations should be judged with this fact in mind. In particular, mean surface fluxes are calculated on the assumption that the flare lasts for the entire exposure time.

Monitoring of active dMe stars for flares has revealed, apart from discrete individual flares, continuous variability in these line fluxes (Butler et al. 1987,

Byrne et al. 1991). This has led to the suggestion that, on the time scales resolvable with *IUE*, the most active dMe's flare continuously at a low level. This raises the question as to whether the measured "quiescent" fluxes of dMe's are "quiet" in the same sense as the solar case.

3 Flares on RS CVn stars

3.1 RS CVn quiescent emission

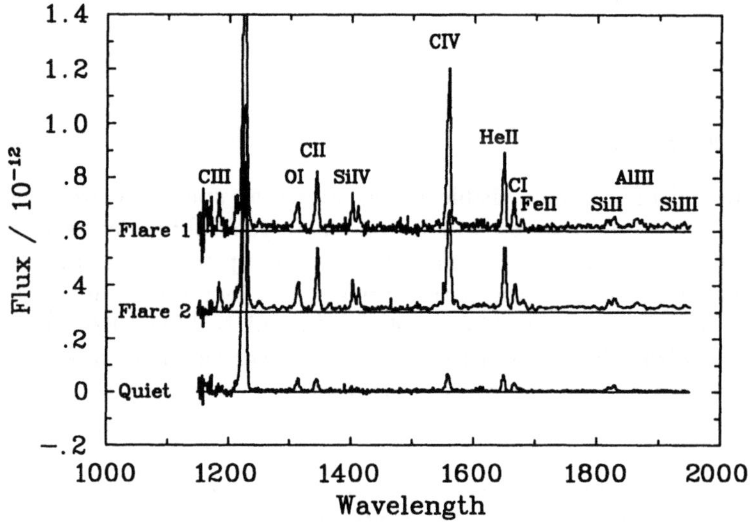

Fig. 2. An example *IUE* SW LORES spectrum of the RS CVn star, II Peg, in quiescence and while flaring (P.B. Byrne, *private communication*). The two flare spectra were separated by ~2 hr in time. Prominent emission lines are indicated. The flare spectra have been displaced to aid their visibility. The horizontal line indicates the zero level for each spectrum. Note the continuum in the flare spectra, which is not detectible in quiescence.

Quiescent ultraviolet line emission from RS CVn systems is qualitatively the same as that from the dMe's except that the sources are more luminous. Some, but not all, of the extra luminosity is accounted for by the greater surface areas of these subgiant or giant stars. The greater brightness of RS CVn's has meant that *IUE*'s HIRES mode can be used, in addition to SW LORES, to study their spectra, at least in the strong Mg II *h*&*k* resonance lines. An example of such spectra will be found in Figs. 2 and 3.

Table 2. Typical mean emission line surface fluxes, in units of $10^4 \mathrm{erg\,cm^{-2}\,sec^{-1}}$, for RS CVn stars in quiesence and during flare, compared to the mean solar quiescent flux in these lines. Figures are also given for continuum surface fluxes in *IUE*'s SW region (1150–1950 Å).

Star Line	II Peg[1] $\log f_q$	 $\log f_f$	V711 Tau[2] $\log f_q$	 $\log f_f$	IM Peg[3] $\log f_q$	 $\log f_f$	λ And[4] $\log f_q$	 $\log f_f$	Quiet Sun
C II	9	49	54	112	8	28	20	43	0.5
C IV	20	104	72	194	13	71	35	105	0.6
N V	2	15	8	40	9	18	12	22	0.1
cont	—	4460	—	—	—	—	270	490	—

[1]Doyle et al. 1989; [2]Linsky et al. 1989; [3]Busazi et al. 1987; [4]Baliunas et al. 1984

3.2 RS CVn flaring emission

As with their quiescent emission, flaring emission in RS CVn's is qualitatively the same as in the dMe stars but RS CVn flares are considerably more energetic. This greater energy arises both from a higher specific flux during the flare and from a much longer flare duration. Indeed RS CVn flares may last up to sizeable fractions of a day, arguing in favour of continual heating on these time scales. As a result *IUE* has been able, in some cases, to observe the time evolution of RS CVn flares in ultraviolet lines (cf. Fig. 2).

Use of LW HIRES spectra to study the behaviour of the Mg II *h&k* resonance lines during RS CVn flares has given important indications of long-lived mass flows in material at chromospheric temperatures. In general, as in optical lines, these velocities are seen as red-shifts, but blue-shifts are also recorded (Doyle et al. 1988, 1989, Linsky et al. 1989). However, reports of representative velocities derived from these observations should be treated with caution. The Mg II *h&k* lines are optically thick and, as a result, there is not a simple mapping of velocity onto the line profile.

A few RS CVn's have been monitored for flaring sufficiently often to allow estimates to be made of the frequency and energy distribution of ultraviolet line flares. For instance Mathioudakis et al. (1992), using earlier *IUE* data from a variety of sources, estimated that the RS CVn system II Peg produced a flare of integrated energy in transition region lines of $\sim 10^{35}$ erg every 10 hours.

4 Physical parameters of flares

Under optically thin (*coronal*) conditions a simple relationship exists between the observed line flux $F_{\rm line}$ and a quantity known as the emission measure (EM) of the plasma, i.e.

$$F_{\rm line} = \xi(N_e^2 dV) = \xi\ EM \qquad (1)$$

where the quantity ξ contains a number of essentially atomic parameters such as the abundance of the element concerned, the relative abundance of the ionic

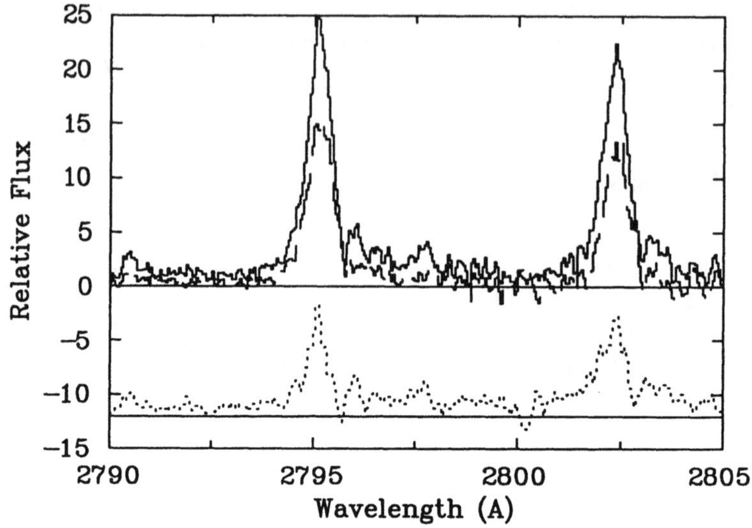

Fig. 3. An example *IUE* LW HIRES spectrum, in the region of the MgII *h&k* chromospheric emission lines, of the RS CVn star, II Peg, in quiescence and while flaring (P.B. Byrne, *private communication*). The *solid line* is the flare spectrum, the *dashed line* is the quiescent spectrum and the *dotted line* is the difference. Note the excess extended emission in the wings, especially the red wing.

species, etc. The validity of the coronal approximation has been tested by, for instance, checking that the ratio of strong resonance doublets is in the ratio of their statistical weights (see e.g. Byrne et al. 1987). Once the overall distribution of EM has been determined from a number of well observed lines formed over a range of temperature, it can be combined with a suitable radiative loss function to determine the *total* radiative losses, including unobserved lines, over the temperature in question (Byrne et al. 1987, Mathioudakis et al. 1991).

It is also possible to determine local electron density, N_e, in the transition region by observing density-sensitive intersystem line ratios, such as between C III] λ1176 Å, C III] λ1908 Å and Si III] λ1892 Å. In the dMe's these lines are extremely weak and the derived densities correspondingly uncertain (Mathioudakis et al. 1991). Furthermore, it is not known whether flare conditions are constant during typical exposure times as explained above. However, values of $N_e \sim 10^{10-10.5}\,\mathrm{cm}^{-3}$ would appear to be indicated. Signals being much stronger for the closest RS CVn's, intersystem line fluxes can be measured with greater confidence (Byrne et al. 1987). The long duration of RS CVn flares also gives us greater confidence in the applicability of the resulting densities. Typical values are $N_e \sim 10^{10.5-12}\,\mathrm{cm}^{-3}$.

Combining EM and N_e yields the volume of the plasma at that temperature via the relation

$$dV = F_{\mathrm{line}} / \xi \, N_e^2 \tag{2}$$

which for dMe's yields values for large flares of $dV \sim 10^{28-29}\,\mathrm{cm}^{-3}$ and for RS CVn flares of $dV \sim 10^{30-32}\,\mathrm{cm}^{-3}$. Assuming that such flares arise within magnetic loops, we can represent such loops as semi-circular in outline with radius R_{loop} and of constant circular cross-section of radius $r \sim 0.1R_{\mathrm{loop}}$. N such loops would have a volume, V_{loop}, given by the following expression.

$$V_{\mathrm{loop}} \sim 0.04\pi^2 R_{\mathrm{loop}}^3/N \qquad (3)$$

Equating this to derived volume leads to a typical loop dimension $R_{\mathrm{loop}}(\mathrm{dMe}) \sim 10^{9-10}\mathrm{cm}$ and dimension $R_{\mathrm{loop}}(\mathrm{RS\,CVn}) \sim 10^{10-11}\,\mathrm{cm}$. Note that R_{loop} is only weakly dependent on N. Such dimensions, at least in the case of the dMe's, are remarkably solar-like. However, we should caution that such dimensions are derived from the volume of material at transition region temperatures only. It is possible, even likely, that the bulk of the flaring plasma may be at X-ray temperatures. Whether the X-ray plasma occupies the same loops as the transition region material is a matter of uncertainty, however, even in the solar case.

5 HST flare observations

As pointed out above, *HST* has disadvantages (spectral coverage, Earth eclipses, etc.) for flare observations, as well as obvious advantages (aperture, spectral and temporal resolution, etc.). Several interesting results have been achieved already. Maran et al. (1994) monitored the dMe flare star, AU Mic, spectroscopically using GHRS in the wavelength region 1345–1375 Å without recording any discrete flares. However, they found that the mean spectrum showed a strong Fe XXI λ1354 Å coronal emission line, whose intensity is comparable to that observed in solar flares. This result would support the conclusion from *IUE* data that the most active dMe's flare almost continually (Butler et al. 1987, Byrne et al. 1991).

Woodgate et al. (1992) also observed AU Mic with the GHRS on *HST* and detected a UV line flare which resulted in a factor 7 increase in the flux of the SiIIIλ1206Å line on the blue wing of Lyα. During this flare they observed a red asymmetry in the wing of Lyα which they attributed to recombining protons from a proton beam.

Perhaps the most dramatic *HST* flare star result to date has been the observation of extended red wings on the C IV λ1548/51 Å resonance lines during the initial stages of a flare on the dMe flare star, AD Leo (Bookbinder et al. 1992). Interpreted as velocities, these imply downflowing material at mid-transition region temperatures of ~2 000 km s^{-1}. Byrne (1993) discussed this result and calculated that the downflowing material's kinetic energy represents ~25 times the energy *radiated* in the C IV lines. Furthermore, the observation of downflows of this magnitude in the optically thin C IV lines lends credence to reports of comparable velocities in optically thick chromospheric lines such as H I Balmer. There is a clear implication that mass flows may be a common and important energy sink in stellar flares.

References

Baliunas S.L., Guinan E.F., Dupree A.K., 1984, ApJ 282, 733

Bookbinder J., Walter F., Brown A., 1992, in Cool Stars, Stellar Systems and the Sun, J. Bookbinder & M. Giampapa (eds.), ASP Conf. Ser., p. 27

Boggess A. et al., 1978, Nature 275, 372

Busazi D.L., Ramsey L.W., Huenemoerder D.P., 1987, ApJ 322, 353

Butler C.J., Doyle J.G., Andrews A.D., Byrne P.B., Linsky J.L., Bornmann P.L., Rodonò M., Pazzani V., Simon T., 1987, A&A 174, 139

Byrne P.B., 1993, in Physics of Solar and Stellar Coronae, J.L. Linsky & S. Serio (eds.), Kluwer, Dordrecht, p. 489

Byrne P.B., McKay D., 1990, A&A 227, 490

Byrne P.B., Doyle J.G., Brown A., Linsky J.L., Rodonò M., 1987, A&A 180, 172

Byrne P.B., Agnew D.J., Cutispoto G., Kilkenny D.W., Neff J.E., Panagi P.M., 1991, in Surface Inhomogeneities on Late-type Stars, P.B. Byrne & D.J. Mullan (eds.), Springer, Berlin, p. 255

Doyle J.G., Butler C.J., Byrne P.B., van den Oord G.H.J., 1988, A&A 193, 229

Doyle J.G., Byrne P.B., van den Oord G.H.J., 1989, A&A 224, 153

Giacconi R. et al., 1982, in The Space Telescope, Space Telescope Institute, Baltimore

Linsky J.L., Neff J.E., Brown A., Gross B.D., Simon T., Andrews A.D., Rodonò M., Feldman P.A., 1989, A&A 211, 173

Maran S.P., Robinson R.D., Shore S.N., Brosius J.W., Carpenter K.G., Woodgate B.E., Linsky J.L., Brown A., Byrne P.B., Kundu M.R., White S., Brandt J.C., Shine R.A., Walter F.M., 1994, ApJ 421, 800

Mathioudakis M., Doyle J.G., Rodonò M., Gibson D.M., Byrne P.B., Avgoloupis S., Linsky J.L., Gary D., Mavridis L.N., Varvoglis P., 1991, A&A 244, 155

Mathioudakis M., Doyle J.G., Avgoloupis S., Mavridis L.N., Seiradakis J.H., 1992, MNRAS 255, 48

Phillips K.J.H., Bromage G.E., Doyle J.G., 1992, ApJ 385, 731

Woodgate B.E., Robinson R.D., Carpenter K.G., Maran S.P., Shore S.N., 1992, ApJ 397, L95

L. Pustil'nik: What is the best time resolution in the UV-range for flare observations? And what is the shortest observed time of a flare itself ($t_{\rm rise}, t_{\rm decay}$)?

P.B. Byrne: This answer must distinguish between dMe's and RS CVn's. In dMe's the exposure time of a typical UV spectrum is in general comparable to, or shorter than, the total duration of the flare. In the case of RS CVn stars, the total duration is much longer and it is possible to take multiple spectra during the evolution of the flare.

On the other hand, durations of flares themselves, as derived from optical continuum data, would suggest that the shortest $t_{\rm rise}$ is about 1 sec. In general $t_{\rm decay}$ is longer by a factor 2 or more. The answer may be dependent on waveband, however. For instance, in soft X-rays, the time scales appear to be longer.

R. Minarini: What about estimates of the amount of mass motions during flares from UV data?

P.B. Byrne: Most indications of mass motions came from chromospheric lines (e.g. Mg II *h&k*) which are very optically thick. Thus it is very difficult to derive unique velocities from line assymetries. The only case of optically thin line evidence comes from the unpublished profiles of C IV during a flare on AD Leo. These indicate velocities up to 1800 $km\,s^{-1}$ decaying to 600 $km\,s^{-1}$ in about 25 sec.

R. Pallavicini: I was surprised to see that these *IUE* observations indicate higher densities for low-gravity RS CVn binaries than for high-gravity dMe stars. Can you comment a little bit more on these density determinations?

P.B. Byrne: This may be in part a result of observational selection. In general dMe flares are much shorter in duration than RSCVn flares. In fact their entire duration is shorter than the typical exposure times. So when we measure densities we measure "averages" over the whole flare. By contrast in the RSCVn's the rise time alone (assumed to be the heating phase) is about equal to the entire duration of a dMe flare. Thus the RS CVn densities may be more meaningful.

Of course, there is a second aspect of flare densities and that is that in flares conditions are far from equilibrium. Thus heating flux and confining magnetic field strength will play a greater role in determining density than gravity.

Post-flare Coronal Activity on AU Mic Detected by EUVE

Maria Katsova[1], *Jeremy Drake*[2], *Moissei Livshits*[3]

[1] Sternberg State Astronomical Institute, Moscow State University, 119899 Moscow, Russia
[2] Center for EUVE Astrophysics, University of California, Berkeley, CA 94720, USA
[3] Institute of Terrestrial Magnetism, Ionosphere, and Radio Wave Propagation, Russian Academy of Sciences, 142092 Troitsk, Moscow Region, Russia

Data of long-duration emission arising after the impulsive rise and decay in a flaring event on the red dwarf star AU Mic are discussed. Intensive EUV emission in the band 65–190 Å was registered by the Extreme Ultraviolet Explorer (*EUVE*) after both impulses during half a day. A similar behavior of the flux in the Fe XVIII 93.9 Å line is detected after the first powerful impulse. The decay of the intensity in the 65–190 Å band and in the Fe XVIII line during this prolonged event is 10 times slower than the time of radiative cooling of coronal loops with a typical flare plasma density. Some difficulties with two explanations of this event proposed earlier are discussed. Explanation (i) – the radiation of dense loops at main phase of the flare, and (ii) – the emission of the low-dense plasma of coronal transients (CME). The temporal behavior of the emission measure is determined for both the 65–190 Å band and the Fe XVIII line fluxes. The total energy emitted in the 1–2000 Å region for the long-duration event lasting almost 12 hours is $3 \cdot 10^{35}$ ergs. The following physical model is proposed to explain the prolonged event (Fig. 1): the source of emission is the system of high coronal loops, the size of which is more than the active region scale, but is less than the stellar radius. The temperature of the plasma in the loops decreases from 10^7 K slowly, during a few hours. The densities in these loops are in the range 10^{13} cm^{-3} to $5 \cdot 10^9$ cm^{-3}. Such systems, when the plasma therein becomes cool, are observed in the Hα line during large solar flares (for instance, June 15, 1991) after CME. Some additional post-flare energy input into this high coronal loop systems can be caused by the reconnection in vertical current sheet, and this post-eruptive energy release provides prolonged and intensive EUV emission.

Apparently, we are faced here with a new kind of surface activity on late-type stars which is intermediate between impulsive flares on red dwarfs and long-duration, powerful events on subgiants, which are components of RS CVn binaries.

The full version of this contribution will be published in Astronomicheskij Zhurnal (Astronomy Reports), 1995, Vol. 72.

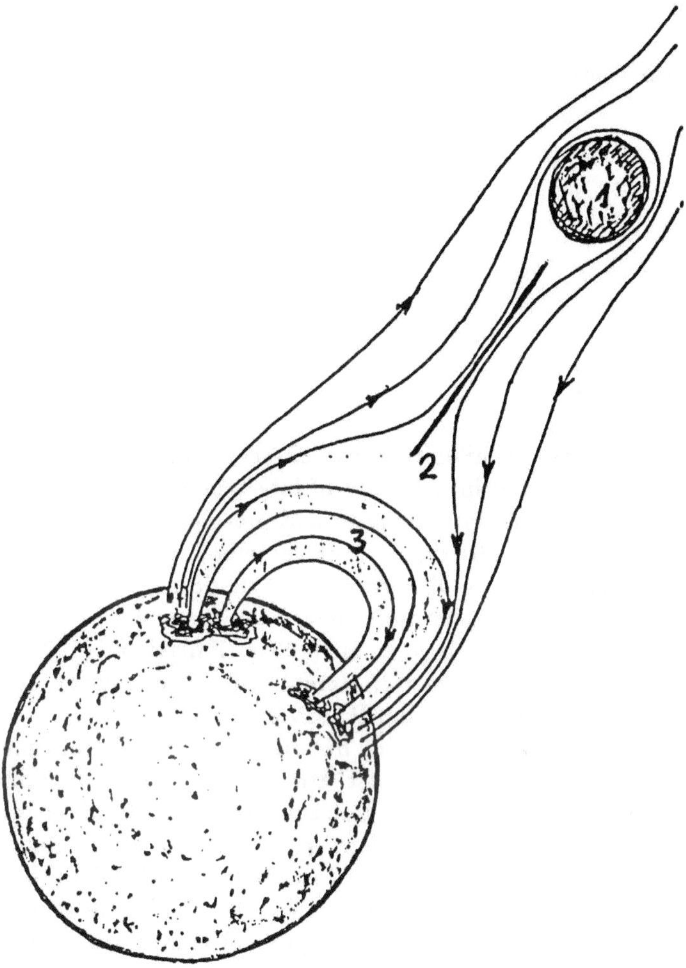

Fig. 1. The event with post-eruptive energy release: (1) – coronal mass ejection (CME), (2) – vertical current sheet, (3) – coronal loops as a source of the EUV radiation

Flares in Late-type Stars: X-Ray

Roberto Pallavicini

Osservatorio Astrofisico di Arcetri, Largo E. Fermi 5, 50125 Firenze, Italy

1 Introduction

Flare-like brightenings, similar to those observed on the Sun but on a larger energy scale, are observed in a variety of late-type stars, including classical M dwarf flare stars, RS CVn-type binaries and pre-main sequence (PMS) objects. These events, which are observable over the whole electromagnetic spectrum, are interpreted as due to rapid release of magnetic energy at the star surface. Comprehensive reviews of stellar flares have been presented by Haisch & Rodonò (1989), Pettersen (1991), Haisch et al. (1991), Pallavicini (1992a), as well as in contributions at this conference. Here I will discuss X-ray emission, with emphasis on flares on M dwarfs and RS CVn binaries.

X-ray observations are important because they provide information on the high-temperature coronal portion of the flare and, at least in principle, could give clues as to the primary energy release and particle acceleration. In the stellar case there are severe limitations of the diagnostic capabilities of X-ray observations, in comparison to the solar case. In fact, stellar flares have up to now been observed only at soft X-ray energies ($\leq$ 10 keV) where thermal processes greatly dominate over non-thermal ones (optical continuum observations have to be used in this case as a proxy for hard X-rays). Moreover, stellar X-ray observations are spatially unresolved and so far of poor spectral resolution. Thus, the observational data that are readily accessible are only light curves, average temperatures T and volume integrated emission measures $EM = \int n^2 dV$, but there is no direct information on parameters such as density, volume and detailed geometry.

2 Observations

When observed in disk-integrated soft X-rays, the Sun is an extremely variable source, and there are clear indications for both transient flare-like events and for more gradual long-term variations associated with rotational modulation and the solar activity cycle (e.g. Pallavicini 1993 and references therein). It may come as a surprise therefore that before *ROSAT* there were only a few X-ray flaring stars which were not part of well-established classes of flare stars like UV Ceti-type stars, RS CVn binaries and PMS objects. The best case was probably a flare observed with *EXOSAT* on the G0V star π^1 UMa (Landini et al. 1986) which released $\sim 2 \times 10^{33}$ erg in the X-ray band. Clearly, the paucity of X-ray flare detections from normal stars was due to the biased nature of most *Einstein* and *EXOSAT* observations, which were pointed primarily at known active stars.

The unbiased observations carried out by *ROSAT* during the All-Sky Survey have shown that virtually all type of late-type stars are flaring sources (Schmitt 1994), thus providing strong support to the notion that magnetic activity is widespread in the cool half of the HR diagram. During the All-Sky Survey, *ROSAT* was scanning great circles perpendicular to the ecliptic plane and each source was observed once every orbit for at least two days (or more depending on its position in the sky). Each observation therefore consisted of short snapshots (typically $\sim$ 20 sec) separated by one orbit period ($\sim$ 90 min) over a time interval of two or more days. Flares have been detected with this technique from stars of all spectral types from F to M: several examples are given in Fleming et al. (1993) and Schmitt (1994). It is worth noting that flares have been detected in this way also from some A and B stars, although in this case it is unclear whether X-ray emission originated from the A and B primaries or rather from unseen late-type companions.

Although the *ROSAT* All-Sky Survey has allowed the detection of flares in an unbiased way, it was limited in the time coverage of the observed events except for long lived flares covering several satellite orbits. For events lasting less that $\sim$ 1 hour, our knowledge of X-ray flare emission is still largely based on previous *Einstein* and *EXOSAT* data as well as on pointed *ROSAT* observations. The long uninterrupted observations carried out by *EXOSAT* are particularly relevant in this respect, since they were not affected by the usual data gaps associated with low-orbit satellites like *Einstein* or *ROSAT*. Observations of X-ray flares by *EXOSAT* have been reviewed by Pallavicini et al. (1990) and Linsky (1991), while *Einstein* flare observations were discussed by Haisch (1983) and Ambruster et al. (1987). The main results obtained by these satellites with regard to flare stars and RS CVn binaries are summaried separately in the following two sections.

2.1 M dwarf flare stars

Short-lived events, with typical durations of minutes to tens of minutes and integrated X-ray energies of $\sim 10^{30}$ to 10^{33} ergs, are commonly observed from late K and M dwarfs of the UV Cet and BY Dra types. These are active red dwarfs with Balmer lines in emission and evidence of strong surface magnetic activity. The light curves of these flaring events resemble strongly those of *compact* flares on the Sun, i.e. flares which are believed to occur in one or more magnetically confined loops. The average temperatures derived from the X-ray data ($\sim 2 - 4 \times 10^7$ K) are similar to those of solar flares, while the volume emission measures are several orders of magnitude larger ($\sim 10^{51} - 10^{53}$ cm^{-3}).

The X-ray emission is typically more gradual and longer lived than the optical continuum emission, consistently with the thermal nature of the X-ray spectra. There is strong evidence that these stellar flares are scaled-up version of solar compact flares: if the physical mechanism is the same, their larger energies imply larger volumes or higher magnetic fields, or both.

In addition to short-lived events, intense long-duration events are occasionally observed from M dwarf stars. Examples are a large flare on YY Gem observed by *EXOSAT* (Pallavicini et al. 1990) and flares on CC Eri and EV Lac observed by *ROSAT* (Pan & Jordan 1994, Schmitt 1994). These flares last from a few to several hours and may release as much as $\sim 10^{34}$ erg in the soft X-ray band. Their long decay time, and sometimes also their long rise time, suggest a gradual energy release process and continued heating during the decay. If an analogy can be drawn with solar flares, these events could be the analog of solar *two-ribbon* flares, where continuous reconnection occurs as open field lines relax back to a closed lower energy configuration (see, e.g., Priest 1995). Note that this analogy is based only on the shape of the X-ray light curve, but there is no proof that the physical mechanism responsible for these long duration events is indeed the same as for solar two-ribbon flares (see discussion in Pallavicini 1992b).

2.2 RS CVn and Algol-type binaries

In contrast to flares in M dwarf stars, long duration flares lasting several hours are the norm rather than the exception for RS CVn and Algol type binaries. These X-ray flares are also much more energetic than flares on M dwarfs, and have typical X-ray energy releases of $\sim 10^{34}$ to 10^{37} erg (i.e. two to five orders of magnitude more than the total energy of the largest solar flares). RS CVn systems are close binaries usually formed by an F or G dwarf and a K subgiant, orbiting each other with periods from less that a day to about two weeks. Algol-type systems, which are nearly-contact binaries with mass transfer between the two components, behave similarly to RS CVn's with regard to the X-ray properties. A large flare on Algol, with a duration of more than 4 hours and a total energy of $\sim 10^{35}$ erg, was observed by *EXOSAT* (White et al. 1986, van den Oord & Mewe 1989) and a somewhat larger flare from the same star (lasting 4 times longer) was observed by the Japanese satellite *Ginga* (Stern et al. 1992). In both cases, the peak temperature derived from the X-ray spectra was $\sim 7 \times 10^7$ K, i.e. larger than the typical temperatures observed in M dwarf flares.

Additional examples of X-ray flares in RS CVn binaries seen by *EXOSAT* have been published by Culhane et al. (1990), White et al. (1990) and Tagliaferri et al. (1991). A flare from II Peg seen by *Ginga* is discussed by Doyle et al. (1991) while Ottmann & Schmitt (1994) and Ottmann (1995) have discussed large flares on AR Lac and Algol seen by *ROSAT*. The latter flares have peak X-ray luminosities of $\sim 2 \times 10^{32}$ erg s^{-1}, peak temperatures approaching $\sim 10^8$ K and total X-ray energies of nearly 10^{37} erg.

In some cases, extremely long flares have been reported for RS CVn binaries. For instance, Tsuru et al. (1989) reported the observation by *Ginga* of the decay phase of a large flare on UX Ari which lasted more than one day and released in the X-ray band more than 10^{37} erg. The peak temperature was 8×10^7 K and the emission measure was nearly 1×10^{55} cm^{-3}. A smaller event (with a total X-ray energy of $\sim 3 \times 10^{35}$ erg) was observed from the same star during the late decay of the major flare. Kürster (1994) reported the observation by *ROSAT* of a flare lasting 9 days from the RS CVn binary CF Tuc. If the light curve is really due to a single event, this is the longest stellar flare ever observed. This raises the question whether small short-lived events (like the ones on M dwarf stars) do exist in RS CVn stars, or whether flares in RS CVn binaries are due to mechanisms fundamentally different from those operating in dMe stars and in the Sun. Note that the much larger quiescent emission of RS CVn binaries makes it difficult to detect small flares, but at least one short lived event (with a rise time of less than 3 min) has been reported from the RS CVn binary σ^2 CrB (van den Oord et al. 1988). It is clear anyway that flares on RS CVn's last typically longer and reach much larger energies than flares on dMe stars.

3 Models

The observed parameters of stellar X-ray flares (both for M dwarf stars and RS CVn binaries) are the X-ray temperature and volume emission measure, the rise and decay time, and the X-ray luminosity and integrated X-ray energy. These can be compared with the analogous quantities measured in solar compact and two-ribbon flares (see tables published by Landini et al. 1986, White et al. 1986, van den Oord et al. 1988, Pallavicini et al. 1990, Linsky 1991). For solar flares, other quantities can also be measured, at least approximately, like the loop length L and height H, the volume V and the density n (the latter usually derived from the observed EM and V). An estimate can also be made of the minimum magnetic field stength B_{min} required to confine the flaring loop. For stellar flares, these quantities must be inferred in an indirect way and the rest of this paper will be devoted to discuss the techniques used for this purpose, and their limitations.

3.1 Order of magnitude estimates

An approximate estimate of flare density and volume can be obtained by considering a simple loop which cools by radiation and conduction, with no additional heating in the decay phase. For simplicity, I assume a semicircular loop of length L and constant cross-section, with an aspect ratio α of loop diameter to loop length (the following considerations can easily be extended to the more general case where the ratio Γ of the loop cross-section at the top and at the base is different from unity). For an ensemble of N identical loops, the volume is given by

$$V = \frac{\pi}{4} L^3 (N\alpha^2) \tag{1}$$

and the emission measure is

$$EM = \frac{\pi}{4} n^2 L^3 (N\alpha^2) \tag{2}$$

The factor $N\alpha^2$ reflects our ignorance of the detailed geometry and is equal to 0.01 for a single loop of aspect ratio $d/L = 0.1$ (as typically observed in the Sun). In the case of an expanding loop, the expressions for V and EM must be multiplied by a factor $(\Gamma + 1)/2$. The height of the loop is given by $H = L/\pi$.

If the flare cools by radiation and conduction the observed decay time will be given by

$$\frac{1}{\tau_{obs}} = \frac{1}{\tau_r} + \frac{1}{\tau_c} \tag{3}$$

where τ_r is the radiative cooling time, τ_c is the conductive cooling time and τ_{obs} is the decay time measured from the observed light curve (more correctly, an effective decay time should be used which takes into account the variation of temperature with time, cf. van den Oord et al. 1988).

The radiative cooling time (in sec) is given by

$$\tau_r = \frac{3kT}{n\Lambda(T)} \tag{4}$$

where $\Lambda(T)$ is the radiative loss function which can be approximated as $\Lambda(T) = \Lambda_o T^{\gamma}$ with $\gamma = 0.25$ for $T \geq 2 \times 10^7$ K and $\Lambda_o = 1.86 \times 10^{-25}$ cgs units.

The conductive cooling time (in sec) is given by

$$\tau_c \sim 1.2 \times 10^{-10} \frac{nL^2}{T^{2.5}} \tag{5}$$

where I have assumed classical Spitzer conductivity.

Note that the ratio τ_r/τ_c is usually unknown and may vary from flare to flare and even in the course of the same flare. In order to determine the flare density and geometry, one has to make further assumptions on the value of this ratio.

A simple approximation (see, e.g., Haisch 1983) is to assume that the cooling mechanisms are equally important, i.e. $\tau_r \sim \tau_c$. Although this assumption is supported by some solar observations close to the flare peak, it remains poorly justified at different times during a flare and when comparing different flares. The obvious advantage is that under this assumption one can determine exactly the flare parameters n, L, and $N\alpha^2$. In fact, by equating the observed decay time to the radiative cooling time, and taking the temperature derived from spectral fits of X-ray data, one can determine the density n. This, combined with the observed EM, gives the volume V. On the other hand, by equating the conductive cooling time to the observed decay time, and using the density derived above, one can also determine the loop length L which, combined with the volume, gives also $N\alpha^2$.

A problem with this approach is that the conductive cooling time (which depends on the actual geometry and the assumed conductivity, i.e. classical or turbulent) is poorly known. An alternative approach (e.g. Pallavicini et al. 1990) is to neglect conductive losses completely, and to derive only upper limits to density (and lower limits to volume) by equating the observed decay time to the radiative cooling time. In this case it is obviously not possible to determine uniquely the loop length L and height H. These parameters can be determined only as a function of $N\alpha^2$, assuming plausible values for this quantity (e.g. $N = 1$ and $\alpha = 0.1$).

The above considerations can be put in more general terms by combining the expression for the volume emission measure (Eq. 2) with those for the radiative and conducting cooling times (Eqs. 4 and 5), and by investigating the dependence of loop height H on $N\alpha^2$. Plots of H vs. $N\alpha^2$, which depend on the temperature, emission measure and observed decay time of individual flares, have been published by White et al. (1986), van den Oord et al. (1988), and van den Oord & Mewe (1989), although with significant differences in the functional expressions adopted by different authors. Depending on which cooling mechanism dominates, one has in fact:

$$H \sim (N\alpha^2)^{-\frac{1}{3}}(\Gamma + 1)^{-\frac{1}{3}} \qquad \text{for } \tau_r \ll \tau_c \tag{6}$$

$$H \sim (N\alpha^2)(\Gamma + 1)^{-1} \qquad \text{for } \tau_r \gg \tau_c \tag{7}$$

$$H \sim (N\alpha^2)^{-1} \qquad \text{for } \tau_r = \tau_c \tag{8}$$

where I have considered the more general case in which the flaring loops have a non constant cross-section (described by the expansion factor Γ). It is clear that assuming equal radiative and conductive losses gives the minimum loop length and height, while the assumption that radiative losses dominate will give a loop length which depends only weakly on the parameter $N\alpha^2$ (for plausible values of Γ and $N\alpha^2$ the loop height will differ from the minimum value by a factor of 2, at most). Much larger uncertainties occur if the flare cools predominantly by conduction: in this case, the derived loop length and height could easily be in error by one order of magnitude or more depending on the actual geometry of the flare (even barring possible departures from classical conductivity). Note however that the large dependence on Γ predicted by Eq. 7 in the case of conductive cooling (which was reported by White et al. 1986 and van den Oord et al. 1988) virtually disappears if account is also taken of the reduction of the conductive flux at the footpoints in a tapered loop (van den Oord & Mewe 1989). In practice, the most reliable parameter is the *minimum* height derived under the assumption of equal radiative and conductive losses. Even in this case, the derived parameters could be easily in error if there is additional heating during the decay.

3.2 Quasi-static cooling

It should be evident from the previous section that in order to determine in a unique way the flare geometry (both L and $N\alpha^2$), one has to introduce an additional relationship which fixes the value of the ratio $\mu = \tau_r/\tau_c$. One such relationship is to assume $\mu = 1$, as discussed above. It has been shown by van den Oord & Mewe (1989) that a more physically meaningful case is the one in which the flare cools quasi-statically through a sequence of constant pressure loops. For a static equilibrium model, in fact, a balance exists between the various energy input and loss terms which in turns implies a constant ratio between the conductive and radiative decay times. It can be shown that for the case of a power-law radiative loss function $\Lambda(T) \sim T^{\gamma}$, this ratio depends only on the exponent γ and is given by $\mu = 0.18$ for $\gamma = 0.25$. Note that the low value of this ratio means that radiative cooling dominates, although conduction is not completely negligible.

For quasi-static cooling with $\mu = 0.18$ and no heating in the decay, simple analytical relationships can be derived for the decay of flare temperature T and radiative energy E. These read:

$$T = T_{\text{o}} \left(1 + \frac{t}{3\tau_{\text{o}}}\right)^{-\frac{8}{7}} \tag{9}$$

$$E = E_{\text{o}} \left(1 + \frac{t}{3\tau_{\text{o}}}\right)^{-4} \tag{10}$$

where $\tau_{\text{o}} = 3kT_{\text{o}}/n\Lambda(T)$ is the radiative cooling time at the start of quasi-static cooling. Note that μ can be rewritten in terms of observable quantities as $\mu \sim T^{13/4}/EM$ and hence the constancy of this ratio throughout the flare decay can be tested observationally from time-resolved spectral data.

By fitting the observed light curve, and using the measured T, one gets τ_{o} and n. From the known value of the ratio μ one also determine L and finally, from the measured value of EM and the derived values of n and L, one determines $N\alpha^2$, in much the same way as for $\mu = 1$. A generalisation is also possible to the case of continuous heating in the decay, although it is not completely clear in this case whether one can solve for both τ_{o} and the heating rate in a unique way. As a matter of fact, in all analyses made up to now of actual flare data (e.g. van den Oord & Mewe 1989, Ottman & Schmitt 1994, Schmitt 1994, Ottmann 1995) the observed light curve was consistent with negligible heating in the decay phase, in spite of the flare long decay time. When no heating is present, there is little difference between the flare parameters derived assuming radiative cooling or quasi-static cooling, except for the parameter $N\alpha^2$ which cannot be determined in the former case. For instance, the density and volume derived by White et al. (1986) for a large flare on Algol assuming radiative cooling were 3×10^{11} cm^{-3} and 1×10^{31} cm^3, respectively, while van den Oord & Mewe (1989) with the quasi-static cooling model derived 2.6×10^{11} cm^{-3} and 1.4×10^{31} cm^3. They also derived a loop height of 5.1×10^{10} cm and $N\alpha^2 = 4.4 \times 10^{-3}$ while White et al. (1986) estimated a height of $(2-5) \times 10^{10}$ cm, within a factor 2 from the value given by the quasi-static cooling model. For a single loop with $\alpha = 0.1$, one would get $N\alpha^2 = 1.0 \times 10^{-2}$, again within a factor of 2 with respect to van den Oord and Mewe (1989). For another flare on Algol which lasted a factor 4 longer, Stern et al. (1992) derived a loop height a factor 4 larger, and a density a factor 5 lower, using again the quasi-static cooling model. The volume was a factor 36 larger and $N\alpha^2 = 3.0 \times 10^{-3}$.

3.3 Reconnection model

All parameters in the previous two sections were derived under the condition (either assumed or inferred) that there is no heating during the flare decay.

Although this possibility cannot be excluded, it appears rather unlikely for long duration flares which last hours and even days. In the Sun, where we know at least approximately the flare density and geometry, the observed decay times of long-duration flares are usually much longer than the radiative cooling time. This proves that there is additional heating in the decay phase, and even more so if conduction losses are not neglible with respect to radiative losses. A model which predicts continuous energy release during the decay phase is the reconnection model of solar two-ribbon flares of Kopp & Poletto (1984) which has been applied to stellar flares by Poletto et al. (1988).

In the reconnection model, a magnetic configuration, torn open by an unspecified explosive event, relaxes back to a lower energy closed configuration. Energy is released gradually as the field lines reconnect at progressively higher altitudes during the flare decay. This process can be modelled analytically in a simple two-dimensional magnetic configuration (described in terms of a single lobe of a Legendre polynomial of degree m). The rate of energy release is given by:

$$\frac{dE}{dt} = A(m, \theta_o) R_\star^3 B_M^2 \frac{y^{2m}[y^{(2m+1)} - 1]}{[m + (m+1)y^{(2m+1)}]^3} \frac{dy}{dt} \tag{11}$$

where $R_\star$ is the radius of the star, $A(m, \theta_o)$ is a known function of the degree m of the Legendre polynomial and of the heliographic latitude of the flare region, B_M is the maximum photospheric field strength in the region, y is the height of the reconnection point expressed in terms of the stellar radius, and dy/dt is its upward velocity. For the latter, one can take:

$$y(t) = \frac{h}{R_\star} = 1 + \left(\frac{H_M}{R_\star}\right)\left[1 - \exp(-\frac{t}{\tau_o})\right] \tag{12}$$

where H_M is the maximum height reached by the reconnection point during its upward movement and τ_o is the time constant of the process.

Analysis of Eq. 11 shows that the shape of the energy release curve depends almost exclusively on the time constant τ_o, while its absolute value depends on both the photospheric magnetic field B_M and the degree m of the Legendre polynomial (and hence on the size of the flaring source, larger values of m corresponding to smaller regions). Clearly, the energy release will be larger for higher photospheric magnetic fields and/or larger regions. Thus, if the energy released in the X-ray band is a constant fraction of the total energy released in the reconnection process, by fitting the observed light curve one can determine the time constant τ_o but cannot separate the source size from the magnetic field strength. The same flare can be reproduced by either a small loop with strong magnetic field, or by a large loop with weaker magnetic field (Poletto 1989, Pallavicini 1992b, Schmitt 1994).

As shown by Schmitt (1994) for the specific case of a long duration flare on the UV Ceti-type star EV Lac, the situation is even more complex. Not only different reconnection models could reproduce the observed light curve (thus leaving essentially unconstrained the loop size), but also models with no heating in the decay (like the quasi-static cooling model discussed above) could also reproduce the same flare. In the latter case, the inferred loop length is 6×10^{11} cm (much larger than the pressure scale height) and the density is 3×10^{10} cm^{-3}, while densities in excess of 10^{12} cm^{-3} are obtained from the reconnection model if the flare region is sufficiently compact. In this case, the inferred loop length is a factor 100 smaller than for the quasi-static cooling model, although reconnection models with larger volumes and lower densities are also acceptable. In the absence of additional constraints (like a direct measurement of the electron density from density sensitive line ratios), it is difficult to discriminate between these possible solutions, although the enormous loop height predicted by the quasi-static cooling model ($\sim$ 10 stellar radii) appears rather unrealistic, if at all acceptable.

4 Conclusions

It is clear from the discussion above that the derivation of flare parameters from spatially unresolved stellar observations (at least of the quality available at present) is a rather tricky business, and that published values should therefore be considered with great caution.

An alternative way to get information on flare loop sizes is to use full hydrodynamic calculations which predict the time behaviour of density, temperature and flow velocity in a magnetically confined loop structure subject to a prescribed heating perturbation (e.g. Peres 1989). For instance, Reale et al. (1988) used a hydrocode to infer a loop length $L = 1.4 \times 10^{10}$ cm for a small flare on Proxima Cen observed by *Einstein*. On the other hand, Cheng & Pallavicini (1991) argued that the inferred loop length was largely the consequence of the assumed absence of additional heating during the decay phase. Reale et al. (1993) have suggested to use the ratio of the decay times of temperature and density as a diagnostics of the loop length. The predicted dependence, however, becomes much flatter when loop integrated quantities are used instead of the values at the loop top, and the predicted ratio is also a strong function of the amount of heating in the decay phase. In practice, rather than a direct diagnostic tool, this ratio is more useful to restrict the range of hydrodynamic models to be explored to fit the observed light curve.

To conclude, any real progress in this area has probably to wait for a major stepforward on the observational side. The Japanese satellite ASCA is already providing spectral data which are of far better sensitivity and resolution of anything else obtained in the past. However, it will be only with *AXAF* and *XMM* at the turn of the century that it should be possible to obtain time-resolved spectroscopy and density-sensity line ratios of the quality needed to constrain stellar flare parameters in an effective way.

References

Ambruster C.W., Sciortino S., Golub L., 1987, ApJS 65, 273
Cheng C.-C., Pallavicini R., 1991, ApJ 381, 234
Culhane J.L., White N.E., Shafer R.A., Parmar A.N., 1990, MNRAS 243, 424

Doyle J.G. et al., 1991, MNRAS 248, 503
Fleming T.A., Giampapa M.S., Schmitt J.H.M.M., Bookbinder J.A., 1993, ApJ 410, 387
Kopp R.A., Poletto G., 1984, Solar Phys. 93, 351
Haisch B.M., 1983, in Activity in Red Dwarf Stars, P.B. Byrne & M. Rodonò (eds.), Reidel, Dordrecht, p. 255
Haisch B.M., Rodonò M. (eds.), 1989, Solar and Stellar Flares, Kluwer, Dordrecht
Haisch B.M., Strong K.T., Rodonò M., 1991, ARA&A 29, 275
Kürster M., 1994, in Cool Stars, Stellar Systems, and the Sun, J.-P. Caillault (ed.), ASP Conf. Series 64, 104
Landini M., Monsignori Fossi B.C., Pallavicini R., Piro L., 1986, A&A 157, 217
Linsky J.L., 1991, in Stellar Flares, B. Pettersen (ed.), Memorie Soc. Astron. Ital. 62, 307
Ottmann R., 1995, this volume, p. 164
Ottmann R., Schmitt J.H.M.M., 1994, A&A 283, 871
Pallavicini R., 1992a, in The Sun: A Laboratory for Astrophysics, J.T. Schmelz & J.C. Brown (eds.), Kluwer, Dordrecht, p. 509
Pallavicini R., 1992b, in Eruptive Solar Flares, Z. Svestka, B.V. Jackson & M.E. Machado (eds.), Lecture Notes in Phys. 399, 291
Pallavicini R., 1993, in Physics of Solar and Stellar Coronae, J.F. Linsky & S. Serio (eds.), Kluwer, Dordrecht, p. 237
Pallavicini R., Tagliaferri G., Stella L., 1990, A&A 228, 403
Pan H.C., Jordan C., 1994, MNRAS, in press
Peres G., 1989, Solar Phys. 121, 483
Pettersen B.R. (ed.), 1991, Stellar Flares, Mem. Soc. Astron. Ital. Vol. 62
Poletto G., 1989, Solar Phys. 121, 313
Poletto G. Pallavicini R., Kopp R.A., 1988, A&A 201, 93
Priest E.R., 1995, this volume, p. 3
Reale F., Peres G., Serio S., Rosner R., Schmitt J.H.M.M., 1988, ApJ 328, 256
Reale F, Serio S., Peres G., 1993, A&A 172, 486
Schmitt J.H.M.M., 1994, ApJS 90, 735
Stern R.A., Uchida Y., Tsuneta S., Nagase F., 1992, ApJ 400, 321
Tagliaferri G., White N.E., Doyle J.G., Culhane J.L., Hassall B.J.M., Swank J.H., 1991, A&A 251, 161
Tsuru T. et al., 1989, PASJ 41, 679
van den Oord G.H.J., Mewe R., 1989, A&A 213, 245
van den Oord G.H.J., Mewe R., Brinkman A.C., 1988, A&A 205, 181
White N.E., Culhane J.L., Parmar A.N., Kellett B.J., Kahn S., van den Oord G.H.J., Kuijpers J., 1986, ApJ 301, 262
White N.E., Shafer R.A., Horne K., Parmar A.N., Culhane J.L., 1990, ApJ 350, 776

J. van Paradijs: You noted that an X-ray flare was observed in the *ROSAT* All-Sky-Survey from the B5 star TT Lupi. Would you expect that?

R. Pallavicini: Of course not, since these stars are not expected to have magnetic activity. However, there are a number of B and A stars detected by *ROSAT* and some of them have been found to flare. One possibility is that the X-ray emission and the flare come from a late-type companion. A flare for instance was also detected by *EXOSAT* (and confirmed by *ROSAT*) on the A star Castor A+B.

Short-Duration X-ray Transients Observed With WATCH on Granat: Are Some of Them Related to Stellar Flares?

Alberto J. Castro-Tirado[1], *Søren Brandt*[2], *Niels Lund*[2], *Igor Lapshov*[3], *Rashid Sunyaev*[3]

[1] Laboratorio de Astrofísica Espacial y Física Fundamental (LAEFF-INTA), P.O. Box 50727, E-28080 Madrid, Spain
[2] Danish Space Research Institute, Gl. Lundtoftevej 7, DK-2800 Lyngby, Denmark
[3] Space Research Institute, Academy of Sciences, Profsoyuznaya 84/32, 117810 Moscow, Russia

Abstract: During 1990-92, the *WATCH* all-sky X-ray monitor on *GRANAT* has discovered 6 short-duration X-ray transients. We discuss their possible relationship to peculiar stars. Only one source, GRS 1100-77 seems to be related to a T Tauri star.

1 Introduction

Amongst the many kinds of sources in the variable X-ray sky, X-ray transients have been observed since the first experiments in the late 60s. According to their duration, it is possible to distinguish between long-duration X-ray transients (lasting from weeks to few months) and short-duration X-ray transients (lasting from hours to very few days).

Long-duration X-ray transients are mainly related to Be-neutron star systems (Be/X-ray Transients) and to soft X-ray transients (a subclass of the low-mass X-ray binaries). The latter include the best black-hole candidates found so far.

Not too much is known with respect to most short-duration X-ray transients. The main reason is the lack of counterparts at other wavelengths for many of them. One subclass, the so-called fast X-ray transients, have been observed with many detectors since the launching of the Vela satellites (Heise et al. 1975, Cooke 1976). Durations range from seconds to less than few hours. Normally, they have been seen only once, and they never have been detected in quiescence, implying high peak-to-quiescent flux ratios ($10^2 - 10^3$) (Ambruster et al. 1983). Spectral characteristics vary substantially, from a hard spectrum in MX 2346-65 ($kT \sim$ 20 keV; Rappaport et al. 1976) to soft spectra with blackbody temperatures from $kT = 0.87$ to 2.3 keV (Swank et al. 1977). In the *Ariel V* database, 27 sources were discovered (Pye & McHardy 1983), and 10 more were detected in the *HEAO 1 A-1* all sky survey (Ambruster & Wood, 1986), implying a fast transient all-sky event rate of 1500 yr^{-1} for fluxes $F_X \geq 3\ 10^{-10}$ erg $cm^{-2}s^{-1}$ in the (2-10 keV) energy band. In the *HEAO 1 A-2* survey, 52 events were found (Connors 1988), but 37 of them were associated with 4 of the brightest X-ray sources in the LMC. Due to the large difference of observational characteristics, it seems that these events are caused by more than one physical mechanism. In several cases, there have been tentative optical identifications on the basis of known sources in the transient error boxes. One suggestion has been that many of the fast X-ray transients are related with stellar flares originated in active coronal sources, like RS CVn binaries or dMe-dKe flare stars.

RS CVn systems peak luminosities are usually $L_X \sim 10^{32}$ erg s^{-1}. The hardest flare yet observed was for the system HR 1099, on 17 Feb 1980. It was detected with *HEAO 2* at energies up to 20 keV (Agrawal & Vaidya 1988). The most energetic X-ray flare was observed with *Ginga* from UX Ari. Its decay time was quite long ($\sim$ 0.5 d).

X-ray flares can be also observed from dMe-dKe stars. In the *Ariel V* sky survey, Rao & Vahia (1984) suggested 7 dMe stars as responsible for X-ray flares that reached peak luminosities $L_X < 10^{32}$ erg s^{-1} (2-18 keV). AT Mic is the dMe star with the largest number of recorded events (4). For a flare observed with *EXOSAT* from YY Gem on 15 Nov 1984, Pallavicini et al. (1990) estimated a total flare energy $E_X = 10^{34}$ erg (0.005-2 keV), and a decay time t_d = 65 min (one of the longest decay times ever mesured for such a flares). The longest flare ever reported was observed for more than 2 h for the X-ray source EXO 040830–7134.7 (Van der Woerd et al. 1989). X-ray flares have also been observed from Algol-type binaries (Schnopper et al. 1976), W Uma systems and young stars. See Linsky (1991) for a comprenhensive review. Theoreticians have suggested other sources like dwarf novae (Stern et al. 1981) or extreme type I X-ray bursts (Lewin & Joss 1981) as origin of some fast X-ray transients.

2 Observations by WATCH during 1990-92

The *WATCH* instrument was the all-sky X-ray monitor onboard the *GRANAT* satellite, launched on 1 Dec 1989. It consisted of four identical units, mounted in a tetrahedral configuration covering the whole sky. Each instrument had an effective area of $\sim$ 30 cm^2. It was based on the rotation modulation collimator principle (Lund 1985). The two energy ranges are approximately 8-15 and 15-100 keV. Usually, the uncertainty for the location of a new and short-duration source was 1 degree in radius (3-σ). A summary of scientific results can be found in Castro-Tirado (1994) and Brandt (1994).

2.1 Fast X-ray transients discovered with WATCH

During 1990-1992, *WATCH* discovered 3 bright fast X-ray transients, a small number compared with the above-mentioned instruments. This was mainly due to the higher low-energy cut off for *WATCH* ($\sim$ 8 keV), implying that only the harder events were detected. The events were immediately noticeable as an increase in the low-energy count rate (Fig. 1). No positive detections were made in the higher energy band. The observational characteristics of the events are summarized in Table 1.

Table 1. *WATCH* Fast X-ray transients.

Source	date	UT	t (min)	α(1950.0)	δ(1950.0)	F_X(Crab)
GRS 1100-77	15 Jan 1992	13:39	90	10h 59m.9	−77.1	0.80
GRS 2037-40	23 Sep 1992	04:35	110	20h 37m.9	−39.8	0.75
GRS 2220-15	19 Sep 1990	03:32	240	22h 20m.4	−14.9	2.10

Note: The UT time refers to the beginning of the event. t is the duration.

2.2 Longer duration X-ray transients discovered with WATCH

Three long events were discovered with *WATCH* during 1990-92, although it is likely that some others are present in the data. The sources reported here were observed to peak at a flux $F_X \leq 0.5$ Crab (8–20 keV), and were discovered by analysis of the corresponding modulation pattern data. Their observational characteristics are summarized in Table 2. A fourth event, GRS 1133+54 (lasting for $\sim$ 1 d on 19-20 Nov 1992) was reported elsewhere (Lapshov et al. 1992).

Table 2. WATCH Longer duration X-ray transients.

Source	date	UT	t (d)	α(1950.0)	δ(1950.0)	F_X(Crab)
GRS 0818-52	10 Oct 1990	14:00	$\sim$ 1	08h 18m.0	−52.0	0.50
GRS 1148-66	10 Oct 1990	12:00	$\sim$ 1	11h 48m.0	−66.5	0.35
GRS 1624-37	24 Sep 1992	03:00	$\sim$ 1	16h 24m.4	−37.5	0.35

Note: The UT time refers to the maximum flux. t is the duration.

2.3 Search for correlated X-ray flares

We have searched for X-ray flares in the *WATCH* data base that could have occurred simultaneously with observed flares at other wavelengths reported elsewhere. The results were negative both for λ Eridani (a flare was observed with ROSAT on 21 Feb 1991, Smith et al. 1993) and AU Mic (a flare was observed with EUVE on 15 July 1992, Cully et al. 1993). In the case of EV Lac, no WATCH data were available for an extraordinary optical flare that occurred on 15 Sep 1991 (Gershberg et al. 1995).

3 Discussion

3.1 The fast X-ray transients

The three events described in 2.1 imply a rate of $\sim$ 5 year^{-1}. If the sources of the three events are galactic, their high galactic latitude would suggest that they are nearby, although no nearby known flare stars (from the list of Pettersen, 1991) have been found in the corresponding *WATCH* error boxes.

GRS 1100–77. The observations implied $L_X = 0.6\ 10^{30}\ (d/1\ \mathrm{pc})^2$ erg s^{-1} (8-15 keV), or $L_X = 2.3\ 10^{30}\ (\mathrm{d}/\ 1\ \mathrm{pc})^2$ erg s^{-1} (2-15 keV) if we assume a Crab-like spectrum. The most interesting object inside the error box is the T Tauri star LH 332-20, from the catalogue compiled by Herbig & Rao (1972). It lies in an anonymous dark nebula at unknown distance. It is known that T Tauri stars may exhibit X-ray flaring activity. In fact, Montmerle et al. (1983) reported a "superflare" in another T Tauri star: ROX-20, in the ρ Oph Cloud Complex. It reached a peak luminosity $L_x = 1.1\ 10^{32}$ erg s^{-1} (0.3-2.5 keV), and an integrated flare X-ray energy $E = 10^{34}$ erg s^{-1}. We propose the T Tauri star LH 332-20 as the most likely candidate for GRS 1100-77.

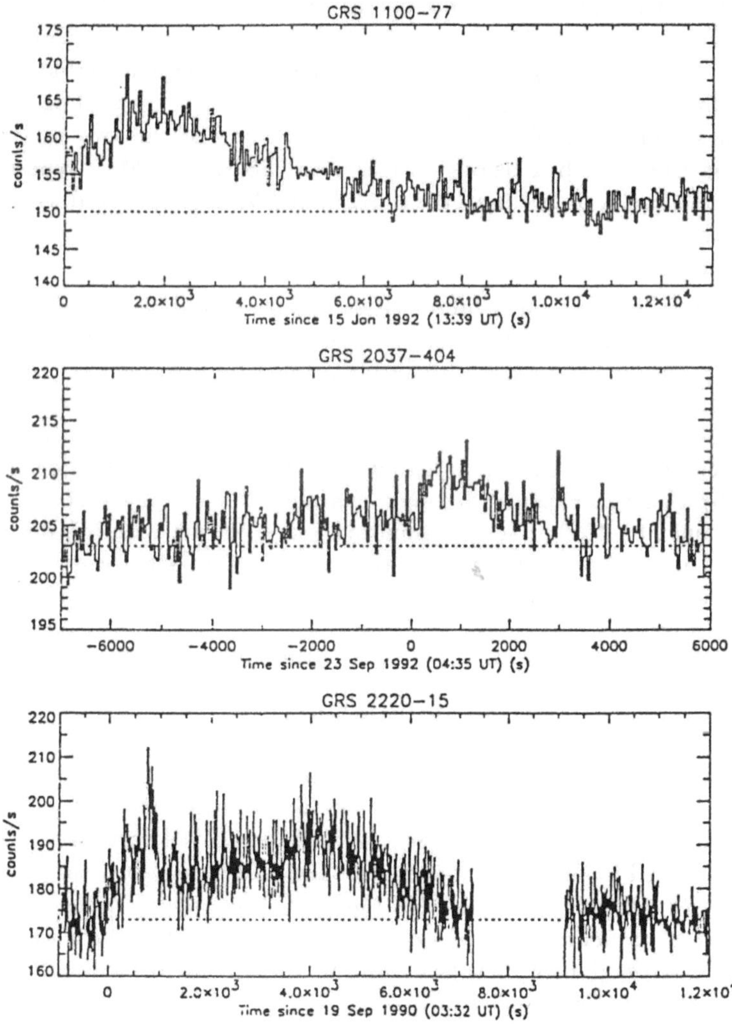

Fig. 1. X-ray light curves in the 8–15 keV energy band of the three fast X-ray transients discovered with *WATCH* on *GRANAT*.

GRS 2037–40. The error box includes the Mira star U Mic. On a plate taken at our request six days after the discovery, this star was near maximum brighteness at $7^{m}_{\cdot}0$. However we do not consider this star as a candidate due to the different time scales of the involved physical processes. Hudec (priv. comm.) proposed another candidate inside the WATCH error box to be further investigated: the variable star RV Mic, discovered by C. Hoffmeister in 1948. Although it is classified as a Mira star, it is apparently a highly variable object. Nevertheless, we carefully examined the Schmidt plate and found no object inside the WATCH error box that differed by more than $0^{m}_{\cdot}5$ compared with the corresponding ESO Sky Survey plate.

GRS 2220–15. The high peak flux (2.1 Crab) was similar to the peak of the fast transient at $20^{h}14^{m} + 30.9^{\circ}$ discovered with *OSO-8* in 1977 (Serlemitsos et al. 1979). For GRS 2220–15, $L_X = 1.5\ 10^{30}$ (d/1 pc)^2 erg s^{-1} (8–15 keV), or $L_X = 6\ 10^{30}$ (d/1 pc)^2 erg s^{-1} (2–15 keV) if we assume a Crab-like spectrum. Clearly, a dMe or a dKe star may be excluded. A possibility would be to consider a RS CVn system at a distance $\leq$ 10 pc, but none of the two variable stars contained in the error box belong to this class: SS Aqr (a Mira-type star) and BW Aqr (an Algol system). If all the flares that BW Aqr can produce are Algol-type (i.e., the peak luminosity is $\sim 5\ 10^{30}$ erg s^{-1}, White et al. 1980), this star can be discarded as a serious candidate for this event.

Considering the high peak luminosities, a flare of a dMe or a dKe star may be excluded in all the three cases. No nearby RS CVn stars from the list of Lang (1992) were found in the error boxes. It may be possible that these events are associated with old isolated neutron stars accreting interstellar matter with unstable nuclear burning (Ergma & Tutukov 1980). According to Zdunik et al. (1992), when the accretion rate is higher than 10^{13} g s^{-1}, the hydrogen burning triggered by electron capture becomes unstable. As the mass of the accreted envelope should be 10^{23} g, several hundred years will be required for accreting this amount of matter. Assuming that the number of isolated neutron stars with such high rate is 10^5 (Blaes & Madau 1993), one should expect 10–100 fast X-ray transients per year.

3.2 The longer-duration X-ray transients

The three longer-duration sources reported in Sect. 2.2. are concentrated near the galactic plane, suggesting that they could be more distant than the three fast X-ray transients described in Sect. 2.1. No nearby dwarf star or RS CVn systems have beed identified in the error boxes. Three variable stars were found inside the *GRS 1148-66* error box: CY Mus, UV Mus and TY Mus. In the case of *GRS 1624-37*, a Schmidt plate was requested on La Silla which was taken on 27 Sep 1992. A previously unknown variable object has been found inside the *WATCH* error box (Castro-Tirado & Lund 1995).

In the *Ariel V* Catalogue of fast X-ray transients, Pye & McHardy (1983) described 10 X-ray transients with durations ranging 0.5–4 d. Some of them were associated with known Be-neutron stars systems, like 4U 0114+65, and other with RS CVn systems, like σ Cen or DM UMa (for which a flare lasting 1.5 d was observed by *HEAO-2*). The long duration events reported here are likely to have originated from compact objects in both LMXBs and HMXBs.

Acknowledgements: We are very grateful to the staff of the Evpatoria ground station in Ukraine and those of the Lavotchkin and Babakin Space Center in Moscow, where *GRANAT* was built. Also to G. Pizarro, A. Smette and R. West (ESO) for the Schmidt plates taken at La Silla.

References

Agrawal P.C., Vaidya G.R., 1988, MNRAS 219, 777
Ambruster C., Snyder W.A., Wood K.S., 1984, ApJ 1984, 270
Ambruster C. et al., 1983, ApJ 269, 779
Ambruster C.W., Wood K.S., 1986, ApJ 311, 258
Blaes O., Madau G., 1993, ApJ 403, 590
Catalano S., 1990, in Active Close Binaries, C. Ibanoglu (ed.), Kluwer, Dordrecht, p. 411
Castro-Tirado A.J., Brandt S., Lund N., 1992, IAU Circ. 5617
Castro-Tirado A.J., 1994, Ph.D. Thesis, Univ. of Copenhagen
Castro-Tirado A.J., Lund N., 1995, in preparation
Connors A., 1988, Ph.D. Thesis, Univ. of Maryland
Cully S.C. et al., 1993, ApJ 414, L49
Ergma E., Tutukov A., 1980, A&A 84, 125
Gershberg R.A. et al., 1995, these proceedings
Gliese W., 1969, Veröffent. Astron. Rechen-Inst. 22
Heise J. et al., 1975, ApJ 202, L73
Herbig H.G., Rao N., 1972, ApJ 104, 401
Lang K.R., 1992, Astrophysical Data, Springer, Berlin
Lapshov I., Sazonov S., Sunyaev R., 1992, priv. comm.
Lewin W.H.G., Joss P.C., 1981, SSR 28, 3
Linsky J.L., 1991, Mem. S. A. It. 62, 307
Lund N., 1985, in X-ray Instrumentation in Astronomy, J.L. Culhane (ed.), SPIE 597, 95
Montmerle T. et al., 1983, ApJ 269, 182
Pallavicini R., Tagliaferri C., Stella L., 1990, A&A 228, 403
Pettersen B.R., 1991, Mem. S. A. It. 62(2), 217
Pye J.P., McHardy I.M., 1983, MNRAS 205, 875
Rao A.R., Vahia M.N., 1987, A&A 188, 109
Rappaport S. et al., 1976, ApJ 206, L139
Schnopper H.W., Delvaille J.P., Epstein A. et al., 1976, ApJ 210, L75
Serlemitsos P., Bunner A., Swank, J.H., 1979, Bull. of the APS 24, 584
Smith M.A. et al., 1993, ApJ 409, L49
Stern R.A., Agrawal P.C., Riegler, G.R., 1981, Nature 290, 573
Swank J.H. et al., 1977, ApJ 212, L73
Van der Woerd H. et al., 1989, A&A 220, 221
White N.E. et al., 1980, ApJ 239, L69
Zdunik et al., 1992, ApJ 384, 129

ROSAT Observation of a Giant X-ray Flare on the Active Binary Algol

R. Ottmann

Max-Planck-Institut für Extraterrestrische Physik, 85748 Garching, Germany

Abstract: ROSAT/PSPC observations of a long-duration, giant X-ray flare on the eclipsing binary Algol are presented. With a total thermally radiated energy of $7 \cdot 10^{36}$ erg, this flare is one of the largest events ever observed. The flare decay is modeled by means of the quasi-static cooling method, which yields a height of the flare loop arcade of $\simeq 1R_K$, and a plasma density of 10^{11} cm^{-3}.

1 Introduction

Active binary systems — like the RS CVn and the Algol type binaries — show the most intense and energetic flares among the late-type stars. Large flares on these systems can reach soft X-ray luminosities of 10^{32} erg s^{-1} and thermally radiated energies of 10^{37} erg, which exceed the luminosities and energies of solar two-ribbon (2R) flares by about 5 orders of magnitude.

Up to now, it is not clear whether large stellar flares can be described by a 2R flare scenario, or whether they involve a different flare type. Even if a 2R flare model is assumed, it is not known whether the large energy release of stellar flares compared to solar ones is the result of a large flaring volume or electron density.

Soft X-ray observations provide the most direct way to investigate the physical properties of the flare plasma. During the last decade, intensive X-ray flares have been detected on Proxima Centauri by the *Einstein* satellite (Haisch et al. 1983), on Algol by the *EXOSAT* satellite (White et al. 1986, van den Oord & Mewe 1989) and by the *Ginga* satellite (Stern et al. 1992), and on σ^2 CrB by *EXOSAT* (van den Oord et al. 1988). Recently, the X-ray satellite *ROSAT* detected large flares with a high S/N ratio on the RS CVn-type binaries AR Lac (Ottmann & Schmitt 1994) and CF Tuc (Kürster & Schmitt 1995), and on Algol (B8 V + K2 IV; $P_{orb} = 2.87d$) (Ottmann 1995).

As the Algol flare has a particularly good coverage of the rise and decay phase, it is well suited for a detailed study of the flare plasma parameters. Further, as the Algol flare occurred in the middle of a long *ROSAT* PSPC pointed observation, we are able to extract the flare-only emission, and to search for a re-structuring of the corona in the course of the flare.

2 ROSAT observation

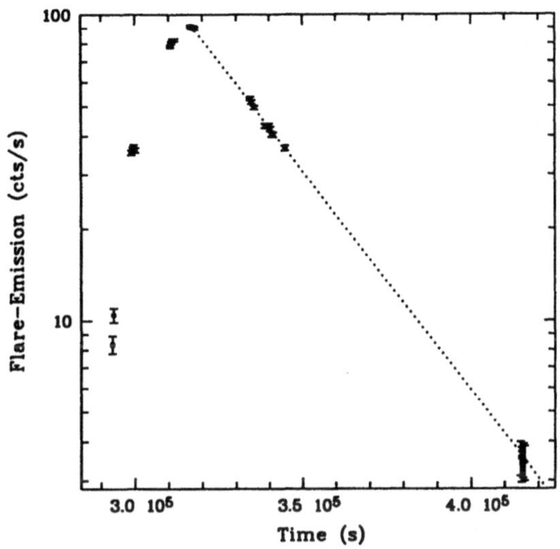

Fig. 1. Flare-only light curve plotted vs. time. The *dotted line* represents the best fit exponential decay with an e-folding decay time $\tau_d = 30400\ s$.

The *ROSAT* PSPC observation of Algol was performed from 1992 August 16 to 23, thus comprising three successive orbital periods (Ottmann 1994). During the second orbit, a giant flare event occurred, lasting for about half the orbital period. Fig. 1 shows the flare-only light curve as a function of time. Thereby, the quiescent contribution (i.e., the count rate level of the first orbit) has been subtracted from the total count rate.

The flare has a total rise time of about 6.6 hours. The flare countrate shows an exponential decay with an e-folding decay time of about 8.3 hours. There is an enhanced emission already 11 hours before the onset of the large flare, continuing for about 36 hours after the exponential decay of thc large flare. The enhanced emission undergoes an orbital modulation. Therefore, it is caused either by the emergence of new magnetic flux, in the course of which also the flare occurs, or by additional heating prior and after the large flare.

The PSPC pulse height spectra are described as the superposition of a two-temperature Raymond-Smith quiescent spectrum and a one-temperature flare spectrum. The parameters $T1$, $EM1$, $T2$, $EM2$ describing the quiescent spectrum are fixed, while the flare temperature $T3$ and emission measure $EM3$ as well as the hydrogen column density are varied. The resulting best fit flare temperature and emission measure are shown in Fig. 2. Obviously, the emission measure follows the X-ray light curve, while the temperature is highest during the early rise phase, reaching about 8 keV or 90 MK.

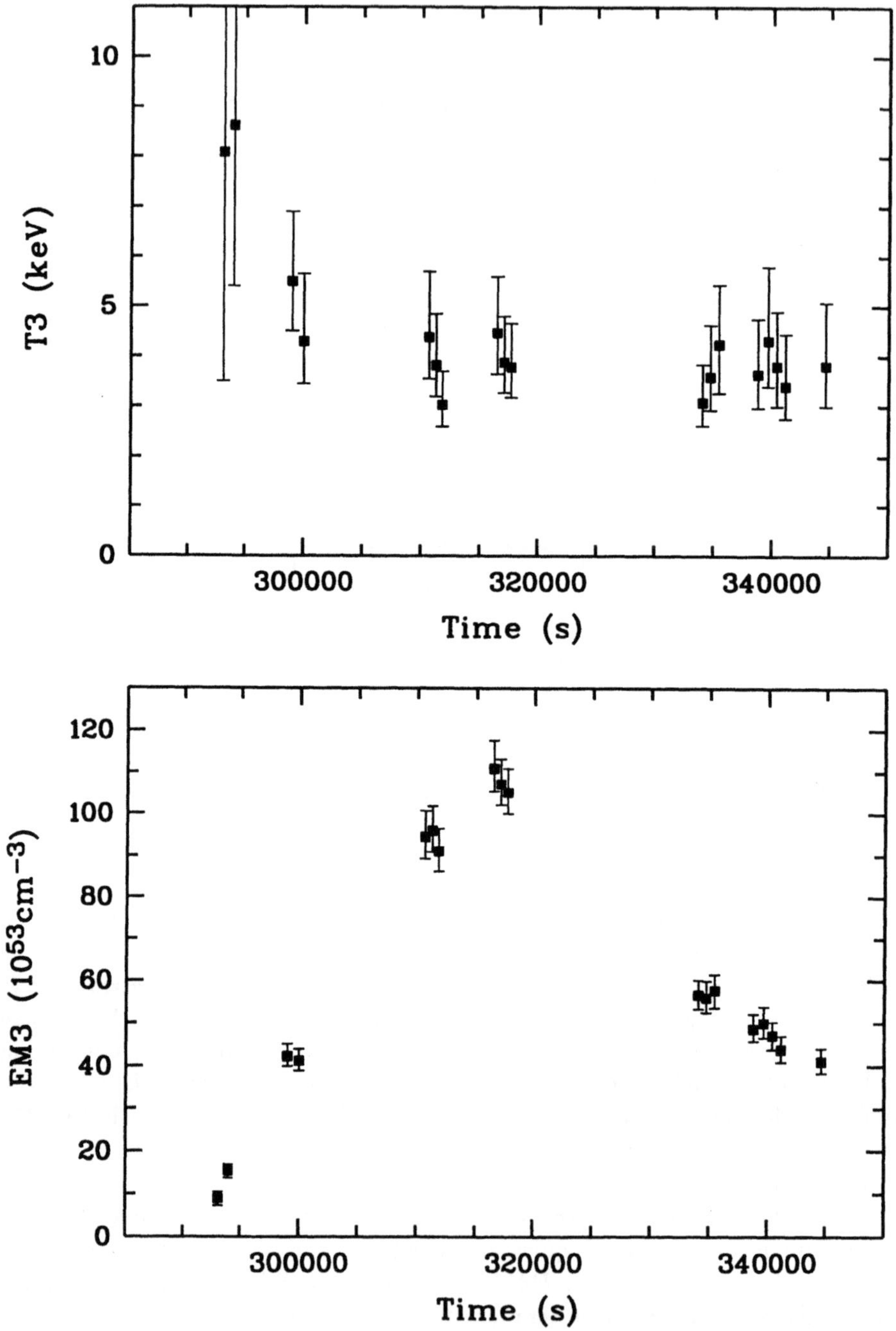

Fig. 2. Temporal evolution of the flare plasma temperature $T3$ and emission measure $EM3$. The errors bars represent 95.4% confidence with one parameter.

3 Physical flare parameters

From the X-ray light curve and spectral results, the physical flare parameters – like the energy content, the spatial extent of the flaring region, and the resulting electron density and gas pressure – can be derived.

For this purpose, the quasi-static cooling formalism (van den Oord & Mewe 1989) is applied to the observed decay of the flare temperature and emission measure. In doing so, we allowed for additional heating during the decay phase. The resulting best fit parameters are as follows. The height of the flare loop arcade amounts to $H = 2.5 \cdot 10^{11}$ cm $\simeq 1R_K$, resulting in a flaring volume of $V = 5 \cdot 10^{33}$ cm^3. This is about 10^5 times the volume of typical solar 2R flares. Additional heating is not present.

For this Algol flare, a peak soft X-ray luminosity of $L_X \simeq 2 \cdot 10^{32}$ erg s^{-1}, and a the total thermally radiated energy of $E_X \simeq 7 \cdot 10^{36}$ erg is estimated, which exceeds the energy release of solar 2R flares by about 5 orders of magnitude. If the solar analogy holds at all, then this Algol flare should be classified as a 2R event on the basis of its long rise and decay time and its large energy release.

From $EM3$ and H, a flare plasma density of $n_e \simeq 10^{11}$ cm^{-3} and a gas pressure of $p \simeq 10^3$ dyne cm^{-2} is derived, which are comparable to the values found for solar flares. Thus, it seems to be the large flaring volume which is responsible for the much larger energy release of stellar compared to solar flares.

Acknowledgements: The *ROSAT* project was supported by the Bundesministerium für Forschung und Technologie (BMFT/DARA) and the Max-Planck-Society.

References

Haisch B.M., Linsky J.L., Bornmann P.L., Stencel R.E., Antiochos S.K., Golub L., Vaiana G.S., 1983, ApJ 267, 280

Kürster M., Schmitt J.H.M.M., 1995 (in preparation)

Ottmann R., Schmitt J.H.M.M., 1994, A&A 283, 871

Ottmann R., 1994, A&A 286, L27

Ottmann R., 1995, A&A (submitted)

Stern R.A., Uchida Y., Tsuneta S., Nagase F., 1992, ApJ 400, 321

van den Oord G.H.J., Mewe R., Brinkman A.C., 1988, A&A 205, 181

van den Oord G.H.J., Mewe R., 1989, A&A 213, 245

White N.E., Culhane J.L., Parmar A.N., Kellett B.J., Kahn S., van den Oord G.H.J., Kuijpers J., 1986, ApJ 301, 262

I. Pustylnik: Is there any contamination of the Algol X-ray data originating from the Perseus Cluster?

R. Ottmann: Due to the high spatial resolution of the ROSAT XRT/PSPC, a source extraction radius of 3.′6 was used. As the distance between Algol and the Perseus Cluster is $> 1^\circ$, a contamination of the X-ray data due to the Perseus cluster can be excluded.

ASCA X–ray Spectra of Quiescent and Flaring Emission from AB Doradus

S. M. White[1], *R. Pallavicini*[2], *J. Lim*[3]

[1] Department of Astronomy, University of Maryland, College Park MD 20742, USA
[2] Arcetri Astrophysical Observatory, Largo E. Fermi 5, I-50125 Florence, Italy
[3] Inst. for Astronomy and Astrophysics, Academica Sinica, Taipei

The Japanese satellite *ASCA* carries X-ray detectors which combine excellent sensitivity and high spectral resolution. We observed the young, rapidly-rotating K dwarf AB Doradus with *ASCA* in November 1993. The star's X-ray flux was essentially steady for the first half of the observation, and then a series of flares occurred during the second half. The flares showed rise times of 30 minutes and decay times of several hours. The quiescent X-ray luminosity was 2×10^{30} ergs s^{-1} (0.5–10 keV). At the flare peaks, $L_X(> 0.5 \text{ keV})/L_{bol} = .002$.

AB Dor has a relatively high count rate in the *ASCA* data and we can obtain high-quality spectra for different periods of the light curve. In Fig. 1 we present spectra corresponding to the period of quiescence, the rising phase and peak of each of the three flares combined, and the decay periods of each of the flares combined. The rise/peak curve (upper) has been multiplied by 2 for purposes of display. We have subtracted the quiescent spectrum from the flare spectra (i.e., used it as a background spectrum) for this display and subsequent analysis.

The detailed differences between these spectra display the importance of the high spectral resolution provided by *ASCA*. One important difference, not readily evident in this figure, is the Fe K line complex at 6.4–6.7 keV. This line, a diagnostic of hot plasma, is strong in the rise/peak spectrum, moderate in the decay spectrum and weak in the quiescent spectrum. A time profile of the counts in this energy range shows little significant emission during the quiescent periods but strong emission at the flare peaks.

It is generally found that two-temperature, variable-abundance models fit *ASCA* data for active cool stars well. We have carried out such fits for the three separate spectra presented above (using the Mewe-Kaastra code, although we do not find major differences when using the Raymond-Smith code). We have only fit for the abundances of those elements whose abundances are reasonably well constrained by the spectrum, and we have forced the abundance to be the same in the hot and cool components. All three spectra show a cool component at 0.6 keV, but the emission measure of this component is much less in the two flare spectra than in the quiescent spectra, indicating (since we used the quiescent spectrum as a "background" for the flare spectra) that there was little significant enhancement in the cool component during the flares. The quiescent spectrum shows a hot coronal component at 1.7 keV with about the same emission measure

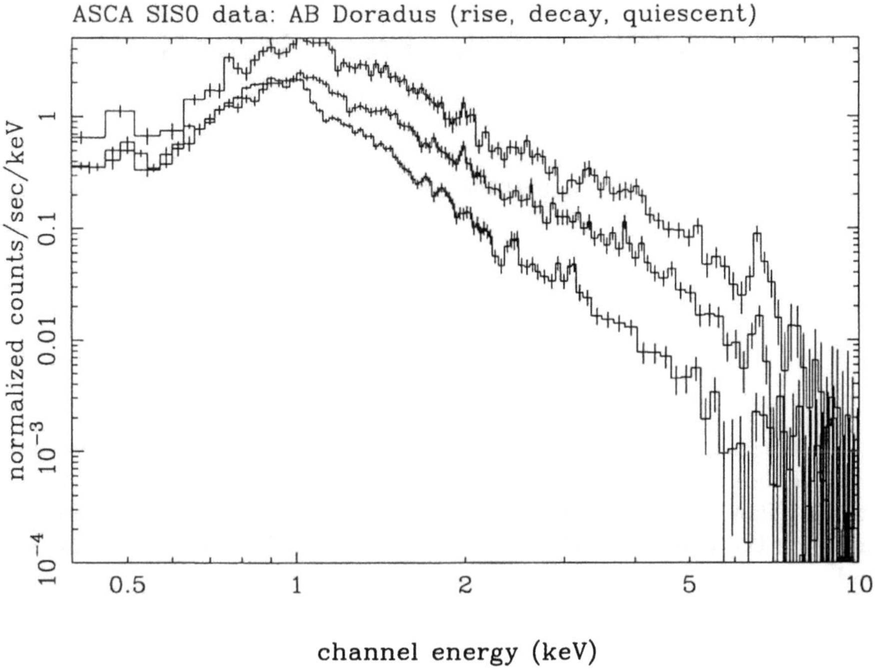

Fig. 1. *ASCA* spectra of AB Dor in its quiescent state (lowest curve), during the rise and peak of three large flares (top curve, count rates multiplied by 2 for clearer display), and during the decay of the flares (middle curve).

as the cool component. The flare rise/peak spectrum has a strong hot component at 3.0 keV; the flare decay spectrum has a hot component with about the same emission measure as in the rise/peak spectrum but a cooler temperature, 2.4 keV.

All three spectra show somewhat low iron abundances: 0.3 (relative to solar) in the quiescent spectrum, 0.8 in the flare rise/peak spectrum and 0.5 in the flare decay spectrum. The low abundance in the quiescent state is supported by aspects of the EUV spectrum obtained in a simultaneous *EUVE* observation (Rucinski et al. 1995). This is a puzzle because in the solar corona Fe, with its low first-ionization potential (FIP), is enhanced. All spectra also show low N abundances, a common feature of *ASCA* spectra which may be partly instrumental (a prominent N line lies at the low-energy cutoff of the SIS detectors).

References

Rucinski S., Mewe R., Kaastra J., Vilhu O., White S. M., 1995, ApJ (submitted)

X-ray Variability of the Flare Star CC Eri

H. C. Pan, C. Jordan

Department of Physics (Theoretical Physics), University of Oxford, 1 Keble Road, Oxford OX1 3NP, UK

The flare/spotted spectroscopic binary star CC Eri (HD 16157) was observed in the periods 1990 July 9-11 and 1992 January 26-27 with the *ROSAT* PSPC detector (Pan and Jordan 1995). These high quality data give the first information on the temporal and spectral variability of CC Eri in the X-ray energy band.

During the *ROSAT* observations the X-ray intensity of the source is variable on timescales from a few minutes to several hours. The X-ray luminosity is in the range $2.5 - 6.8 \times 10^{29}$ erg s^{-1}, which is similar to values found from previous measurements with the *Einstein* IPC and *EXOSAT* LE. On 1990 July 10 an X-ray flare-like event was detected with an exponential rise time of about one hour and a decay time of about two hours. The X-ray spectrum of CC Eri can be described by current thermal plasma codes with two temperature components or with a continuous temperature distribution. The spectral results show that plasma at $T_e \sim 10^7$ K exists in the corona of CC Eri. We find that the X-ray spectrum is also variable and the variations of the emission measure, and to a lesser extent of the temperature, are correlated with the source intensity.

The X-ray variability of CC Eri may in principle be caused by flaring events and/or by the rotational modulation of active regions. However, rotational modulation of a spot related active region requires an unphysically large X-ray flux in a concentrated area. Thus it seems more likely that some sort of flare was observed. The variation in the observed source flux can be reproduced by adopting a magnetic reconnection model. Comparisons with an unheated model, towards the end of the flare, suggest that the area and volume of the flare are substantially larger than in a solar two ribbon flare, while the electron pressure is similar. The larger flare area on CC Eri, compared with the Sun, may not be surprising given the larger optical starspot area. Thus a two ribbon flare provides one possible explanation of the variations observed. The emission measure and temperature of the non-flaring emission, interpreted as the average corona, lead to an electron pressure similar to that in a well-developed solar active region.

References

Pan H. C., Jordan C., 1995, MNRAS 272, 11

X-ray Observations of the dMe Star EQ1839.6+8002 in 1980-1993

H.C. Pan[1], *C. Jordan*[1], *K. Makishima*[2], *R.A. Stern*[3], *K. Hayashida*[4], *M. Inda-Koide*[5]

[1] Department of Physics (Theoretical Physics), University of Oxford, UK
[2] Department of Physics, University of Tokyo, Japan
[3] Lockheed Palo Alto Research Laboratory, USA
[4] Department of Earth and Space Science, Osaka University, Japan
[5] Department of Electronics and Information, Toyama University, Japan

1 Introduction

During *Ginga* observations of galaxy 3C390.3 on 1991 February 14, a large impulsive flare was detected between 22:30-23:00 UT (Inda et al. 1994). Inda et al. (1994) suggested that the flare was associated with EQ1839.6+8002, an M4Ve star at a distance of 15.2 pc (Fleming et al. 1988). The star has been observed previously with the *Einstein* IPC and *EXOSAT* LE (Gioia et al. 1990; Giommi et al. 1991), and more recently with the *ROSAT* PSPC.

In this paper we present the results of an analysis of the observations of EQ1839.6+8002 with *Ginga* and of the data obtained with the *Einstein* IPC (between 1980 January and May), the *EXOSAT* LE (in 1984-1986), and the *ROSAT* PSPC instruments (in 1991 March, 1992 April and 1993 April).

2 The Ginga flare

Figure 1 shows the (2-20 keV) light curve obtained with the *Ginga* LAC between 20:50-23:03 on 1991 February 14. After the start of the flare (at $\sim$ 22:30), the flux increases rapidly from 28 to 172 counts s^{-1} in about two minutes, and then rises more slowly to 206 counts s^{-1} by 22:34. At the end of the observation, the flux is about 40 counts s^{-1}, nearly a factor of 2 above the pre-flare level.

The *Ginga* spectra, after subtracting the pre-flare X-rays, have been fitted with a one temperature Raymond-Smith (RS) model. Fig. 2 shows the time history of the derived plasma temperature T_e and volume emission measure $EM(V)$. The temperature reaches a maximum of $\sim 10.7 \times 10^7$ K within 2 minutes and then decays rapidly with an e-folding time $\sim$ 350 s until 22:37. After that it slowly decreases to $\sim 3.8 \times 10^7$ K by the end of the observation. The emission measure, however, continues to increase after the temperature peak, and reaches a maximum two minutes after the temperature. After the peak, $EM(V)$ gradually decreases (decay time $\sim$ 910 s) and has not returned to its pre-flare level by the end of the observation. The total energy released in the 2-20 keV energy band during the flare, after subtracting the pre-flare quiescent flux, is $\sim 10^{34}$ erg. At the flare peak, the total luminosity is $\sim 10^{31}$ erg s^{-1}.

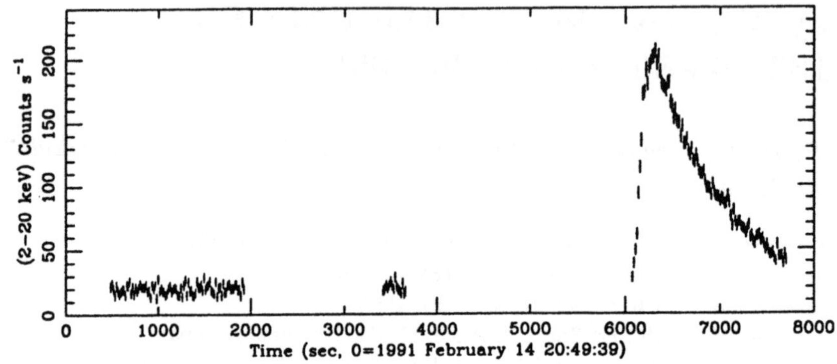

Fig. 1. The X-ray light curve before and during the flare on EQ1839.6+8002.

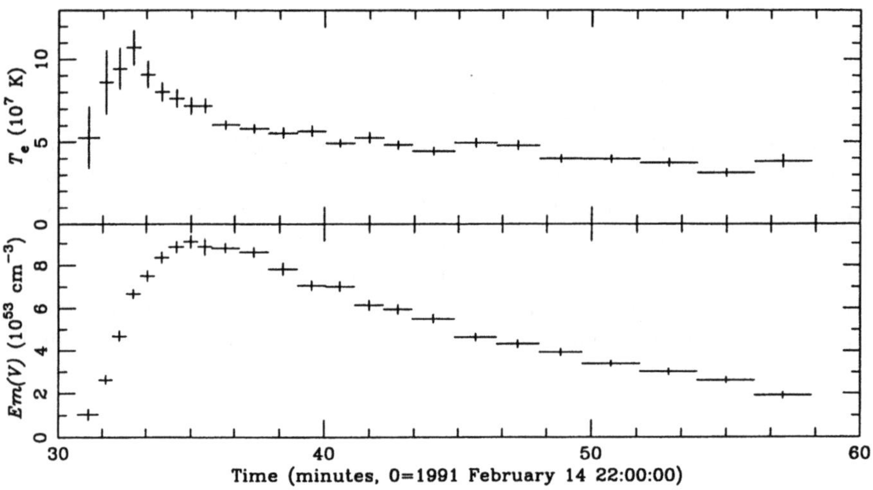

Fig. 2. The evolution of the flare temperature and emission measure.

The time lag between the peak temperature and emission measure illustrated in Fig. 2 strongly suggests that the flaring region of EQ1839.6+8002 relaxes back to the quiescent state by the 'evaporation/condensation' mechanism (e.g. Cheng & Pallavicini 1991). The temperature in the flaring loop increases when excess heating is being supplied. The high temperature results in an excess conductive flux into the transition region, causing chromospheric material to expand up the loop. Thus the emission measure can continue to increase after the heating has stopped or decreased. The density rise in the corona increases the radiative cooling since the radiative losses scale as the square of the density. The decrease in the emission measure may result from a decrease in the density as material flows back to the chromosphere, or to material cooling to temperatures not recorded by the 2-20 keV energy band.

3 The soft X-ray emissions

Figure 3 shows the (0.2-2 keV) luminosity of EQ1839.6+8002 derived from the observations made with the *Einstein* IPC, *EXOSAT* LE and *ROSAT* PSPC. The luminosity varies between 1.3×10^{27} erg s^{-1} to 2.4×10^{29} erg s^{-1}. The source was in a flaring state on 1980 April 8-10. The ratio of the peak to the quiescent flux is ~ 23. At the peak of the flare the star is the brightest X-ray source in the field of view. On 1985 November 7 EQ1839.6+8002 was the brightest X-ray source in the LE image. The source stayed in the high luminosity state ($\sim 5.7 \times 10^{28}$ erg s^{-1}) for over one and half hour without significant variation, which implies that a long duration flare might have occurred.

The soft X-ray spectra obtained with *Einstein* IPC and *ROSAT* PSPC have been modelled with a two temperature RS model. There is no significant differences between the temperatures of the flare and quiescent spectra. However, the emission measures of the low and high temperature components of the flare spectrum are about a factor of 10 larger than the quiescent emission measures. The mean spectral parameters in the quiescent state are $T_1 = 1.5 \times 10^6$ K, $EM_1 = 4.3 \times 10^{50}$ cm^{-3}, $T_2 = 7.6 \times 10^6$ K, and $EM_2 = 6.3 \times 10^{50}$ cm^{-3}.

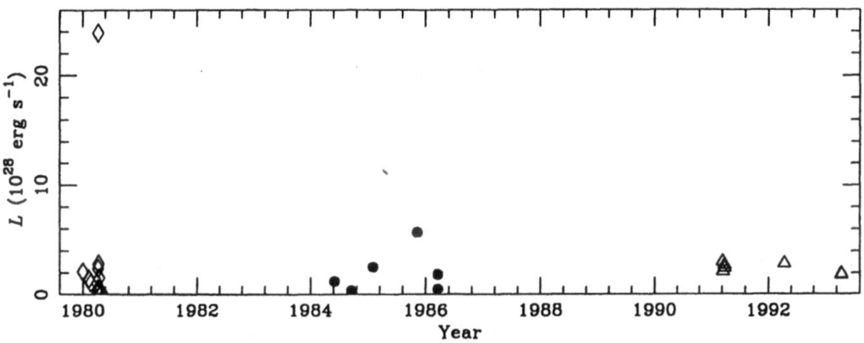

Fig. 3. The X-ray (0.2-2 keV) luminosity (L) of EQ1839.6+8002 from the measurements of the *Einstein* IPC, *EXOSAT* LE and *ROSAT* PSPC

4 Discussion

Using *EXOSAT* data (mostly from the LE instrument), Pallavicini et al. (1990) find that the X-ray quiescent luminosity and bolometric luminosity of dMe stars are correlated in terms of $\log L_X = -9.83 + 1.21 \log L_{Bol}$. The quiescent X-ray luminosity of EQ1839.6+8002 during the *ROSAT* observations (see Fig. 3) is $\sim 2.4 \times 10^{28}$ erg s^{-1}, which is consistent with the calculation using the above relationship and $L_{Bol} = 3.9 \times 10^{31}$ erg s^{-1}.

On the basis of the short rise time, the *Ginga* flare can be classified as an impulsive stellar flare, whose solar counterparts are associated with confined

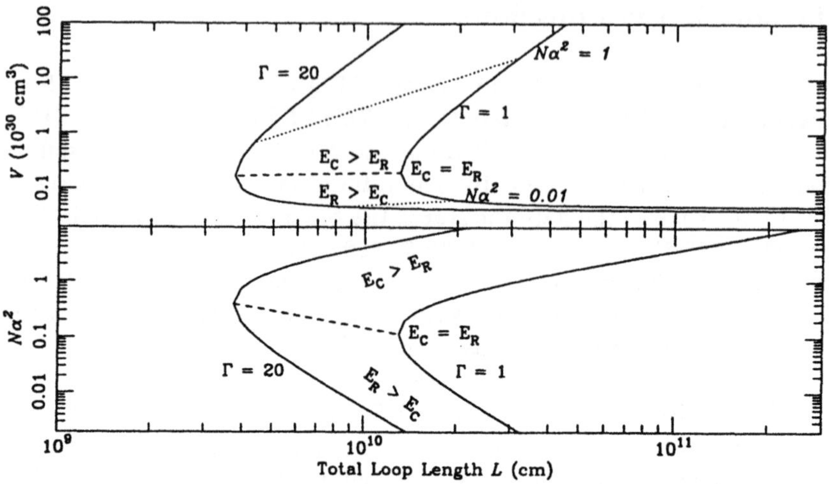

Fig. 4. The loop volume V and $N\alpha^2$ as a function of L for a loop with constant cross section, $\Gamma = 1$, and a loop with cross section expanding towards the apex, $\Gamma = 20$

magnetic loops. Using the observed time variability of the emission measure and temperature, we can put constraints on the loop volume and length.

We assume that the flare is generated by N identical flaring loops, each having a length L and expanding linearly towards its apex. If the loop aspect ratio is α and the ratio between the apex and footpoint cross-sections is Γ, the volume of the N loops can be expressed as $V = (\Gamma + 1)\frac{\pi}{8}N\alpha^2L^3$. In the case of the Sun, $N\alpha^2$ can range from 0.01, for a single loop with $\alpha = 0.1$, to nearly 1, for a very compact flare made of many loops (e.g. Pallavicini et al. 1990).

If there is no heating in the flare decay phase and the loop plasma loses energy mainly by radiation (E_R) and conduction (E_C), relationships between $N\alpha^2$, V and L can be derived from the loop energy equation. Figure 4 shows the derived value of $N\alpha^2$ and V as a function of the loop length for $\Gamma = 1$ and 20. If $N\alpha^2$ is in the range of 0.01-1, similar to that of a solar flare, we find 3.7×10^9 cm $< L < 3.1 \times 10^{10}$ cm, and 4.7×10^{28} cm^3 $< V < 2.4 \times 10^{31}$ cm^3 for $1 \leq \Gamma \leq 20$.

References

Cheng C.C., Pallavicini R., 1991, ApJ 381, 234
Fleming T.A., et al., 1988, ApJ 331, 958
Gioia I.M., et al., 1990, ApJS 72, 567
Giommi P., et al., 1991, ApJ 378, 77
Inda M., et al., 1994, ApJ 420, 143
Pallavicini R., Tagliaferri G., Stella I., 1990, A&A 228, 403

THEORY OF FLARES

Physical Processes During Impulsive Solar and Stellar Flares

Maria Katsova[1], *Moissei Livshits*[2]

[1] Sternberg State Astronomical Institute, Moscow State University, 119899 Moscow, Russia
[2] Institute of Terrestrial Magnetism, Ionosphere and Radio Wave Propagation, Russian Academy of Sciences, 142092 Troitsk, Moscow Region, Russia

1 Introduction

Investigations of impulsive flares on both the Sun and red dwarf stars during more than 30 years allow us to arrive at quite definite conclusions. Here we will consider impulsive events; on the Sun the impulsive phase of a flare is observed as a hard X-ray burst with the emission of photons with energies $E > 30$ keV up to the γ-ray range. At the same time microwave radio bursts, and sometimes UV and optical continuum bursts are registered. Typical durations of these processes are ~1–3 min. In this time interval other kinds of flare emission like soft X-ray (2–10 keV) emission, meter radio bursts and Balmer line emission begin to rise, but their maxima occur later on, in the gradual (thermal) phase of the flare.

Impulsive stellar flares are often observed as a significant increase in optical continuum, especially in the U-band, of similar duration (1–3 min), and this time interval is, like in the solar case, the rise phase of the soft X-ray emission.

Modern observations demonstrate that both the impulsive phase of a flare or an impulsive flare develops in low-lying loops. Earlier only indirect evidence existed in optical and radio data. Recently, however, the heights of the hard X-ray sources in impulsive solar events were determined directly from *YOHKOH*'s HXT (Kosugi 1994, Masuda 1994) (Fig. 1a). Statistically, the height of the hard X-ray source in the 14–23 keV range is 9700 ± 2000 km above the photosphere, and this height reduces to 6500 km in the 53–93 keV range. Besides two hard X-ray sources in the loop footpoints, a third hard X-ray source exists at the top of the loop at least in some cases. The authors of this experiment suppose that the appearance of this loop-top source is due to reconnection in the impulsive phase. Note that the reconnection begins close to the apex of the loop, when this loop is filled by hot plasma that evaporated from both footpoints.

Thus, the impulsive phase of large flares takes place at low chromospheric heights (< 10000 km), often close to a spot (Fig. 1b). In the process electrons are accelerated and impact on the chromosphere adjacent to the footpoints of the low-lying loop. The response of an atmosphere to the action by primary accelerated particles and by thermal and radiation fluxes is referred to as secondary processes in a flare and will be discussed separately below. The impulsive phase

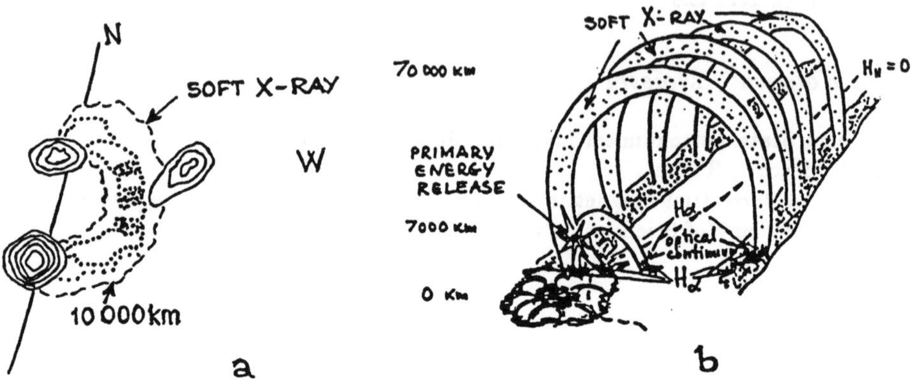

Fig. 1. (a) The brightness distribution in the 14–23 keV and the soft X-ray ranges according to Kosugi (1994) and Masuda (1994); (b) The system of flare loops: low-lying impulsive source and high coronal loop.

lasts 1–3min and later accelerated particles and hot plasma go into the higher coronal loops, and the main soft X-ray source is formed (Katsova & Livshits 1991).

During evolution of magnetic fields or the motion of one spot relative to others, or when new spots emerge, energy is stored gradually in the non-potential part of the magnetic field. In large flares magnetic tensions are relaxed and the magnetic field lines return toward a potential configuration as the excess energy is released. The process can be repeated up to a few times leading to series of homologous flares while active regions are on the solar disk.

In some cases the motions of magnetic hills or the emergence of magnetic flux can be observed, but as a rule it is difficult to detect these effects directly. It is clear, however, that the impulsive phase begins close to spots, often in their penumbrae, near to photospheric bridges or to a neutral line of longitudinal magnetic field. Zirin & Wang (1993a) emphasize that big flares occur in regions with very strong transverse magnetic field (1000 – 2000 G, e.g. on 4 June 1991 – 4000 G). The motion of photospheric footpoints does form extremely narrow elongated channels of intense horizontal field (Zirin & Wang 1993b). In such regions the strongest impulsive solar flares take place, like the events of June 1991.

Magnetic fields of an unipolar spot are more squeezed to the photosphere in the direction where they confine the points of the umbrae and bright points of the chromospheric network, i.e. in these azimuths the "canopy effect" is observed (Zhang 1994). Magnetic force lines of other azimuths of an unipolar spot extend up to large heights, like in the case of the field of a vertical dipole. It is quite possible that flare activity is enhanced in those places where this "canopy" effect is present.

At the onset of an impulsive flare particles, particularly electrons, are accelerated. This primary acceleration takes place in a very small volume. The total energy of fast particles exceeds the energy of the magnetic field in this volume, therefore the question remains (see also van den Oord 1993): how does the globally stored energy dissipate locally?

Energy storage can occur in large-scale current flows. The current intensities and changing of electric conductivity and other parameters provide the power, and the duration of the acceleration can be estimated using an equivalent circuit model (Spicer et al. 1986, Stepanov & Zaitsev 1992, Melrose, 1994). However, it is required that this theory be able to describe such a process in the real solar atmosphere, not only in ideal circuits.

The details of the process of primary energy release are still unknown. X-ray data provide evidence for extremely fast and very effective particle acceleration. From a theoretical point of view a trigger mechanism is required and can be connected with an instability of a ring current (like the pinch-effect); it is necessary to provide such a mechanism of particle acceleration operating for a few seconds at least.

At present, it is becoming clear that both in the Earth's magnetosphere and in the magnetospheres of solar active regions several mechanisms of particle acceleration should be taken into account. Strong non-linear plasma effects, like solitons, can be significant.

Outstanding problems are: to give a theory of the evolution of particle beams in coronal magnetic tubes, to explain the radio spectra in the entire range from meter to millimeter, and to explain the flat spectra of microwave type-IV radio emission. If we could know how the energetics and the angular distribution of particles change in a magnetic trap, we could compute the gyrosynchrotron radiation of accelerated particles to explain brightness temperatures as high as $3 \cdot 10^9$ K.

2 Accelerated particles in impulsive solar flares

There are three main types of observations from which we can derive the properties of accelerated particles: hard X-ray and γ-ray bursts and direct registration of accelerated particles in interplanetary space and at the Earth. The differential spectrum of the X-ray emission with $E > E_1 = 15$ keV can be described as

$$I(h\nu) = I_0(h\nu)^{-\delta} \quad \text{photons/cm}^2\text{s keV} \tag{1}$$

where the values of the fluxes I, I_0 and the index δ of the power law depend on the time.

The solution of the Abel integral equation in the Born-approximation for the differential cross-section of bremsstrahlung radiation allows us to determine the number of electrons in the emission source:

$$N_e(E,t) \approx 10^{42} \frac{I_o}{n} \frac{(\delta-1)\Gamma(\delta-1/2)}{\Gamma(\delta-1)Z^2} E^{-(\delta-1/2)} \quad \text{electrons/keV} \tag{2}$$

where n is mean ion density in the emission source, $Z^2 \approx 1.8$ for cosmic plasmas and $\Gamma(x)$ is the Γ-function.

From the solution of the equation of continuity for accelerated particles the particle spectrum in the region of the acceleration is

$$F_o(E,t) \approx 5 \cdot 10^{-9} \delta n E^{-3/2} N_e(E,t) \quad \text{electrons/s keV} \tag{3}$$

These equations are based on Brown's (1971) pioneering paper and some later considerations (Syrovatskij & Shmeleva 1972, Korchak 1976).

Using Eqs. 1–3 it is possible to derive the flux of electrons with $E > 15$ keV as 10^{36} electrons/s for a typical solar hard X-ray burst; the total number of electrons accelerated in a large impulsive solar flare is 10^{38}. From here the electron energy input to the low-lying chromosphere is $1 - 3 \cdot 10^{11}$ erg/cm^{2}s, for bright points of area $\sim 10^{17}$ cm^2.

During some flares significant numbers of electrons are accelerated to 10MeV, as observed with the gamma-spectrometers on *SMM* and *YOHKOH*. Besides bremsstrahlung radiation of accelerated electrons, a number of nuclear lines are detected with the same *SMM* and *YOHKOH* equipment. To explain the appearance of these lines it is necessary to accept that 10^{33} protons are accelerated up to energies $\geq$30 MeV (Ramaty 1986). These same observations give evidence, apparently, for the existence of a cut-off in the proton spectrum beyond 700 MeV (Alexander et al. 1994).

Analysis of solar cosmic ray increase supports both the above conclusions concerning the proton spectrum. So, large number of protons, up to $2 \cdot 10^{33}$ with $E > 30$ MeV were accelerated in the solar flare on 24 May 1990 and in the impulsive phase of the flare on 15 June 1991 (Belov & Livshits 1995, Belov et al. 1994). These data contain the only hint of the presence of a cut-off of the proton spectrum at 700 MeV.

The proton spectrum in the flare of 29 September 1989 which occurred behind the solar limb does extend up to very high energies. Here the impulsive phase could not be observed directly and the registered accelerated particles originated from the very high corona. For other flares with an expressed impulsive phase, the proton spectrum beyond 700MeV turns out to be sharper (Fig. 2 from Belov & Eroshenko 1995). At the moment this problem is under more detailed investigation.

However, it is possible to propose the following preliminary conclusion for the most powerful solar flares. The length of the current sheet even in the most powerful impulsive solar flares is not sufficient to provide acceleration of particles up to energies of a few GeV, while in some events, after the pulse, the current sheet can reach a solar radius in length. Then the particles can be accelerated up to energies of 10^{10} eV.

Modern information on accelerated particles is sufficient for a correct estimation of the production of light elements during flares.

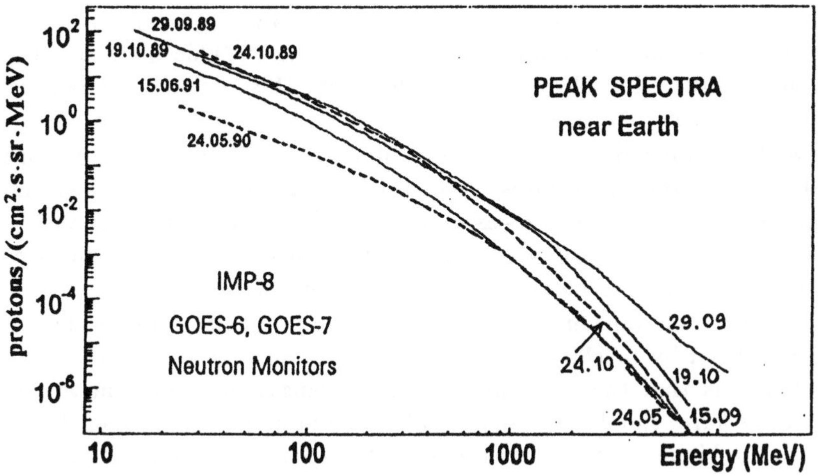

Fig. 2. The proton spectrum observed near the Earth at the maximum of the largest enhancement of solar cosmic rays during solar cycle XXII.

3 The response of solar and stellar chromosphere to impulsive action

The first discussion of the analogy between solar and stellar flares was that of Gershberg & Pikel'ner (1972). Attention was drawn to the problem of the primary energy release which on both the Sun and the stars is the result of magnetic field evolution. The paper by Kostyuk & Pikel'ner (1975) was fundamental in the serious study of secondary processes in solar flares. This question was further investigated in detail by Fisher (1986, and references therein), and general results of gas-dynamical simulations were reproduced by ten groups (Kopp et al. 1989). As a source of heating one more often considers accelerated electrons, although both protons and EUV or soft X-rays are able to make significant contributions in some cases.

The different influences of particle beams and the heat fluxes from action of X-rays need to be emphasized: the accelerated particles and heat fluxes propagate along magnetic field lines and lead to the appearance of soft X-ray and optical emission in separate bright points (close to the loop footpoints). Flaring coronal X-ray emission illuminates all the low-lying area and results in a diffusion "halo" source of the UV and optical radiation.

The process of chromospheric response to impulsive heating distributed over column mass, i.e. the process of explosive evaporation, develops on the Sun in the following way. In the first tenth of a second the upper chromosphere is heated suddenly without noticeable change of total density, because motions are still absent. Two disturbances propagate, upwards and downwards, from the resulting zone of high pressure. One second after the beginning of heating the quasi-stationary plasma flow moves downwards and a shock wave forms. The gas behind this radiative shock front is compressed by about a factor of 100 and a geometrically thin emission source, with $T = 10^4$ K and high density, forms. This source is called "the low-temperature condensation". This disturbance propagates downwards. Hot plasma evaporates from this region, and becomes the source of soft X-ray emission.

The effects predicted by the theory of gas-dynamic processes are observed directly in impulsive solar flares. First, the expected outflow of hot gas into the corona during the hard phase was detected with *SMM* (Antonucci et al. 1982). Second, the properties of the Hα emission in bright points in the impulsive phase agree well with those calculated in models (Bruevich 1995).

The first calculations of explosive evaporation in the atmosphere of red dwarfs were carried out by Livshits et al. (1981) and Katsova et al. (1981). Their main conclusion was that the low-temperature condensation formed in this process during large flares should emit in optical continuum. This feature of the solution is due to the higher density of red dwarf atmospheres as compared to the solar case.

Recently we carried out a new modeling of this gas-dynamical process for flares on AD Leo, using modern numerical methods (Boiko & Livshits 1995, Katsova et al. 1995). The system of equations of gas-dynamics for $2T$- and one-fluid plasma includes heating flux, radiative energy losses, thermal conduction flux and the function of the energy change between the electron and ion component of plasma. Previous experience showed that the usual Lagrangian coordinates should not be used, and we adopted the motions of the sites of the numerical grid in order to compute exactly the shock wave and the thermal conduction fronts. This numerical method of solution is similar to the mixed Euler-Lagrangian approach.

As an initial model we used the Hawley-Fisher model for the chromosphere and transition region and then added coronal layers. We supposed that the chromosphere is heated by an electron beam of energy of $3 \cdot 10^{11}$ erg/cm^{2}s and the temporal profile of the heating flux has a rectangular form. The computations have been carried out for one elementary burst lasting 10 sec. Real stellar flares can be represented as a set of such bursts. The propagation of the low-temperature condensation downwards through the red dwarf chromosphere is shown in Fig. 3.

This gas-dynamical computation allows us to explain many features of simultaneous observations of flares in different spectral ranges. Optical continuum emission of the low-temperature condensation has a thermal origin. This is supported at the present by the UV-continuum observations. The area of optical flares is about of $(1 - 5) \cdot 10^{18}$ cm^2. The temperature of the source of optical emission is in the range 8000–15000 K. It is worth noting that the temperature of the condensation as a whole should be determined by taking into account the absorption therein of the soft X-ray emission. This effect was analyzed by Hawley & Fisher (1992).

Some features such as, e.g., the evolution of the Balmer decrements and the appearance of broad components of the line profiles, can be explained in the framework of this model. Part of the line emission is caused by radiation of the

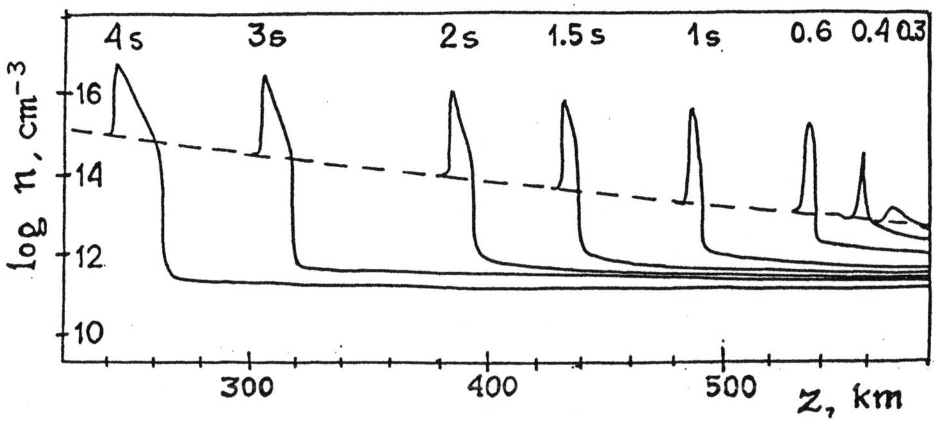

Fig. 3. The density distribution versus the height in the stellar chromosphere. The null height is photospheric level where $\tau(5000\text{Å}) = 1$. The time in seconds after the beginning of heating is indicated. (Katsova et al. 1995).

low-temperature gas above the flare region.

UV (1000-2000 Å) emission should accompany the beginning of each elementary process of explosive evaporation (Katsova & Livshits 1989). Furthermore, we found one new feature of our solution: regions with high pressure arise from time to time close to the thermal conduction front due to specific conditions of the heat propagation. The motions restore the previous flow. However, the process of isruption of the hot-cool gas boundary should be accompanied by additional UV emission. Thus, the EUV light curve should show a set of spikes of short duration.

It is also worth noting that the evolution of the source of soft X-ray emission can be explained as well.

Intergrated over the duration of a flare, energies emitted in the X-ray, UV and optical spectral range should be close to each other. This theoretical conclusion is in agreement with observations.

4 Concluding remarks

We are happy to understand something about the nature of solar and stellar flares, for instance, the thermal origin of the optical continuum emission. However, a number of questions are far from a final solution. Among other things:
– the real relationship between magnetic field motions in or above spots and the impulsive flare onset,
– high energy processes and the interpretation of radio emission of stellar flares,
– verification of the presence of short UV spikes in solar and stellar flares.

References

Alexander D., Dunphy P.P., MacKinnon A.L., 1994, Solar Phys. 151, 147
Antonucci E., Gabriel A., Acton L., et al., 1982, Solar Phys. 78, 107
Belov A.V., Eroshenko E.A., 1995, Nuclear tracks and radiation measurements (in press)
Belov A.V., Eroshenko E.A., Livshits M.A., 1994, Proc. 8th Int. Symp. on Solar-Terrestrial Physics, Sendai, Japan Pt. 1, 26
Belov A.V., Livshits M.A., 1995, Sov. Astron. Lett. 21, 1
Boiko A.Ya., Livshits M.A., 1995, Astron. Zh. 72, (in press)
Brown J., 1971, Solar Phys. 18, 489
Bruevich E.A., 1995, Astron. Zh. 72, No. 1
Fisher G.H., 1986, Radiation Hydrodynamics in Stars and Compact Objects, Lect. Notes in Physics, 255, 53
Gershberg R.E., Pikel'ner S.B., 1972, Comments Ap. Space Phys. 4, 113
Hawley S.L., Fisher G.H., 1992, ApJS, 78, 565
Katsova M.M., Livshits M.A., 1989, Sov. Astron. 33(2), 155
Katsova M.M., Livshits M.A., 1991, Sov. Astron. 35(1), 65
atsova, M.M., Boiko A.Ya., Livshits M.A., 1995, (in prep.)
Katsova, M.M., Kosovichev A.G., Livshits M.A., 1981, Astrofiz. 17, 285
Kopp R.A., Fisher G.H., MacNice P., et al., 1989, Energetic Phenomena on the Sun, Kundu M.R., Woodgate B., Schmahl E.J. (eds.), Kluwer, Dordrecht, p. 601
Korchak A.A., 1976, Astron. Zh. 53, 370
Kostyuk N.D., Pikel'ner S.B., 1975, Sov. Astron. 18, 590
Kosugi T., 1994, New Look at the Sun with Emphasis on Advanced Observations of Coronal Dynamics and Flares. Proc. of Kofu Symp. NRA Rep. No. 360, Enome S. & Hirayama T. (eds.), p. 11
Livshits M.A., Badalyan O.G., Kosovichev A.G., Katsova M.M., 1981, Solar Phys. 73, 269
Masuda S., 1994, PhD Thesis. Univ. of Tokyo
Melrose D.B., 1994, New Look at the Sun with Emphasis on Advenced Observations of Coronal Dynamics and Flares, Proc. of Kofu Symp. NRA Rep. No. 360 eds. Enome S. & Hirayama T. (eds.), p. 235
van den Oord G.H.J., 1993, Advances Space Res. 13, No. 9, 143
Ramaty R., 1986, Physics of the Sun, Sturrock, P. et al. (eds.) Reidel, Dordrecht, Vol. 2, 291
Spicer D.S., Mariska J.T., Boris J.P., 1986, Physics of the Sun, Sturrock, P. et al. (eds.) Reidel, Dordrecht, Vol. 2, 181
Stepanov A.V., Zaitsev V.V., 1992, Solar Phys. 139, 343
Syrovatskij S.I., Shmeleva O.P., 1972, Astron. Zh. 49, 334
Zirin H., Wang H., 1993a, Solar Phys. 144, 137
Zirin H., Wang H., 1993b, Nature 363, 426
Zhang H., 1994, Solar Phys. 154, 207

The Role of Flare Stars in Cosmic-ray Origin

Maurice M. Shapiro

University of Maryland, College Park, MD, USA*

Abstract: Supernovae and their expanding shock fronts are evidently the main agents of cosmic-ray acceleration. The thermal gas in the interstellar medium has been regarded as the reservoir of seed particles destined to become cosmic-ray nuclei. This assumption is, however, at variance with the source composition of galactic cosmic rays. In an alternative hypothesis, the seed particles are injected into the interstellar material as suprathermal seed ions, and it has been surmised that flare stars provide the initial boost. We find that the dMe and dKe stars are probably the principal sources of cosmic-ray seed particles. Most stars in the Galaxy are red dwarfs and many of these flares much more powerfully and frequently than solar flares. Augmenting the optical data, recent X-ray and far-ultraviolet observations now permit a better estimate of the energy budget. Altogether, dMe and dKe stars seem to be the most promising class of cosmic-ray injectors.

1 Introduction

According to well-founded theory, most of the galactic cosmic rays (GCR) are accelerated to their high energies by shock waves in the interstellar medium (ISM). This theory has generally assumed that the shock fronts, generated mainly by supernova explosions, pick up the particles directly from the thermal pool of ions in the ISM gas. This seems to imply that no special injection of suprathermal seed particles is required prior to their main acceleration. There are compelling reasons to question this view. These are based on a comparison of the element abundances in the ISM gas with the source composition inferred from (a) observations of the "arriving" GCR composition in the vicinity of Earth, and (b) analysis of CR propagation in the Galaxy.

We represent an alternative model – one that yields better agreement with observations (Shapiro 1987, 1990). It invokes the injection of energetic seed ions from stellar flares, and especially those of dMe and dKe stars, as a precursor to the main accelerating process. The model is supported by examination of the relevant physical parameters.

* Address for correspondence: 205 Yoakum Pkwy 2-1514, Alexandria, VA 22304

In Sect. 2, we summarize some of the earlier ideas on cosmic ray origin (Fermi 1949, Ginzburg 1951, Shklovsky 1953, Shapiro 1962, Ginzburg and Syrovatsky 1964). Then we discuss the more recent theory of shock-wave acceleration. Sect. 3 explains why injection of suprathermal seed particles is required. Sect. 4 describes the hypothesis of injection from stellar flares, comparing these to solar flares. It focuses on the dMe and dKe dwarfs, as these are likely to be the dominant contributors of energetic seed ions for shock acceleration to GCR.

Sect. 5 discusses the energy budget of the cosmic-ray reservoir in the Galaxy, and the available supply of stellar seed ions to feed shock acceleration. It attempts to evaluate the various parameters that determine the adequacy of the particle-energy supply from stellar flares. Sect. 6 deals with the question as to whether emitted particles will meet a supernova shock. Sect. 7 considers the possible role of other types of flare stars as injectors. It also mentions the possible contribution of stellar winds. Finally, Sect. 8 discusses some related issues, and summarizes our conclusions.

2 Magnetohydrodynamic theories of cosmic-ray origin

Fermi's seminal proposal for cosmic-ray acceleration (Fermi 1949, Chandrasekhar & Fermi 1953) were based partly on principles of magnetohydrodynamics developed by Alfvén (1981) and others. In Fermi's first scheme – the statistical mechanism – acceleration is produced by collisions of particles with magnetized clouds in the ISM. A notable success of this "second-order theory" is that it gave a power-law energy spectrum. It proved, however, to be too slow; moreover, it did not work for heavy CR nuclei. Fermi then developed another theory involving special magnetic-field configurations in the ISM; these resembled the mirror-machines subsequently developed for controlled fusion devices.

Central to Fermi's concepts, as well as to recent schemers of shock acceleration, was the mechanism of energization by particle collisions with moving magnetic "walls" that reflect and accelerate ions. Hence subsequent theories (e.g. Krymsky 1977, Bell 1978, Eichler 1979, Axford et al. 1977, Blandford & Ostriker 1978) are still called "Fermi acceleration" even though he did not invoke shocks.

Given a strong shock front that compresses the gas in its path, and the diffusion of particles between scattering centers in the ISM, ions are repeatedly reflected back and forth, gaining energy at each step. The resulting energy spectrum conforms to a power law E^{-2}, consistent with the observed CR spectrum, after correction for the preferential escape of particles with the higher rigidities (Cesarsky 1983). A high-energy cutoff ($\approx 10^{14}$ eV) is predicted for shock acceleration in the Galaxy (Lagage 1983).

It was formerly supposed that CR are literally born in processes of stellar collapse that produce supernovae (SN), i.e. that the CR precursor particles are picked up from material ejected into the SN remnant (Ginzburg 1951, Shapiro 1962, Ginzburg & Syrovatsky 1964). Today it still seems possible that some relativistic nuclei and electrons could arise in this way. (Here we shall be dealing with the CR nuclei.) For nearly two decades acceleration theory has continued to regard SN explosions as the principal "machine", i.e. as the energy source, through the action of shock fronts at the leading edge of expanding SN shells. However, in current models for the origin of the preponderant bulk of cosmic-rays (say, up to 10^{13} eV/amu[2]) the atomic nuclei that are ultimately promoted to CR are assumed to reside in the ISM, and there to be accelerated by the shocks they encounter[3].

In Sect. 1 we mentioned that in the theories of acceleration by shock waves it has been tacitly (if not always explicitly) assumed that the seed-ions reside in the background of thermal gas in the ISM, i.e., that they are directly picked up from the thermal pool by the shock fronts. While this possibility cannot yet be ruled out, we are here arguing that injection is ordinarily required for the acceleration of the GCR. As will be seen, we invoke a promising source of suprathermal ions for the seed particles.

3 The observed composition of GCR constrains the source of seed ions

The source composition of the GCR is well correlated with the preferential selection – prior to injection or acceleration – of those heavy ions with relatively lower (< 10 eV) first ionization potentials (FIP). This mechanism works well at temperatures in the atmospheres of M and K dwarfs. The ions are transported to the corona, where they are energized to MeV and keV energies in flares. Selection by FIP favors a scenario of origin of the CR progenitor ions in stellar envelopes, and their subsequent injection into the ISM, rather than direct shock acceleration of thermal particles in the ambient ISM. Most of the volume of the ISM is in the hot phase, at $\approx 10^6$ K – a temperature too high for the observed FIP selection.

Finally we note that while shocks can energize thermal ions such as those in the ISM, the efficiency for picking up nuclei with MeV energies is greater than that for thermal ones. This, by itself, is not a cogent argument for injection. It is rather the compositional constraints cited above that indicate a need to inject seed particles from another source.

4 Injection from stellar flares

Solar flares produce energetic particles that have, in fact been called solar cosmic rays. Here we shall refer to them as solar energetic particles (SEP); they differ significantly from the GCR, especially in their energy spectra, although the

[2] It is noteworthy that $< 10^{-4}$ of the CR nuclei have energies exceeding 10^{12} eV/amu. The electron component, which is not discussed here, contributes less than 2% of the arriving CR.

[3] Of course, it is recognized that most of the heavy atoms ("metals") originally came into the ISM, and thence into second-generation stars, from the debris of SN explosions

typical SEP composition has much in common with that of the CR sources. The latter have a relatively flat spectrum, whereas the SEP spectra are very steep, with a preponderance of MeV particles. It has long been known that late-type main sequence stars, notably M and K dwarfs, display prolific flaring activity. In particular, the dMe and the dKe stars, so designated because of their prominent emission lines, have frequent flares (Gershberg 1988, Shakhovskaya 1995). The exceptionally strong flares that occur only once in several years in the Sun seem commonplace among dMe and dKe stars. We shall usually be referring to these dwarfs when we use the terms "flare stars" or "stellar flares", although there are other active stars – far less numerous – that show conspicuous flaring activity. As to solar flares, it is noteworthy – and hardly the result of pure coincidence – that their SEP exhibit the same FIP selection as described in Sect. 3 for the GCR (Meyer 1985).

5 Energy budget for injection

In assessing the adequacy of stellar flares as injectors, we compare the power density $W - \mathrm{i}$ available for injection from flares, with the power density W_{cr} of the GCR. In so doing, we confront the following questions: What is the rate W_{em} of energy emission in electromagnetic radiations (UV, X-ray, and visible light) per unit volume of the ISM? How much of the flare energy is likely to go into the production of energetic nuclei? What fraction f_{i} of these escape into the ISM? If SN shocks (and other shocks) boost the ions to CR, does the rate of supply of injected particles suffice to meet the requirements imposed by the observed CR energy density, lifetime, and mean particle energy? Further, once the ions escape from a flare, what is their potential slowing-down time t? Within that time, is it likely that the particle will encounter at least one shock front in the ISM?

Many of these questions do not yet have firm answers; only rough estimates are available for some of the parameters, while reasonable limits can be assigned to others. Despite disparities in published data (a dilemma not unknown to astrophysics) we try, by adopting conservative values and/or lower limits, to arrive at tenable conclusions.

Let us estimate the power density W_{i} introduced above. For the *average* flare, we define:

E_{em} is the energy in the form of electromagnetic radiation (visible + UV + X-ray),

ν_{f} is the corresponding flare frequency,

E_{p} is the flare energy that goes into ion production,

E_{i} is the energy in ions that escape from a flare,

f_{i} is the ratio $E_{\mathrm{i}}/E_{\mathrm{p}}$.

Observations and theory suggest that particle acceleration is intimately tied to the emission of radiation. Hence, lacking definitive data, we *assume* in our model that

$$E_{\mathrm{p}} \approx E_{\mathrm{em}} \,. \tag{1}$$

Now let W_p be the power density for the production of fast ions in flares. Let W_i be the power density in fast ions *escaping* from flares, and n_d be the mean number density of dMe and dKe stars in the Galaxy.

Based on data in the flare literature (some of which allow considerable leeway), we provisionally adopt values of the parameters that probably yield an underestimate[4] of W_i. These values are based in part on the extensive flare observations of Gershberg (1988) and Shakhovskaya (1995):
$n_d = 0.04$ pc^{-3},
$E_{em} = 10^{34}$ erg,
$\nu_f = (100 \text{ h})^{-1} = (3.6 \times 10^5\text{s})^{-1}$.
$f_i = 0.15$.
The power density in emitted energetic ions is:

$$W_i = n_d \, E_{em} \, \nu_f \, f_i = 1.7 \cdot 10^{26} \text{ erg pc}^{-3} \text{ s}^{-1} \,. \tag{2}$$

Let W_{cr} be the power density in GCR.

$$\begin{aligned} W_{CR} &= \text{energy density/lifetime} \\ &= 1 \text{ eV cm}^{-3}/1.3 \cdot 10^7 \text{ y} \\ &= 1.15 \cdot 10^{29} \text{ erg pc}^{-3} \text{ s}^{-1} \,. \end{aligned} \tag{3}$$

Define $\Gamma \equiv W_i/W_{CR}$. Then,

$$\Gamma \approx 1.5 \cdot 10^{-3} \,. \tag{4}$$

Thus, the power required – and available for injection – is a small fraction of that needed for replenishment of the GCR as these escape from the CR reservoir in the Galaxy.

6 Will the emitted particles meet a shock?

Next we estimate the probability P that the ions will meet a strong shock, i.e. one sufficiently powerful to accelerate them to CR energies. This depends on the time t_S available for such encounters, and on the fraction of the Galactic volume that contains the expanding envelopes of SN remnants.

In evaluating t_S, we use the range-energy tables of Northcliff & Schilling (1970). First, however, we note that the mean energy of the emitted ions is obtainable from the value of Γ in Eq. 4. Let w_i be the mean particle energy of the seed ions, and w_{cr} be the mean energy of a GCR nucleus, ≈ 6 GeV. Then $w_i = \Gamma w_{CR}$. With Eq. 4, $w_i \approx 9$ MeV. Further, to compute the range, we adopt a mean value of 0.1 H atom cm^3 for the number density in the ISM, most of which consists of rarefied gas in the hot phase. The slowing-down time t_s for a 9 MeV ion will then be ≈ 12 million years – about the same as the lifetime of the CR against escape from the Galaxy.

[4] In Sect. 8 we refer to an independent study which suggests that our provisional estimate of W_i is a lower limit.

With a SN rate of $(50\ \mathrm{y})^{-1}$ in the Galaxy, there will be 2.4×10^5 SN outbursts in $1.2 \cdot 10^7$ years. The total volume V_{sn} filled by their remnants, each within a radial distance r_s of 70 pc is $4.3 \cdot 10^{11}$ pc^3. (At $r > r_s$, the shocks become weak).

To evaluate P, we estimate the volume V_g of the Galaxy with a disk thickness of 10^3 pc. This yields $V_g \approx 5 \cdot 10^{11}$ pc^3. Within the uncertainties in the values of V_{sn} and V_g, P is of the order unity. Thus it appears that the seed ions are likely to meet a shock.

7 Other flare stars; stellar winds

We have devoted this paper mainly to CR injection from dMe and dKe stars, owing to their overwhelming numbers and to the frequency of their intense flares. Other flaring stars are widely discussed (e.g. Haisch & Rodonò 1989, Haisch et al. 1991, Wallerstein 1990), and their features are described elsewhere in these Proceedings. Their numbers are small compared with those of red dwarfs: yet they may add to the pool of seed particles, provided their atmospheres are conductive to ion selection yielding the FIP effect. We make no effort here at quantitative estimates of their net effect.

Stellar winds consist of much slower particles than the ions from stellar flares. Nevertheless, it seems possible that, under appropriate conditions, they might also provide some CR seed particles. This topic is beyond the scope of the present discussion.

Probably we have not established a conclusive argument for the adequacy of the red dwarf flares as the principal injectors of GCR (see remarks on the uncertainties below). It should be noted, however, that the case for requiring a suitable source of injected particles is compelling, as shown in Sect. 3.

8 Discussion and summary

The idea that there might be a connection between CR and stellar emissions is not new[5] (Unsöld 1957, Edwards & McQueen 1977). That flares on M and K stars might directly produce the lower-energy ($< 10^9$ eV) CR was proposed by Lovell (1974). Later, Mullan (1979) concluded that flare stars cannot generate the *total* CR energy density in the Galaxy. Our analysis supports his result. However, this conclusion by no means precludes the more modest role that flare stars seem to play as *injectors*. Based on early X-ray observations, Gorenstein (1981) favored a hypothesis similar to Lovell's. Arguments were adduced by Cassé ((183) and Meyer 91985) to show that the nature of the CGR source composition is inconsistent with the direct acceleration of thermal ISM material. As to the stellar injection of GCR, the first attempt to go beyond speculation, and to treat the problem quantitatively, was made by Shapiro (1990). The present paper is an updated version, and a fuller treatment of that earlier work.

[5] Today it seems curious that the Sun itself was once regarded by a few eminent physicists as *the* source of CR.

We have avoided assigning figures of merit to the parametric values used here. In the present state of our knowledge of M-dwarf flares, estimates of error for most of the parameters would be rather subjective. As an example of the difficulties in observing and measuring the properties of faint stars like red dwarfs, consider the value assigned to the electromagnetic energy emission from dMe and dKe stars. The value in optical emission is based mainly on extensive observations (Gershberg 1988, Shakhovskaya 1995). However, our tentative estimate that UV and X-ray emission each contributes energy comparable to that in the optical channel is based on a limited number of *simultaneous* detections of flares at the various wavelengths. Clearly the simultaneous observations from space telescopes required for the X-ray and UV emissions from flares occurring at unpredictable times on faint stars, present logistical problems; the data are accumulating slowly. Moreover, our assumption (Eq. 1) that the energy going into production of flare particles is comparable to that in electromagnetic radiation is based largely on theoretical considerations rather than on available data. Our value for f_i, the fraction of ions that escape from a flare is perforce very uncertain. The more reliable figures we have used are those for the GCR, e.g., their energies and lifetimes. At worst, our numerical estimates may be subject to Ginzburg's famous "law of astrophysics", i.,e. $1 \approx 10$.

On the other hand, in discussing our earlier paper (Shapiro 1990), Matthews et al. (1991) asserted that we had underestimated the power density available in injected ions by a very large factor. They based their opinion on analysis of ROSAT X-ray data. If they are correct, energetic seed ions from flaring red-dwarf stars would exceed the requirements for CR injection, irrespective of the uncertainties cited above. It would certainly bolster my belief that the numerical estimates I adopted have been conservative.

In *summary*, we conclude that dMe and dKe stars contribute significantly to the injection of GCR. The output of their flares can provide seed particles for acceleration to CR energies, mainly by SN shocks in the ISM. These red dwarfs are unlikely to be the only injectors of the GCR. Yet they seem to be important sources of seed ions since (a) they flare frequently and powerfully, emitting radiations in the various electromagnetic channels; (b) their atmospheres can provide FIP selection that satisfies the GCR source composition; (c) the required energy budget for injection is modest; and (d) once emitted, the ions are likely to meet a strong shock in the ISM. Accordingly, given a need for injection of the GCR, we believe that we have made a plausible case for red dwarf stars as injectors.

Acknowledgments: Among colleagues who have kindly shared their expertise with me are R.E. Gershberg, N. Shakhovskaya, E.N. Parker, M. Rodonò, R. Pallavicini, R. Rosner, and D. Reames. To R. Silberberg I am particularly indebted for helpful discussions.

References

Alfvén H., 1981, Cosmic Plasmas, Reidel, Dordrecht

Axford W.I., Lear E., Skadron G., 1977, Proc. 15th International Cosmic Ray Conf. 11, p. 132

Bell A.R., 1978, MNRAS 182, 147
Blandford R.D., Ostriker, J.P., 1978, ApJ 221, L129
Cassé M., 1983, in Composition and Origin of Cosmic Rays, M.M. Shapiro (ed.), Reidel, Dordrecht, p. 193
Cesarsky C.J., 1983, in Composition and Origin of Cosmic Rays, M.M. Shapiro (ed.), Reidel, Dordrecht, p. 161.
Chandrasekhar S., Fermi E., 1953, ApJ 118, 113
Edwards P.J., McQueen M., 1971, Proc. 12th International Cosmic Ray Conf. 1, p. 323
Eichler D. 1979, ApJ 229, 415
Fermi E. 1949, Phys. Rev. 75, 1169
Gershberg R.E., 1988, in Solar and Stellar Flares, Proc. IAU Coll. 104, Kluwer, Dordrecht, p. ...
Ginzburg V.L., 1951, Dokl. Akad. Nauk SSSR 76, 377
Ginzburg V.L., Syrovatsky S.I., 1964, The Origin of Cosmic Rays, Pergamon, London
Gorenstein P., 1981, in Proc. 17th International Cosmic Ray Conf. 12, p. 99
Haisch B.M., Rodonò M. (eds.), 1989, Solar and Stellar Flares, Proc. IAU Coll. 104, Kluwer, Dordrecht
Haisch B.M., Strong K.T., Rodonò M., 1991, ARAA 29, 295
Krymsky G.F., 1977, Dokl. Akad. Nauk SSSR 234, 1306
Lagage P.O., 1983, in Composition and Origin of Cosmic Rays, M.M. Shapiro (ed.), Reidel, Dordrecht, p. 169
Lovell B., 1974, Phil. Trans. Roy. Soc. London A277, 489
Matthew L., Quenby J.J. Sumner T.J., 1991, Proc. 22th International Cosmic Ray Conf. 2, p. 459
Meyer J.-P., 1985, ApJS 57, 173
Mullan D., 1979, ApJ 234, 588
Northcliff L.C., Schilling R.F., 1970, Nuclear Data Tables A7, 233
Shakhovskaya N.I., 1995, these proceedings (and priv. comm.)
Shapiro M.M., 1962, Science 135, 175
Shapiro M.M., 1987, Proc. 20th International Cosmic Ray Conf. 2, p. 260
Shapiro M.M., 1990, Proc. 21th International Cosmic Ray Conf. 4, p. 8
Shklovsky I.S., 1953, Dokl. Akad. Nauk SSSR 90, 983
Unsöld A., 1957, in Radioastronomy, Proc. IAU Symp. 4, H.C. van de Hulst (ed.), Cambridge Univ. Press, Cambridge, p. 242
Wallerstein G. (ed.), 1990, Cool Stars, Stellar Systems, and the Sun, Astr. Soc. Pacific Conf. Series.

Flare Activity in Algol-type Binaries

Izold Pustylnik

Institute of Astrophysics and Atmospheric Physics, Toravere, EE2444, Estonia

Two sites of flare activity have been identified in Algol-type binaries. One is associated with the shock region from the accretion on B-A type primary, another, predominant one is apparently connected with coronal mass ejections (CME) from the late-type, Roche lobe filling secondary. The relative contribution from both sources of activity is still to be found. According to Peters & Polidan (1984), a moderately hot (10^5 K) and low density $n_e \simeq 10^9$ cm^{-3} high turbulent accretion region (HTAR) surrounding the hot component has been discovered. Typical dimensions of HTAR agree with the scale height of the X-ray source $H_X \simeq 1.2 \cdot 10^{11} \cdot T_7$ cm (Harnden *et al* 1977), where T_7 is the temperature in units of 10^7 K. But the decay time of the HTAR, comparable to the orbital period, more favourably agrees with the idea of CME as deduced from *Ginga* observations of X-ray flares on Algol (Stern et al. 1992). One solution of the controversy as to the contribution from HTAR would be to concentrate on X-ray observations of systems like u Her and U CrB, where the secondary is possibly earlier than F0 (thus, is not expected to possess an extensive convection zone) and on the other hand on systems like U Cep, where during periods of high activity $\dot{M} \simeq 10^{-6}\, M_\odot$ per year. Thus, the accretion luminosity must be at least by two orders of magnitude higher than in Algol. At present the analysis of X-ray flares is based upon the solar type CME assuming different scenarios of radiative and/or conductive cooling (see Stern et al. 1992). However, the current models of CME do not take into account the binary nature of the flaring object.

We shall consider now a model of an anisotropic wind in a moderately close binary system where the wind from one component overwhelms the wind from the other. Our purpose is to estimate the effects of the anisotropy due to the displacement of a sonic point caused by the gravitational attraction of a companion. For a stationary flow in a radially expanding wind, and ignoring Coriolis forces

$$\frac{1}{2}\left(1-\frac{u_s^2}{u^2}\right)\frac{\mathrm{d}\,u^2}{\mathrm{d}\,R}=\frac{2}{R}\,u_s^2-\frac{\mathrm{d}\,u_s^2}{\mathrm{d}\,R}-\frac{\mathrm{d}\,\Phi}{\mathrm{d}R}\,. \tag{1}$$

Here u is the velocity of gas, u_s is the local sound velocity and Φ is the gravitational potential effectively reduced by radiation pressure both in continuum and in spectral lines. At the sonic point

$$R=R_s\;,\;\; u=u_s\;,\qquad \frac{2}{R_s}\,u_s^2-\frac{2\,k}{m}\,\frac{\mathrm{d}\,T}{\mathrm{d}\,\mathrm{R}}\bigg|_s-\frac{\mathrm{d}\,\Phi}{\mathrm{d}R}\bigg|_s=0\,. \tag{2}$$

For the evaporative wind we use the condition of conservation of mass flux $J = J_0 \exp(-\Phi_s / k T_s) r^2$ along the streamline. We estimate the rate of anisotropy by assuming $T_s = T_{s0} + (\mathrm{d}T/\mathrm{d}r)\,\Delta r_s$, where $\mathrm{d}T/\mathrm{d}r$ is taken from Eq. (2) and Δr_s is the displacement of the sonic point. We assume $T = T_s\,(r_s/r)^m$ to estimate Δr_s. Let us consider now a case of a plane-parallel thin conduction zone. The equation of heat conduction with the energy loss by radiation is

$$\frac{\mathrm{d}}{\mathrm{d}r}\left(K_0 T^{5/2} \frac{\mathrm{d}T}{\mathrm{d}r}\right) = j_0\, n_e^2\,, \tag{3}$$

where K_0 is rather insensitive to the density and the temperature (Hearn 1975). Substituting (2) for the temperature gradient into (3) one can estimate the angular dependence of the emission measure $j_0\, n_e^2$. The results are indicated in Fig. 1 for some values of $\chi = G\,(M_1 + M_2)/k\,T$ ($\mu = 1$ corresponds to the direction towards the centre of the accretor). The mass flux $J(\mu)$ also turns out to be directionally dependent by a factor of 3–5 being maximum for $\mu = 0.6-0.7$. Thus, the anisotropy of the wind should affect the standard picture of the shock region (Richards 1992). Secondly, phase variations of $\mathrm{d}T/\mathrm{d}r$ stemming from the binary nature of the source should be taken into account in future analysis.

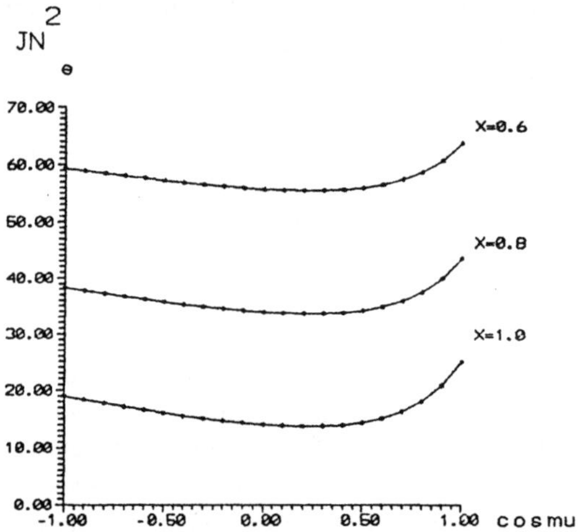

Fig. 1. An example of the angular dependence of the emission measure $j_0\, n_e^2$ for χ=0.6,0.8,1.0, χ being the ratio of the escape velocity to the gas velocity at the sonic point, $\Gamma_s = 0.9$,(Γ_s being the ratio of radiation pressure to the gravitational force of the mass losing component), mass ratio $q = 1, r_s = 0.3, \mu = 1$ corresponds to the direction towards the centre of the companion star.

References

Harnden Jr. F.R. , Fabricant D., Topka K., et al., 1977, ApJ 214, 418
Hearn A.G., 1975, A&A 40, 355
Peters G.I., Polidan R.S., 1984, ApJ 283, 745
Richards M., 1992, ApJ 387, 328
Stern R.A., Uchida Y., Tsunata S., Nagase F., 1992, ApJ 400, 321

Influence of Coronal Mass Ejections from the Red Dwarf Component on the Accretion Pattern in CVs and LMXBs

Roberto Minarini[1,2], *Grigory Beskin*[2]

[1] Dipartimento di Astronomia, Università di Bologna, via Zamboni 33, 40126, Bologna, Italy
[2] Special Astrophysical Observatory of the Russian Academy of Sciences, Nizhnij Arkhyz, Karachaevo-Cherkessiya, 357147, Russia

1 Introduction

Low-mass main sequence stars show a magnetic activity similar to the Sun and as a consequence they lose mass in the form of a variable stellar wind. In the latest spectral types (red dwarfs) the activity and the mass loss rate appear to increase by a large factor of $\sim 10^3$ with respect to the solar case, reaching $\dot{M} \sim 2 \cdot 10^{-11}$ $M_\odot$/yr (Badalyan & Livshits 1992, Katsova 1993). The same happens for coronal mass ejections (CMEs), which are the most relevant transient events of mass loss in these objects. In the Sun, these appear as bubbles of coronal material, with dimensions of some fraction of the solar surface, mass $M \simeq 2 \cdot 10^{14} - 2 \cdot 10^{16}$ g and ejection velocity $v \simeq 3 \cdot 10^7 - 2 \cdot 10^8$ cm/s, with an instantaneous mass loss rate $\dot{M} \sim 10^{-13} - 10^{-11}$ $M_\odot$/yr (Wagner 1984). In red dwarfs, as recently observed, the ejection velocities are higher, up to $v \simeq 3 \cdot 10^8$ cm/s and the mass loss rate can reach the value $\dot{M} \simeq 10^{-8}$ $M_\odot$/yr (Mullan *et al.* 1989, Houdebine *et al.* 1990). In both cases, the observations suggest that a bubble expands, once ejected, with a velocity of several hundreds of km/s.

2 Possibility of an accretion event

We asked ourselves what would happen if a bubble, expelled by the red dwarf component of a cataclysmic variable or a low-mass X-ray binary, is accreted by the compact object. The ejection velocity of the bubble (some thousands of km/s) is one order of magnitude larger than the escape velocity from the red dwarf and the orbital velocity of these systems. Thus, if the bubble travels toward the compact object, it is accreted directly. It is possible in this way to take into account a second stream of accretion in these binary systems. The transfer time scale is of the order of some minutes, much smaller than the accretion disk viscous time scale. It means that a transient mass loss event on the red dwarf can rapidly be followed by a transient event of mass accretion by the compact object.

The bubble can be expelled from the star in any direction. The accretion probability can be approximated by the ratio between the cross section of the bubble itself and the surface of the sphere with radius equal to the distance between the two components. We estimate $P \sim 10^{-2}$. This must be combined with the production rate of CMEs, which is about 1/day in the Sun and not yet determined but probably larger in red dwarfs.

3 Estimates of the emitted luminosity

Taking the typical values for the mass, density and velocity of a bubble and for the parameters of a binary system, we estimated the amount of energy emitted during the accretion process.

First of all, the accretion radius is of the same order of the radius of the bubble, so a relevant fraction of the gas is effectively accreted. We calculated the accretion luminosity by the formula $L = GM\dot{M}/R$, where $\dot{M}$ is the amount of matter which enters the accretion surface per unit time interval. For a CV we get $L \sim 10^{31} - 10^{33}$ erg/s. If this energy is thermalized in the white dwarf photosphere, it is blackbody emission in optical-UV. It will be bremsstrahlung emission in X-rays from the accretion column of a magnetized white dwarf. From a LMXB we get $L \sim 10^{33} - 10^{36}$ erg/s, in X-rays considering bremsstrahlung emission from a neutron star, magnetized or not.

These luminosities are significant and a burst can be observed as a consequence of a CME, also if the system already emits accretion luminosity. The duration will be of some minutes up to some tens of minutes, corresponding to the time a bubble needs to cross the compact object's accretion sphere. We conclude that the described phenomenon can be a cause of some bursting behaviour in CVs and LMXBs.

References

Badalyan O.G., Livshits M.A., 1992, Sov. Astron. 36, 70
Houdebine E.R., Foing B.H., Rodonò M., 1990, A&A 238, 249
Katsova M., 1993, in Inside the Stars, IAU Coll. 137, W.W. Weiss & A. Baglin (eds.), ASP Conf. Ser. Vol. 40, p. 654
Mullan D.J., Sion E.M., Bruhweiler F.C., Carpenter K.G., 1989, ApJ 339, L33
Wagner W.J., 1984, ARA&A 22, 267

Magnetic Field Reconnection as a Possible Source of Nonthermal Processes in Accreting Relativistic Systems

L.A. Pustil'nik

Observatory of the Jordan Valley Regional College, 15132, Israel

I. Recent observations of the *EGRET*-telescope aboard the Compton Gamma Ray Observatory *(CGRO)*, results from Čerenkov shower telescopes, and fast variability studies have demonstrated the dominant role of nonthermal processes for some accreting relativistic objects. This is perhaps an unexpected result, because the basic assumption of the standard accretion models (Shakura & Sunyaev 1973, Sunyaev & Titarchuk 1980, Galeev et al. 1979) is the thermal nature of all processes that determine the energy output. At the same time, the recent observations argue that the source of the emission is not the thermal plasma as a whole, but a very small fraction of charged particles accelerated to very high energy ($\varepsilon /mc^2 > (30\text{–}1000)$ with a "universal" energy spectrum $n_\varepsilon \sim \varepsilon^{-\alpha}$. These observational results suggest a revison of the global approach to the accretion process to find any new (or forgotten "old") energy conversion mechanisms with predominately nonthermal output (Lynden-Bell 1969, Ikhsanov & Pustil'nik 1994).

II. The nature of the energy source and mechanisms of energy conversion to either plasma heating ("thermal" mode) or to acceleration of charged particles ("nonthermal" mode) is the main question for accretion theories.

The primary source of the energy in both approaches is the same – gravitational energy of the accreted plasma or kinetic energy of the rotation of the central body is converted into kinetic energy of the plasma. The distinction between "thermal" and "nonthermal" models arises at the next stage – conversion of the plasma kinetic energy as a whole to energy in emitted charged particles.[1]

The "thermal" concepts assume that the main mechanism of kinetic energy dissipation is plasma heating by friction or by compression. The "nonthermal" approach is based on the existence of magnetic fields (either "frozen" in the accreted plasma, or external fields in the magnetosphere). It is assumed that any dynamo-processes would convert kinetic energy to magnetic field energy. Exactly these fields and their currents are the physical base of acceleration processes in the nonthermal models of accretion. Evolution of these current structures is

[1] I would like to note here that the standard list of emission mechanisms (bremsstrahlung, cyclotron or synchrotron radiation, Compton scattering, and plasma waves...) operates for both thermal plasmas and beams of ultrarelativistic particles.

very similar to the situation for solar and stellar flares and to the numerous laboratory experiments: global fast MHD-instability of the current structures lead to the formation of singular lines and current pinches, local dissipative instabilities lead to current disruption with fast magnetic field reconnection and development of high plasma turbulence. As a consequence numerous electrostatic "double layers" develop and effective acceleration of charged particles takes place in these regions. Emitted structures in the "nonthermal" models are the magnetic corona above the accretion disc and/or polar Z-pinch structures formed above the central object (Ikhsanov & Pustil'nik 1994).

III. Expected observational manifestations for thermal and nonthermal accretion models, possible "experimentum crucis", results of recent observations:

A choice among these two theoretical possibilities could be made only on the basis of the results of an "experimentum crucis". In other words, we would have to know the critical values of any observational parameters (energy, spectral slope, variability, etc.) which is impossible for most models [2]. We analyze from this point of view the expected manifestations of "thermal" and "nonthermal" classes of models and the comparisons to observations.

1. Energy spectra.

(a) "Thermal accretion" – the spectrum consists of the sum of Planck spectra from different parts of the accretion disc or accretion column. An additional spectral distortion may be caused by inverse Compton scattering of thermal emission of high energy electrons in the corona (resulting in the generation of a power spectrum), cyclotron scattering in a magnetized plasma (spectral breaks and lines could be generated).

The "first principles" limitation of particle energies (and, correspondingly, photon energies) for thermal models is the rest energy of particles. This limit is caused by the fact that the source of thermal energy is the kinetic energy of falling matter. The maximum kinetic energy is

$$E_{\max} = 0.5\ m_p\ c^2 = (0.3\ \mathrm{MeV} - \text{for electrons};\quad 0.5\ \mathrm{GeV}\ - \text{for protons})$$

In reality, the upper limit for the emitted spectrum, predicted by numerous models and numerical simulations, is not more than $\sim$ 100 keV. However, we will use as the absolute limit of energy for "thermal emission" $\varepsilon_{cr}^{th} = 0.5 - 1$ GeV as model-independent.

(b) "Nonthermal accretion models" assume high-energy particle acceleration by electric fields with a single power law spectrum with a wide range of energies (or the sum of power law components). Maximum energies for these particles (and, correspondingly, photons) are determined by potential drops in the acceleration region $\varepsilon = eE_* l_*$. For estimating the upper limit, using the size of the acceleration region $l_* = \zeta_l r_g$ and the induced electric field determined by the regular magnetic field value $E = \zeta_E H_0$, we obtain the critical value

$$\varepsilon_{cr}^{nth} = 10^{13}\mathrm{eV}\ (\text{for AGN}) - 10^{17} eV\ (\text{for neutron stars})\ .$$

[2] If models are tuned too much in order to fit all observational data through "games with free parameters", it would not be science altogether.

The emission spectrum is determined by energy loss processes (synchrotron radiation, Compton scattering and plasma waves, creation of an e^+e^- pair plasma, inelastic scattering of high energy particles with π –mesons and γ - quanta.
(c) Observational results. Observations of *EGRET* aboard *CGRO* (Hartman et al. 1992, Hurley 1994) show that many accreting objects exhibit "universal" power law spectra $n_\varepsilon \sim \varepsilon^\alpha$ from 0.1 GeV to $\varepsilon \geq$ 10-30 GeV. The spectrum of many γ-ray bursts is also a similar power law up to this energy. The slope of the spectra for all these objects is nearly the same: $\alpha \simeq 1$. The bulk of total emitted energy emerges in the form of ultrarelativistics photons. Observations with VHE (Very High Energy) telescopes have detected numerous sources with emission in the energy range up to $\varepsilon \geq 10^{13}$ eV (Meintjes et al. 1992).
2. Variability.
(a) "Thermal accretion" models assume as a source of variability some kind of hot plasma motion (rotation, free fall) and excitation processes (shock waves, thermal instabilities). The minimum time scales of these processes are determined by hydrodynamic or thermal velocities, or radiation loss time scales: $\Delta t_{mine} = (r/v_{ff}; r/c_s; nkT/L_*)$ As the size of the emitting region for these objects is limited by $r_{min} = (2-3)\ r_g$, we may estimate the minimal time of "elementary flares" for "thermal" modes of accretion:

$$\Delta t_{\min} = (10 - 100\text{ms } for\ M_{bh} = 10 M_\odot; 1 \text{ day} - 1\text{week } for\ M_{b.h.} = 10^8 M_\odot)$$

(b) The "nonthermal accretion" approach assumes that the flare process is caused by ultrarelativistic particle propagation and emission. The velocity of flare front motion for these models is c – the velocity of light. The minimum time of a flare in this picture is $\Delta t_{min} = d/c$. This time interval is 1-2 orders of magnitude shorter than that in the "thermal accretion" picture. Brightness temperature is also an essential parameter:

$$T_B = \frac{\Delta F_\nu}{c^2\ \nu^2\ d^2} \geq T_{B_{\min}} \frac{\Delta F_\nu}{c^4\ \nu^2\ \Delta t^2} \quad .$$

The value of kT_B sets the lower limit for the energy of emitted particles, $\Delta t_{\min}$ determined from observations of variability give a unique method for estimating particle energies.
(c) Observational results. The search for fast variability in the optical and X-ray emission from accreting compact objects demonstrates for several sources that this phenomenon is real and may be explained only by "nonthermal" accretion. As an example we may refer to the observations of ultra-short X-ray flares of Cyg X-1 up to 0.5 msec (Rothschild 1974), shot-noise component in its emission with $\tau \simeq 3msec$ (Meekins 1984). But the best evidence for the nonthermal nature of flares is provided by the optical emission from X-ray novae A0620-00, Nova Per 1992, and also from X-ray bursters of Type II (Beskin et al. 1983, Bartolini et al. 1994). The duration of flares (from 1 msec up to 100 msec) corresponds to brightness temperatures ranging from 10^9 K to 10^{10} K. These time scales and brightness temperatures are only expected in the nonthermal mode of accretion.
IV. Possible acceleration processes in the accretion plasma may be divided into two groups:

(a) Regular acceleration in large scale electric fields in the inner reconnection region or a place of current disruption (Dreicer 1959). The energy spectrum for this acceleration process has an exponential form. This process assumes free particle motion in the regular electric and magnetic field – an assumption that is not likely to be correct in reality, if we take into account processes of particle-plasma wave scattering.
(b) Stochastic acceleration resulting from diffusion processes in regions with large scale electric fields as well as diffusion in momentum-energy space.
The second process is based on inelastic "particle-plasmon" scattering, and is more less as effective as the first process (based on elastic collisions). In some intermediate region particle acceleration becomes "transparent" to inelastic scattering (for lower energies) and also for elastic scattering. The energy spectrum derived from stochastic acceleration is a power law. This fact is a consequence of Fermi's arguments (1949) – if the acceleration processes and particle dissipation have the same origin, the spectrum will be a power law with slope determined by the ratio of particle life time to the time scale of acceleration. This ratio is constant and does not depend on particle energy [3]. While the universal power law of the emission spectrum is understood as a natural consequence of diffusive acceleration, for now the origin of the "universal" slope ($\alpha \simeq 1$) is not so obvious. It may be result of the nonlinear self-agreement of the efficiency of acceleration and space diffusion from the acceleration region for situation with extremely high electric fields (compared with Dreicer's limit) in the very strong plasma turbulence (compared with thermal energy). Exactly this state has to arise in the region of current structure disruption. As we have shown in previous work (Ikhsanov & Pustil'nik 1994) the distribution of particle energies for this mechanism are in good agreement with observations.

References

Bartolini C., Piccioni A., Beskin G., Neizvestny S., 1994, ApJS 92, 455
Beskin G.M., et al.: 1983, Sov. Astron. Lett. 14, 156
Dreicer H., 1959, Phys. Rev. 115, 238
Fermi E., 1949, Phys. Rev. 75, 1169
Galeev A.A., Rosner R., Vaiana G.S., 1979, ApJ 229, 318
Hartman R.C., et al., 1992, ApJ 385, L1
Hurley K., et al., 1994, Nature 372, 652
Ikhsanov N.R., Pustil'nik L.A., 1994, ApJS 90, 959
Lynden-Bell D., 1969, Nature 233, 690
Meintjes P.J., 1992, ApJ 401, 325
Meekins J.F., 1984, ApJ 278, 288
Rothschild R.E., Bodt E.A., Holt S.S., Serlemitsos P.J., 1974, ApJ 189, L13
Shakura N.I., Sunyaev R.A., 1973, A&A 24, 337
Sunyaev R.A., Titarchuk L.G., 1980, A&A 86, 121

[3] Exactly this physics lead to paradoxical power spectrum formation by Compton scattering of thermal Planck radiation of Maxwellian electrons (Sunyaev & Titarchuk 1980).

FLARES IN T TAURI STARS

Flares in T Tauri Stars

Gösta F. Gahm

Stockholm Observatory, S-133 36 Saltsjöbaden, Sweden

1 Introduction

This review concerns rapid variability, on time-scales of a few hours or less, occurring on T Tauri stars (TTS). There are several recent reviews on observed properties of TTS (e.g. Appenzeller & Mundt 1989, Bertout 1989, Gahm 1990a, Kuhi & Cram 1989), some with more emphasis on "flare-like" activity (Feigelson et al. 1991, Gahm 1990b, Montmerle 1991, Montmerle et al. 1993), and the ambition below has been to summarize more recent work. We will also give a compilation of published works on observations of optical variability on these time-scales and a related statistical overview. Some early, very interesting and extensive studies of this kind were made here at the Sonneberg Observatory, where for instance Götz & Wenzel (1967) concluded that the light-curve of RW Aurigae contains several components: quasi-periodic fluctuations over days, 'waves' over several hours, rapid outbursts with symmetric light-curves, rapid outbursts with asymmetric (flare-like) light-curves (see also Fürtig & Wenzel 1964) and small fluctuations with very small amplitudes, which they postulated could be due to changes in the emission line fluxes. Modern photometric results of higher sensitivity and time-resolution confirm the existence of these different types of rapid variations in classical TTS (CTTS), having strong emission (lines and continuous, so called veiling) superimposed on the photospheric absorption line spectrum. As described in the subsequent chapters there are new concepts for the interpretation of the short-lived fluctuations. Concerning the long-term quasi-periodic variations we can now usually relate them to the stellar rotational period (bright or dark spots), let be that RW Aur still is an uncertain case. There are other slow changes most likely related to variable circumstellar extinction in circumstellar dust in the line-of-sight to the star. The long-term optical changes were most recently discussed by Herbst et al. (1994).

2 X-ray bursts

A number of X-ray bursts were observed with *Einstein* (see the reviews above). Some extremely energetic events were recorded, the most extreme one on the

WTTS doublet ROX 20 with a peak luminosity of 10^{32} erg s^{-1} and a total X-ray energy of $E_x \geq 10^{35}$ ergs (Montmerle et al. 1983). An even more powerful burst on LkHα 92 with 5 10^{32} erg s^{-1} (peak) and $E_x = 4\ 10^{36}$ ergs was detected by *ROSAT* (Preibisch et al. 1993) and the spectrum of what could be a TTS burst in the ρ Oph region was traced to energies above 10 keV by Koyama et al. (1992) from *Ginga* observations. The X-ray bursts are flare-like with short rise-times and exponential declines and they last for a few hours. From spectrally resolved events temperatures of a few 10^7 K are inferred (model dependent) and if the X-ray emitting volumes are single loops rooted at the stellar surface, then they typically extend to up to about 1 stellar radius. These results are summarized and compared to corresponding events on the Sun and flare stars in Table 1, most of which is based on Preibisch et al. (1993), which should be consulted for references on ranges in the values – not given here. Data for post-T Tauri candidates (PTTS), older than the TTS but presumably still in contraction, come from Tagliaferri et al. (1988, HD 560B) and Pakull (1981, AB Dor: in parenthesis).

It has always been assumed that these X-ray bursts in fact are flares, similar to what is seen on the Sun and flare stars but much more powerful. In view of recent theoretical developments concerning possible interactions between a circumstellar disk and the star, one must also ask whether hot transient phases can develop from inhomogeneous mass accretion or cometary bodies flying into a stellar wind and disintegrating. On the basis of models (Camenzind 1990, Königl 1991; see also Ushida & Shibata 1984, 1985 and Bertout et al. 1988) with accretion along magnetic loops connecting a circumstellar disk and the star, Gullbring (1994) showed, assuming inhomogeneous mass accretion, that X-rays will be produced if the optical depth in the plasma of infalling blobs is small. However, the temperature is determined by the potential energy extracted from the almost free falling plasma and can only reach a few 10^6 K. Hence, the hotter, strong X-ray bursts are most likely connected to magnetic surface activity and not to accretion. This does not exclude the presence of X-ray events generated by accretion. Strong X-ray flaring may be a rather rare phenomenon, however. Gagné & Caillault (1995) have made a detailed examination of all the *Einstein* IPC and HRC fields in the region centred at the Orion nebula. From 45 hours of time-resolved observations only 2 X-ray flares with fluxes $\geq 2.5\ 10^{30}$ erg s^{-1} were found among the over 200 X-ray sources. 1/4th of the sources vary on time-scales of 1 day to 1 year.

3 Radio bursts

Radio emission at cm wavelengths has been detected from several TTS and many sources are known to vary with time. This radiation can be thermal and arise in massive stellar winds (Panagia & Felli 1975) or non-thermal. The CTTS have radio spectra consistent with thermal emission while for the WTTS the evidence of sources of gyrosynchrotron radiation is increasing: First, the brightness temperatures derived are high, $10^{7.5} - 10^9$ K, much higher than those inferred from

Table 1. X-ray flare characteristics

Object	E_x (10^{30} erg)	T (10^6 K)	l (10^{10} cm)
Sun, 3b flare	10	10	1
dMe (EV Lac)	10^4	30	2
TTS	10^4	20	2
ROX 20	10^6	15	10
LkHα 92	$4\ 10^6$	50	8
PTTS	$2 \cdot 10^4$ ($4 \cdot 10^6$)	50	

X-ray data (Phillips et al.J1991). Second, the spectral index is flat or negative in many stars as was pointed out already by Bieging et al. (1984) and discussed by e.g. Stine et al. (1988). The VLA and VLBI studies made so far show that for some WTTS, the sources are very extended. Third, circular polarisation is present in many stars (O'Neal et al. 1990, Skinner 1993).

Considerable flux changes occur on time-scales of several hours (see e.g. White et al. 1992). Hence, the radio characteristics provide the best evidence of magnetic activity in WTTS. It is not straightforward to model the flares. Efforts in this direction have given magnetic field strengths of B $<$ 1 kGauss and single loop lengths of one up to ~15 stellar radii as envisioned by e.g. Montmerle & André (1988). It should be pointed out that a lot of attention has been given to the most extreme cases DoAr 21 and HD 283447, which may be peculiar in the WTTS group.

The first simultaneous multi-frequency observations of TTS start to appear. Feigelson et al. (1994) observed what could be the declining part of a radio flare on (again) HD 283447 over a few hours when the X-ray, optical and emission line fluxes (Hα, Mg II) stayed constant. In this paper several possible origins of TTS activity are discussed, including solar type activity, star-disk interactions, magnetic fields in the disk and binarity. More studies of this type are under way. Finally we note that a rapid decrease in the radio flux of T Tau (S) (a possible protostar?) was discovered by Phillips et al. (1993).

4 Short-term optical variability

Optical monitoring of TTS has a long history. In the review by Gahm (1990b) it was pointed out that: 1. *U*-band 'flaring' with changes by more than 20% in flux is rare; 2. CTTS most often show symmetric rather than asymmetric (flare-like) light-curves, like observed for RW Aur by Götz & Wenzel (see Chapter 1); 3. Assuming black-body radiation, the events are cool ~7000K, different from flares on the Sun and flare stars; 4. These flares may originate from inhomogeneous mass accretion rather than from magnetic surface activity. More published results have been added to the data bank since that time and we can now make an overview of short-term variability of the stars sorted in CTTS and WTTS in

Table 2. Stars in the data bank

CTTS
Taurus: RY, T, DF, DG, DH, DK, GG, GI, GK, DL, CI, AA, DN, HP, DO, DR
Chameleon: CV, SY, TW, VW
Corona Austrina: S, VV, Wa/1, Wa/2, Wa/3
Other: RW Aur, GW Ori, R Mon, LHα 332-17, Sz 68,
GW Lup, GQ Lup, EX Lup, SR 4, SR 12, SR 9, V853 Oph,
Haro 1-16, IX Oph, KK Oph, LkHα 118, DI Cep, BM And, TW Hya
WTTS
Taurus: DI, UX-A, HP/G2, V 773, V819, V827, V826, V830, V836
Orion: CO, V370, KM, KN
Other: SU Aur, AB Dor, AK Sco, SZ Cha, T Cha, ROX 3, DoAr 21, DoAr 51

Data for these stars are taken from the references marked with an asterisk.

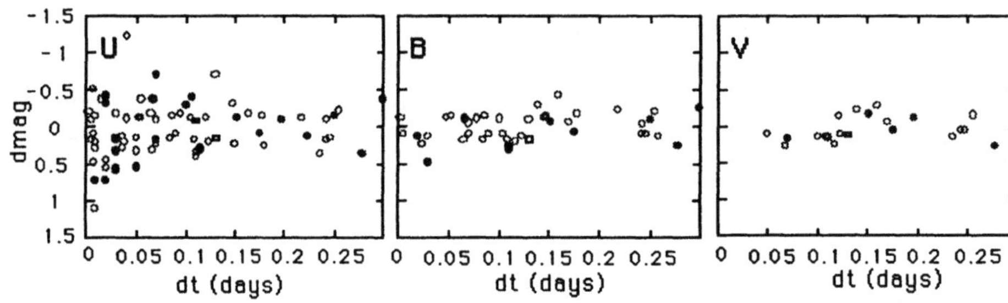

Fig. 1. Rises ($d(mag.)$ negative) and declines with total amplitudes larger than $0^m\!.05$ in U, B and V versus total duration, as compiled from the literature

Table 2 (designations from Herbig & Bell 1988: HBC), and ordered according to increasing HBC *number*). We follow strictly the definition of an WTTS as having an equivalent width of the Hα emission line of W(α) $\leq$ 10Å, although some of these stars are known to carry circumstellar dust. Only series of observations with $\geq$ 2 photometric observations per night including the U-band enter. This compilation, which is based on the references marked with an asterisk in the reference list, covers a total of about 2000 hours of monitoring.

For the statistical analysis we proceed as before by measuring d(magnitude), the total change in any filter over an observed time interval dt. If the direction of the change is the same over several measured points, then the total interval in magnitude and time is stored. Slow but significant changes can be missed by automatic procedures due to noise and we also check all nights with several observed points to restore all gross trends during a night.

Fig. 1 shows all recorded changes with $|d(\text{mag.})| \geq 0.05$ in U, B and V where the material has been divided into CTTS (subgroups: spectral types F&G and K&M) and WTTS. It should be noted that depending on telescope and observing conditions some of the points of low amplitude may be spurious. The diagrams show that: A. Rapid changes of $0^m_.2$ ($\sim$ 20% in flux) or more are exceptional even in the U-band. B. Rapid increases (upwards in the diagrams) do not dominate over rapid declines, which would be the case if the fluctuations were flare-like. In this respect there is not a significant difference between the CTTS and WTTS as was found from a smaller sample of Strömgren photometric observations by Gahm (1990b); C. In spite of the long total patrole time, no changes larger $0^m_.4$ in V and only two $> 0^m_.8$ in U have occurred and since eruptions with amplitudes of much more than 1^m. has been reported from earlier photographic work we can conclude that such variations, if real, must be extremely rare indeed. Along the same lines as before we have also derived the frequencies of rapid fluctuations, that is the probability that an observer a given night will find the star to undergo a major and rapid change ($t \leq 0.1$ day). Regarding all changes with $|dU| \geq 0^m_.1$ as real, we find that for the CTTS (type K&M) this happens only 3% of the time and for the F&G stars no event was recorded, which can be explained mainly as a contrast difference in the background U-flux between the two subgroups. The corresponding frequency for the WTTS is 2%.

The data bank is based on published individual measurements, but some major events have only been shown as figures in the literature. This applies to the observations made in Sonneberg referred to above and also events shown by e.g. Rydgren & Vrba (1983: on FK X-2 = V827 Tau) and Zaitseva et al. (1985: on RY Tau). Strong U-band flaring on two WTTS during minima in the spot modulated light-curve, was inferred from single (not time-resolved) observations by Vrba et al. (1993). Not included in the data bank is a recent, simultaneous spectroscopic and photometric study of short-term variability on 2 CTTS, 3 WTTS and one PTTS candidate by Gahm et al. (1995a). The general results from the statistical study above are confirmed. Moreover it was possible to show that *all* the rapid events on VW Cha (CTTS) were due to variable veiling. However, on the WTTS most events were truly flare-like, starting with a sudden increase in the Balmer line- and continuous emission. It was possible to confirm the postulate by Götz & Wenzel (1967) about small changes in B due to changes in the Balmer line fluxes. Also slower changes due to variable extinction appear to be present. Accounting for the larger radii of TTS compared to flare stars, one can conclude that the WTTS possess a magnetic surface activity comparable to the flare stars (as overviewed by Gershberg & Shakhovskaya 1983), while there is not a single flare-like event recorded for the CTTS.

Also the profiles of emission lines can vary on short time-scales. For a review on what has been accomplished from spectroscopy at high spectral resolution see Gahm et al. (1995b).

5 Concluding remarks

The work in Sonneberg has been pioneering in pointing out the different components of optical variability in TTS. It is now evident that the TTS are not exceedingly active on time-scales of hours. Even at detection levels of milli-magnitudes the stars are mostly constant in brightness or vary only very slowly. Regarding the rapid fluctuations, some appear to be connected to exceptional magnetic surface activity – especially on the WTTS – but most of the CTTS activity comes from variable veiling, which according to current models is a consequence of mass accretion from a disk to the star, possibly along the dipole stellar magnetic field. Hence, inhomogeneous mass accretion controls the optical activity, something apparent also from night-to-night studies by e.g. Basri & Bathala (1990), Hartigan et al. (1991) and Valenti et al. (1993). In addition, Guenther & Hessman (1993) found accompanying variable, inverse P Cygni emission line profiles on DR Tau (CTTS). A considerable fraction of the stellar surface of a CTTS can be hidden by foreground dust and rapid infall may inhibit normal surface activity in another region around the visible pole. A possibility is that the lack of surface flaring on CTTS is due to the combination of these effects. Also, if the suggestion by Bouvier et al. (1994) and Edwards et al. (1993) is correct, namely that the CTTS rotate slower than the WTTS, then the smaller dynamo effect on the CTTS could lead to a smaller surface activity on these stars compared to the WTTS. The fact that a considerably larger fraction WTTS are detected at X-rays, compared to the CTTS, renders support to this idea (Neuhäuser et al. 1995).

However, soft X-ray events can be produced at the poles. Many X-ray flares indicate plasma temperatures which are larger than those expected from infall, and particularly on the WTTS powerful magnetic activity, sometimes very extended, is present with radio characteristics typical for flares. Simultaneous X-ray, optical and radio monitoring provides an important tool in the diagnosis of TTS eruptions.

Acknowledgement: My thanks go to Erik Gullbring for stimulating discussions and help with the analysis of the data bank.

References

Abuladze O., Vardanian R.A., Kovalenko V.M., Kumsishvili Ja., Melikian N.D., Mironov A.V., Oshchepkov V.A., Stepanian J.A., Totochava A., Cherepashchuk A.M., Shanin G.I., Shpychka I.V., Shcherbakov A.G., 1975, Variable Stars 20, 47*

Alcala J.M., Covino E., Franchini M., Krautter J., Terranegra L., Wichmann R., 1993, A&A 272, 225*

Appenzeller I., Mundt R., 1989, A&AR 1, 291

Basri G., Batalha C., 1990, ApJ 363, 654

Bastian U., Mundt R., 1979, A&AS 36, 57*

Bertout C., 1989, ARA&A 27, 351

Bertout C., Basri G., Bouvier J., 1988, ApJ 330, 350

Bieging J.H., Cohen M., Schwartz P.R., 1984, ApJ 282, 699

Bouvier J., Bertout C., Bouchet P., 1988, A&AS 75, 1*

Bouvier J., Cabrit S., Fernández M., Martin E.L., Matthews J.M., 1993, A&A 272, 176

Camenzind M., 1990, in Reviews in Modern Astron. 3, G. Klare (ed.), p. 234

Cohen M., Schwartz R.D., 1976, MNRAS 174, 137*

Covino E., Terranegra L., Franchini M., Chavarria-K C., Stalio R., 1992, A&AS 94, 273*

Edwards S., Strom S.E., Hartigan P., Strom K.M., Hillenbrand A., Herbst W., Attridge J., Merrill K.M., Probst R., Gatley I., 1993, AJ 106, 372

Evans L., 1987, South African Astron. Obs. Circ. No 11, p. 73*

Evans L., Koen M.C.J., 1987, South African Astron. Obs Circ. No 11, p. 21*

Feigelson E.D., Giampapa M.S., Vrba F.J., 1991, in The Sun in Time, C.P. Sonett, M.S. Giampapa M.S. Matthews (eds.), Univ. of Arizona Press, p. 658

Feigelson E.D., Casanova S., Montmerle T.,Guibert J., 1993, ApJ 416, 647

Feigelson E.D., Welty A.D., Imhoff C.I., Hall J.C., Etzel P.B., Phillips R.B., Lonsdale C.J., 1994, ApJ 432, 373

Fürtig W., Wenzel W., 1964, Sterne 40, 35

Gagné M., Caillault J.-P., 1995, ApJS, in press

Gahm G.F., 1990a, in Formation of Stars and Planets and the Evolution of the Solar System, B. Battrick (ed.), ESA SP-315, p. 43

Gahm G.F., 1990b, in Flare Stars in Star Clusters, Associations and the Solar Vicinity, IAU Sym. 137, L.V. Mirzoyan, B.R. Pettersen, M.K. Tsvetkov (eds.), Kluwer, Dordrecht, p. 193

Gahm G.F., Gullbring E., Fischerström C., Lindroos K.P., Lodén K., 1993, A&AS 100, 371*

Gahm G.F., Lodén K., Gullbring E., Hartstein D., 1995a, A&A, in press

Gahm G.F., Petrov P., Tuominen I., Gullbring E., Ilyin I., 1994b, in Proc. 4th MUSICOS Workshop, Multi-site Continuous Spectroscopy, Beijing June 1994, in press

Gershberg R.E., Shakhovskaya N.I., 1983, Ap&SS 95, 235

Götz W., Wenzel W., 1967, Mitteilungen über Veränderliche Sterne 4, 71

Grinin V.P., Efimov Ju.S., Krasnobatsev V.I., Shakovskaya N.I., Shcherbakov A.G., Zaitseva G.V., Kolotilov E.A., Shanin G.I., Kiselev N.N., Gjulaliev Ch.G., Salmanov I.R., 1980, Variable Stars 21, 247*

Guenther E., Hessman F.V., 1993, A&A 268, 192

Gullbring E., 1994, A&A 287, 131

Hartigan P., Kenyon S.J., Hartmann L., Strom S.E., Edwards S., Welty A.D., Stauffer J., 1991, ApJ 382, 617

Herbig, G.F., Goodrich, R.W. 1986, ApJ 309, 294*

Herbig G.H., Bell K.R., 1988, Lick Obs. Bull., No 1111

Herbst W., Holtzman J.A., Klasky R.S., 1983 AJ 88, 1648*

Herbst W., Booth J.F., Koret D.L., Zajtseva G.V., Shakovskaya N.I., Vrba F.J., Covino E., Terranegra L., Vittone A., Hoff D., Kelsey L., Lines R., Barksdale W., 1987 AJ 94, 137*

Herbst W., Herbst D.K., Grossman E., Weinstein D., 1994, AJ 108, 1906

Kardopolov V.I., Filipev G.K., 1985, Variable Stars 22, 103*

Kilkenney D., Whittet D.C.B., Davies J.K., Evans A., Bode M.F., Robson E.I., Banfield R.M., 1985, South African Astron. Obs. Circ. No 9, p. 55*

Kilyachkov N.N., Shevchenko V.S., 1976, AZh Lett. 2, 494*

Kolotilov E.A., 1986, AZh 63, 298*

Königl A., 1991, ApJ 370, L39

Koyama K., Asaoka I., Kuriyama T., Tawara Y., 1992, PASJ 44, L255

Kuhi L.V., Cram L.E., 1989, In: Cram L.E. & Kuhi L.V. (eds.) FGK stars and T Tauri stars, NASA SP-502, p.99

Lindroos K.P., 1989, private comm.*

Montmerle T., 1991, in The Physics of Star Formation and Early Stellar Evolution, NATO ASI Ser. C 392, C.J. Lada N.D. Kylafis (eds.), p. 675

Montmerle T., André P., 1988, in Formation and Evolution of Low Mass Stars, NATO ASI Ser. C 241, A.K. Dupree, M.T.V.T. (eds.), p. 225

Montmerle T., Koch-Miramond L., Falgarone E., Grindlay J.E., 1983, ApJ 269, 182

Montmerle T., Feigelson E.D., Bouvier J., André P., 1993, in Protostars and Planets III, E.H. Levy, J.I. Lunine (eds.), Univ. of Arizona Press, p. 68

Mundt R., 1979, A&A 74, 21*

Neuhäuser R., Sterzik M.F., Schmitt J.H.M.M., Wichmann R., Krautter J., 1995, A&A, in press

O'Neal D., Feigelson E.D., Mathieu R.D., Myers P.C., 1990, AJ 100, 1610

Pakull M.W., 1981, A&A 104, 33

Panagia N., Felli M., 1975, A&A 39, 1

Phillips R.B., Lonsdale C.J., Feigelson E.D., 1991, ApJ 382, 261

Phillips R.B., Lonsdale C.J., Feigelson E.D., 1993, ApJ 403, L 43

Preibisch Th., Zinnecker H., Schmitt J.H.M.M., 1993, A&A 279, L 33

Pugach A.F., 1975, Variable Stars S. 2, 195*

Rucinski S.M., 1983, A&AS 52, 281*

Rucinski S.M., 1988, Inf. Bull. Variable Stars (IAU), No 3146*

Rucinski S.M., Krautter J., 1983, A&A 121, 217*

Rydgren A.E., Vrba F.J., 1983, ApJ 267, 191*

Rydgren A.E., Zak D.S., Vrba F.J., Chugainov P.F., Zaitseva G.V., 1984, AJ 89, 1015*

Shaimeva A.F., Shutiomova N., 1985, Variable Stars 22, 176*

Shevchenko V.S., Shutiomova N.A., 1981, Afz 17, 286*

Simon T.,Vrba, F.J., Herbst W., 1990, AJ 100, 1957*

Skinner S.L., 1993, ApJ 408, 660

Stine P.C., Feigelson E.D., André P., Montmerle T., 1988, AJ 96, 1394

Tagliaferri G., Giommi P., Angelini L., Osborne J.P., Pallavicini P., 1988, ApJ 331, L 113

Ushida Y., Shibata K., 1984, PASJ 36, 105

Ushida Y., Shibata K., 1985, PASJ 37, 515

Valenti J.A., Basri G., Johns C.M., 1993, AJ 106, 2024

Vrba F.J., Rydgren A.E., Chugainov, P.F., Shakhovskaya N.I., Zak D.S., 1986, ApJ 306, 199*

Vrba F.J., Herbst W., Booth J.F., 1988, AJ 96, 1032*

Vrba F.J., Rydgren A.E., Chugainov P.F., Shakovskaya N.I., Weaver W.B., 1989, AJ 97, 483*

Vrba F.J., Chugainov P.F., Weaver Wm.B., Stauffer J.S., 1993, AJ 106, 1608*

White S.M., Pallavicini R., Kundu M.R., 1992, A&A 259, 149

Zaitseva G.V., 1978, Afz 14, 17*

Zaitseva G.V., 1982, Afz 18, 67*

Zaitseva G.V.,1988, Afz 31, 489*

Zaitseva G.V., Lyuti V.M., 1976, Variable Stars 20, 255*

Zaitseva G.V., Lyuti V.M., 1976, Variable Stars 20, 266*

Zaitseva G.V., Kolotilov E.A., Petrov P.P., Tarasov A.E., Shenavrin V.I., Shcherbakov A.G., 1985, AZh Lett. 11, 271

H. Zinnecker: The missing optical surface flares on classical T Tauri stars which you emphasized suggest that these objects differ fundamentally (maybe even genetically) from the weak-line T Tauri stars in their magnetic field properties; or do you see another explanation?

G. Gahm: Yes, and we can only speculate on its course. Maybe the strong dipole filed connecting the disk and the star could have the effect of surpressing the development of at least very strong surface magnetic flares.

P.B. Byrne: Why, if T Tauri stars have deep convective zones, is the proposed dipole field not shredded and locally intensified as in the sun and, presumably, active late type stars?

G. Gahm: This is a difficult question and I have no input to give on this topic.

P.B. Byrne: Is it not possible that the CTTS are more slowly rotating and so possess overall weaker fields than the active PTO?

G. Gahm: I think this postulate is very good and should be followed up.

Properties of X-ray Flares on Young Stars

Th. Preibisch[1], *R. Neuhäuser*[2]

[1] Universität Würzburg, Astronomisches Institut, Am Hubland, D-97074 Würzburg, Germany
[2] Max-Planck-Institut für Extraterrestrische Physik, D-85740 Garching, Germany

Abstract: The interest in the giant flares found on young stars results from the fact that their energetics is vastly different from solar flares. The recent *ROSAT* observations of young clusters yield a good basis for a systematic study of these flares.

We have investigated 18 very deep pointed observations of star forming regions in IC348, Chamaeleon, ρ Ophiuchi, and Orion. To study the age dependence of the flare properties we also investigated the observations of the young clusters IC2391, Pleiades, and Hyades. In this way we have evaluated the X-ray emission of several hundred young stars with ages ranging from about 10^6 years to $7 \cdot 10^8$ years. We have found 36 large flares on identified cluster members and have determined the flare properties with a simple flare model.

The flares show the same relation between the total X-ray flare energy and the quiescent X-ray luminosity as the dM star flares. The total X-ray flare energies decrease smoothly from the very high values found for very young stars to solar values with increasing cluster age. Our results indicate that the large flares on young stars are similar to scaled-up solar-like activity rather than to accretion processes.

1 Introduction

Since the observations of the *Einstein* satellite young stars are known to show giant X-ray flares. While on the Sun the total flare energies radiated in X-rays usually do not exceed 10^{31} erg even for the largest flares, values of up to $4 \cdot 10^{36}$ erg have been found for flares on T Tauri stars (Montmerle et al. 1983, Preibisch et al. 1993). The vastly different energetics of the flares on the Sun and on young stars raise the question of the origin of these large flares: are these giant flares just the result of scaled up solar-like coronal activity commonly found on young stars, or are they somehow related to accretion processes (see Gahm 1995, these proceedings) or to a disk corona? To address this question, we have made a systematic search for X-ray flares in deep *ROSAT* observations of several young clusters.

2 ROSAT observations and data analysis

We have used ROSAT observations[3] of the Orion nebula region, the star-forming regions in Chameleon, ρ Oph, IC348 and the young clusters IC2391, Pleiades, and Hyades. While the age of the stars in the first four regions is about a few million years, the age of the Pleiades is $7 \cdot 10^7$ years and the age of the Hyades $7 \cdot 10^8$ years. The flares on stars in the Orion nebula region have been detected by Gagné et al. (1995) and re-analyzed by us.

We have investigated the light curves of all identified cluster members that showed variability in their X-ray emission and we have found 36 large flares. In the subsequent analysis we used only flares where the pre-flare count rate was high enough to extract and fit a pre-flare spectrum and where the decay phase was covered well enough to derive the decay time (the low orbit of the satellite causes frequent gaps in the light curves).

We extracted an X-ray spectrum in the pre-flare phase to determine the quiescent X-ray luminosity and emission measure. The flare spectrum was obtained by subtracting the pre-flare spectrum from the spectrum at flare maximum. The spectra were fitted with Raymond-Smith plasma emission models (Raymond & Smith 1977) including interstellar absorption. The best fit parameters were determined using a χ^2 method. We used several spectral models: an isothermal model, a two temperature model, and a model with a power-law temperature distribution (see Schmitt et al. 1990 for a discussion of these models). From the fit to the flare spectrum we obtained the maximum X-ray luminosity and the temperature and emission measure of the flaring plasma. In nearly all cases we found increased plasma temperatures during the flare. The plasma temperatures at flare maximum were always around or above $3 \cdot 10^7$ K.

This detailed spectral modeling was possible for about one third of the flares. For the rest of the flare sources, where the signal-to-noise ratio of the X-ray data did not allow such an analysis, we could only fit the pre-flare spectrum. In these cases we calculated the peak X-ray luminosity and emission measure from the ratio of the maximum and the pre-flare count rate. For the flare plasma we assumed a temperature of $3 \cdot 10^7$ K.

3 Flare modeling

To estimate the flare parameters we used the same analysis as Pallavicini et al. (1990) for their study of dM star flares. We estimated the plasma density by assuming only radiative cooling and the minimum magnetic field strength by equating magnetic pressure and gas pressure. We estimated the loop length by assuming the flaring plasma to be trapped in a loop with a ratio of diameter to length of 1:10. We are aware that this analysis may give only rough estimates, but it should give at least the right orders of magnitude.

[3] Most of the data were taken from the *ROSAT* data archive.

4 Results and discussion

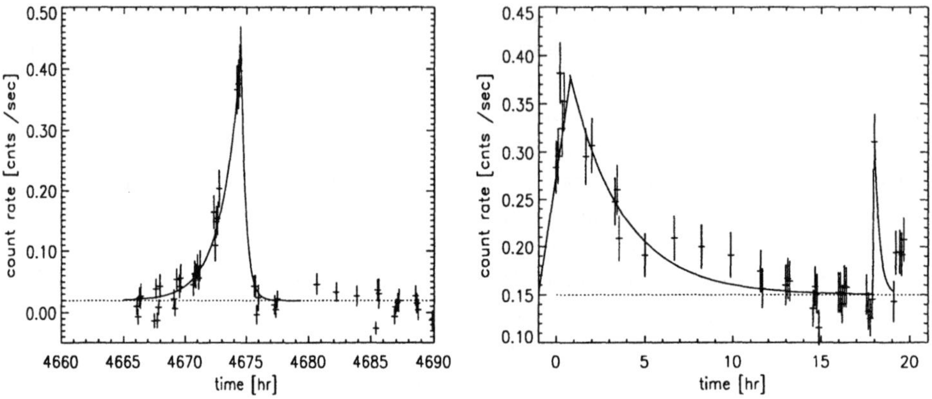

Fig. 1. Lightcurves of the flares on HGC 144 (left) and Hz 2411 (right). The solid lines represent fitted light curves.

The decay times range from a few minutes to several hours. For most flares the rise time is clearly shorter than the decay time with one remarkable exception: the flare found on the Pleiades member HGC 144 (Fig. 1) shows an exponential rise phase with a rise time (1.8 hours) considerably longer than the decay time (0.4 hours).

A couple of flare stars show more than one flare during the *ROSAT* observation. A particular example is the Hyades member Hz 2411, a well-known dMe flare star, for which Gurzadyan (1980) finds a mean frequency of optical flares of 1 every 11 hours. In the 20 hour patrol time of the *ROSAT* observation there are two large X-ray flares (Fig. 1).

A comparison of the flare properties of our sample with those of solar and dM star flares yields the following results: the total X-ray flare energies range from $3 \cdot 10^{32}$ erg to $4 \cdot 10^{36}$ erg (for the flare on LHα 92) and are much larger than the ones found in solar flares. The plasma density ($10^{11} - 10^{12}\,\mathrm{cm}^{-3}$) and magnetic field strength ($\approx 100 - 1000$ G) are in the same range as for solar flares. The much larger energy output of flares in young stars is related to the much larger volumes and loop lengths of the flaring plasma: the loop lengths range from 10^{10} cm to $6 \cdot 10^{11}$ cm and are much larger (l/R_* up to 8.5) than for solar flares (l/R_* usually below 0.1).

In their study of dM star flares Pallavicini et al. (1990) found a good correlation between the total X-ray energy of the flares and the quiescent X-ray luminosity. In our sample we find the same correlation, only shifted towards higher flare energies and luminosities. This may indicate that the same flare mechanism is at work in both samples.

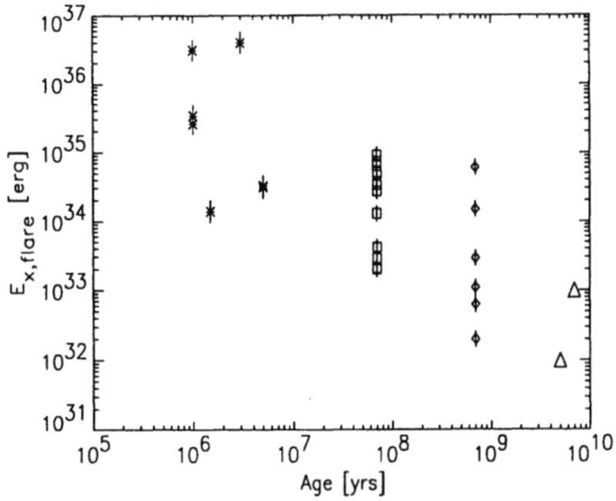

Fig. 2. Total X-ray flare energy vs. cluster age; the triangles in the right corner indicate the maximum values found for solar (left) and dM star (right) flares.

In Fig. 2 we have plotted the total X-ray flare energy vs. the cluster age. One can see a smooth decrease from the very high values for the youngest stars towards solar values. This is very similar to the decrease of the X-ray luminosity with age, probably caused by the decrease of the rotation velocity during early stellar evolution (see Neuhäuser et al. 1995). The same age dependence is found for the loop length.

These results suggest that a relation of the flaring process to accretion processes or a disk corona seems improbable: in this case we would expect to find a sharp decrease of the flare energies at the age of a few million years, when the disks are getting dispersed. In contrast to this, we find a smooth decrease extending over more than 10^9 years. Our results can be best explained by the decrease of scaled-up solar-like coronal activity usually found for young stars.

References

Gagné M., Caillault J.-P., Stauffer J.R., 1995, ApJ (in press)

Gurzadyan G.A., 1980, Flare Stars, Pergamon Press, Oxford, p. 179

Montmerle T., Koch-Miramond L., Falgarone E., Grindlay J.E., 1983, ApJ 269, 182

Neuhäuser R., Sterzik M.F., Schmitt J.H.M.M., Wichmann R., Krautter J., 1995, A&A (in press)

Pallavicini R., Tagliaferri G., Stella L., 1990, A&A 228, 403

Preibisch Th., Zinnecker H., Schmitt J.H.M.M., 1993, A&A 279, L33

Raymond J.C., Smith B.W., 1977, ApJS 35, 419

Schmitt J.H.M.M., Collura A., Sciortino S., Vaiana G.S., Harnden F.R., 1990, ApJ 365, 704

X-ray Flare Rate of T Tauri Stars in Taurus

R. Neuhäuser[1], *Th. Preibisch*[2]

[1] Max-Planck-Institut für Extraterrestrische Physik, 85740 Garching, Germany
[2] Universität Würzburg, Astronomisches Institut, Am Hubland, 97074 Würzburg, Germany

Abstract: We study the X-ray emission of several hundred (young, low-mass, late-type, pre-main sequence) T Tauri stars (TTS) in the Taurus T association, a nearby well-studied region of ongoing star formation. We report on X-ray emission variability of TTS as observed with the flux-limited *ROSAT* All-Sky Survey (RASS). Since RASS observations are spatially unbiased, we can investigate the X-ray flare rate of TTS on a large sample. We find that large flares are very rare (once per year), while medium-size flares can occur once in ~ 40 days.

1 Introduction

TTS are fast rotators with periods down to a few days (e.g. Bouvier et al. 1993). A strong correlation between X-ray surface flux and stellar rotation first observed by Bouvier (1990) indicates that weak-emission line TTS (WTTS) are intrinsically more X-ray active than classical TTS (CTTS), because WTTS rotate faster. The rapid stellar rotation sustains the internal magnetic field. A dynamo effect produces X-ray flares, and magnetic field lines emerge out of the stellar surface containing hot X-ray emitting plasma. TTS X-ray emission has first been observed with the *Einstein* Observatory. As expected from the model given above, irregular variability as well giant flares have been found in the X-ray light curves of TTS (see Montmerle et al. 1983).

2 ROSAT observations

As reported in Neuhäuser et al. (1995), CTTS are rarely (11%) found with RASS, while most WTTS (66%) are detected as strong X-ray sources; TTS X-ray emission can be modelled with one-temperature Raymond-Smith (Raymond & Smith 1977) spectra with energy peaking at ~ 1 keV ($\sim 10^7$ K); WTTS and CTTS luminosity functions are significantly different; X-ray luminosity and surface flux are correlated with rotation period. Also, RASS observations confirm the irregular variability of TTS X-ray emission. However, due to low exposure time and, hence, statistical errors in RASS X-ray count rates, light curves are in many cases also not inconsistent with constant X-ray emission.

The X-ray emission model outlined above also predicts large flares. Preibisch et al. (1993) summarize properties of X-ray flares and report on the largest such flare detected to date among TTS, the X-ray flare of LHα 92, observed during a *ROSAT* pointing on the Perseus OB association. The X-ray telescope *ROSAT* has surveyed the sky in great circles, i.e., each object has been observed once in 90 minutes for up to $\sim$ 30 seconds. In Taurus–Auriga, each TTS has been observed on average for 575 seconds, on average in 27 scans of average exposure time of 21 seconds. As typical TTS X-ray flares last several hours (e.g. Preibisch et al. 1993), i.e., longer than the separation between any two scans, RASS can detect any sufficiently large flare during RASS patrol time (start of first until end of last scan). The average RASS patrol time for a TTS in Taurus–Auriga is 41 hours.

In evaluating the X-ray light curves, so far we have recognized flares just by visual inspection, accepting as flares those events with short rise time, e-folding decay time of several hours, and peak count rate significantly above quiescent level.

3 Results

We have detected one large X-ray flare on the single WTTS 040234+2143. Also, we have found eleven medium-size flares including three peaks in GV Tau. In total, we have detected twelve flares among 256 TTS observed (including both TTS known prior to the *ROSAT* mission and newly discovered TTS). Eleven medium-size flares in 256 TTS observed (on average for 41 hours) correspond to roughly one flare in 40 days. The fact that we have observed only one large flare during this patrol time means that such large flares are very rare (once per year). These flare rates are very rough estimates and need to be improved by implementing objective flare recognition techniques on a larger sample including TTS in other star forming regions. Extending the X-ray flare luminosity versus flare rate plot (Montmerle et al. 1983) down to solar-type small and frequent flares can help distinguish whether TTS X-ray emission consists of many small flares or whether there is a truly quiescent level.

References

Bouvier J., 1990, AJ 99, 946

Bouvier J., Cabrit S., Fernandez M., Martin E.L., Matthews J.M., 1993, A&A 272, 176

Montmerle T., Koch-Miramond L., Falgarone E., Grindlay J.E., 1983, ApJ 269, 182

Neuhäuser R., Sterzik M.F., Schmitt J.H.M.M., Wichmann R., Krautter J., 1995, A&A (in press)

Preibisch Th., Zinnecker H., Schmitt J.H.M.M., 1993, A&A 279, L33

Raymond J.C., Smith B.S., 1977, ApJS 35, 419

Middleterm Flares in T Tau Stars of the Tau – Aur Complex

Alisher Hojaev

Ulugbek Astronomical Institute, Tashkent 700052, Uzbekistan

1 Introduction

Light variations of different kinds appear to be one of the characteristic features of T Tau stars. They are mainly caused by changes in the continuum radiation. The availability of large observational databases makes it possible to investigate different patterns of variability such as stochastic, quasiperiodic and quasicyclic ones with typical time scales from days to years, as well as sporadic flare and fuorlike outbursts. Recent attempts in classification of the photometric variability of PMS-stars were made by Herbst (1994). In this paper, we study the observational results on flares in T Tau and related stars belonging to the Taurus-Auriga complex, which resemble that of ordinary flare stars.

2 Observational data and results

A census of T Tau stars with homogeneous observations of short and middleterm flares is presented in Table 1. Our selection criteria were: pre- and postflare levels of brightness ought to be quasiquiescent; the amplitude has to be more than 3 times the value of standard error; the duration of brightening does not exceed a few hundred minutes; the lightcurve shape should be similar to that of a typical flare. For convenience, all observational data in Table 1 are divided into 3 subgroups.

The light curves of all flare events cannot be not distinguished from those occuring in cluster flare stars. Morphological consideration shows a presence of multiple structure (two of the FY Tau flares) as well as a simple one (another FY Tau flare, flares in V927 Tau, FZ Tau, CI Tau etc.). Estimated flare energies released in T Tau stars (on the average 10^{36} erg) are greater than in flare stars (up to 10^{35} erg). The $V/B-V$ diagram for the above stated stars is shown in Fig. 1. Flaring T Tau stars are apparently located somewhat lower and spread over a larger area than other objects and lie generally between T Tau stars without detected flares (or with smaller frequency of flare activity) and cluster flare stars, which occupy the dotted field.

Table 1. Flares in T Tau stars of the Tau-Aur complex

Star name	Flares number	JD 2400000+	Magnitude	Ampl. [mag]	Duration [min]	Notes/Remarks
FY Tau	3	44634.29	16.7	1.5	60	tt(1),vb
		45013.21	16.8	2.1	215	
		45286.44	17.1	1.6	100	
VY Tau	2	45644.32	16.4	1.5	60	tt(1),b(2)
		45678.28	16.4	1.1	50	
V955 Tau	2	44689.25	16.1	1.2	40	tt(1),b(2)
		45680.45	16.3	0.7	15	
V927 Tau	1	45683.49	16.1	2.1	55	tt(1),b(2)
FZ Tau	1	45368.19	15.7	1.2	50	tt(1),vb
GI Tau	1	45286.59	15.4	1.7	35	tt(1),vb,b(2)
CI Tau	1	45368.26	15.2	2.3	90	tt(1)
HP Tau	1	45013.18	15.5	0.7	40	tt(1)
DP Tau	1	45644.38	16.1	1.0	30	tt(1)
FF Tau	1	35041.79	15.8	1.4	61	wtt(1)
HV Tau	1	44880.47	17.1	1.4	50	wtt(1),b(2)
B 13	1	44994.31	17.1	2.3	>50	wtt(1),b(2)
B 41	2	45289.38	16.1	1.3	25	tt?,e-a
		45311.23	16.3	2.1	>65	
HQ Tau	1	45644.29	14.6	1.6	20	in
GN Tau	1	45724.25	17.3	1.4	20	in,e-a
SVS 1099	1	45373.17	14.5	1.3	170	in
V590 Tau	1	44668.20	16.8	1.0	30	in

Comments: (1) Herbig & Bell (1988); (2) Leinert et al. (1993);
b – binary; vb – vis. binary; in – Orion variable; e-a – Hα in emission

Besides the indicated stars in the Tau-Aur complex, references on flares and flare-like events are available which were detected in T Tau, RY Tau, DF Tau, V410 Tau, SU Aur, DG Tau, and DP Tau, which belong to the tt- and wtt-types. T Tau is a relatively frequently flaring T Tau star ($f \approx 30$/h). Flares of DF Tau may be due to active plages on the stellar surface. Krautter et al. (1994) have shown that in Tau-Aur the ratio WTTS/CTTS = 1.3 and close that of the Orion Nebula (1.5). These complexes are similar as concerns the flare activity of inner stars. A large percentage of flaring T Tau stars exists in the Orion Complex, but only a few of these stars are found in the NGC2264 and NGC7000 aggregates.

Possible mechanisms of flare activity of T Tauri stars are in accordance with the evolutionary scenario outlined by Montmerle and Andre (1989): In classical T Tau stars, flares seem to be caused by outbursts on the stellar surface,

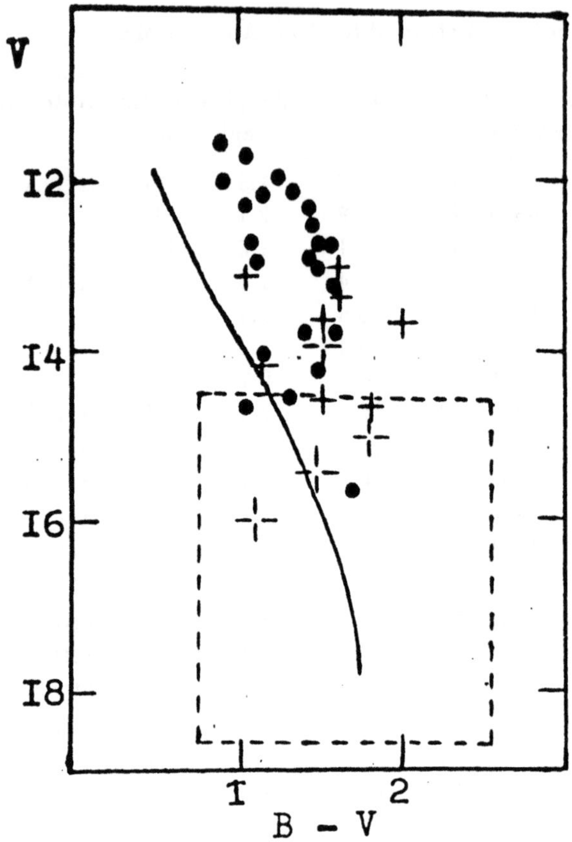

Fig. 1. The $V/B-V$ diagram for the T Tau stars given in Tab. 1: + marks T Tau stars with detected flares, – T Tau stars without flares.

but the initial radiation is distorted by circumstellar obscuration as well as by interaction processes with the surrounding accretion disk and the dust+gas envelope (sporadic outflow and infall of matter). Weak T Tauri stars have lost their circumstellar envelopes and their flares are presumably caused by the same mechanism which is dominant in typical flare stars.

References

Herbig G., Bell K., 1988, Lick Obs. Bull. No. 1111, p. 90.

Herbst W., 1994, in Nature and Evolutionary Status of Herbig Ae/Be stars, PASP Conference Series Vol. 62, p. 35

Krautter J., et al., 1994, Rev. Mex. Astr. Astrofis. 29, 41

Leinert Ch., et al., 1993, A&A 278, 129

Montmerle T., Andre P., 1989, in Low Mass Star Formation and Pre-Main Sequence Objects, ESO Conf. and Workshop Proceedings No. 33, p. 407

Activity on T Tauri Stars

Erik Gullbring[1], *Gösta F. Gahm*[1], *Heinz Barwig*[2], *Peicheng Chen*[3]

[1]Stockholm Observatory, S-133 36 Saltsjöbaden, Sweden
[2]Universitäts-Sternwarte München, Scheinerstrasse 1, 81679 München, Germany
[3]Yunnan Observatory, P.O. Box 110, Kunming, China

1 Observations

We have made a detailed investigation of the short-term variability of the classical T Tauri star BP Tauri in *UBVRI*. Data were collected from the Wendelstein Observatory in 1991, 1992 and 1993 with time resolutions down to 1 sec. The 0.8m telescope was equipped with a fiber-fed fifteen channel high-speed photometer (Barwig et al. 1987). Observations (in *UBV*) were also collected in China at the Yunnan and Shanghai Observatories to get a long base line in time. To search for differences in the properties between the brightness variations of classical T Tauri stars (CTTS) and T Tauri stars with weak emission lines (WTTS) we performed simultaneous photometry (in the *UBV* and Strömgren systems) and spectroscopy of 6 young stars during two observing periods at ESO La Silla. The study concerned mainly short-term variability on time-scales of minutes to a few hours. The sample contained two CTTS, SY Ori and VW Cha; three WTTS, San 1, SZ Cha and ADA 481 and one post-T Tauri candidate, HD 70309B.

2 Results

We found no pronounced flare activity on BP Tauri. Most of the time the star was constant or varied only very slowly in brightness. The few distinct events observed had amplitudes of $0\overset{m}{.}1$ to $0\overset{m}{.}3$ in the U-band and temperatures < 8000 K. The optical energy release of these events ranges from 10^{35} to 10^{36} ergs and there was a striking lack of events with lower energies. No brightness variations with time durations of < 0.6 hours were observed. Normally the rise time of an event was comparable to the decay time. One active night of BP Tauri is shown in the figure. The activity of BP Tauri can most easily be understood in terms of inhomogeneous accretion from a disk towards the star. The short time-scales of the variations favour models with magnetically controlled accretion.

The simultaneous spectroscopic and photometric observations showed that the short-term brightness variations of CTTS and WTTS behaved differently. We could distinguish between two types of fast events: (a) Rapid bursts of durations

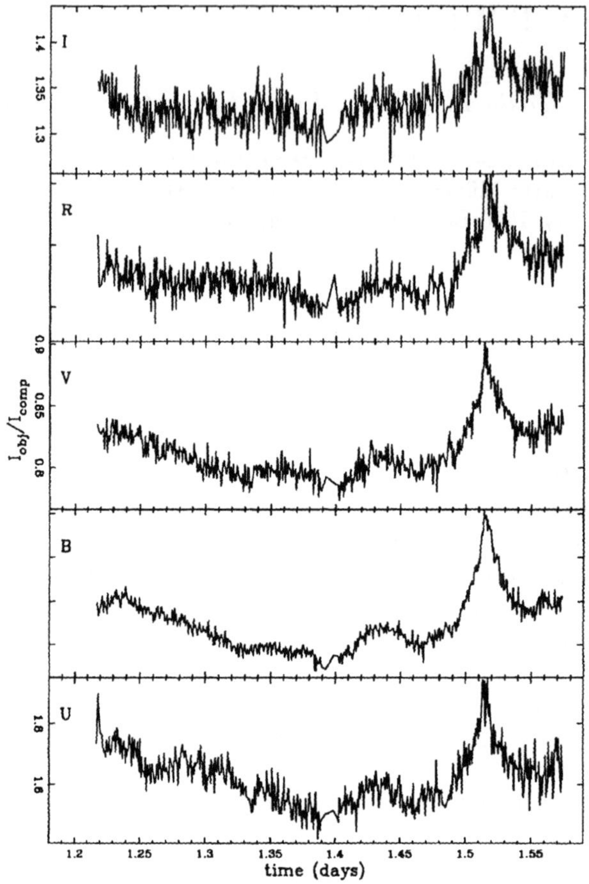

Fig. 1. High-speed photometry of BP Tauri. The time is given at JD 2449004+.

less than an hour, manifested solely in the Balmer continuous and line emission. Such events were found only on the WTTS. (b) Slow and smooth changes occurring over several hours in particular for VW Cha, where we could show that variable veiling dominates. These investigations show that the mechanisms behind most of the activity on CTTS and on WTTS are different. Accretion from circumstellar disks are probably responsible for the observed varibility on CTTS, while magnetically active regions cause the short-term variability on WTTS.

References

Barwig H., Shoembs R., Buckenmayer C., 1987, A&A 175, 327

Gahm G.F., Lodén K., Gullbring E., Hartstein D., 1995, A&A (in press)

Gullbring E., Barwig H., Chen P.S., Gahm G.F., Bao M.X., Chen Y.K., 1995, A&A (in prep.)

A Very Active T Tauri Star in NGC 7129

Evgeni Semkov

Institute of Astronomy, Bulgarian Academy of Sciences, 72 Tsarigradsko Shose, BG-1784 Sofia, Bulgaria

1 Introduction

The star forming region NGC 7129 was first studied by Herbig (1960), who discovered two bright (BD+65°1637 and LkHα234) and some weaker Hα emission stars. Subsequent investigations in this region led to the detection of five Herbig-Haro objects (Strom et al. 1974, Gyulbudaghian et al. 1978, Eiroa et al. 1992) and one T Tauri star, V350 Cep (Gyulbudaghian & Sarkissian 1977). Some non-stable stars in NGC 7129 are surrounded by reflection and cometary nebulae.

Our Hα-emission survey (Semkov & Tsvetkov 1986) revealed an interesting irregular variable star in the dark clouds near the emission nebulae. On objective prism plates the star shows very strong Hα-emission and was included in our list of such stars as No. 7. An identification chart and the coordinates can be found in Semkov (1993a). The observations reported here include long-time photographic photometry and CCD spectroscopic and photometric observations.

2 Observations

The photographic UBV observations (Semkov 1993a) were made with the 50/70/172cm Schmidt telescope of the Rozhen Astronomical Observatory of the Bulgarian Academy of Sciences. The star brightness ranges from $14\overset{m}{.}9$ to $17\overset{m}{.}0$ in B-light (Fig. 1) with periods of enhanced activity. The colour index $B-V$ varies around the value $1\overset{m}{.}5$ and $U-B$ around $-0\overset{m}{.}6$, hence the star has a strong ultraviolet excess.

Spectroscopic CCD observations of star No. 7 were made in September 1992 with the Focal Reducer and P8603/B camera of the Max-Planck-Institute for Aeronomy attached to the Rozhen 2 m RCC telescope. The spectra were taken with a grating prism at a dispersion of 0.47 nm/pixel (Fig. 2). They exhibit strong emission lines of H I, O III, Fe II, Mg I and other metals (Semkov 1993b). Hα is very intense and surpasses the frame of Fig. 2.

Photometric CCD observations of the star were made with the SBIG ST-6 camera installed in the RC focus of the 2m RCC telescope. This camera was

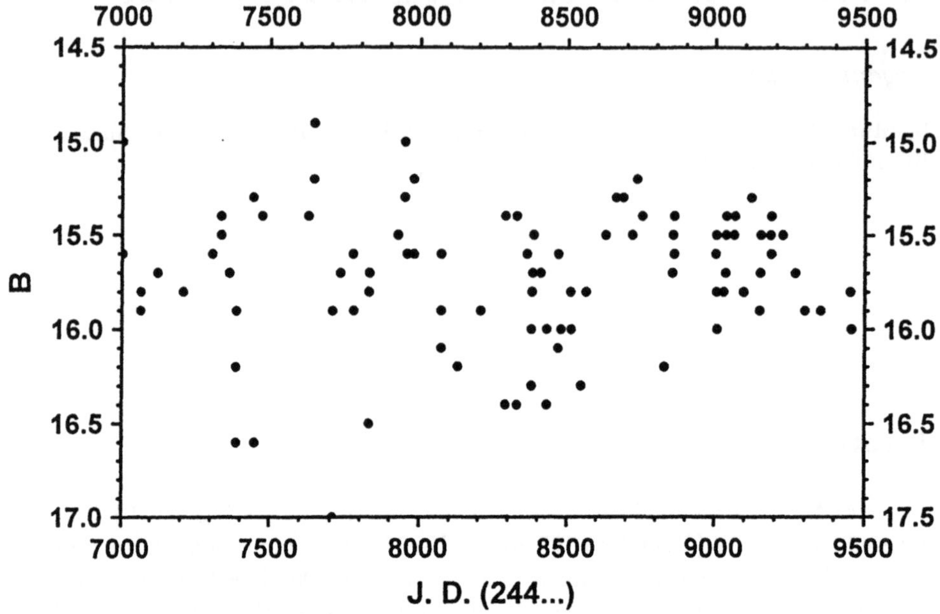

Fig. 1. Light curve of the new T Tau star during the period July 1987 – April 1994 in *B*-light.

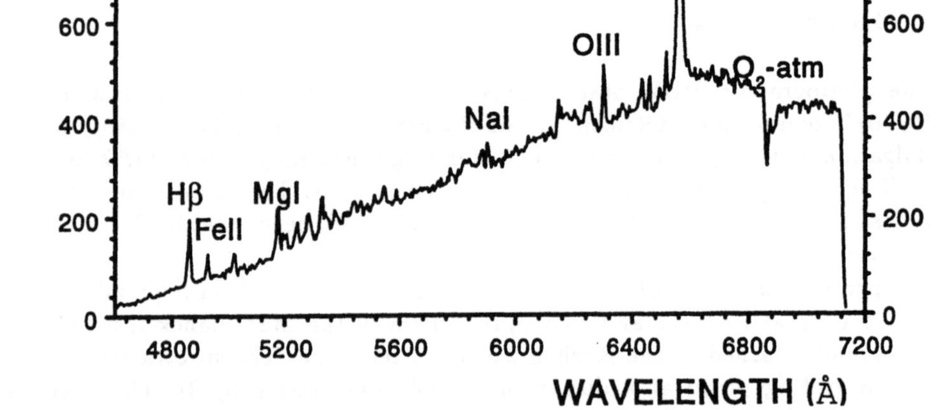

Fig. 2. CCD spectrum of the new T Tau star.

given as a present to the Rozhen Astronomical Observatory from the European Astronomical Society. The observations were made in the period August 1993 – August 1994 in the standard *BVRI* system (Table 1). The CCD photometry suggests that the star is now in a period of decreasing activity (the amplitude is about $0^m.3$). CCD frames through broad band filters *I* and *R*, obtained on Desember 18, 1993, are given in Fig. 3a and 3b. In both figures North is at the top, East is on the left and the scale is 0.5 arcmin on a side. The CCD observations showed that there is a small cometary nebula around the star, which is not visible on the Palomar Observatory Sky Survey prints. To the East from the star a compact object is allocated which is visible better on the infrared frame and is absent on the *B* and *V* frames. Preliminary photometry leads to a colour index $R - I = 1^m.3$ for this object.

Table 1. CCD photometry of No. 7a.

Date	J.D. (244000+)	V	$B-V$	$V-R$	$V-I$
12/13.08.1993	9212.425	14.34	1.30	0.85	
13/14.08.1993	9213.378	14.38			
14/15.08.1993	9214.369	14.33			
18/19.12.1993	9340.322	14.39	1.26	0.86	1.82
19/20.12.1993	9341.267	14.41	1.22	0.81	1.69
7/8.08.1994	9572.398	14.53		0.98	1.83
8/9.08.1994	9573.435	14.42	1.11	0.97	1.79
10/11.08.1994	9575.436	14.21	1.13	0.88	1.70

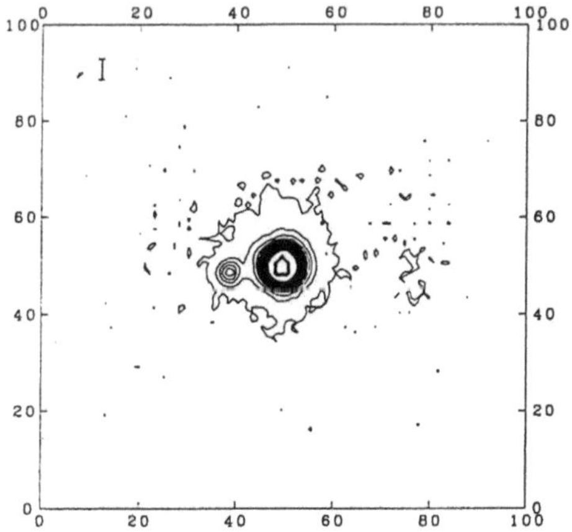

a

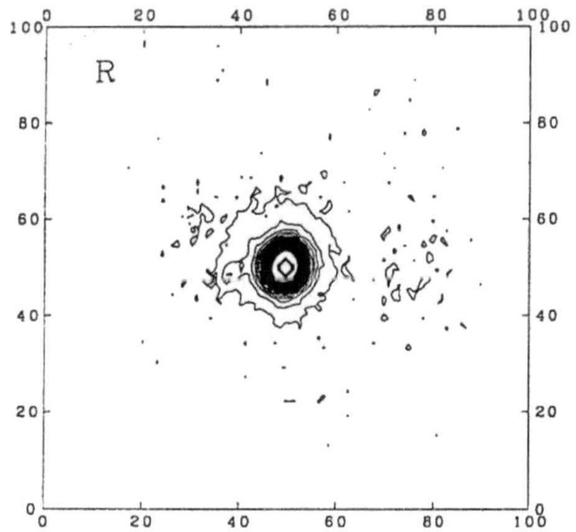

b

Fig. 3. CCD frames of the new T Tauri star.

3 Discussion

The spectral observations confirm our preliminary assumption that the star has T Tau type variability: such a spectrum is typical for very active T Tau stars connected with nebulae according to Herbig's classification (1962). The distance between the two components of star No. 7 is about 3". If we assume that the object is at a distance of 1 kpc (Racine 1968), the real distance between the two components must be at least 3000 AU. For the second object (No. 7b) we must reject our first assumption that it is a jet from the star or a Herbig-Haro object. The $R-I$ colour index is typical only for deeply embedded infrared sources (for very young stars). Our conclusion is that No. 7a and 7b are members of a wide double system of pre-main sequence stars in different stages of evolution.

Acknowledgements: I am thankful to the IAU and the Deutsche Forschungsgemeinschaft for a grant to attend the IAU Colloquium 151 in Sonneberg. Our research has been partly supported by the Bulgarian National Science Foundation (grants F-311/93 and F-340/93).

References

Eiroa C., Gomez de Castro A.I., Miranda L.F., 1992, A&AS 92, 721
Gyulbudaghian A.L., Sarkissian R.A., 1977, Astr. Tsirk. 972, 5
Gyulbudaghian A.L., Glushkov Yu.I., Denisyuk E.K., 1978, ApJ 244, L137
Herbig G.H., 1960, ApJS 4, 337
Herbig G.H., 1962, Adv. in Astr. and Astrophys. 1, p. 47
Racine R. 1968, AJ 73, 233
Semkov E.H., Tsvetkov M.K., 1986, in Star Clusters and Associations, G. Szecsenyi-Nagy (ed.), Publ. Astr. Dep. Eotvos Univ., Budapest, 8, p. 141
Semkov E.H. 1993a, IBVS 3870
Semkov E.H. 1993b, IBVS 3919
Strom S.E., Grasdalen G., Strom K. M., 1974, ApJ 191, 111

Polarimetric Outbursts in Young Herbig Ae/Be Stars and in Peculiar B[e] Stars

R. V. Yudin

Central Astronomical Observatory of the Russian Academy of Sciences at Pulkovo, 196140 Saint-Petersburg, Russia

1 Observations

Quasi-simultaneous optical and IR photometric (*BVRIJHK*) and optical polarimetric (*BVRI*) observations were carried out at the 1 m telescope of Assy Observatory (Institute of Astrophysics of the Kazakhstan Academy of Sciences). The photometer-polarimeter of the Pulkovo Observatory (Bergner et al. 1988) was used for the study. All our measurements were done with apertures of 26″ and 52″.

2 Results

HDE 259431. Our data show a strong polarimetric variability $\Delta p = 2\% - 3\%$ and $\Delta\Theta = 30°$ (especially in UV and optical region). Although the general nature of polarization is connected with the dust disk around the star (Dzhakusheva et al. 1988), the polarimetric variability probably has an another origin. This is due to the fact that we have observed a strong polarimetric variability on a time-scale of up to several hours (Dzhakusheva et al. 1988). Incidentally, this rapid polarimetric variability has no correspondence in photometry.

MWC 137. Optical polarimetric observations (9 R-band measurements) were obtained for the first time. A very large polarization degree (up to 9%), together with a large variability ($\Delta p = 4\%$) was detected. It is unusual that the increase of the degree of polarization is accompanied by an increase of the brightness in the UV.

MWC 342. In our data, a periodic photometric and polarimetric variability was detected. The period of photometric variability in the optical spectral region is 132 days, but the period of the polarimetric variability is only half this value (about 66 days). Moreover, the maximum of the degree of intrinsic polarization falls on phases 0.2 and 0.8 of the photometric period. According to theoretical calculations (Dolginov et al. 1979), such a conformity between the photometric and polarimetric periods is predicted for binary systems having a compact X-ray source. This effect may be explained by the existence of a hot spot on the surface of the Be star which is connected with the conversion of X-rays from the compact component. Using the data from Amnuel et al. (1979), we have found an X-ray source near MWC 342. The coordinates of the X-ray source ($\alpha = 20^h 19^m 00^s$, $\delta = +39°30'00''$). almost coincide (within the error bars) with those of MWC 342 ($\alpha_{1950} = 20^h 21^m 14\overset{s}{.}6$, $\delta_{1950} = +39°20'09''$)

Z CMa. *UBVRIJHK* photometry and *BVRI* polarimetry were obtained in 1985–1990. During the optical photometric outburst in 1986–1987 a significant increase of the polarization degree (up to 5%) was detected. We explain the polarization behaviour during the outburst in terms of the increase of the orientation degree of nonspherical dust grains in polar "lobes". This model is discussed in more detail by Miroshnishenko & Yudin (1993).

MWC 349. We obtained 27 polarimetric measurements (mainly in the *R*-band). During the observations reported here, both the degree of polarization and the position angle are variable ($\Delta p = 8\%$ and $\Delta\Theta = 15°$). The typical time-scale of variability is about 1–2 days. During 12–13.09.90 five polarization observations were made and a variation of the degree of polarization by 3% during 7 hours was detected. No correlations were found between the degree of polarization and the brightness in all photometric bands.

Some of our observational results are as follows:
- Some objects from our program show a synchronous increase of polarization degree and brightness (for example MWC 137 and Z CMa).
- Typically, polarimetric variability is observed on a time-scale of days or less.

3 Conclusions

The difference in the polarimetric properties for some Herbig Ae/Be stars and B[e] stars is connected with different physical conditions in their circumstellar shells. This can be specifically connected with the different evolutionary stages of their circumstellar shells.

References

Bergner Yu.K., Bondarenko S.L., Miroshnichenko A.S., Moralev Yu.D., Schumacher A.V., Yudin R.V., Yutanov N.Yu.: 1988, Izv.GAO AN SSSR, vol. 205, p. 142

Dzhakusheva K.G. et al., 1988, Sov. Astr. Lett. 14, 317

Dolginov A.Z., Gnedin Yu.N., Silant'ev N.A., 1979, Dissemination and Polarization of Radiation in Space, Nauka, Moscow

Amnuel P.R., Guseinov O.H., Rakhamimov Sh.Yu., 1979, ApJS 41, 327

Miroshnichenko A.S., Yudin R.V., 1993, Astrophys. Space Sci. Library 186, p. 173

Photometric Behaviour of the EXors V1118 Orionis and V1143 Orionis

C. Paul[1], *P. Kroll*[2], *T. Lehmann*[3]

[1] Astrophysikalisches Institut und Universitäts-Sternwarte Jena, Schillergäßchen 2, 07740 Jena, Germany
[2] Sternwarte Sonneberg, 96515 Sonneberg, Germany
[3] European Southern Observatory, Casilla 19001, Santiago 19, Chile

Abstract: We investigated the long-term photometrical behaviour of the EXors V1118 Ori and V1143 Ori on the basis of the Sonneberg Plate Archive. About 2800 plates were checked for outbursts of these stars. We found several dozens of plates where the stars appeared to be distinctly brighter than the plate limit. To get magnitudes we have scanned those plates and applied a special method of brightness determination which is currently used at Sonneberg Observatory. Because of the small number of plates we were able to extract only crude estimates of some typical time scales of outburst and quiescent phases.

1 Introduction and brightness determinations

EXors are a small group of very active T Tauri stars. Their main characteristics are strong outbursts of up to 5^{rmm}, but the stars normally show only small amplitude irregular variations. The typical recurrence time scale of these events is of the order of a decade. FU Orionis stars (FUors) are another group of young low-mass stars with even stronger outbursts, which may last for a century.

Herbig (1977) pointed out that FU Orionis eruptions with strong intensities occur in T Tauri stars, while eruptions are absent on the main sequence, and there should exist similar objects with minor activity. EX Lupi was the first object and prototype (that's why they are called EXors).

Two recently found EXors are V1118 Ori (Chanal's object) and V1143 Ori (Sugano's object), which we studied on the basis of the Sonneberg plate archive.

For the derivation of magnitudes we have scanned the plates on which the stars are visible. V1143 Ori is an object without bright stars in its neighbourhood. Therefore the fitting method of Kroll & Neugebauer (1993) was suitable to obtain reliable magnitudes. The accuracy σ of these values is about $0\overset{\text{m}}{.}2$ to $0\overset{\text{m}}{.}3$ is mainly due to the rather poor plate quality (e.g., due to short exposure times). The star V1118 Ori could not be measured with the fitting method. It is embedded in an H II region and is located very near to the bright star V372 Ori. In this case the program fails, and the traditional method of Argelander was used, the stellar magnitude was estimated visually using a plate microscope. Since the comparison stars are also embedded in the H II region, the variable had to be linked to stars not only with similar magnitude but also with similar background level. We estimate the accuracy to be $\Delta m \sim 0^{m}.5$.

2 Results

We examined about 2 800 photographic patrol plates, taken between 1928 and 1993, for visibility of the investigated stars. In most cases, both stars were not visible because of the insufficient plate limit of $m_{pg} \sim 12^{m}.5 \ldots 14^{m}$ of the short-focus (patrol) cameras.

From the ~50 astrograph and Schmidt plates ($f \sim 160$ cm, $m_{pg/lim} \sim 17^{m} \ldots 18^{m}$) on which V1143 Ori was found and from the patrol plates, it appeared that this star never flared brighter than $m_{pg} = 12.5$ during this time. The variable V1118 Ori, however, owing to the vicinity of the bright star V372 Ori, could not be found on patrol plates at all.

For V1118 Ori, discovered by Chanal in 1984, only a few pre-discovery observations were published (Gasparian et al. 1987). However, with the data of the Sonneberg plates we found that there were further flares in the past. Fig. 1 shows these few data. Chanal's object was visible at the following dates: 1939 Feb. 24 with $m_{pg} \sim 14^{m}.0$, 1956 Feb. 5 with $m_{pg} \sim 14^{m}.0$ and 1961 Feb. 16 with $m_{pg} \sim 14^{m}.0$. The other data confirm the outburst in 1982, which probably had its maximum in 1983. Further data were not available. It is interesting that V1118 Ori flared on 1983 Dec. 5 to a photographic brightness of $m_{pg} \sim 12.5$. Compared with the literature that seems to be the brightest value ever recorded (assuming $m_{pg} \simeq m_{B}$).

V1143 Ori showed also an outburst in 1982/83 and, after several brightenings, probably a second one in 1989. This star was active in the past inb a similar way as V1118 Ori (Fig. 2).

Our data confirm the outburst in 1983. Further results are a brightening of about 2 mag (photographically) in 1939 (Feb. 22, $< 17^{m}.7$; Feb. 23, $17^{m}.1$; Feb. 24, $15^{m}.8$) and a possible outburst between ~ 1966 and ~ 1970 (1965 Oct. 22, $< 16^{m}.5$; 1967 Feb. 6, $14^{m}.2$; 1969 Oct. 6, $14^{m}.0$; 1972 Feb. 15, $< 14^{m}.7$ - photographic magnitudes). This conclusion is in agreement with the lightcurve published by Gasparian et al., showing a possible post-outburst activity in September 1970. On the other hand, Chavira et al. (1985) claimed that V1118 Ori showed no outburst from 1944 to the beginning of 1982.

3 Discussion

V1118 Ori and V1143 Ori exhibited outbursts or brightenings in the last few decades. However, owing to our sparse material it is not possible to derive rising times or duration of maxima.

On the other hand, according to the published data (mainly from the electronic AFOEV database) it is obvious that the timescales of these events reached several years at most.

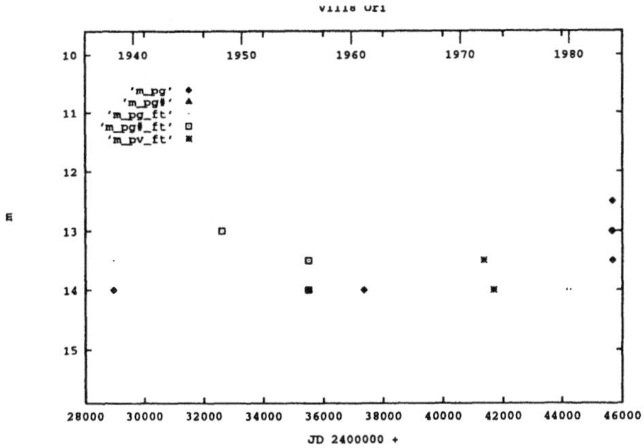

Fig. 1. Light-curve of V1118 Ori (m_{pg} – photographic magnitudes, $m_{pg\#}$ – uncertain photogr. mag., other – fainter than).

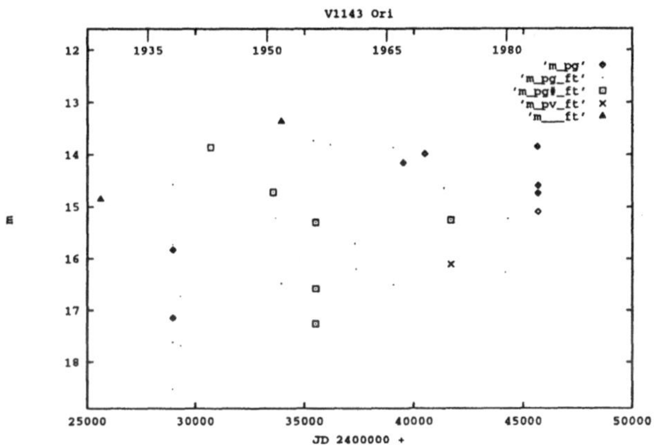

Fig. 2. Light-curve of V1143 Ori (m_{pg} – photographic magnitudes, other – fainter than).

Acknowledgement: This work was supported by grant 2JN12A of Bundesministerium für Forschung und Technologie (Paul) und by grant 05 2S0 524 of DARA/ DESY (Kroll).

References

Kroll P., Neugebauer G., 1993, A&A 273, 341
Gasparian K.G., et al., 1987, IBVS 3024
Chavira E., et al., 1985, IBVS 2746
Herbig G.H., 1977, ApJ 217, 693

Observations of Short-term Dips in the Orion Variables DD Serpentis and AB Aurigae

I.A. Verlyuk

Main Astronomical Observatory, National Academy of Sciences of Ukraine, 252650 Kiev, Ukraine

Abstract: Short-term fadings (dips) in the range of a few seconds with amplitudes from $0\overset{m}{.}3$ to $1\overset{m}{.}2$ were discovered in the two T Orions type variables DD Ser and AB Aur. Similar features were observed in T Tau and R CrB by other observers.

1 Introduction

The Orion variables (INA) of spectral types B–A are characterized by occasional sudden Algol-like fadings with amplitudes of about $0\overset{m}{.}5 - 1\overset{m}{.}0$ in the range from a few hours to several days. This kind of activity may by interpreted in terms of episodic condensation of grains in a stellar wind. Our observations reveal that similar features take place on time scales a short as several seconds.

2 Observations

The observations were obtained at the 60 cm telescope at the Mount Terskol Observing Station in North Caucasus (3 100 m above sea level). They were obtained with a high-speed two-channel photometer (Zhilyaev et al. 1992). Simultaneous measurements of a reference star was performed to accurately determine atmospheric conditions during the time of observations. High time resolution monitoring of AB Aur and DD Ser was made in the *UBV* bands in autumn of 1992 and 1993. Integration times in the range 0.1 to 0.5 sec were used.

3 Results

In DD Ser, few dips were detected with amplitudes and durations in the range from $0\overset{m}{.}3$ to $0\overset{m}{.}7$ and 4 – 5 sec, respectively. One short-lived dip was detected in AB Aur with an amplitude of the order of $1\overset{m}{.}2$. The duration of the latter was 1.7 sec. All dips show more or less symmetrical smooth light curves. Figs. 1–2 show that the reference star remained constant. Similar dips features were observed in T Tau and R CrB. As noted by Zajtseva (1980) the light curve of T Tau on 1977 October 30, exhibits a few sudden fadings up to about 1^{m} on

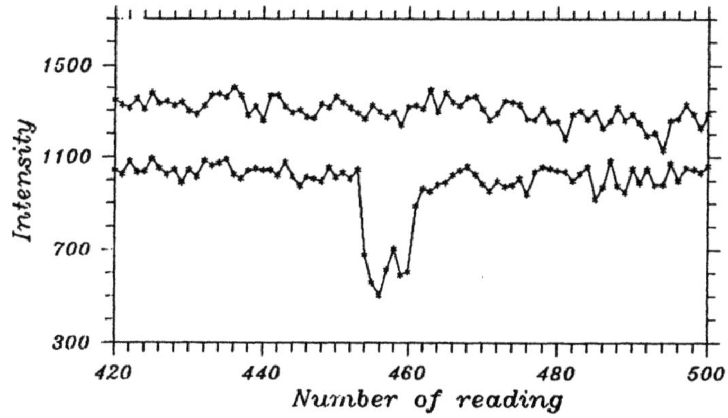

Fig. 1. DD Ser on 1992, July 6. B band. Time resolution: 0.5 sec. Top: reference star; bottom: variable.

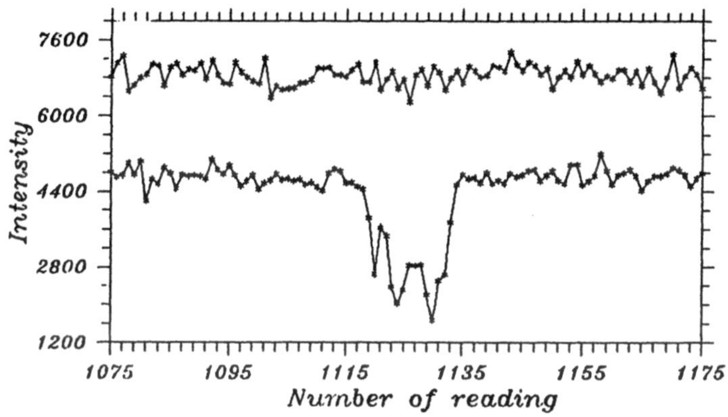

Fig. 2. AB Aur on 1992, August 21. Same parameters as Fig. 1

a time scale of 10 sec in the *U* band. Similar phenomena had been observed in R CrB by Totochava (1977) during the deep minimum passage in 1976. One can imagine that dips observed in the cases of DD Ser and AB Aur are quite similar to those observed on T Tau and R CrB.

Acknowledgements: I wish to thank Dr. B.E. Zhilyaev for his kind assistance.

References

Totochava A.G., 1977, Photometric study of the RCB type stars, Ph.D. thesis, Odessa State University

Zajtseva G.V., 1980, in Flare Stars, Fuors and Herbig-Haro Objects, L. Mirzoyan (ed.), Yerevan, 1980, p. 61

Zhilyaev B.E., Romaniuk Ya.O., Svyatogorov O.A., 1992, AZh 69, 895

CATACLYSMIC VARIABLES:

ERUPTIONS AND FLICKERING

Hydrogen and Helium Flashes in Cataclysmic Variables

Mariko Kato

Keio University, Hiyoshi, Kouhoku-ku, Yokohama 223, Japan

Abstract: Light curve analysis for the decay phase of novae gives estimates of the white dwarf mass, the luminosity (or distance) and the chemical composition of ejecta. Differences in these parameters among nova subclass, i.e. fast, slow and recurrent, are briefly summarized to speculate on their cause. Helium shell flashes occur in mass-increasing white dwarfs, their theoretical light curves are given in UV and visual wavelength bands. An evolutional scenario for fast, slow and recurrent, and slow novae through a helium nova/EUV source-stage toward a type Ia supernova/neutron star formation-stage is given.

1 Hydrogen shell flashes in cataclysmic variables

A nova explosion is a thermonuclear runaway event (TNR) on the white dwarf of a close binary system. A hydrogen shell flash on the white dwarf causes the outburst in which the star quickly brightens up and the hydrogen-rich shell greatly extends. The envelope is eventually blown off. This extended development of a nova outburst can be modelled by the optically thick wind theory which has been so far the only method in reproducing nova light curves. The optically thick wind is a continuum-radiation driven mass loss in which the acceleration occurs deep inside the photosphere. The structure of the envelope is obtained by solving the equations of motion, continuity, radiative transfer in diffusion approximation, and energy conservation. Steady-state and spherical symmetry is assumed. The new OPAL opacity is used. The nova decay phase is described as a sequence of the steady wind solutions. The numerical methods are summarized in detail by Kato & Hachisu (1994).

2 Light curve analysis of the decay phase of novae

Novae can be divided into three subclasses, i.e. fast, slow and recurrent novae. Fast novae brighten by more than 10 magnitudes within a few days and after the luminosity peak, they almost return to their pre-outburst magnitude with time scales of months to a year. The chemical composition of ejecta deviates very much from the solar abundance and is abundant in He, C, O and other heavy elements. Slow novae show a decline time of several months to years. Helium enrichment is observed in some slow novae. Recurrent novae have a short decline time of 10 days to several months and usually a relatively small amplitude. The outburst repeats every few decades. No heavy element enhancement has been reported.

The first example of light curve fitting is for the fast classical nova V1668 Cyg (Nova Cyg 1978). Fig. 1 shows the theoretical light curves in optical and UV regions as well as the observed data. TNRs on massive white dwarfs show a rapid evolution because of the small envelope mass. Both the optical and UV data show a good agreement with the model of a 1.0 $M_\odot$ white dwarf. The distance to the star can also be estimated from the comparison between observed apparent magnitude and the theoretical absolute magnitude and turns out to be 2.9 kpc (UV) and 3 kpc (optical). The theoretical expansion velocity at the photosphere is also consistent with *IUE* data. Details are published by Kato (1994) and Kato & Hachisu (1994).

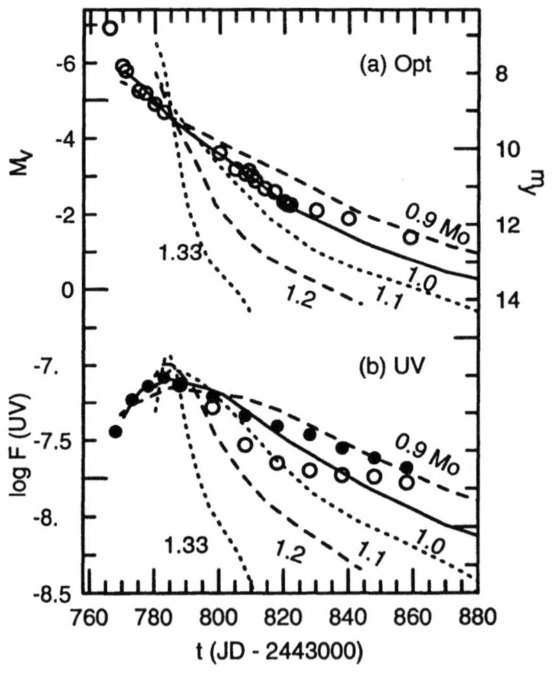

Fig. 1. Light curve fitting for the fast classical nova V1668 Cyg. Theoretical curves are denoted by thick curves, which are labelled by the white dwarf masses. (a) Optical light curves. The left ordinate shows the absolute visual magnitude for the theoretical curves and the right ordinate the apparent y-magnitude for observed data. Data are taken from Gallagher et al. (1980). (b) UV light curves. Open circles denote the UV flux (1140 Å- 3290 Å) and filled circles give the summation of the UV and IR fluxes (> 12000 Å) in units of erg/cm^2 sec^{-1} (Stickland et al. 1981). In theoretical UV flux, $F = L_{\rm UV}/4\pi D^2$, the distance to the star $D = 2.88$ kpc is assumed.

Another example of a classical nova is GQ Mus (Nova Muscae 1983). Kato (1995) shows that the light curves of UV, optical, and IR bands are well fitted by models with the white dwarf of mass 0.5–0.6 $M_\odot$. These results indicate that the white dwarfs in classical novae be not always as massive as the Chandrasekhar limit as has been assumed in many papers.

As an example for the light curve fitting of a slow nova, Kato (1995) investigated RR Pic and gives an estimate of the white dwarf mass of 0.8 – 0.9 $M_\odot$.

Recurrent novae show very rapid decline rates and short recurrence periods. Such a rapid evolution can be obtained only in very massive white dwarfs. Fig. 2 shows the light curve fitting of V394 CrA. This figure shows theoretical light curves with different sets of chemical composition for a 1.377 $M_\odot$ white dwarf as well as observational data.

Kato (1995) shows that three other recurrent novae, U Sco, V745 Sco and T CrB, have very similar light curves to V394 CrA so that light curve fitting gives also very massive white dwarfs in these objects.

From these results of light curve fitting and observational characters, we can summarize differences in physical parameters among nova subclasses. Fast classical novae have relatively small white dwarf masses as small as 1.0 $M_\odot$ or less with heavy element enrichment in ejecta. Metal enrichment causes strong wind even on less massive white dwarfs. Slow classical novae show a long decay time, which is explained by models with small white dwarf mass and almost solar abundance of heavy elements. Rapid evolution in recurrent novae is explained by models of very massive white dwarfs, close to the Chandrasekhar limit.

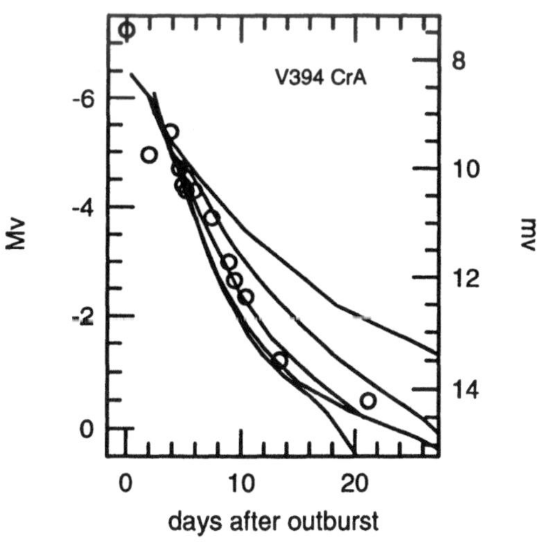

Fig. 2. Light curve fitting for V394 CrA. Dots denote the observational data by Duerbeck (1988). Theoretical curves show the dependence on the chemical composition; from upper to lower, (X,Z)=(0.5, 0.002), (0.1 0.004), (0.1,0.01), (0.7, 0.02), and (0.1, 0.02) are assumed. The white dwarf mass of 1.377 $M_\odot$ is assumed.

3 Does the white dwarf mass increase or decrease?

Now, we will consider whether the white dwarf mass increases or decreases after many cycles of hydrogen shell flashes. The envelope mass, accreted from the companion, will be lost in part by the wind during the nova outburst, and the rest of it remains on the white dwarf. From the wind models fitted to observational light curves, we can estimate the amount of mass ejected by the wind and the amount of helium processed by nuclear burning. The ratio of matter that remains on the white dwarf after one cycle of hydrogen shell flash to the matter accreted is estimated to be a few percents for fast and slow novae, whereas it is a few tens of percents for recurrent novae.

This ratio further decreases if we include the effect of mixing of white dwarf matter into the hydrogen-rich envelope. There are two kinds of mechanisms proposed for mixing; One is hydrogen diffusion into the white dwarf matter (Prialnik & Kovetz 1992) and the another is shear mixing due to differential rotation (Fujimoto 1993). Such mechanisms mix the white dwarf matter into the upper envelope during the quiescent phase and then hydrogen ignites deep inside the envelope where heavy elements are enriched. Convection carries such material upward and mixes it into the upper envelope that will be ejected by the wind. In such a case the white dwarf losts more mass that it had accreted.

The heavy element enrichment observed in fast classical novae is, therefore, the evidence of dredge up of white dwarf material. In these objects the white dwarf mass will decrease after it suffers many cycles of nova outbursts. It is to be noted that neon novae, i.e. novae with neon rich ejecta, not necessarily contain a massive white dwarf, although stellar evolution theory concludes ONeMg white dwarfs should be massive ($> 1.35\ M_\odot$) at their birth.

In the case of slow classical novae, the helium enhancement in ejecta suggests that the white dwarfs have a helium layer under the hydrogen-rich envelope, and a part of the helium layer is dredged up by the same mechanisms as in the previous class.

In the case of recurrent novae, there is no evidence of heavy-element enrichment in ejecta. This suggests that no white dwarf matter is dredged up. Helium enhancement can be explained in part by hydrogen burning that is strong on very massive white dwarfs, and in part by dredged up material from the helium layer. When these observational aspects are combined with light curve fitting, we can conclude that the white dwarf keeps a part of the accreted material and will grow in mass.

4 Helium shell flashes

When a part of the accreted matter remains after the hydrogen shell flash, the white dwarf develops a helium layer under the hydrogen burning zone. This helium layer will grow in each hydrogen shell flash and when the mass of the helium layer reaches a critical value, unstable helium burning occurs to trigger a nova-like phenomenon. Energy generated by helium burning is consumed partly by the gravitational work to push matter upward and is partly radiated away. When most of the helium layer is burnt or blown off, the star becomes faint and the helium shell flash finishes.

Figure 3 shows light curves of helium flashes obtained by the optically thick wind theory. The radius of a hot white dwarf is assumed (Iben 1982). The light curve of helium novae is similar to that of typical novae but the development is very slow. When the helium shell flash is weak because of small ignition mass, the light curve starts from the middle part of curves in Fig. 4. In this case the star is observed as a bright UV source but it is faint at optical wavelengths.

Observational identification of a helium shell flash is important for the study of a relation between novae and type Ia supernovae. A helium shell flash occurs only on mass increasing white dwarfs, and will be observed as helium nova or a bright UV source with a composition which is highly deficient in hydrogen.

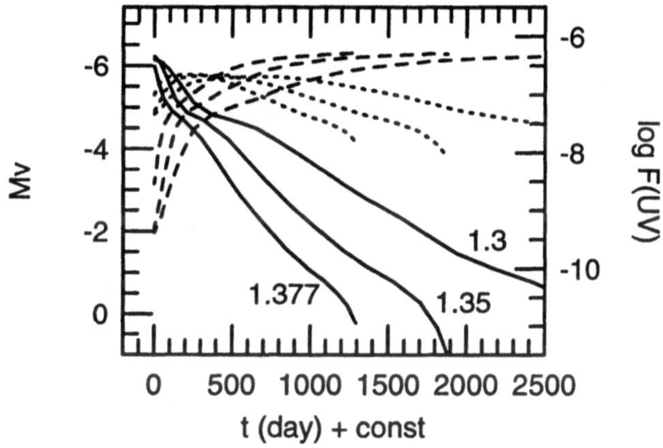

Fig. 3. Theoretical light curves of helium shell flashes on a white dwarf. Each curve is labelled by the corresponding white dwarf mass. Light curves in the visual, ultraviolet, and extreme ultraviolet bands are denoted by solid, dotted and dashed curves, respectively. In UV and EUV fluxes, a distance of 2.2 kpc is assumed.

5 Connection between novae and type Ia supernovae

Figure 4 summarizes the fate of white dwarfs in nova systems. In fast classical novae, the white dwarf has no helium layer under the hydrogen envelope and the white dwarf mass decreases after a nova outburst.

For slow classical novae, the white dwarf has a helium layer, a part of which will be lost after one nova outburst cycle. When all of the helium layer will be gone after many outburst cycles, the white dwarf becomes naked and then a part of the white dwarf matter will be mixed into the envelope and blown off during the next outburst. Therefore, this outburst is a fast classical nova outburst.

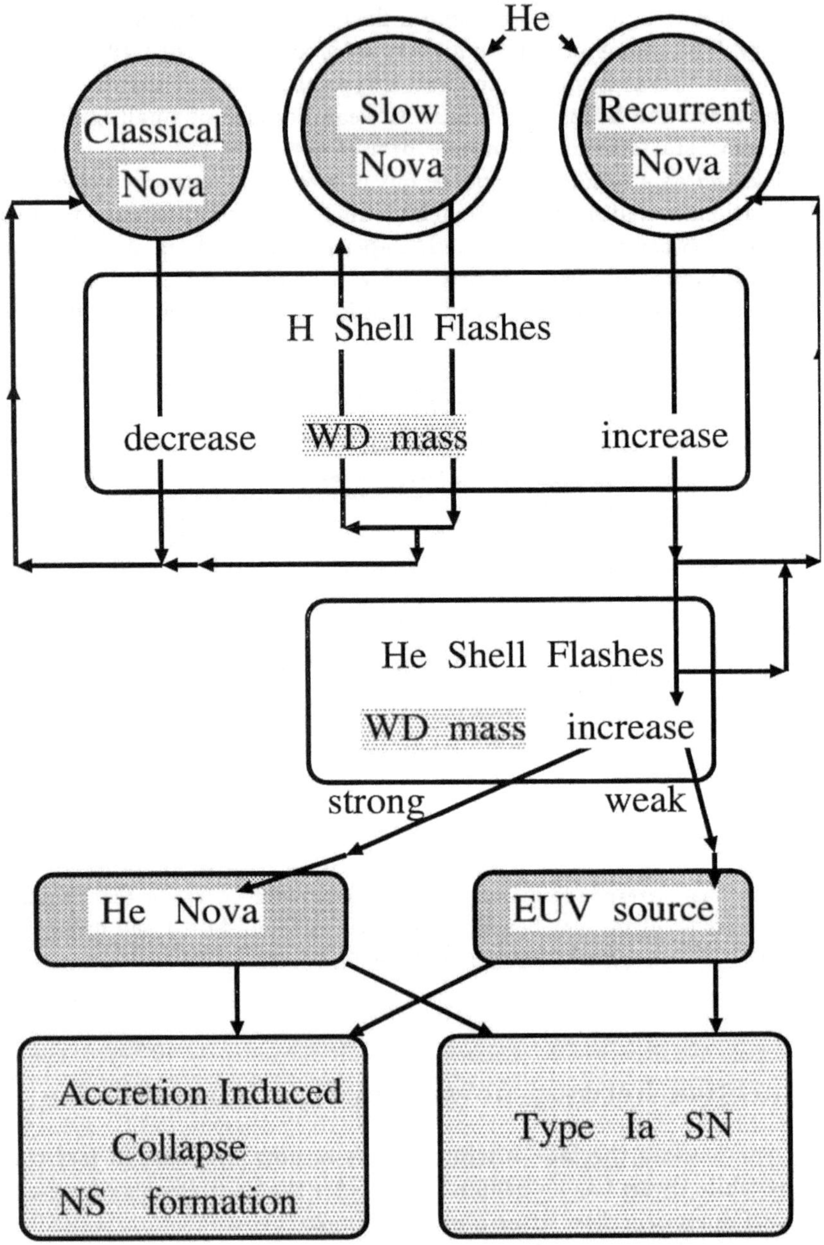

Fig. 4. Connection between novae and type Ia supernova/neutron star formation

In recurrent novae, on the other hand, a part of the envelope mass remains on the white dwarf after shell flashes and the white dwarf grows in mass. A helium layer develops under the hydrogen burning zone and when its mass reaches a critical value, an unstable helium shell flash occurs to trigger a helium nova. If the shell flash is strong enough, it will be observed as a bright slow nova-like object with hydrogen deficient abundance, and if the shell flash is weak, it will be a bright EUV/UV source with a faint optical counterpart. In such a object, the white dwarf grows to close the Chandrasekhar limit and will be a type Ia supernova or a neutron star via accretion induced collapse depending on the initial and internal conditions of the white dwarf (Nomoto & Kondo 1991).

Acknowledgement: This work has been supported in part by the Grant-in-Aid for Scientific Research (05640314) of the Japanese Ministry of Education, Science and Culture and also in part by Yamada Science Foundation.

References

Duerbeck H.W., 1988, A&A 197, 148

Fujimoto M.Y., 1993, ApJ 419, 768

Gallagher J.S., Kaler J.B., Olson E.C., Hartkopf W.I., Hunter D.A., 1980, PASP 92, 46

Hassall B.J.M., Snijders M.A.J., Harris A.W., Cassatella A., Dennefeld M., Friedjung M., Bode M., Whittet D., Whitelock P., Menzies J., Lloyd Evans T., Bath G.T., 1990, in Physics of Classical Novae, A. Cassatella and R. Viotti (eds.), Springer, Berlin.

Iben I., 1982, ApJ 259, 244

Kato M., 1994, A&A 281, L49

Kato M., 1995, in Proceedings of the Abano Terme CV conference, in press

Kato M., Hachisu I., 1994, ApJ 437, 802

Nomoto K., Kondo Y., 1991, ApJ 367, L19

Prialnik D., Kovetz A., 1992, ApJ 385, 665

Stickland D.J., Penn C.J., Seaton M.J., Snijders M.A.J., Storey P.J., 1981, MNRAS 197, 107

M. Shapiro: In connection with Dr. Kato's proposal of the relation of recurrent novae to supernovae of type Ia, the following may be of interest: To explain the relative deficiency of H and He in the cosmic rays, I suggested in 1985 (ICRC Proceedings, La Jolla, USA) that one component of the cosmic rays, mainly the "heavy nuclei" ($Z > 5$) originate in large part from SN Ia, since these show virtually no H and He. This idea was developed more thoroughly by Hayakawa et al. (ICRC Proceedings, Adelaide, Australia), but the cosmic ray composition they deduced was rather contrived, depending on uncertain parameters.

L. Pustil'nik: You talked about novae with surface thermonuclear detonation. But a new type of objects of completely different nature (unstable accretion on a black hole), manifest themselves in the optical as recurrent novae, but show up as strong X-ray sources. Have you any test in the light curve (from your models) which may be used to distinguish between absolutely different objects when only the optical light curve has been observed (e.g. in the past when no X-ray observation was available)?

M. Kato: Some of the X-ray novae show very similar light curves to those of recurrent nova. To distinguish these two kinds of objects, we need more observational information such as expansion velocities, detection of X-rays, light curves in UV, EUV, soft X-ray etc.

J. van Paradijs: Apart from your proposed He burning massive white dwarfs there is another potential progenitor of SN Ia's and AIC formed neutron stars, i.e. the supersoft sources detected with *ROSAT*. They are generally considered to be steadily hydrogen burning massive white dwarfs. What are the expected relative formation rates of the two different progenitors?

M. Kato: To estimate the number of H burning white dwarfs and He burning white dwarfs is complicated and not yet be done. It is closely related to the binary evolution which include mass loss due to the optically thick wind and the angular momentum loss of the system. The observational properties of H burning white dwarfs and He burning white dwarfs may be similar, so that a part of the supersoft X-ray sources is possibly from the contribution of He burning white dwarfs.

PU Vul: 15 Years of Studies, Important Findings and Remaining Mysteries

R.E. Gershberg

Crimean Astrophysical Observatory, Nauchny, Crimea, Ukraine

In autumn 1979, the young W. Wenzel – at this time he was only 50 years old – came to the Byurakan meeting on flare stars and informed the participants about a recently discovered flaring object: the Kuwano-Honda object = PU Vul. During that time, FU Ori and related stars were the most intriguing variable objects, and PU Vul had shown some similar features such as the rate of brightening and the spectrum at maximum. It turned out to be an extremely interesting object, and the Crimean team has studied it for the last 15 years. Over this time, this team included T.S. Belyakina, N.I. Bondar, K.K. Chuvaev, Yu.S. Efimov, R.E. Gershberg, V.I. Krasnobabtsev, E.P. Pavlenko, P.P. Petrov, I.S. Savanov, N.I. Shakhovskaya, N.M. Shakhovskoy, A.G. Shcherbakov and V.A. Shcherbakov, and we collaborated with V.I. Shenavrin from Russia, with D. Chochol, J. Grygar and L. Hric from former Czechoslovakia and with V. Piirola, I. Tuominen, M. Poutanen and J. Huovelin from Finland.

Now the number of publications on PU Vul approaches one hundred. It was established that PU Vul is a symbiotic nova, the object among the known variables of this type, whose explosion became observable most recently (with the possible exception of Wakuda's object in Sgr, Grebel et al. 1994), and many details have been observed from the very beginning of the flare. I intend to list the most interesting findings and the most intriguing remaining mysteries of PU Vul.

1 Important findings

The binarity of PU Vul was established by two independent observations: (i) in 1979 during brightness maximum a hot stellar spectrum was seen at optical wavelengths, but IR photometry simultaneously showed radiation with a colour temperature of 3 200 K (Bensammar et al. 1980), and (ii) at the descending branch to the 1980–81 brightness minimum (Fig. 1) the spectrum changed from F to M without the appearance of any intermediate spectral classes (Gershberg et al. 1982) and the same M spectrum was found 20 years before the flare (Stephenson 1979). Now it is clear that PU Vul consists of a normal late M giant and a rapidly evolving burst component.

A strong similarity between the spectrum of PU Vul and that of a normal F supergiant was observed in 1979–1983, during the prolonged brightness maxi-

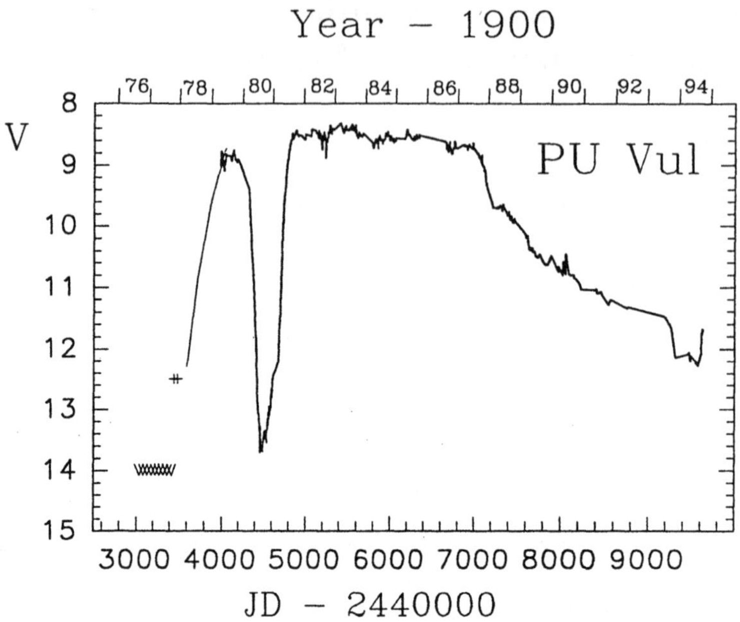

Fig. 1. *V* lightcurve of PU Vul over the last 15 years.

mum. This similarity was noted from the energy distribution from UV to near IR (Nakagiri & Yamashita 1982, Friedjung et al. 1984, Belyakina et al. 1985, 1990, Vogel & Nussbaumer 1992) as well as from relative intensities of absorption lines (Hric et al. 1980, Gershberg et al. 1982, Yamashita et al. 1982, 1983, Kolotilov 1983a, Iijima & Ortolani 1984) and from quantitative analysis by the curve of growth method (Belyakina et al. 1985, Kanamitsu 1991). During these years, PU Vul could not be distinguished from a normal F supergiant by its luminosity, effective temperature, physical parameters and chemical composition of its atmosphere. This fact leads us to consider the existence of a non-homogeneity of high luminosity F stars, and I consider the enigmatic ϵ Aur a probable candidate for such a mimicry (Gershberg 1990).

A general accordance between observable features of PU Vul and those of theoretical models for the explosion of an accreting white dwarf in a binary system was found, when observations were compared with Iben's (1982) computations (Fig. 2). Iben did not know about the existence of PU Vul when he calculated the models, but Fig. 2 shows the excellent agreement between the observed and the computed luminosity, the effective temperature at brightness maximum, and the rate of brightening before maximum. Theory predicts that the exploded object should describe a loop in the (luminosity – effective temperature) plane near brightness maximum, and in Fig. 3 this loop is presented in the (magnitude – colour index) plane from our data.

The 1980–81 brightness minimum. In the beginning of 1980, the gradual development of PU Vul along the Iben model track ceased because of a strong brightness decrease. In August-September 1980, PU Vul passed a bright-

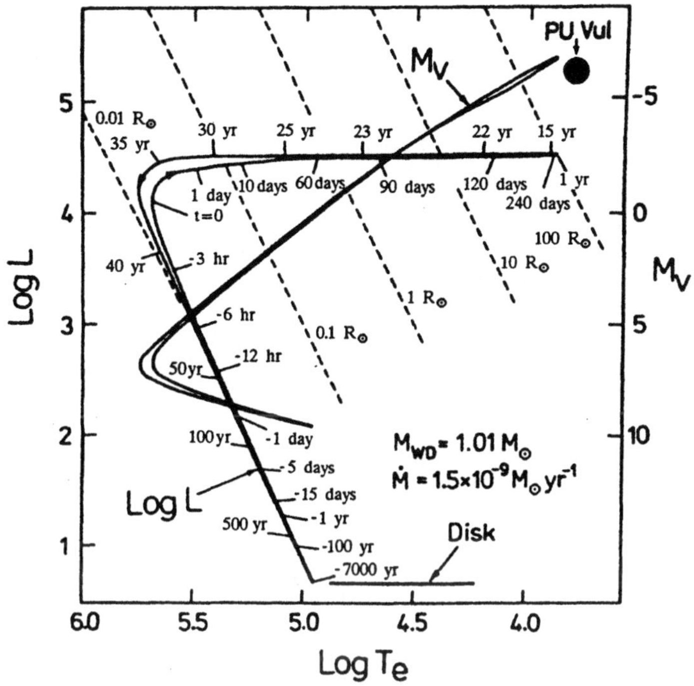

Fig. 2. The theoretical track of an accreting white dwarf (Iben 1982) and the position of PU Vul at brightness maximum in the (M, $\log T$) plane (Gershberg 1986)

ness minimum with $\Delta V = 4^{\rm m}.8$, and in September 1981, it had recovered from this decrease. Belyakina et al. (1982b) suggested two possible interpretations of this 500-day minimum: an eclipse of the outbursting component by an M giant, or the formation of a dust envelope around it, with a subsequent dissipation of the envelope. At that time both these schemes had weak points: Accordingly to Belyakina et al. (1985), the observed duration of the PU Vul brightness decrease and the known size of an M giant lead to an expected orbital period of 200 – 300 years and a distance between the components of tens of AU. It is unlikely that in such a wide binary a high accretion rate, necessary for a nova-like flare, would be realized. On the other hand, the occurrence of an eclipse of the bursting component in such a system requires that the angle between the line of sight and the orbital plane does not exceed 1° and that the PU Vul flare took place near conjunction phase. It is obvious that a coincidence of these independent circumstances has an extremely low probability. Recently Vogel & Nussbaumer (1992) attempted to reanimate the eclipse scheme. However, they have carefully considered the problem from the photometric point of view only and did not take into account the geometrical and dynamical aspects mentioned above, which apparently force to reject this scheme.

The scheme of a non-stable dust envelope was proposed independently by Yamashita et al. (1982) and supported by Friedjung et al. (1984) and Men'shchikov

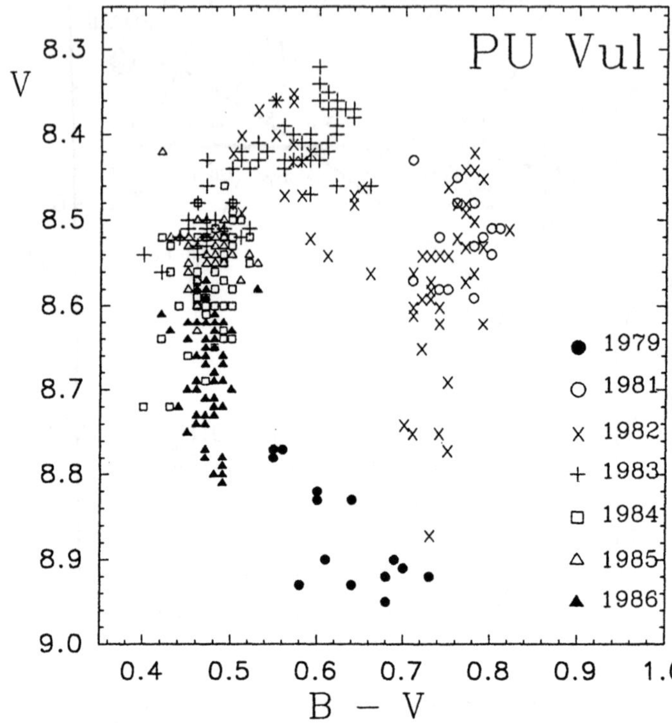

Fig. 3. The observed loop traced by PU Vul in the magnitude – colour index plane near brightness maximum (from Belyakina et al. 1990).

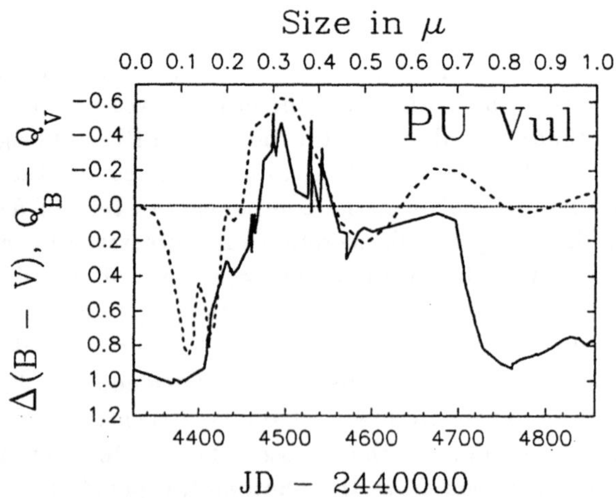

Fig. 4. The observed $\Delta(B - V)$ curve of PU Vul in the deep minimum in 1980–81 (solid line) and the calculated $Q_B - Q_V$ curve for a spherical silicate dust formation process (dotted line). The lower scale is JD, the upper one is a characteristic size of growing particles (from Efimov 1988a).

et al. (1985). However, a strong bluening of colour indices at brightness minimum (Belyakina et al. 1982a, Kolotilov & Belyakina 1982) contradicts the standard idea of the reddening of a star when its brightness decreases due to interstellar dust absorption. This has resulted in critics on the dust scheme (Kenyon 1986, Iijima 1989, Vogel & Nussbaumer 1992, Sion 1993). Meanwhile, 6 years ago, Efimov (1988a) showed that the observed variations in the colour indices at brightness minimum do not contradict the dust scheme but are a decisive argument in favour of it. To realize this fact, one should consider dust formation as a particle growth process. When conditions for dust formation appear within a circumstellar medium, this process evolves rather quickly, and a given time corresponds to a certain characteristic size of particles. In such a case, brightness minima at different wavelengths should take place at different times, earlier at shorter wavelengths and later at longer ones. Such a sequence was indeed observed. As the result of a monotonous growth of characteristic size of particles, variations of colour indices near brightness minimum should be described by sign-reversing functions. It means that at a certain time a star occulted by a forming dust envelope will be even bluer than before the dust appeared – see Fig. 4. (This concept was successfully applied by Efimov (1988b) to R CrB variables as well, where the cause of non-stable dust envelopes for stellar brightness minima is beyond any doubt.)

Another objection against the dust model of the 500 day brightness minimum of PU Vul are the small variations of IR radiation of during this event, while during nova flares accompanied by dust envelope formation, the IR emission usually increases strongly (Kenyon 1986). However, this objection does not account for an essential difference between the considered bursting objects. In novae, a powerful IR excess appears (and is easily detectable) when the donor-star is of low luminosity. In PU Vul, the total IR radiation of the system is mainly determined by the M giant, and the smallness of the IR flux variation does not contradict the dust model (Belyakina et al. 1985). Moreover, these small variations may be even of different signs, depending on their main cause: either due to an intrinsic radiation of dust or to the M giant light reflected from dust – in these cases the total IR radiation of the system would increase, or due to a veiling of the M giant by the dust – in this case the total IR radiation of the system would decrease. As known, the JHK band fluxes from PU Vul decreased by $0^{m}_{.}7 - 0^{m}_{.}4$ (Belyakina et al. 1985), while $\Delta V = 4^{m}_{.}8$.

Thus, objections against the dust model of the brightness minimum in 1980–81 are rejected and the above model can be accepted. A general scheme of the evolution of PU Vul, including this event, was made by Kanamitsu (1991) and is presented in Fig. 5.

The outflow and the entering to the nebular phase was traced during the slow evolution of PU Vul with unprecendent details. Several stages of this process can be distinguished. The P Cyg profile of Hα – the first evidence of an outflow – was found just after discovery (Mochnacki 1979). Simultaneously detected weakened metallic lines were also considered as due to numerous weak emission lines filling in respective absorptions. These features were observed

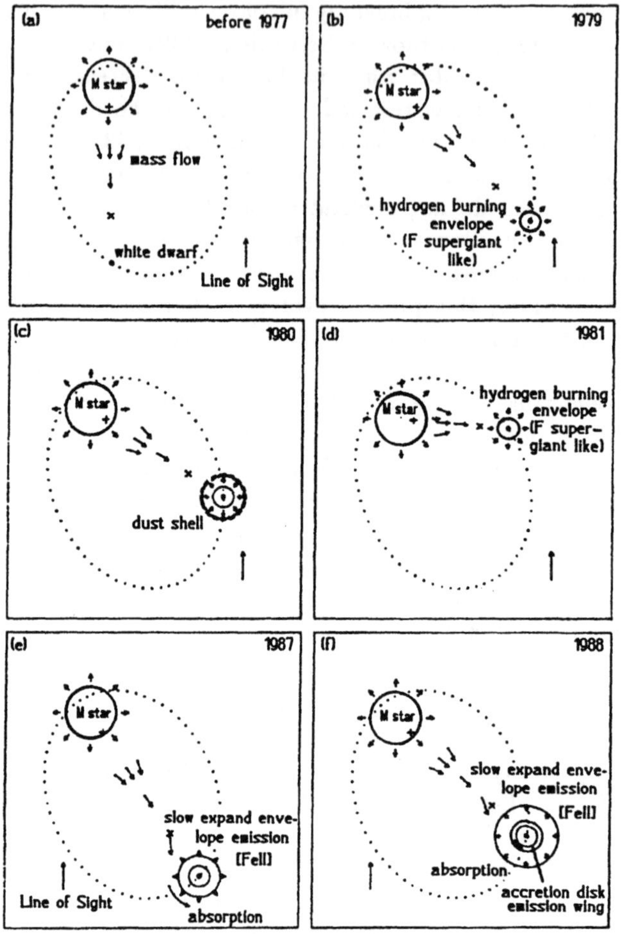

Fig. 5. Schematic plot of the PU Vul binary system in different years (Kanamitsu 1991)

during 6 months (Honda et al. 1979, Yamashita et al. 1982, Kanamitsu et al. 1991). The second episode of the absorption line weakening began in November 1982, two weeks after the brightening of the system by $\Delta U \sim 0\overset{m}{.}8$ and with the true appearance of emission lines of multiplet series of many elements. In this rather long episode, the line weakening effect was found to be dependent on ionization potentials of respective species. The effect was strongest in July 1984 (Yamashita et al. 1983, 1984, Iijima & Ortolani 1984, Kolotilov 1984, Kanamitsu et al. 1991). Then, similar weak-line events with hydrogen and Fe II emission lines of different intensities took place in August 1985, in May 1986 and in April 1987 (Kanamitsu 1991, Kanamitsu et al. 1991, Klein et al. 1994).

The last weak-line episode was followed by the appearance of P Cyg line profiles of many species during several months and by the fast flaring of strong emissions of H, He, Fe II and [Fe II] lines in December 1987. For two years the H and Fe emission lines became the main features of the spectrum. This was the beginning of the entering in the nebular phase (Iijima 1989, Kanamitsu 1991, Kanamitsu et al. 1991, Klein et al. 1994). During this stage a sequence of appearance of P Cyg line profiles of other species and the flaring of lines of different species depending on their ionization potentials was seen. Then, absorption components of P Cyg profiles weakened and the spectrum turned into a pure emission spectrum. In November 1989, only the interstellar NaI D lines were seen in absorption.

The nebular phase of PU Vul showed further systematic shifts of emitting species to higher ionization potentials, due to continuing contraction of the burst component in accordance with the increase of effective temperature. During 1988 the $\lambda < 1900$ Å flux increased by a factor 2 (Vogel & Nussbaumer 1992), HeI emission became noticeably stronger (Klein et al. 1994) and very wide wings of H and He lines appeared (Iijima 1989, Sion et al. 1993). From May 1989, the strong UV HeII, SiIV, CIV, SiIII], CIII] and AlIII emission lines became visible (Vogel & Nussbaumer 1992; Sion et al. 1993; Klein et al. 1994). In the middle of 1990, emission lines characteristic of WR stars as well as many forbidden lines were registered (Tomov et al. 1991).

2 Remaining mysteries

The comparison of the main observational features with theoretical calculations by Iben (1982) for a quasi-static approximation of hydrogen burning on an accreting white dwarf and calculations by Livio et al. (1989) for a hydrodynamical approximation of this process permit to understand the general nature of the PU Vul flare and the evolutionary status of the variable. Accordingly to Duerbeck's (1981) classification, PU Vul corresponds to the Eddington novae, and later it was counted among the symbiotic novae (Viotti 1988). However, these general conclusions leave many open questions for understanding the rich variety of observational data. I intend to list and discuss some of these mysteries.

1. What is the meaning of the small-amplitude quasi-periodical brightness variations in 1979? In that year PU Vul showed such variations with $\Delta V = 0^{m}_{.}15$ and a quasi-period of about 78 days (Chochol et al. 1981). After the 1980–81 minimum, similar variations of a smaller amplitude and with a shorter time scale were also suggested (Kolotilov & Belyakina 1982); however, this suggestion was not confirmed (Kolotilov 1983b). On the other hand, nonlinear self-excited oscillations of envelopes of low-massive stars of high luminosity lead to dust formation (Fadeyev 1982, 1983). Is this the case in PU Vul?

2. What is the cause of the small-amplitude brightness variations with unusual colour anomaly $\Delta U < \Delta B \sim \Delta V$, detected in autumn 1982 (Purgathofer & Schnell 1983; Belyakina et al. 1985)?

3. What is the cause (or causes) of the polarisation of the light of PU Vul? In the long series of polarimetric measurements one can find episodes when the polarization was apparently due to the dust envelope – see, e.g., the data for the 1980–81 minimum. However, when three models – light scattering in a dust envelope, light scattering on free electrons in a magnetized envelope and a spotty photosphere – are used to interpret the complete polarimetric data, none of them fits the whole variety of observed wavelength dependence and none of them can be completely rejected. The most intriguing facts are polarimetric data in November 1982 and in September 1983, when during brightenings in the *U* band the polarization angles essentially changed simultaneously in all the *UBVRI* bands, while the degrees of polarization were practically constant, and in October–December 1984, when essential variations were detected in p_{UBVRI} without noticeable changes in θ_{UBVRI} (Belyakina et al. 1985, 1990).

4. Where and why originated the emission spectrum in March–April 1981? At the ascendant branch of the light curve after the deep minimum, the spectrum was quite unique: emission lines of Sc II, La II, Y II and Nd II were dominant while Fe emission was rather scanty. Only the solar chromosphere (Pierce 1968) and η Car and RR Tel (Meinel et al. 1969) show comparable numbers of lines of mentioned ions (Belyakina et al. 1984, 1985). Why was this spectrum so short-lived?

5. Why was the chemical composition of the photosphere so close to that of the normal F supergiant γ Cyg in 1979 and 1981? In both cases, there are overabundances of Y, Zr, Ba and La (Belyakina et al. 1985). Does it mean that the accreted hydrogen-rich wind from the M giant did not interact with the upper layers of the hydrogen-depleted accreting white dwarf? What is the cause of chemical abundance variations after 1981? At this time, the PU Vul atmosphere became very similar in this respect to atmospheres of "metallic" stars: excesses of the iron group (excluding Fe itself) and of heavier elements appeared, these excesses being higher with atomic numbers, Ca and Sc showed essential deficits (Belyakina et al. 1990, Kanamitsu 1991). Are these variations a real effect? Or are they results of an improper application of the curve of growth method? Or results of a real stratification of chemical abundances along the depth of the quasi-photosphere? Or the consequence of weak-line events?

6. And finally, what are the spatial and kinematic structures of PU Vul? What numbers should be added to Kanamitsu's (1991) scheme to convert this qualitative plot into a quantitative model? To answer this question, we have one of the richest observational data sets and the most scanty conclusive results. Numerous measurements of radial velocity have been carried out (Hric et al. 1980, Yamashita et al. 1982, 1983, 1984, Kolotilov 1983a, 1984, Iijima & Ortolani 1984, Belyakina et al. 1985, Kanamitsu et al. 1991, Tomov et al. 1991). However, in the very beginning of the measurements an essential heterogeneity was found in the results (Hric et al. 1980). Chochol & Grygar (1987) have carefully selected lines to calculate an orbit of the binary, but the resulting orbit was not confirmed directly by later observations (Belyakina et al. 1990, Kanamitsu 1991). However, the estimated orbital period of about 8 years turned out to be in agreement with gradual variations of the polarization angle (Belyakina et al. 1990); then, there is observational evidence that in 1981 the bursting component was nearer to us than the M giant (since the dust envelope around the bursting component decreased the IR radiation of the M giant) while in 1989–91 the wind from the M giant produced absorption structures in the spectrum of the bursting component (Sion et al. 1993).

Many measurements of radial velocity differences between absorption and emission components of P Cyg profiles have been performed (Mochnacki 1979, Yamashita et al. 1982, Vogel & Nussbaumer 1992, Tomov et al. 1991, Klein et al. 1994). However, these differences were found to be dependent both on time and on the emitting species.

The existing confusing picture of radial velocities is probably partly due to the filling in of absorptions by respective emission lines as well as by the weak-line events which are probably of the same nature. For instance, in June 1984 a monotonous change of radial velocities was found along the Balmer series from 0 km/s for Hγ to +55 km/s for H20 (Belyakina et al. 1990).

In spite of the uncertainty mentioned above, the observations before the nebular phase were in agreement with theoretical predictions by Livio et al. (1989), that the envelope expansion velocity is about 40 km/s. At the nebular phase very wide wings of emission line profiles were discovered, and this required new emitting structures: a WR hot wind from the contracting burst component (Tomov et al. 1991) or an accreting disc around this component (Kanamitsu 1991).

In any case, the problem to separate a regular orbital motion and an expansion of the envelope of variable optical thickness is still far from a solution. Several weak-line events in the epoch of the absorption spectrum and very fast – time-scales of several days – profile variations in the nebular phase (Tamura et al. 1992; Efimov et al. 1995) lead to suggest a non-spherical envelope, similar in structure to that of N Cyg 1975 (V1500 Cyg) (Boyarchuk & Gershberg 1977). However, this makes the search for the structure of PU Vul even more difficult.

Acknowledgement: This research was made possible in part by Grant R2Q000 from the International Science Foundation.

References

Belyakina T.S., Efimov Yu.S., Pavlenko E.P., Shenavrin V.I., 1982a, Astron. Zh. 59, 1

Belyakina T.S., Gershberg R.E., Efimov Yu.S., et al. 1982b, Astron. Zh. 59, 303

Belyakina T.S., Bondar N.I., Chochol D., Chuvaev K.K., Efimov Yu.S., Gershberg R.E., Grygar J., Hric L., Krasnobabtsev V.I., Petrov P.P., Piirola V., Savanov I.S., Shakhovskaya N.I., Shakhovskoj N.M., Shenavrin V.I., 1984, A&A 132, 12

Belyakina T.S., Bondar N.I., Gershberg R.E., Efimov Yu.S., Krasnobabtsev V.I., Petrov P.P., Piirola V., Savanov I.S., Chuvaev K.K., Shakhovskaya N.I., Shakhovskoy N.M., Shenavrin V.I., 1985, Bull. Crimean Astrophys. Obs. 72, 3

Belyakina T.S., Bondar N.I., Gershberg R.E., Hric L., Grygar J., Efimov Yu.S., Krasnobabtsev V.I., Piirola V., Poutanen M., Savanov I.S., Tuominen I., Chochol D., Chuvaev K.K., Shakhovskaya N.I., Shakhovskoy N.M., Shenavrin V.I., Shcherbakov A.G., 1990, Bull. Crimean Astrophys. Obs. 81, 28 (A shortened version appeared in A&A 223 (1989) 119)

Bensammar S., Friedjung M., Assus P., 1980, A&A 83, 261

Boyarchuk A.A., Gershberg R.E., 1977, Astron. Zh. 54, 488

Chochol D., Hric L., Papousek J., 1981, IBVS 2059

Chochol D., Grygar J., 1987, ApSS 131, 487

Duerbeck H.W., 1981, PASP 93, 165
Efimov Yu.S., 1988a, Astron. Zh. 65, 807
Efimov Yu.S., 1988b, Astron. Zh. 65, 979
Efimov Yu.S., et al., 1995, this volume, p. 255
Fadeyev Yu.A., 1982, ApSS 86, 143
Fadeyev Yu.A., 1983, ApSS 95, 357
Friedjung M., Ferrari-Toniolo M., Persi P., Altamore A., Cassatella A., Viotti R., 1984, in The future of ultraviolet astronomy based on six years of IUE research, J.M. Mead R.D. Chapman and Y. Kondo (eds.), NASA CP-2349, Washington, p. 305
Gershberg R.E., 1986, Commun. Konkoly Obs. 86, 351
Gershberg R.E., 1990, Astron. Zh. 67, 76
Gershberg R.E., Krasnobabtsev V.I., Petrov P.P., Chuvaev K.K., 1982, Astron. Zh. 59, 6
Grebel E.K., Duerbeck H.W., Greiner J., Richter G.A., 1994, IBVS 4019
Honda M., Ishida K., Noguchi T., Norimoto Y., Nakagiri M., Soyano T., Yamashita Y., 1979, Tokyo Astron. Bull. Second Ser. N 262, 2983
Hric L., Chochol D., Grygar J., 1980, IBVS 1835
Iben I., 1982, ApJ 259, 244
Iijima T., 1989, A&A 215, 57
Iijima T., Ortolani S., 1984, A&A 136, 1
Kanamitsu O., 1991, PASJ 43, 225
Kanamitsu O., Yamashita Y., Norimoto Y., Watanabe E., Yutani M., 1991, PASJ 43, 523
Kenyon S.J., 1986, AJ 91, 563
Klein A., Bruch A., Luthardt R., 1994, A&AS 104, 99
Kolotilov E.A., 1983a, Astron. Zh. 60, 324
Kolotilov E.A., 1983b, Pisma Astron. Zh. 9, 486
Kolotilov E.A., 1984, Pisma Astron. Zh. 10, 284
Kolotilov E.A., Belyakina T.S., 1982, IBVS 2097
Livio M., Prialnik D., Regev O., 1989, ApJ 341, 299
Mochnacki S., 1979, IAU Circ. 3350
Meinel A.B., Aventi A.F., Stockton M.W., 1969, Catalog of emission lines in astrophysical objects, Univ. Arizona, Tucson
Men'shchikov A.B., Tutukov A.V., Shustov B.M., Ergma E.V., 1985, Pisma Astron. Zh. 11, 529
Nakagiri M., Yamashita Y., 1980, Tokyo Astron. Bull. Second Ser. N 263, 2993
Pierce A.K., 1968, ApJS 17, 1
Purgathofer A., Schnell A., 1983, IBVS 2264
Sion E.M., 1993, AJ 106, 298
Sion E.M., Shore S.N., Ready C.J., Scheible M.P., 1993, AJ 106, 2118
Stephenson C.B., 1979, IAU Circ. 3356
Tamura S., Kanamitsu O., Yamashita Y., 1992, PASJ 44, 543
Tomov T., Zamanov R., Iliev L., Mikolajewski M., Georgiev L., 1991, MNRAS 252, 31
Viotti R., 1988, in The symbiotic phenomenon, J. Mikolajewska, M. Friedjung, S.J. Kenyon and R. Viotti (eds.), Kluwer, Dordrecht, p. 269
Vogel M., Nussbaumer H., 1992, A&A 259, 525
Yamashita Y., Maehara H., Norimito Y., 1982, PASJ 34, 269
Yamashita Y., Norimito Y., Yoo K.H., 1983, PASJ 35, 521
Yamashita Y., Norimoto Y., Watanabe E., 1984 (priv. comm.)

Spectral Variability and Photometric Behaviour of the Symbiotic Nova PU Vul in 1985–1994

Yu.S. Efimov, R.E. Gershberg, I.S. Savanov, A.G. Shcherbakov, V.A. Shcherbakov

Crimean Astrophysical Observatory, Nauchny, Crimea, Ukraine

We present preliminary results of observations of the Hα, He I 5876 and Na I D lines profiles in spectra of PU Vul. The variability of profiles is confronted with the photometric behaviour. The observations were carried out with a CCD camera, installed in the coudé spectrograph of the 2.6 m Shajn telescope. 68 spectra of the Hα line and 25 spectra in the region of the He I and Na I lines were obtained.

Fig. 1 (left panel) presents the evolution of the Hα line from 1985 to 1994, displaying selected typical profiles in different years. The profiles are shown in a heliocentric wavelength scale and in logarithms of intensity. Variations of line intensity, line width, as well as radial velocity of the centre of weight of the total emission and absorption components are observed.

Stronger and more complex variability is observed in the region of the He I and Na I D lines. The strong P Cyg type absorption in the Na I D lines during 1985-1987 disappeared in later spectra as a result of total ionization of Na I (Fig. 1., right panel). Contrary to that behaviour, He I 5876 shows variations from a pure absorption profile to a very strong and wide emission profile.

Nevertheless, even more drastic fast changes of Hα emission line were observed in 1987 and 1989. Tamura et al. (1992) found an abrupt change of the Hα profile in September 1989. On September 10, the Hα line had an extremely wide wing and showed only a trace of the central absorption feature which was very strong on July 26. In 1990 August, the Hα emission revealed a similar profile to that of 1989 July. In Fig. 2 (right panel) Hα profiles compiled from Tamura et al. (1992) together with our observations are shown. It is seen that the profile obtained by us on July 21, practically coincides with the profile on July 26 (Tamura et al. 1992, dashed line). We succeeded to reduce the time interval of these drastic changes in Hα. Our subsequent observations were obtained on 1989 October 24. Already at this time, the Hα profile had returned to its former shape (Fig. 2). According to Van Winckel et al. (1993), the former shape of Hα profile was observed only 5 days after 1989 September 10. Thus, this dramatic event lasted not more than 5 days. Similar fast changes of the profile were observed in 1987 (Fig. 2, left panel). The shape of the profile changed from a double-peaked to a quasi-symmetrical one in less than 37 days and returned to the normal shape in 6 days. On 1987 August 8 only a weak trace of absorption in the blue wing of

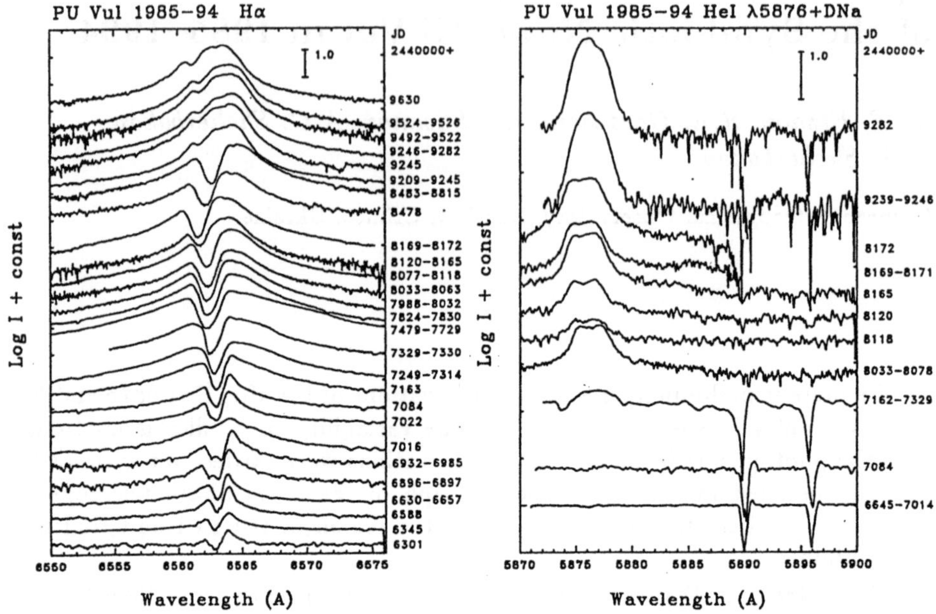

Fig. 1. Selected profiles of the Hα line (left) and of He I 5876 and Na I D (right).

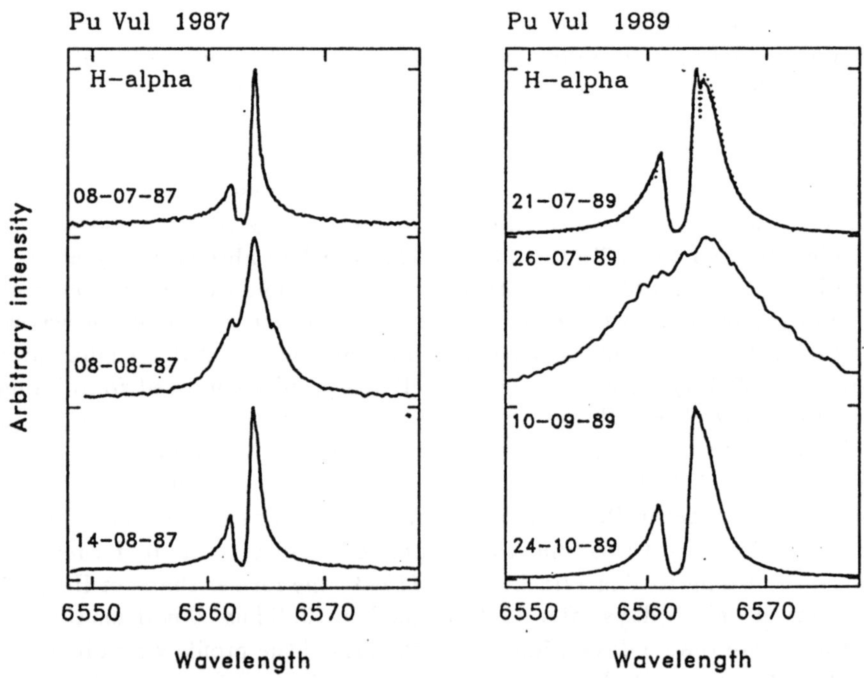

Fig. 2. Abrupt changes of the Hα profiles in 1987 and 1989

the line is seen as compared with absorption in profiles, obtained on July 8 and on August 14. A similar weak absorption is also seen in the red wing of the line.

For a quantitative study of the character of the line variability, all observed profiles (except abnormal profiles, shown in Fig. 2) were approximated by Voigt-type profiles, as shown in Fig. 3 (left panel). Three types of profiles are displayed: a moderately weak and narrow profile in 1985, a strong and very wide profile in 1987–1992, and a very strong and moderately narrow asymmetric profile in 1993–1994. These profiles were approximated using three criteria: the shape of the blue and red wings of the line and the gradient of the growth of intensity on the red part of the profile, which is not disturbed by the absorption.

The following parameters of a line were measured: radial velocity of absorption component of the observed profile (RVabs), radial velocity of the centre of weight of emission Voigt profile (RVem), full width of the line at half maximum level (FWHM), measured on the Voigt profile, and full width of the line at the 10 % of maximum level (FW0.1M).

In Fig. 3 (right panel), a comparison of measured parameters of the line and the V-brightness is presented. Two vertical dashed lines mark the times of drastic changes of the line profile in 1987 and 1989, as seen in Fig. 2. All measured parameters show slow changes from year to year, as well as fast variations, correlated with variations of stellar brightness.

The FWHM and the FW0.1M vary synchronously and indicate a strong growth of emission width between 1987 and 1989. The broadening of the line began at the initial stage of transition to the nebular phase, when the brightness began to drop abruptly. The width of the emission smoothly decreases while the star progressively enters the nebular phase.

The radial velocity of the total emission, corresponding to the movement of an envelope in the binary system, also shows variations from year to year with an amplitude of more than 10 km/s. The orbital elements of the system have not yet been established with certainty. According to Kanamitsu (1991), the orbital period of PU Vul must be of the order of 10 years. We have analysed our series of radial velocity measurements to search for a possible period. Several methods were applied. The highest peak in the power spectrum obtained by the Deeming method gives a period near 9.3 years. A refinement by Lafler-Kinman's and the string-length method yield a value of about 9.1 ± 0.4 years. The variability was also modelled by a least squares fitting of observed data by trigonometric polynomials. The period of the model was iteratively refined by the Levenberg-Marquardt algorithm (Pelt 1992, Press & Teukolsky 1988). The result of the estimate – the dashed curve in the third panel in Fig. 5 – gives an optimal period of changes of the radial velocity of about 11.2 ± 2.2 years.

The fast changes of the radial velocity of the total emission are surprising. These variations cannot be explained in the framework of a simple model of an expanding envelope. It was found, however, that such variability is caused by variations in the wings of the line, by their asymmetry. The model of a binary system, in which a wind outflow plays a significant role, as proposed by Kanamitsu (1991), could explain the observed variations as result of interaction

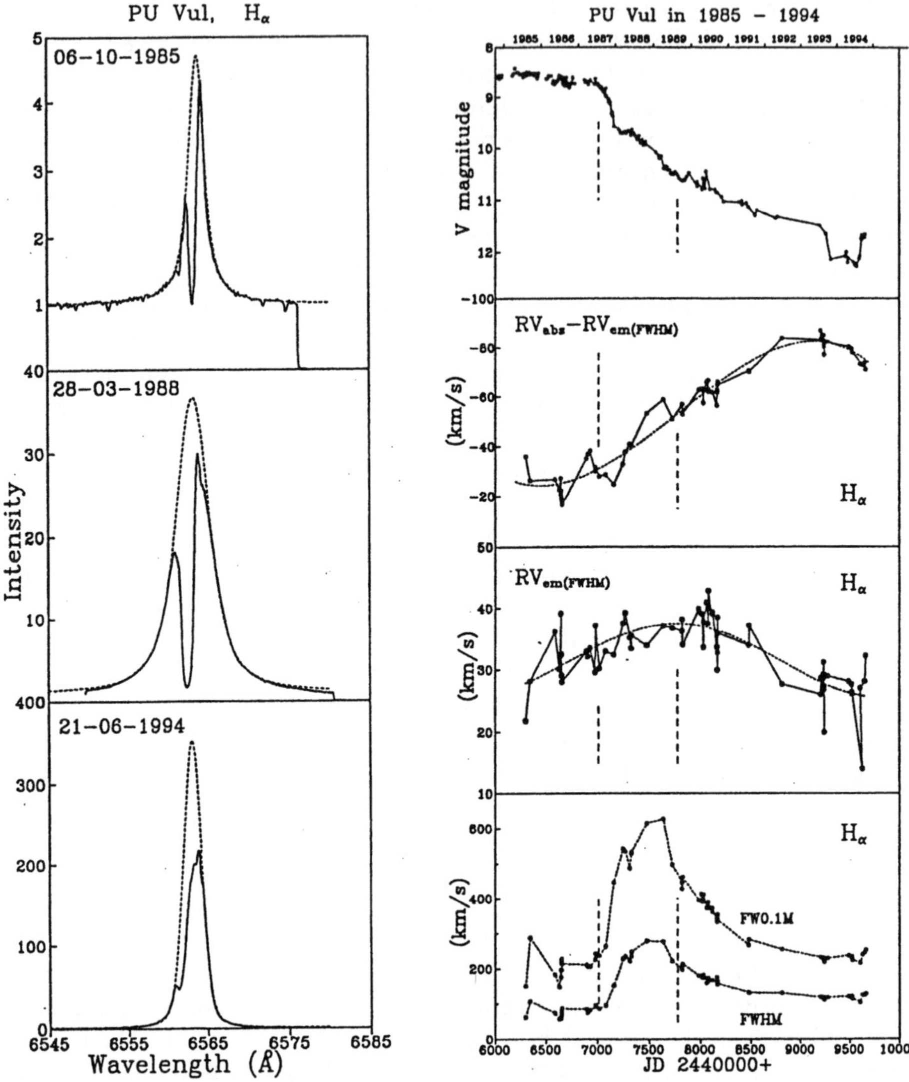

Fig. 3. Approximation of observed Hα line by Voigt profiles (left panel), and comparison between measured parameters of the Hα profiles with the *V* band brightness of the star.

between a non-stationary wind with an expanding envelope. The correctness of such a model is confirmed by the indication of a bipolar mass outflow found by Bensammar et al. (1991), by the onset of a Wolf-Rayet wind outflow (Sion et al. 1993), and the Wolf-Rayet features observed in the spectrum of PU Vul (Tomov et al. 1991). Similar unexpected fast variations of the radial velocity were observed in the spectrum of the enigmatic binary system ϵ Aur (Gershberg 1990).

The difference between the radial velocities of the absorption component of the line and the total emission as fitted by a Voigt profile (RVabs-RVem) determines the velocity of expansion of the envelope. Fig. 5 shows that the velocity of expansion during the time of our observations changed from 20 km/s up to 80 km/s (dashed curve). In 1985–1986 the velocity varied only by a few km/s, from 1986 to 1992 the envelope expanded with acceleration, and since 1992, evidence of retardation is observed. Besides the slow increase of radial velocity, rapid variations with timescales of a year and less are observed. Attention should be paid to the fact that both episodes of drastic changes of the Hα profile were followed by retardation. Tamura et al. (1992) interpret the abrupt change of the Hα profile as a result of interaction between the hot stellar wind from the hot contracting star that underwent an explosion with the cold wind from an M star. However, the episodes of retardation of the envelope in 1987 and 1989 indicate that the drastic changes of the Hα profile could be as well the result of the passage of a shock wave in the expanding envelope.

Acknowledgement: This research was made possible in part by Grant R2 Q000 from the International Science Foundation.

References

Bensammar S., Friedjung M., Chauville J., Letourneur N., 1991, A&A 245, 575
Gershberg R.E., 1990, Astron. Zh. 67, 76
Kanamitsu O., 1991, PASJ 43, 225
Pelt J., 1992, Licenced Package "Irregularly spaced data analysis", Helsinki, Finland
Press W.H., Teukolsky S.A., 1988, Computers in Physics 2, 77
Sion E.M., Shore S.N., Ready C.J., Scheible M.P., 1993, AJ 106, 2118
Tamura S., Kanamitsu O., Yamashita Y., 1992, PASJ 44, 543
Tomov T., Zamanov R., Iliev L., Mikolajewski M., Georgiev L., 1991, MNRAS 252, 31
Van Winckel H., Duerbeck H.W., Schwarz H.E., 1993, A&AS 102, 401

Photometry of CH Cygni During 1991–1994*

K.P. Panov[1], *J.S.W. Stegert*[2], *G. Hildebrandt*[3]

[1] Institute of Astronomy, Bulgarian Academy of Sciences, Zarigradsko Chaussee 72, 1784 Sofia, Bulgaria
[2] Sternwarte der Universität Bonn, Auf dem Hügel 71, 53121 Bonn, Germany
[3] Astrophysikalisches Institut Potsdam, An der Sternwarte 16, 14482 Potsdam, Germany

CH Cygni has been observed since about one decade at the Bulgarian National Astronomical Observatory Rozhen. *UBV* photometry was obtained with the 60 cm telescope. The computer controlled photometer is equipped with an EMI 9789 QB multiplier, a set of *u,b,v* filters and a diaphragm of 28″. The reduction includes corrections for dead-time and background, differential extinction and transformation to the standard *UBV* system.

Observations between 1991 and 1994 are displayed in Fig. 1 and show the new outburst, which is still going on. The previous major outburst lasted until 1984, and CH Cyg declined slowly between 1985 and 1988. In 1988, it reached its minimum in brightness and activity. The cause for the activity is probably mass transfer from the M giant component onto the hot companion. In this model, only the M star is seen during activity minimum. In July 1988, the mean brightness was $V = 8.50$, $B - V = 1.39$ and $U - B = 1.48$. Since 1989, some erratic activities appeared, including a hot variable continuum, broad and variable emissions in the Balmer lines and rapid light variations (Leedjarv 1990, Tomov and Mikolajewski 1992, Kuczawska et al. 1992, Panov and Ivanova 1992). Since 1991, a trend of increasing activity is observed, which persisted also in 1994. The brightening of CH Cyg of about 5^{m} in U in 1994 with respect to 1991 shows that it undergoes another major outburst. Moreover, it exhibits light variations with different other time scales, minutes (flickering), days and weeks, probably resulting from the mass-transfer. Fig. 2 showns an example of a short time variation on October 1, 1994. The flickering of CH Cyg was very pronounced in the U, B and V filters and the amplitudes reached several $0\overset{\mathrm{m}}{.}1$ in about half an hour. More observations of this interesting system are planned.

References

Kuczawska E., Mikolajewski M., Kirejczyk K., 1992, IBVS 3806
Leedjarv L., 1990, IBVS 3474
Panov K.P., Ivanova M.S., 1992, IBVS 3817
Tomov T., Mikolajewski M., 1992, IBVS 3721

* Based on observations collected at the National Observatory, Rozhen, Bulgaria

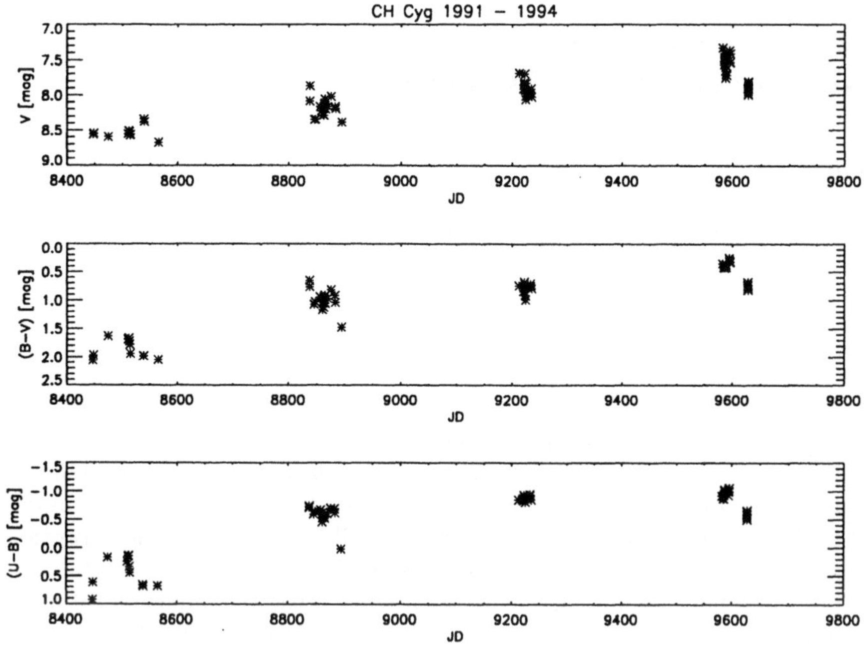

Fig. 1. UBV photometry of CH Cygni between 1991 and 1994

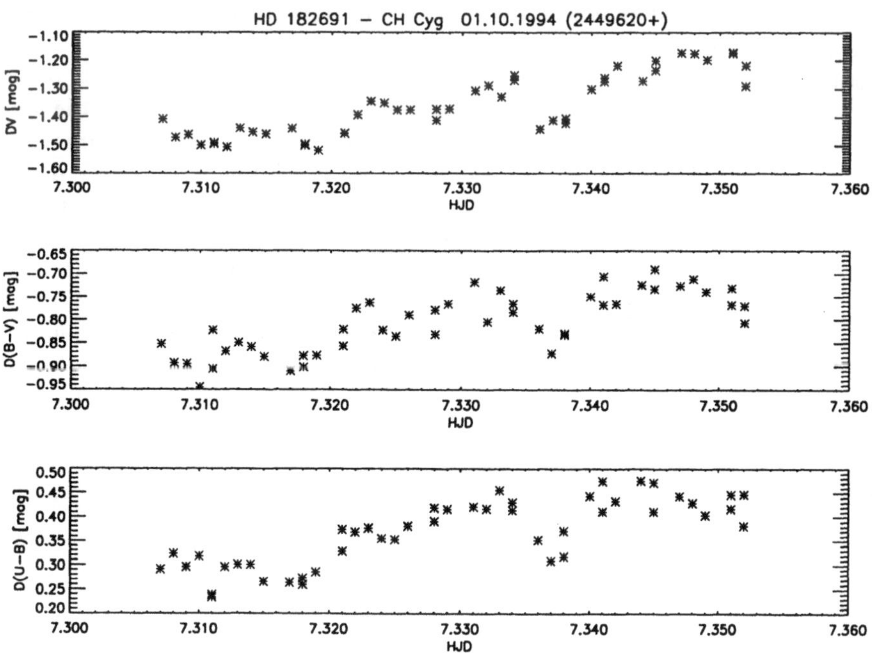

Fig. 2. Short time light variations of CH Cygni on October 1, 1994.

Long-term Photometry of the Symbiotic Star AG Draconis

R. Luthardt

Sonneberg Observatory, 96515 Sonneberg, Germany

Photoelectric *UBV*-observations of the well known symbiotic star AG Draconis were obtained between 1982 and 1984 with the 60 cm photoelectric reflector II of Sonneberg Observatory. The light curve shows two large optical outbursts: (at JD = 244 5000...5500 and JD = 9000...9500) and two smaller eruptions (at JD = 6200, 6600). The recent outburst is the largest one since 1952.

During the quiet stage a significant variability with an amplitude of about one magnitude is only visible in the *U*-band. L. Meinunger (1979) detected a period of 554 days and calculated the following ephemeris:

$$U_{max} = 243\,8900 + 554 \cdot E$$

All outbursts and smaller eruptions are correlated with the period of the U-maxima, the outburst maxima always coincide with U-maxima of the quiet stage, calculated by the above formula.

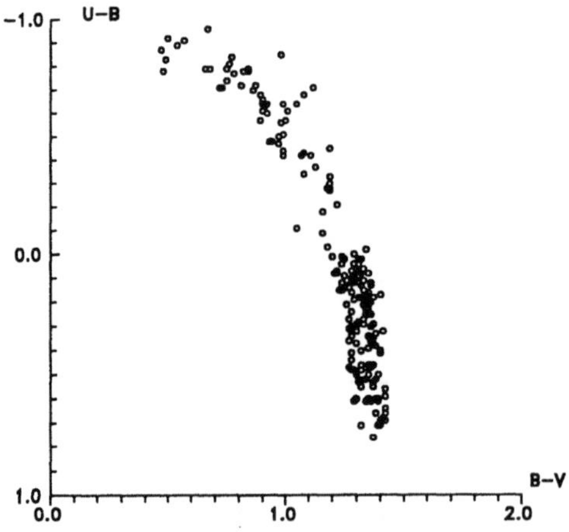

Fig. 1. Correlation between the color indices $(B-V)$ and $(U-B)$ of AG Dra.

The amplitudes of the outbursts differ also very significantly in color: the amplitude of the last outburst amounts to $\approx 5^{m}$ in U but only $\approx 1\overset{m}{.}5$ in V. $(U-B)$ varies from $\approx +0.8$ during the quiet stage to ≈ -1.0 during the large outburst maximum (Fig. 2). The variations of the color indices $(B-V)$ and $(U-B)$ are always strongly correlated (Fig. 1).

During the quiet stage the V-light curve shows small periodical variations of $\approx 0\overset{m}{.}15$ with a period of ≈ 356 days. This period differs significantly from the U-band period. The cause of this variability is still unknown.

References

Meinunger L., 1979, IBVS 1611

Luthardt R., 1983, Mitt. Veränd. Sterne 9.5, 130

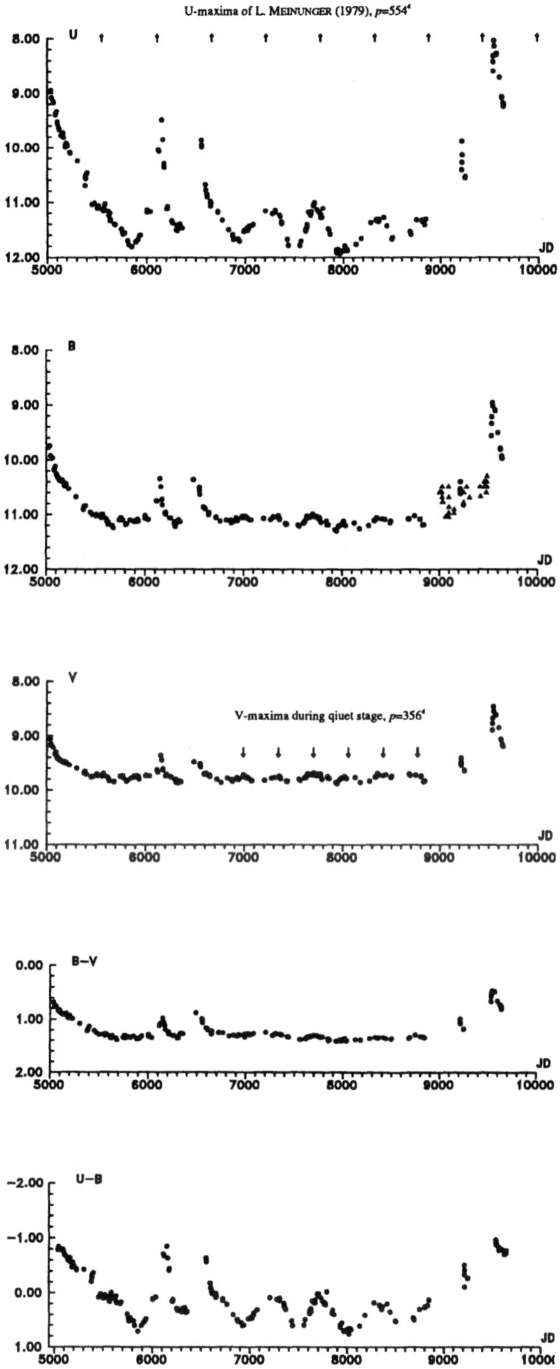

Fig. 2. *UBV* light curve of AG Dra (top panels) and the variation of the color indices $(B-V)$ and $(U-B)$ as obtained at Sonneberg Observatory over the last 12 years.

Space Densities of Cataclysmic Variables and Nova Recurrence Times

H.W. Duerbeck[1,2], *R. Covarrubias*[1]

[1] Astrophysics Group, P. Universidad Catolica de Chile, Santiago 22, Chile
[2] Astronomisches Institut der Universität Münster, D-48149 Münster, Germany

Abstract: Space densities of novae, dwarf novae and novalike stars are derived on the basis of the most recent catalogue of cataclysmic variables and improved values of absolute magnitudes. Assuming that these objects represent different stages of the same type of variable, changing over short-term cycles, we derive an average time interval of 10 000 years between two subsequent nova eruptions. During this time, the cataclysmic variable remains almost 2 000 years in the novalike stage and the remaining time in the dwarf nova stage, in which a total of nearly $3 \cdot 10^{-5}\ M_{\odot}$ is accreted on the white dwarf primary. This amount is comparable to the average mass of ejected nova shells.

1 Introduction

In 1983, one of the present authors determined the space densities of novae, dwarf novae, and related systems, and drew attention to the fact that for realistic outburst intervals, quiescent novae in the form of 'novalikes' appear to be underrepresented among cataclysmic variables (CVs) (Duerbeck 1984). Similar suggestions were voiced independently by Patterson (1984). This led to the revival and extension of the concept, first put forward by Vogt (1981), that novae spend their time between eruptions as novalike stars *and* as dwarf novae (see also Vogt 1989). Furthermore, the hibernation scenario was suggested by Shara et al. (1986), in which novae not only change into novalikes and dwarf novae, but may spend many thousands of years in a detached state, with a dramatic reduction of the accretion rate ('deep hibernation').

In the meantime, our knowledge of CVs and their physical properties has increased considerably. A recent catalogue of novae, dwarf novae, novalike stars and related systems is at hand (Downes and Shara 1993, henceforth D+S), and Warner (1987, henceforth W87) has determined improved absolute magnitudes for different types of CVs. It thus seemed worthwhile to check whether the 'deficit' of quiescent novae is still felt as strongly as in the past, and whether the 'deep hibernation' scenario is still needed.

2 Procedure

From D+S, we take all dwarf nova (DN) and novalike (NL) systems, which are situated in a cylinder of 250 pc radius, centered on the Sun and perpendicular to the galactic plane. Their $m_{\max}$ is obtained from the catalogue, an average $M_{\max}$ is assigned to each system (see below), interstellar extinction is taken into account with $A = 1^{\mathrm{m}}/\mathrm{kpc}$, when the radius of vision is inside a slab of thickness 100 pc above and below the galactic plane, and assumed to be $A = 0$ otherwise, and the height d_z above the plane is calculated.

Following W87, we assigned absolute magnitudes $M_V = +4.5$ to all dwarf novae at maximum, except those of SU UMa type, for which we assumed $M_V = +5.2$. Novalike stars show large scatter in their absolute magnitudes. We thus chose the average absolute magnitude in the sample of W87, i.e. $M_V = +4.1$. Distances were also calculated under the assumptions that M_V differs by ± 0.7 mag from this value.

The number of objects found in z-bins of 25 pc were fitted with the simplex algorithm to the function

$$\rho(z) = \rho_0 \exp\left(-|z|/H\right), \tag{1}$$

where ρ_0 is normalized to the space density per cubic parsec in the galactic plane, and H is the scale height.

3 Results

Fitting the distribution of all dwarf novae by equation (1) yields:

$$\rho(z) = (1.85 \pm 0.07) \cdot 10^{-6} \exp\left(-|z|/(98 \pm 5)\right) [\mathrm{pc}^{-3}] .$$

For the U Gem/Z Cam-subclasses one obtains

$$\rho(z) = (1.15 \pm 0.06) \cdot 10^{-6} \exp\left(-|z|/(96 \pm 6)\right) [\mathrm{pc}^{-3}] .$$

For the SU UMa-subclass the result is

$$\rho(z) = (0.70 \pm 0.03) \cdot 10^{-6} \exp\left(-|z|/(102 \pm 7)\right) [\mathrm{pc}^{-3}] .$$

Within the error margins, both groups have the same space distribution, but the SU UMa-stars are roughly half as frequent as the U Gem/Z Cam-stars.

The space distribution of novalike stars was derived, assuming the absolute magnitudes given above. The exnova DQ Her was included in the sample since it fulfills all criteria for a NL in the solar neighbourhood, for which 'by chance' a nova eruption was observed. The following results for novalike stars are obtained, assuming $M_V(\max) = 3.4$

$$\rho(z) = (0.28 \pm 0.06) \cdot 10^{-6} \exp\left(-|z|/(97 \pm 32)\right) [\mathrm{pc}^{-3}] ,$$

assuming $M_V(\max) = 4.1$

$$\rho(z) = (0.42 \pm 0.06) \cdot 10^{-6} \exp\left(-|z|/(97 \pm 20)\right) [\mathrm{pc}^{-3}],$$

assuming $M_V(\max) = 4.8$

$$\rho(z) = (0.66 \pm 0.06) \cdot 10^{-6} \exp\left(-|z|/(70 \pm 9)\right) [\mathrm{pc}^{-3}].$$

$M_V(\max) = 4.8$ leads to a strong concentration of NL towards the galactic plane which appears unlikely. In the following, the result for $M_V(\max) = 4.1$ is used.

For novae, we consider a cylinder with a radius $R_{x,y} = 1\,500$ pc in the galactic plane. Since the compilation of the list of galactic novae in the solar vicinity (Duerbeck 1984), no bright nova has appeared within this cylinder, but many of the listed novae have now revised distances (Cohen 1985, W87). We take observed novae which erupted in the said volume during the 20th century to derive the space density of nova eruptions per cubic parsec per year. The effective 'monitoring time', however, was set to $T_{\mathrm{obs}} = 70$ years, because novae may have been overlooked due to seasonal or other effects. With this assumption, the error of the result should not be larger than 25%. We obtain the space/time density $\rho^*(z)$:

$$\rho^*(z) = (0.23 \pm 0.02) \cdot 10^{-9} \exp\left(-|z|/(120 \pm 24)\right) [\mathrm{pc}^{-3}\ \mathrm{yr}^{-1}].$$

The above results show that the average scale height of all types of cataclysmic variables is about 110 pc. The number of objects in the solar neighbourhood, however, is too scarce to test whether they are a mixture of disk and bulge objects, as seems to be the case for novae (see Duerbeck 1990 and Della Valle at al. 1992).

4 Discussion

If we assume in a straightforward way that all dwarf novae and novalikes (except the highly magnetized polars, whose space density is possibly much lower), are manifestations of the same type of variable – a nova between two outbursts –, we can draw the following conclusions:

(1) The fact that DN are by a factor 4 more frequent than NL signifies that novae spend 4 times more time in the DN than in the NL stage.

(2) Assuming the space/time density of novae to be $0.23 \cdot 10^{-9}\ \mathrm{pc}^{-3}\ \mathrm{yr}^{-1}$, and the space density of DN plus NL $(1.85 + 0.42) \cdot 10^{-6}\ \mathrm{pc}^{-3}$, the *average outburst interval* Δt of novae is (Duerbeck 1990):

$$\Delta t = \frac{\rho^* \cdot T_{\mathrm{obs}} + \rho_{\mathrm{DN}} + \rho_{\mathrm{NL}}}{\rho^* \cdot T_{\mathrm{obs}}} \cdot T_{\mathrm{obs}} \approx \frac{\rho_{\mathrm{DN}} + \rho_{\mathrm{NL}}}{\rho^*} = \frac{2.27 \cdot 10^{-6}}{0.23 \cdot 10^{-9}}$$

or $\Delta t \approx 10\,000$ years. If we assume that conclusion (1) is valid, a nova between eruptions spends, on the average, 2000 years in the NL stage (including the so-called exnova-stage), and 8000 years in the DN stage. If accretion in the first stage is $10^{-8}\ M_\odot\ \mathrm{yr}^{-1}$, and in the second stage $10^{-9}\ M_\odot\ \mathrm{yr}^{-1}$, a total of about

$3 \cdot 10^{-5}\ M_{\odot}$ is transferred, to be ejected in the subsequent explosion. This agrees with the average shell mass of novae, $3 \ldots 4 \cdot 10^{-5}\ M_{\odot}$, as determined by Cohen and Rosenthal (1983).

The above estimates can at best be taken as a guideline for further studies, because, e.g., the mass spectrum of accreting white dwarfs might have to be taken into account (e.g. Politano et al. 1990). Since the relation between critical mass and white dwarf mass seems to be still quite unclear, however, we refrain from discussing this problem here. A fainter stage ('deep hibernation') with a negligible accretion rate is possible, but not necessary, unless the 'average' mass transfer rate of about $3 \cdot 10^{-9}\ M_{\odot}\ \mathrm{yr}^{-1}$ is found to be too high on the basis of theoretical arguments.

Acknowledgements: We are grateful to Rolf Duemmler, Münster, for making his simplex function fitting program available to us, as well as to Ron Downes and Mike Shara (STScI) for the updated version of their catalogue. The stay of HWD with the Astrophysics Group of P.U.C. was made possible by a grant of the Deutscher Akademischer Austauschdienst.

References

Cohen J., 1985, ApJ 292, 90
Cohen J., Rosenthal A., 1983, ApJ 268, 689
Downes R.A., Shara M.M., 1993, PASP 105, 127 (D+S)
Della Valle M., Bianchini A., Livio M., Orio M., 1992, A&A 266, 232
Duerbeck H.W., 1984, ApSS 99, 363
Duerbeck H.W., 1990, in IAU Coll. 122, Physics of Classical Novae, A. Cassatella and R. Viotti (eds.), Springer, Berlin, p. 34
Patterson J., 1984, ApJS 54, 443
Politano M., Livio M., Truran J.W., Webbink R.F., 1990, in IAU Coll. 122, Physics of Classical Novae, A. Cassatella and R. Viotti (eds.), Springer, Berlin, p. 386
Shara M.M., Livio M., Moffat A.F.J., Orio M., 1986, ApJ 311, 163
Vogt N., 1981, Mitt. Astr. Ges. 57, 79
Vogt N., 1989, in Classical Novae, M.F. Bode and A. Evans (eds.), Wiley, Chichester, p. 225
Warner B., 1987, MNRAS 227, 23 (W87)

Extension of the Radio Spectrum of AE Aqr to the Sub-millimetric Range

Meil Abada-Simon[1,2], *Tim S. Bastian*[3], *Jay A. Bookbinder*[4], *Monique Aubier*[2,5], *Gordon Bromage*[6], *George A. Dulk*[2], *Alain Lecacheux*[2]

[1] Sterrekundig Instituut, Postbus 80000, 3508 TA Utrecht, The Netherlands
[2] Observatoire de Paris, 92190 Meudon, France
[3] National Radio Astronomy Observatory, Socorro, NM 87801, USA
[4] Smithsonian Astrophysical Observatory, Cambridge, MA 02138, USA
[5] Université Paris 6, France
[6] Lancashire Polytechnic, Physics Department, Preston, PR1 6AD, UK

1 Introduction

AE Aquarii is a magnetic cataclysmic variable containing a white dwarf and a K3-K7 star which lies slightly above the main sequence. The white dwarf is the most rapidly rotating known ($P_{\rm rot} \simeq 33.08$ s, Patterson 1979), and it is the most strongly asynchronous with its revolution ($P_{\rm orb} = 9.88$ hr). The white dwarf accretes matter from the K star, which approximately fills the Roche lobe. AE Aqr exhibits flares in the soft X-rays, the ultra-violet, and almost continuously in the visible and the radio regimes. Rapid optical and TeV γ-ray bursts have also been discovered, which are modulated with the period of the white dwarf and at half of this period (de Jager & Meintjes 1993). This modulation, also found in X-rays, is interpreted as the accretion of matter onto the white dwarf's magnetic poles. The strength of the white dwarf's magnetic field is not well-determined, it is estimated to be $\sim 6.10^4 - 10^5$ G (Lamb & Patterson 1983, Cropper 1986) at the white dwarf's surface. Eracleous et al. (1994) recently suggested that the magnetic dipole axis lies close to the equatorial plane ($\sim 20°$). De Jager et al. (1994) discovered a rapid spin down of the white dwarf leading to a spin down power which exceeds the accretion power. They suggest that a significant fraction of the spin down power may be converted to the acceleration of particles, which may explain the radio and the γ-ray emissions. Both the characteristics of the optical flares and the existence of TeV γ-rays suggest a relation with the non-thermal radio flares.

AE Aqr was discovered by Bookbinder & Lamb (1987), and by Bastian et al. (1988, hereafter BDC) as a strong radio source from 1.4 to 22.5 GHz (21 to 1.3 cm), highly variable on time scales of minutes, hours, and days. BDC also showed that AE Aqr's time-averaged spectrum is a remarkably constant power-law increasing with frequency as $S_\nu \propto \nu^\alpha$, where $\alpha \simeq 0.35$. Similar flares were also observed at 88 GHz (3.4 mm) and AE Aqr's radio spectrum was found to increase up to 240 GHz (1.25 mm), at which the flux density was ~ 48 mJy, and $\alpha \simeq 0.35 - 0.60$ (Abada-Simon et al. 1993). A detection at 250 GHz (1.2 mm) gave ~ 24 mJy (Altenhoff et al. 1994). These results suggest that the turnover frequency is ≥ 250 GHz, maybe in the sub-millimetric range.

Assuming that the radio source is not larger than the binary separation ($\sim 2.10^{11}$ cm) the inferred brightness temperature is $T_b \geq 10^{10}$ K, implying a non-thermal emission process. The lack of circular polarization and of rapid time variations are in favour of an incoherent process. BDC argue, for several reasons, that the radio flares cannot arise from the K star.

BDC attribute the radio flares to the superposition of synchrotron radiation from "clouds" of MeV electrons which are presumably accelerated at times of disruptions of the accretion flow near the white dwarf. According to BDC, as the "plasma clouds" expand, they change from optically thick to optically thin. The initial maximum flux density corresponds to an optical thickness $\tau \simeq 1$, and to a turnover frequency ν_{m0} which decreases with time. In their model, ν_{m0} is directly related to the initial magnetic field strength B_0 in a "plasmoid". Under several assumptions, among which an energy distribution for the electrons in the form of a power-law with spectral index $\delta = 2.5$, the measurement at 230 GHz then provided $B_0 \geq 250$ G, for a source radius $r_0 \simeq 10^{10}$ cm.

From the IRAS survey, which was not sensitive enough to detect AE Aqr, one can infer that its flux density is lower than 1 Jy at 100 μm, and lower than 200 mJy at 12, 25 and 60 μm: these upper limits are too high to constrain the observed radio emission.

In this paper we report the first detections of AE Aqr at 1.1 and 0.8 mm made with the James Clerk Maxwell Telescope (JCMT, Mauna Kea, HI) a few hours after microwave observations made with the Very Large Array (VLA). These observations were performed in the frame of a coordinated multi-wavelength campaign on AE Aqr, whose main part was held around 20-21 October 1993. In Sect. 2 we report our observations, and in Sect. 3 we discuss our results.

2 Observations

2.1 Observing Techniques

Microwave observations were performed with the VLA in a time sharing mode between 4.9, 8.4, 15 and 22.5 GHz (6 to 1.3 cm) on 17 and 18 October 1993, from about 20:30 to 25:30 UT.

Millimetric observations were made at 264 and 394 GHz (respectively 1.1 and 0.8 mm), with the UKT14 bolometer of the 15 m diameter mm/sub-mm JCMT on 18 and 19 October 1993, from 05:00 to 07:30 UT. The "beam-switching" observational technique was used, i.e. the source was placed alternatingly in the signal and reference beam of the chop cycle. This provided integrated scans of $\sim 3 - 9$ min on AE Aqr.

2.2 Observational Results

At centimetric wavelengths, AE Aquarii was, as usual, detected, with flares lasting for a few 10 min to several hours. They were almost continuously present, but stronger in the second night (18-19 October 1993) than in the first night. The average flux densities at 3 and 6 cm are reported for the two nights in Fig. 1.

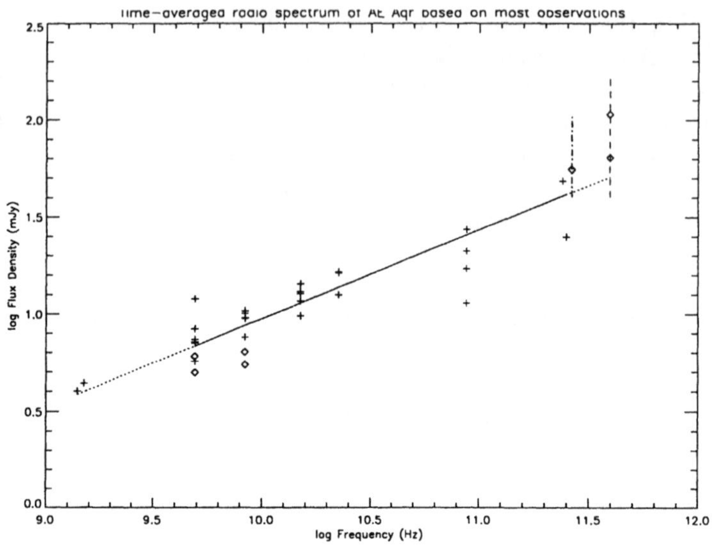

Fig. 1. Time-averaged radio spectrum of AE Aqr, based on most observations since 1984. The "+" symbols are average flux densities previously reported in the literature, and the lozenges are the values measured by the VLA and the JCMT in October 1993. The vertical dashed bars indicate the range of values measured at 1.1 and 0.8 mm with their 1-σ uncertainty.

On Mauna Kea, although the weather conditions were not good, the observations of AE Aqr made with the JCMT at 1.1 and 0.8 mm show some consistent detections, especially at 0.8 mm over the two consecutive nights. On 18 October 1993, the average flux densities at 0.8 and 1.1 mm were respectively 107.2 mJy (3.5σ) and 55.8 mJy (3.7σ), and on 19 October, an average value of 64.3 mJy (4.5σ) was obtained at 0.8 mm. These values are plotted in Fig. 1, with a vertical (dashed) "error" bar indicating the uncertainties inferred from all the significant ($S/N \geq 3$) scans (minimum to maximum: 40 to 138 mJy at 0.8 mm, and 50 to 73 mJy at 1.1 mm). The inclined dotted line represents the increasing power law obtained from fitting all the plotted flux densities, which gives a spectral index $\alpha \simeq 0.46$.

3 Discussion

The uncertainties on the flux densities measured by the JCMT lie mainly above the plotted power law line; it is clear that the turnover frequency of AE Aqr's radio spectrum is higher than 394 GHz. Applying BDC's model with all their hypotheses, we find that the initial magnetic-field strength in a plasmoid of radius $r_0 \simeq 10^{10}$ cm is 270 G $\leq B_0 \leq$ 1000 G. This upper limit is inferred from BDC's estimate of the $B_0 - r_0$ domain in which losses due to betatron deceleration dominate over radiative synchrotron losses, for a plasmoid expanding into a uniform medium with an initial velocity $v_0 \simeq 1000$ km s^{-1}. The other alternative proposed by BDC was an expansion into a stellar wind, then they find that $v_0 \simeq 200$ km s^{-1} is required; in this case, it is interesting to find out that the detections at frequencies $\geq$ 300 GHz infer that B_0 is restricted to $\sim$ 300 G. These results suggest to re-investigate BDC's model with other possible values for the assumed parameters.

AE Aqr's radio flares, and more particularly its time-averaged spectrum, increasing with frequency over more than two decades, up to an unknown frequency, remain intriguing. Its emission at frequencies $\geq$ 400 GHz, up to the far-infrared regime, is unexplored. More sensitive observations at several mm and sub-mm wavelengths need to be performed, not only to detect AE Aqr at wavelengths shorter than 0.8 mm, but also to study its flares and their correlations with the microwave domain (the only observations of AE Aqr ever made simultaneously in these two regimes lasted for less than 2 hours). More precise investigations of the BDC model could then be made. There is no observational constraint on the source size yet. The way electrons can be accelerated also needs to be investigated quantitatively. Finally, investigating correlations of AE Aqr's radio flares with the optical and with the TeV γ-rays should also provide constraints on the radio emission process.

References

Abada-Simon M., Lecacheux A., Bastian T.S., Bookbinder J.A., Dulk G.A., 1993, ApJ 403, 692

Altenhoff W.J., Thum C., Wendker H.J. 1994, A&A 281, 161

Bastian T.S., Dulk G.A., Chanmugan, G. 1988, ApJ 234, 431

Bookbinder J.A., Lamb D.Q., 1987, ApJ 323, L131

Cropper M., 1986, MNRAS 222, 225

Eracleous M., Horne K., Robinson E.L., Zhang E.-H., Marsh T.R., Wood J.H., 1994, ApJ 433, 313

De Jager O.C., Meintjes P.J., 1993, A&A 268, L1

De Jager O.C., Meintjes P.J., O'Donoghue D., Robinson E.L., 1994, MNRAS 267, 577

Lamb D.Q., Patterson J., 1983, in Cataclysmic Variables and related Objects, M. Livio & G. Shaviv (eds.), Reidel, Dordrecht, p. 229

Patterson J., 1979, ApJ 234, 978

The Origin of Flaring Activity in AE Aquarii

Nazar Ikhsanov

Pulkovo Observatory, 196140 St. Petersburg, Russia

1 Introduction

1.1 General properties

AE Aqr is a close non-eclipsing binary system with an orbital period $P_{\rm orb} \approx 9.88$ hr, eccentricity of the orbit $e = 0.02$ (Chincarini & Walker 1981) and a mass ratio $q = 1.14$ (Reinsch & Beuermann 1994), situated at the distance of ~ 84 pc (Van Paradijs et al. 1989). The secondary is a K3 red dwarf on or close to the main sequence (Bruch 1991). The primary cannot be observed directly. However, the stable photometric oscillations at a period of 33 s (and at half this period) in the optical (Patterson 1979), UV (Eracleous et al. 1994), X-rays (Patterson et al. 1980) and, probably, VHE γ-rays (Meintjes et al. 1992, Bowden et al. 1992) leave no doubt that the primary is a rapidly rotating magnetized compact object. The observed rate of deceleration of the primary rotation is $\dot{P} = 5.64 \times 10^{-14}\,{\rm ss}^{-1}$ (De Jager et al. 1994), and its mass is $m_1 sin^3 i = (0.56 \pm 0.03) M_{\odot}$ (Reinsch & Beuermann 1994). The lack of eclipses allows to put a lower limit to the mass of the primary $m_1 \geq 0.62 M_{\odot}$.

AE Aqr emits detectable radiation in all parts of the electromagnetic spectrum. In radio and γ-rays it is a powerful non-thermal source. However, in the optical, UV and X-rays the radiation is likely to be thermal and can be well explained by gas accretion onto a compact star (Patterson 1979, Patterson et al. 1980). The observed impulse profile of the pulsating component (Eracleous et al. 1994) looks similar to that of the accreting X-ray pulsars.

1.2 Flaring component

The main peculiarity of AE Aqr is its extraordinary flaring activity. Active and quiet phases alternate at a time scale of a few hours without any noticeable transitions. The persistent characteristic of the flare is flickering, clearly showing up in the power spectra during the active phase in contrast to the quiescence (Bruch 1991).

A major part of the thermal energy release in the flares is radiated in the UV part of spectrum. The light curves of AE Aqr during the active phase show a great diversity of flaring events and non-trivial fine structure. This raises the the question how meaningful average characteristics of the flares are. The most commonly adopted classification of flares has been based on their spectral properties (Chincarini & Walker 1981, Bruch 1991):

1. pure continuum flares,
2. pure emission lines flares,
3. combination of line and continuum radiation.

With respect to their temporal characteristics three types of flaring components can be distinguished:

A: *An active phase* with durations of 3-5 hr, rise times of 1-1.5 hr and amplitudes in the U-band $\Delta m \sim 1^m - 2^m$ (Chincarini & Walker 1981, Bruch 1991). An active phase can be considered as a superposition of flares of shorter duration with the following characteristics:

B: *Medium time scale flares* lasting 10-30 min with rise times of 2-10 min (Van Paradijs et al. 1989). Some flares show multi-component fine structure:

C: *Short time scale flashes* with duration of 2-10 min and rise time of few seconds (De Jager & Meintjes 1993).

The state of flaring activity of AE Aqr does not correlate with orbital phase (Van Paradijs et al. 1989). As a rule, an active phase begins with a sequence of flares of type **B** with increasing amplitude.

Near the maximum of the active phase two independent groups detected γ-ray emission from AE Aqr with energies $\sim$ TeV and the luminosity $L_\gamma \sim 1.2 \times 10^{32}$ $\mathrm{erg\,s^{-1}}$ (Meintjes et al. 1992, Bowden et al. 1992). Modulation of detected γ-ray radiation with a period $P_{1/2} = 16.5\,\mathrm{s}$ gives evidence that it is generated in the vicinity of the compact object in AE Aqr.

Observations in the UV (Eracleous et al. 1994) have shown that the amplitude of the 33 s oscillations depends weakly on the state of the system (active/quiet). Finally, Abada-Simon et al. (1993) reported about the flaring character of the radio emission from AE Aqr. However, there is no evidence for a correlation between radio and optical flares.

2 The primary and the evolutionary status of AE Aqr

According to the standard model, proposed by Patterson (1979) the primary in AE Aqr is a rapidly rotating magnetized white dwarf undergoing disk accretion. Hence, AE Aqr can be classified as an intermediate polar, belonging to the DQ Her group. However, recent observational data force us to doubt the validity of this model. As is shown by Ikhsanov (1995a) any attempt to explain the totality of observational data on AE Aqr in the framework of the standard model leads to a paradox: the white dwarf should be simultaneously in three different states: *ejector*, *propeller* and *accretor*. This is in obvious contradiction with the modern theory on evolution of close binary systems. At the same time, this paradox does not arise under the assumption that the primary of AE Aqr is a neutron star. According to Ikhsanov (1995a) the state of the neutron star can be classified as *accretor-propeller*. A principal feature of the new model is a low mass loss rate of the secondary during the quiet state $\dot{M} \sim 10^{12}\,\mathrm{g\,s^{-1}}$. This is close to the mean loss rate of the late type stars in the form of stellar wind (Pottasch 1970). Chincarini & Walker (1981) and Bruch (1991) pointed out that the degree of the secondary filling its Roche lobe might be insufficient to realize the mechanism of matter outflow through the first Lagrangian point L_1. Thus, according to this picture the evolutionary status of AE Aqr is determined as a pre-low-mass X-ray binary pulsar (Ikhsanov 1995a).

According to calculations of Davies & Pringle (1980), a quasi-spherical type of accretion should be realized in this system. That is why we will discuss further a scenario of flaring activity in AE Aqr in the framework of the model of quasi-spherical accretion onto a magnetized neutron star.

3 Main points of a flare scenario for AE Aqr

Bruch (1991) pointed out that the flaring behaviour of AE Aqr does not fit the model of flaring activity of the secondary component with respect to either its temporal or its energetic characteristics. Thus, the flares are more likely to be connected with a non-stationary accretion process. The reasons for this can be as follows:

- variations of the secondary mass loss rate,
- instabilities in the accretion flow,
- interaction of the accretion flow with the magnetosphere of the primary.

Taking into account the temporal characteristics of type **B** (see 1.2), Van Paradijs et al. (1989) suggested that the trigger of the flares is localized at the Alfvén surface of the primary and is not connected with disk instabilities. They proposed a model of "magnetospheric gating", assuming that the state of the primary is quasi-periodically changing from propeller to accretor and back. However, this model cannot explain the temporal characteristics of the active phase **A** as a whole (see 1.2) as well as spectral differences between flares.

Starting from a model of quasi-spherical accretion onto a neutron star instead of the standard (accreting white dwarf) model can shed new light on the problem of flaring activity in AE Aqr..

First of all, neither the surface of a neutron star nor the accretion column can be the source of the observed thermal energy release during a flare. Otherwise we should detect flares in hard X-rays, rather than optical or UV flares. Thus, we can assume that the flare energy release is localized at the Alfvén surface of the neutron star or/and at the surface of the secondary. In both cases, a flare proves to be connected with changes in the atmosphere of the secondary. Really, in the binary system with practically circular orbit and quasi-spherical type of accretion quasi-periodical increase of the rate of accretion onto the Alfvén surface can be connected only with change of the mass loss rate of the secondary. If we exclude the mechanism of the secondary magnetic activity (Bruch 1991), we can suppose that the state of the secondary may be changed due to interaction of the secondary and the primary. Such interaction may occurr via the illumination of the secondary by the radiation of the neutron star.

Models connected with the irradiation of the secondary by UV and X-rays (Osaki 1985, Hameury et al. 1986) cannot work in AE Aqr because of the very low luminosity in these spectral regions (Eracleous et al. 1994; Patterson et al. 1980). At the same time, we can suppose that perturbation of the red dwarf atmosphere may occur due to the non-thermal mode of the neutron star energy release (accelerated particles and γ-rays). Under certain conditions, heating of the red dwarf atmosphere by the beam of accelerated protons ($E_p \sim$ GeV) results in the increase of the secondary mass loss rate up to $\dot{M} \sim 10^{-10} M_\odot$ yr^{-1} on a dynamical time scale $\sim$ 100 s (Ikhsanov 1995b). Accretion of this gas onto the Alfvén surface of the neutron star leads to UV flare with the blackbody temperature $T_{\rm eff} \sim 16000 - 18000$ K. The rise time of such a flare is of the order of $b/V_{\rm ff}(b) \sim 1 - 1.5$ hr, where b is the orbital separation, and $V_{\rm ff}(b)$ is the free-fall velocity at this distance.

The different spectral behaviour of flares in this case can be classified according to the values of the flux of accelerated particles at the surface of the red dwarf. This value depends on the relation between the main axes of the system and the degree of collimation of the particle beam. If the flux reaches a critical value $S \geq S_{\rm cr}$, the model of explosive evaporation and matter accretion onto the Alfvén surface will be realized. Otherwise ($S < S_{\rm cr}$), the mass loss rate of the red dwarf does not change much, and the flares are very similar to the secondary processes of flares on flaring stars (Livshits et al. 1981). A detailed scenario of flares will be presented in a subsequent paper (Ikhsanov 1995b).

References

Abada-Simon M., et al., 1993, Ap.J. 406, 692
Bowden et al., 1992, Astroparticle Phys. 1, 47
Bruch A., 1991, A&A 251, 59
Chincarini G., Walker M.F., 1981, A&A 104,24
Davies R.E., Pringle J.E., 1980, MNRAS 191, 599
de Jager O.C., et al., 1994, MNRAS 267, 577
de Jager O.C., Meintjes P.J., 1993, A&A 268, L1
Eracleous M. et al., 1994, Ap.J. 433, 313
Hameury J.M., et al., 1986, A&A 162, 71
Ikhsanov N.R. 1995a, A&A in press
Ikhsanov N.R., 1995b, in preparation
Livshits M.A., et al., 1981, Solar Phys. 73, 269
Meintjes P.J., et al., 1992, ApJ 401, 325
Osaki Y., 1985, A&A 144, 369
Paradijs J. van et al., 1989, A&AS 79, 205
Patterson J., 1979, ApJ 234, 978
Patterson J., et al., 1980, ApJ. 240, L133
Pottasch S.R., 1970 in Interstellar Gas Dynamics, H.J. Habing (ed.), Reidel, Dordrecht
Reinsch K., Beuermann K., 1994, A&A 282, 493

UBVRI Photometry of AE Aquarii in July-August 1994

Nina Beskrovnaya[1], *Nazar Ikhsanov*[1], *Albert Bruch*[2]
Nikolaj Shakhovskoy[3]

[1] Pulkovo Observatory, 196140 St. Petersburg, Russia
[2] Astronomisches Institut, Wilhelm–Klemm–Straße 10, D-48159, Münster, FRG
[3] Crimean Astrophysical Observatory, 334413 Nauchnyi, Crimea, Ukraine

The most striking feature of the close binary system AE Aqr is its unique photometric behaviour – irregular alternations of quiet and active phases without noticeable transition periods. Although extensive photometric observations of AE Aqr have been carried out for more than forty years (see Bruch 1991 and references therein), the mechanism of its flaring activity is still poorly understood and the star – although generally classified as a cataclysmic variable – does not fit well into any of the subclasses.

According to us, much insight into this peculiar object can be gained by an investigation of the connections between the flaring activity of AE Aqr and the strictly periodic oscillations at 33^s, identified as the spin period of the primary. The latter have been seen in optical (Patterson 1979) and UV light (Eracleous et al. 1994), in X-rays (Patterson et al. 1980) and in TeV γ rays (Meintjes et al. 1992). There are contradictory reports about the relation between their amplitude and the flaring activity in AE Aqr : Patterson (1979) noted that the amplitude of the pulsations increases during active phases, while Eracleous et al. (1994) observed that their characteristics are not strongly affected by the flares.

During July–August 1993 the oscillations in AE Aqr vanished below a detectable limit (Bruch et al. 1994), but neither brightness and colour of the system nor its flaring activity showed any dramatic peculiarities. In this contribution we present a preliminary report on results of five-channel time resolved photometry of AE Aqr in 1994, performed to study further the optical oscillations and the flaring behaviour of the system.

AE Aqr was observed during four nights in July–August 1994 with the photometer-polarimeter of Helsinki University (Piirola 1987) at the 1.25 m telescope of the Crimean Astrophysical Observatory. In the photometry mode the instrument provides simultaneous observations of the star and the sky background in five broad passbands close to the standard *UBVRI* system. A new version of the photometer control and signal recording system based on CAMAC modules and a PC AT-286 was used (Kalmin & D. Shakhovskoy 1995).

An integration time of 5^s was chosen. The variable star was monitored continuously for $\sim 30^m$ (with the use of offset guiding) through a $10''$ diaphragm. Between these series of integrations a comparison star (situated $10'$ to the west form AE Aqr) was observed. The measurements were tied into the *UBVRI* system by observations of the nearby photometric standard HD 196395 (Barnes & Moffet 1979).

Power spectra of all light curves were calculated. Only in one night (July 10) could the 33^s oscillations confidently be detected in the U, B and V bands. On July 9 some quasi-periodic oscillations close to the frequency of the 33^s signal are possibly seen in the B band. However, this detection is marginal. The first harmonic of the 33^s oscillations, often appearing strongly in power spectra of AE Aqr, remains undetectable in our data. In order to assess the reliability of our non-detections we added tracer signals of various amplitudes to the data: an oscillation with an amplitude of a few millimagnitudes would have been easily detected (except in the U band). Possible reasons for the cessation of the 33^s oscillations in AE Aqr were discussed by Beskrovnaya et al. (1995).

While the coherent signal was thus largely absent during our observations the irregular activity was quite normal. AE Aqr showed a strong flaring behaviour interspersed with shorter periods of relative calmness. The time scale of the large outbursts and the rise time of rapid flares is of the order of 1 hour. This is much longer than the characteristic time scale ($\sim 100^s$) for the rise of flares in the model of magnetospheric gating (Van Paradijs et al. 1989). However, the observed time scales are comparable with the free-fall time of the matter flowing from the secondary star to the compact primay. It may thus be suggested, that the origin of the large flares is connected with variations of the secondary mass loss rather than with variations of the accretion rate in the vicinity of the primary. This idea is supported by the fact, that the amplitude of the 33^s oscillations is not affected by the large flares (Eracleous et al. 1994). A corresponding physical model is currently being worked out by one of us (N. Ikhsanov).

Acknowledgements: We thank D. Shakhovskoy for his assistance during the observations. This study was partly supported by RFFR grant 94-02-05343.

References

Barnes T.G. & Moffet T.J., 1979, PASP 91, 289

Beskrovnaya N., Ikhsanov N. & Bruch A., 1995, Proc. of the Padova Conference on CVs, in press

Bruch A., 1991, A&A 251, 59

Bruch A., Beskrovnaya N., Ikhsanov N., Borisov N., 1994, IBVS 3996

Eracleous M., Horne K., Robinson E.L., Zhang E.-H., Marsh T.R., Wood J.H., 1994, ApJ 433, 313

Kalmin S. & Shakhovskoy D., 1995, Kinematika i fizika nebesnykh tel, in press

Meintjes P.J., Raubenheimer B.C., de Jager O.C., Brink C., Nel H.I., North A.R., van Urk C., Visser B., 1992, ApJ 401, 325

Patterson J., 1979, ApJ 234, 978

Patterson J., Branch D., Chincarini G., Robinson E.L., 1980, ApJ 240, L133

Piirola V., 1988, in: Polarized Radiation of Circumstellar origin, G.V. Coyne (ed.), Vatican, p. 735

Van Paradijs J., Kraakman H., Van Amerongen S., 1979, A&AS 79, 205

Flares and Flickering in the Cataclysmic Variable AE Aquarii

William F. Welsh[1,4], *Keith Horne*[2], *Richard Gomer*[3]

[1] Keele University, Dept. of Physics, UK
[2] University of St. Andrews, School of Physics & Astronomy, UK
[3] Dept. of Biochemistry and Cell Biology, Rice University, Texas USA
[4] Space Telescope Science Institute, Maryland, USA

1 Introduction

Detected from the radio to the γ-ray bands, AE Aquarii is an extraordinary cataclysmic variable. It is a relatively bright ($V \sim 11.4$), non-eclipsing binary with an orbital period of $9^{\rm h}\!.9$. The mass-donor secondary star is of approximate spectral type K5V and contributes about $\sim$ 80% of the total flux in the optical. While the mass ratio is well determined at $q = M_2/M_1 = 0.646$, the individual stellar masses are poorly known since the inclination is unknown. AE Aqr's flickering behavior is unique among CVs, alternating from periods of very violent activity to near total quiescence. Large amplitude flares have been detected in the radio, optical and UV. Previous work has shown the optical flare spectrum to be similar to stellar flare spectra.

2 The data (the "cable experiment")

Four nights of exactly simultaneous Hα spectroscopy and photometry were obtained on the 2.5 and 1.5m telescopes at Mt. Wilson on 27–30 July 1982. The simultaneity allowed us to renomalize the spectroscopy to agree with the photometry, giving us spectrophotometric data (spectral resolution $\sim$ 50 km s^{-1}, time resolution $\sim 6^{\rm s}$). A total of $\sim 17^{\rm h}$ of data were obtained, resulting in 10,098 spectra.

3 Preliminary findings

Large flares in both the line and continuum are easily seen. Changes of a factor of two in the continuum and a factor of three in the integrated line flux can occur on a timescale of a few hours. The variations in the Hα emission are highly structured and it is clear that the Hα emission is not simply a linear function of the continuum emission (the light curves are not scaled versions of each other). It appears that the Hα variations contain more power on shorter timescales than the continuum variations.

The Hα line does not act as a single entity; rather the wings and core can exhibit different behaviour (e.g. a flare was seen in the red wing but is absent in the core, blue wing and continuum). If we assume the flares arise in an accretion disk (perhaps not a good assumption given AE Aqr's peculiarities!), this localization in velocity space corresponds to localization in position space, implying the flares come from localized regions in the disk.

The rise time of a particularly clean flare was seen to be about 6 min, with an exponential-like decay. The decay from the peak takes longer for Hα than for the continuum. Closer scrutiny suggests that there can be a delay between the maxima, with the continuum leading the line emission. Thus it appears that the Hα emission is a delayed and smeared version of the continuum.

"Cross–correlations images" were constructed (ccfs between the mean continuum light curve and the light curves for each pixel in the spectrum, displayed as an image). The ccfs do not remain the same across the line profile, but rather show a striking change in behavior. *The fact that correlation strength between the continuum and the line changes dramatically across the line profile implies that the variations are not global, but rather are localized in velocity space.*

Night 4 exhibits a somewhat different ccf image than the other nights, showing a noticeable trend across the core of the line profile — the peak of the ccf occurs at larger lags moving from negative to positive velocities. This behavior can be interpreted as meaning the continuum and blue wing vary synchronously, while the core and red wing lag behind, with the lag increasing to positive velocities. A maximum delay of $\sim 25 - 30$ min is present.

The Hα flux was plotted versus the mean continuum flux for each night. While Night 1 shows some structure, Nights 2 and 3 show little more than what appears to be random fluctuations. On Night 4 a linear correlation at low continuum flux levels is seen, as well as an apparent Hα saturation effect during flares. A loop structure was seen, indicating a strong hysteresis effect in the large flares on this night.

4 Summary

The relationship between the continuum fluctuations and the Hα fluctuations is complicated and varies from night to night. Flickering and flaring in the wings and the line core do not behave in identical fashion. This localization in velocity space may correspond to localization in position space, implying the flares come from localized regions in the accretion flow. The data suggest that the Hα emission decays slower than the continuum emission after a flare, additional evidence that the flares in CVs are analagous to those of stellar flares. No 33 sec oscillation was seen in Hα.

Time Dependence of the UV Resonance Lines in the Cataclysmic Variable Star VW Hyi*

M.T. Özkan[1], *T. Ak*[2], *A.T. Saygac*[2], *H.H. Esenoglu*[2] *and S. Guler*[2]

[1] Istanbul University Observatory Research and Application Center, 34452 University-Istanbul, Turkey

[2] Istanbul University, Science Faculty, Department of Astronomy and Space Sciences, 34452 University-Istanbul, Turkey

Abstract: In this work, we study orbital variations in the ultraviolet resonance lines of the SU UMa type dwarf nova VW Hyi examining spectra in the data archive of the International Ultraviolet Explorer (*IUE*). We show that VW Hyi exhibits an orbital and long term variability in terms of UV line profiles and continuum flux at 1450 Å during superoutburst.

1 Introduction

To learn more about the nature of accretion disks, winds and dwarf novae outbursts, it is useful to study UV observations of cataclysmic variables. These observations reveal that one of the most striking properties of the wind-formed lines observed in dwarf novae is related to the variation of their profiles. There are two types of regular variability in the resonance line profiles. These are dependent on (i) outburst phase, and (ii) orbital phase. It has been found that there are about twenty CVs exhibiting more or less orbital change in their resonance lines (Friedjung et al. 1982, Szkody & Mateo 1986, Verbunt et al. 1987, Wood et al. 1990, Drew et al. 1991, Wood et al. 1992, Harlaftis et al. 1992, Knigge et al. 1994, van Paradijs et al. 1994, Mansperger et al. 1994, Özkan et al. 1995). They are generally dwarf novae (but one of them, MU Cen, does not show any variation). In addition to the three subgroups of dwarf novae, orbital variations were also discovered in two nova-like systems and one old nova. In this paper, we study the UV line variations associated with the orbit of the SU UMa type dwarf nova VW Hyi. Although VW Hyi has been studied over a large wavelength range through the outburst cycles (Pringle et al. 1987), we believe that it is necessary to consider the wind-formed line profiles in more detail again. We have chosen eight consecutive SWP spectra covering 3.88 orbits during superoutburst. Similarly, a second selection includes a total of ten SWP spectra belonging to three different normal outbursts. We also consider all SWP spectra at quiescence. All spectra are corrected for any wavelength shifts using the centroid of the geocoronal Lyα line.

---

* Based on the spectra taken from VILSPA IUE database

2 Results

Superoutburst: N V λ1240, Si III λ1300, Si IV λ1400, C IV λ1550 profiles and line centroid velocities of three resonance lines examined at increasing orbital phase suggest the presence of regular orbital variations. Although the variability pattern in the C IV, N V lines seems to be similar, Si IV shows somewhat different variability. Continuum fluxes at 1450 Å are also variable during the orbital cycle.

Normal outbursts: Although there is an insufficient phase coverage in the first and third outburst, the data exhibit long-term-scale variations in their UV resonance lines and continuum during 50 and 35 orbits, respectively. On the other hand, the conclusion reached for the second outburst is that the orbital variations in the wind-formed lines and the UV continuum may be present during 2.6 orbits, but the number of spectra is insufficient to confirm the orbital nature of these variations.

Quiescence: Examination of the continuum fluxes at 1450 Å versus JD and orbital phase, reveals two different groups with respect to JD. Every group includes five spectra. No significant variation is found for the UV continuum in the first group. However, the data in the second group lead to a 65% variation.

In summary, we have determined the variability of the UV resonance lines N V λ1240, Si IV λ1400, C IV λ1550 observed during superoutburst as a function of orbital phase. This confirms a similar result associated with variations of the C IV line in a study of VW Hyi by Verbunt et al. (1987). We detected no significant orbital variation in the normal outburst fluxes compared to the UV superoutburst flux modulation. The same behaviour can be seen in another SU UMa type cataclysmic variable star, Z Cha (Harlaftis et al. 1992). The mean equivalent widths of the UV absorption lines during superoutburst seem to be weaker than during normal outbursts while the continuum flux distribution during superoutburst is stronger than during normal outbursts.

Acknowledgements: We thank the AAVSO and Janet Mattei for obtaining photometric data. M.T. Özkan also thanks the IAU for supporting his travel and accomodation for this colloquium. This work is partly supported by TUBITAK.

References

Drew J.E., Hoare M.G., Wood J.A., 1991, MNRAS 250, 144
Friedjung M., Andrillat Y., Puget P., 1982, A&A 114, 351
Harlaftis E.T., et al., 1992, MNRAS 259, 593
Knigge C., et al., 1994, MNRAS 269, 891
Mansperger C.S., et al., 1994, PASP 106, 858
Özkan M.T., et al., 1995, XXIInd General Assembly of the IAU, poster paper (in press)
Pringle J.E., et al., 1987, MNRAS 225, 73
Szkody P., Mateo M., 1986, ApJ 301, 286
van Paradijs J., et al., 1994, MNRAS 267, 465
Verbunt F., et al., 1987, MNRAS 225, 141
Wood J.A., et al., 1990, MNRAS 245, 323
Wood J.A., et al., 1992, MNRAS 255, 237

The Benefit of Archival Research in Investigating Dwarf Novae

C. la Dous

ESA/IUE Observatory, Villafranca del Castillo, Apartado 50727, 28080 Madrid, Spain

1 Introduction

Dwarf novae and nova-like stars are a sub-group of the class of cataclysmic variables. Most of these stars show high and low brightness states (maxima or outbursts and minima or quiescent states, respectively). They all are short-period interacting binaries consisting of a white dwarf primary, a late-type Roche-lobe filling secondary, and an accretion disk around the white dwarf. For details on observations of cataclysmic variables and their theoretical explanations and modelling, the interested reader is referred to recent reviews.

The main problem when trying to understand these systems is the wealth of patterns presented by the whole group: no two objects are nearly identical and it is hard to say which the 'typical' features are. As a result of this, from an individual observation, or from observations of a few systems only, it is not possible to decide which of the observed features are characteristic of the entire class, and which are specific of individual members. The only way is to investigate many different members of the same class, arrange them according to their different physical parameters, and try to determine what the general characteristics are.

So far, data archives of the size and accessibility required for statistical investigations do not exist in the optical. Fortunately, from the very beginning of the *IUE* project, all observations were archived in a homogeneous way, together with the tools for a proper reduction and calibration; since many years the low-resolution UV spectra are available through ULDA, basically on-line, in a fully calibrated form. For dwarf novae and nova-like stars, more than 2000 spectra of over 100 objects have so far been accumulated (la Dous 1990). With this data set, which covers nicely the parameter space, it is possible to carry out a statistical investigation like the one discussed above. An additional help that proved to be crucial is the series of visual observations made by members of the American Association of Variable Star Observers (AAVSO) which allow a rather exact phasing of the observations in the outburst light curve.

2 Results

Without going into details, in the following I will summarize the most important results of the statistical analysis of low-resolution *IUE* spectra of dwarf novae and non-magnetic nova-like stars. Details of these investigations are given in la Dous (1991; 1994; 1995, in preparation).

2.1 The continuum radiation at maximum

For dwarf novae at maximum, and likewise for nova-like stars in the high brightness state, it is assumed that the accretion disk is stationary. Thus, assuming a value for the mass-transfer rate and for some other less important parameters, the continuum spectrum of the radiation emitted by the disk can be computed. If the continuum flux distributions of dwarf novae and nova-like stars at the peak of an outburst and in the high brightness state are compared with each other and with theoretical models, the following picture emerges: some 80% of the dwarf novae have approximately identical continuum flux distributions which are systematically redder than predicted for a stationary disk. In the case of nova-like stars, the internal agreement between the flux distributions of the different members of the class is not as good as in the case of dwarf novae, but with very few exceptions all of them are considerably redder, i.e., cooler, than expected on theoretical grounds. The reasons for this disagreement are not yet clear, but it is obvious that some mechanism is extracting energy from the continuum radiation of the accretion disk. Possible candidates are winds or magnetic loops forming above the disk, but the question remains open.

2.2 The line radiation at maximum

The pattern of line radiation emitted by accretion disks in the high brightness state ranges from strong emission lines to strong absorption lines with P Cygni profiles, to about featureless continua. The most obvious physical parameter on which a dependence is to be expected is the inclination angle of the system. If the line strengths of all observed systems are arranged according to this angle, and if we accept that one system — even if it is the brightest and best-known (SS Cyg) — does not follow the general trend, a very clear pattern emerges: low-inclination systems (where the disk is seen essentially pole-on) show strong absorptions of the UV resonance lines and strong P Cygni profiles, in particular in C IV 1550 Å; as the inclination increases, the absorption lines, as well as both the emission and the absorption components of the P Cygni lines, gradually disappear. In the case of dwarf novae, around an inclination of some 60° or 70° a basically featureless continuum is left; only at very high inclinations, when the disk is seen edge-on, very strong and broad emissions of the resonance lines appear. In the case of non-magnetic nova-like stars the pattern is essentially identical, the only difference being that the emission components already appear when the absorptions have not yet totally disappeared. The immediate interpretation of this behaviour is that, in addition to the optically thick disk at outburst maximum, a wind exists which is ejected, to first approximation, perpendicularly to the disk. In addition, there must be an extended stationary corona above the disk which is responsible for the strong emission lines seen at high inclinations; this corona must be considerably larger in nova-like stars than in dwarf novae.

2.3 The line radiation at minimum

If we look for a similar dependence on inclination of the line radiation of dwarf novae at minimum, the emerging pattern is that at low inclinations the UV resonance lines are seen as strong emissions; as the inclination increases, the line strength gradually decreases, leaving again an essentially featureless continuum at high inclinations. The obvious interpretation is that the quiescent disk is at least partially optically thick, in particular in the hot inner areas.

2.4 The continuum radiation at minimum

When the continuum radiation is considered as a function of the inclination angle, it becomes obvious that at low inclinations the continuum flux consists of two components, a blue one at short wavelengths and a red one at long wavelengths; the wavelength at which the major contribution changes from one to the other varies from object to object. Furthermore, if different spectra of the same system are inspected for this feature, it results that the turn-over point moves to longer wavelengths as the overall flux level decreases. From another investigation (la Dous in preparation) it results that the temperature of the white dwarf, and whether it is of type DA or DB, can be determined with rather high accuracy from the short-wavelength *IUE* spectrum alone. Thus, in the case of dwarf novae at minimum, supposing that the bluer one of the two continua is that of the white dwarf primary – an assumption which is supported by the presence of a strong Lyα absorption in several systems – the temperatures and types of these stars can be determined. If their contribution to the emitted flux is subtracted from the spectrum, what remains must be the spectrum of the accretion disk alone, which turns out to be that of an optically thick emitter. In all cases, the overal shape of the supposed disk spectrum is the same, except for the fact that the amount of the excess continuum flux decreases as the inclination angle increases, in agreement with the disk being basically a two-dimensional object which contributes progressively less to the overall flux as is seen more and more edge-on.

References

la Dous C., 1990, SSR 52, 203
la Dous C., 1991, A&A 252, 100
la Dous C., 1994, SSR 67, 1

Do We See Magnetic Effects in Dwarf Nova Outbursts?

N. Vogt[1,2], *E. Meyer-Hofmeister*[1], *F. Meyer*[1]

[1] Max-Planck-Institut für Astrophysik, 85740 Garching, Germany
[2] Universidad Catolica de Chile, Santiago, Chile

Many observations indicate that fast rotating late type stars show magnetic activity. We therefore argue that some of the secondary stars in cataclysmic binaries might also have magnetic fields. Such magnetic fields would reach over the accretion disk around the white dwarf primary. We investigate their effect on dwarf nova outbursts. The magnetic field lines will penetrate the disk and remove angular momentum. This shifts the accumulation of mass towards the inner disk, closer to the white dwarf, and therefore leads to a different outburst behaviour, which can be recognized in observed light curves of dwarf novae. If a magnetic field of the order 50 - 100 gauss is acting on the accretion disk, we expect narrower and more frequent outbursts as compared to the non-magnetic case. Outburst records of three dwarf novae above the period gap ($P > 3$h), whose long-term light curves are well covered, were used to search for traces of magnetic activity. All three cases display a pronounced bimodality in the distribution of their outburst width, i.e. either narrow or wide outbursts occur. We found evidence for a cyclic behaviour in one case, SS Aur: possible "magnetic episodes" repeating every 18 years reveal epochs with abnormally frequent narrow outbursts and nearly or totally missing wide ones. There are also indications for a similar behaviour of SS Cyg with a 7-years cycle, but with less pronounced periodicity. The third case, U Gem, does not show clear evidence of magnetic activity although we found some indications for a transitory oscillation of the width of wide outbursts after 1926, with a period of 13.6 years. The behaviour of SS Aur and SS Cyg resembles the theoretical predictions, there is, however, an important difference: Throughout the entire cycle, marked by the "magnetic episodes", neither wide nor narrow outbursts alter their mean light curves: magnetic fields seem to affect only the observed proportion in the frequency of both outburst types without altering the light curves of individual outbursts. Remarkably, most of the "anomalous outbursts" (which are characterized by an abnormally slow rising branch to an outburst) occur also at or near the "magnetic episodes".

The complete version of this paper has been submitted to Astron. & Astrophys. (E. Meyer-Hofmeister, N. Vogt and F. Meyer, "The influence of a magnetic field of the secondary star on dwarf nova outbursts", MPA Preprint no. 843, Dec. 1994).

Flares in Dwarf Novae

Frank M. Bateson

Astronomical Research Ltd., P.O. Box 3093, Greerton, Tauranga, New Zealand

Abstract: This report provides a summary of flares observed visually in the dwarf novae V442 Cen, OY Car, TU Men and Z Cha. The flares precede an outburst by an average of around ten days.

1 Introduction

Members of the Variable Star Section, Royal Astronomical Society of New Zealand, have closely monitored most southern dwarf novae over the past five decades. The many hundreds of thousands of visual estimates include a number of definite observations of short lived flares which in the majority of cases preceded an outburst by a few days.

Bateson (1991a) suggested that these flares should be termed precursors to outbursts. He discussed in detail problems inherent in visual estimates and showed that he only accepted as flares those estimates that conformed to high standards of reliability.

Observations of V442 Cen, OY Car and TU Men were the first dwarf novae chosen to search the records for flares because these stars had minima well below the thresholds of the instruments used. A summary of the results are given below.

2 Results

<u>V442 Cen:</u> Type SS Cyg. Mean cycle 27.13 days (Bateson et al. 1991). Narrow and wide maxima differ in duration by little in mean brightness. Rise to outburst is steep and rapid. Minima are fainter than 16.0 mag.

A total of 19 flares were listed ranging in magnitude from 12.7 to 14.3, all fainter than the mean outburst magnitude of 12.0. These flares on average preceded an outburst by 8.8 days (range 4–14 days).

<u>OY Car:</u> SU UMa type. Superoutbursts have mean maximum magnitude of 11.35; mean width 11.5 days. Corresponding values for normal maxima are 12.38 and 1.8 days. Mean super cycle 328 days but this changes abruptly and varies within wide limits. Normal maxima follow superoutbursts at intervals of about 171 days (Bateson and Dodson 1983). Minimum magnitude is 17.5.

Bateson (1991a) listed 11 possible flares of which two may have been normal outbursts. The magnitudes of the remainder ranged from 12.3 to 14.3. These apparent flares preceded outbursts by intervals between 16 and 174 days excluding two long intervals probably caused by an outburst being unobserved. One flare preceded an outburst by 45 days and was then recorded at magnitude 12.3 fading to fainter than 13.5 within 5 minutes. This was probably an eclipse.

TU Men: SU UMa type. The mean cycle length, maximum magnitude and outburst duration have all undergone abrupt and dramatic changes (Bateson 1991b). These followed two prolonged superoutbursts with durations of 20 and 24.6 days and maximum magnitudes of 11.8 and 11.6.

Seven possible flares were listed with maxima 13.2 to 14.2. These occurred from 20 to 222 days before an outburst.

Z Cha: SU UMa type. Superoutbursts have a mean cycle of 217.6 days; mean maximum magnitude 12.1; mean duration 9.8 days. Corresponding values for normal outbursts are 51.3 days, 12.6 mag and 0.92 days (Bateson 1991c). Minimum magnitude is 17.2.

Bateson (1991d) listed 26 possible flares of which two may have been of poorly observed normal maxima.

The magnitudes of the remainder ranged from 12.3 to 14.0. The intervals which these flares preceded an outburst were short, ranging from 2 to 73 days with one exception of 117 days.

3 Conclusions

There appears to be reliable evidence that flares occur in dwarf novae. The majority precede an outburst by an average of around 10 days.

References

Bateson F.M., 1991a, Publ. Var. Star Section, R. astr. Soc. N.Z. No. 16, 75
Bateson F.M., 1991b, Publ. Var. Star Section, R. astr. Soc. N.Z. No. 16, 40
Bateson F.M., 1991c, Publ. Var. Star Section, R. astr. Soc. N.Z. No. 17, 11
Bateson F.M., 1991d, Publ. Var. Star Section, R. astr. Soc. N.Z. No. 17, 74
Bateson F.M., Dodson, A.W., 1983, Publ. Var. Star Section, R. astr. Soc. N.Z. No. 11, 1
Bateson F.M., McIntosh, R., Brunt, D. 1991, Publ. Var. Star Section, R. astr. Soc. N.Z. No. 16, 4

Flickering in Cataclysmic Variables: Inventory and Perspectives

Albert Bruch

Astronomisches Institut, Wilhelm–Klemm–Str. 10, 48149 Münster, F.R.G.

Abstract: Stochastic variations on short time scales called flickering are a characteristic feature in the light curves of CVs. Their amplitudes indicate that their origin must be related to one of the principle light sources of these systems. But the physical mechanism behind the flickering is unknown. New attempts to quantify the properties of flickering are presented, and a consistent scenario for its origin is outlined.

1 Phenomenology of flickering

Cataclysmic variables (CVs) are interactive binary stars where a white dwarf primary receives matter via an accretion disk from a late type secondary star. In most systems the optical and ultraviolet light is dominated by the accretion disk. But a considerable fraction of the total radiation is emitted by a unsteady light source which so far has not been identified with certainty and which is responsible for the phenomenon of flickering in CVs.

Any light curve of a CV in any photometric state (except for classical and recurrent novae in outburst) with a time resolution of the order of seconds reveals more or less strong irregular variations, appearing as a continuous series of flares with a duration of seconds to minutes, depending on their amplitudes. These variations are called flickering. As an example, Fig. 1 shows an unfiltered light curve of the novalike variable TT Ari.

While the total amplitude of the flickering is approximately constant for a given system in a given state, it can differ strongly for different CVs. In general, most novae and the members of the UX UMa subtype of novalike variables reveal modest amplitudes of the order of $0\overset{m}{.}1$ (in white light), while the VY Scl stars, i.e. those novalike systems which occasionally drop into a low state, appear to have systematically a more violent flickering of several tenths of a magnitude. The flickering in outbursting dwarf novae resembles that of UX UMa stars, while in quiescence they exhibit a large total range of amplitudes. In some cases it is so violent, that flares of more than a magnitude peak brightness rise and decay within just a few minutes.

Although being typical for CVs, flickering is not unique to cataclysmic variables. It is also observed in some symbiotic stars where a large part of the

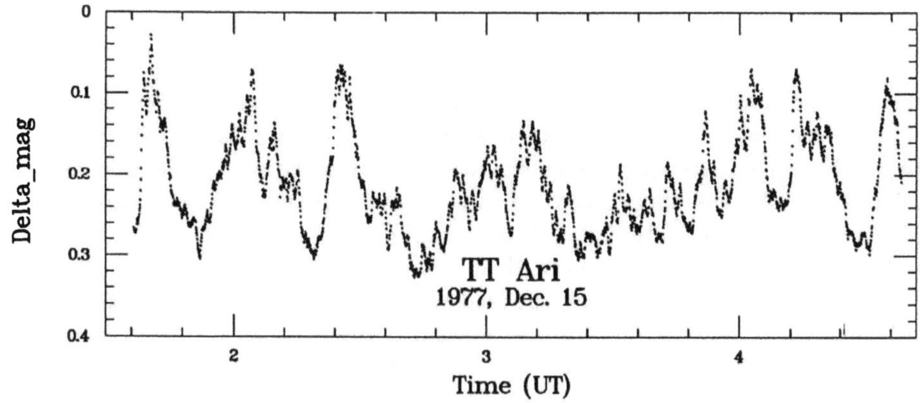

Fig. 1. Unfiltered differential light curve of the novalike variable TT Ari on 1977, Dec. 15.

luminosity is suspected to be powered by accretion. Other candidates showing this phenomenon are low-mass X-ray binaries where, however, with the exception of the case of Sco X-1 it is by far not as well documented as in CVs. Light curves of T Tau stars with accretion disks also show irregular variations (Gahm 1995) which resemble very much the flickering in CVs and can be regarded as its counterpart. It appears thus that the accretion process is ultimately responsible for flickering.

The often considerable amplitudes of the flickering indicate immediately that a significant fraction of the total visual light of a CV – sometimes half of it or even more – is emitted by the flickering light source. I therefore want to make a point saying that *we cannot claim to understand the physics of a CV unless we understand what flickering is.* In spite of this obvious fact, and although the phenomenon is known for more that 40 years, there are scarcely any systematic investigations of the flickering. Most of the corresponding statements in the literature (with very few exceptions) are restricted to a mere qualitative description or at most to the derivation of some rather superficial statistical parameters.

The first encompassing attempt to quantify flickering, to derive general properties and to find constraints for a physical model were performed by Bruch (1992). It was somewhat hampered by the fact that the observational material available to him was not really sufficient neither in quantity nor in quality. In order to improve the situation, no *new* observations are required, but one can draw upon the numerous light curves of CVs observed in the past which were never investigated with respect to the properties of the flickering. With the kind support of colleagues all over the world I could collect a data base of about 1400 light curves suitable for flickering studies, representing about 140 days and nights of continuous observing time. This material is currently being investigated, and the first results are just beginning to emerge.

2 The quantification of flickering

One can choose different approaches to quantify flickering. On the one hand a flickering light curve can be regarded as an ensemble of data, of which one can obtain global characteristics. This approach involves the study of power spectra and the like. On the other hand we can try to identify individual flares in the light curves, perform statistics with them and deduce properties of the flickering in this way. Both approaches have their benefits.

The most serious difficulties opposed to a systematic study of flickering are problems which affect many physical measurements, but which are particularly annoying in the present connection:

- *Noise:* Just as noise, flickering is a stochastic process. Moreover, its amplitude is the smaller, the shorter the regarded time scales are. Therefore, flickering below a certain limit in time scale and amplitude level – which depends on the noise level in the data – is practically indistinguishable from noise. At the same time, just because noise and flickering cannot be distinguished, this limit is not easily quantifiable. This means also that flickering in different light curves with different noise levels is difficult to compare.
- *Time resolution:* Flickering occurs on time scales up to about 20^{m} – 30^{m}. A lower limit is unknown. The high frequency part of the flickering can only be studied if the time resolution of the measurements is adequately high. A sufficient time resolution is particularly important if individual flares are to be recognized. Again, it is obviously difficult to compare the flickering in light curves observed with different time resolutions.

In the general case the problems due to noise and time resolution get mixed, which certainly does not facilitate the quantitative analysis!

The problem of noise can in certain cases be solved by a Fourier analysis of the light curves. As an example, Fig. 2 shows a power spectrum of a light curve of KR Aur on a double logarithmic scale. It is typical for power spectra of flickering light curves. There is a linear drop from low to high frequencies, which reflects that larger events in the flickering occur on longer time scales. At a certain frequency, however, the power becomes constant. This is not due to flickering having no power at high frequencies but to white noise which prevails over flickering here. To estimate the noise level, one can compare this power spectrum to that of a simulated constant signal, subjected to Poisson noise. If the count rate of the constant signal is chosen such that the high frequency parts of both power spectra coincide, the corresponding Poisson noise is equal the noise in the real data.

But the power spectra of flickering light curves serve also other purposes. The slope of the decreasing part contains information about the flickering. Similarly, a wavelet analysis can tell us something about the properties of the flickering. Corresponding investigations are currently being carried out. It is, however, too early to report about results.

In another approach flickering can statistically be regarded as a Markov process. The individual data points in a light curve are transferred into symbolic

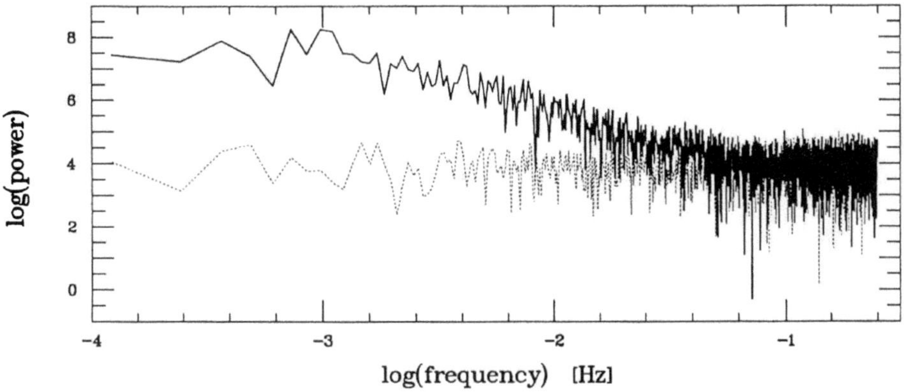

Fig. 2. Power spectrum of the light curve of KR Aur of 1977, Nov. 14 (solid line) and of an artificial constant signal, subjected to Poisson noise, chosen such that its high frequency part coincides with that of the real light curve (broken line).

observations which are then elements of a Markov chain. At any point in the light curve the system is considered to be in one of several states. Going to the next data point there is a certain (unknown) probability to remain in the same state or to make a transition to another state. This probability depends only on the state the system is currently in. In the first step, we investigate the simplest case of three states: (1) getting brighter, (2) remaining constant (within the noise limits), or (3) getting fainter.

Using the observations it is then possible to calculate a statistical model for the process driving flickering. Comparing such models for different light curves, one can define a measure for the distance of the models, i.e. for the similarity of the light curves. Light curves of similar statistical properties have a small, such with strongly differing characteristics a large distance. While we do not yet understand well how to interpret these differences physically, for the time being we regard them as an empirical measure for the similarity of different light curves.

Before statistics with individual flares in a light curve can be performed, it is necessary to devise a method to identify flares in the presence of noise and in view of the available time resolution. There is no unique way to do this, and therefore any method will be arbitrary to a certain degree. But this can be suffered as long as it is secured that this arbitrariness introduces only a minimum of *systematic* errors into light curves of different qualities. A prescription for flare identification has been given by Bruch (1992). To avoid problems with noise, it purposely disregards flares below a pre-defined amplitude limit.

The goal of all efforts to quantify flickering must be to gain a physical understanding of the underlying mechanism (or mechanism<u>s</u>?). A way to learn something about it is to look for correlations among the flickering properties and other system parameters. A most interesting result of this kind is a rela-

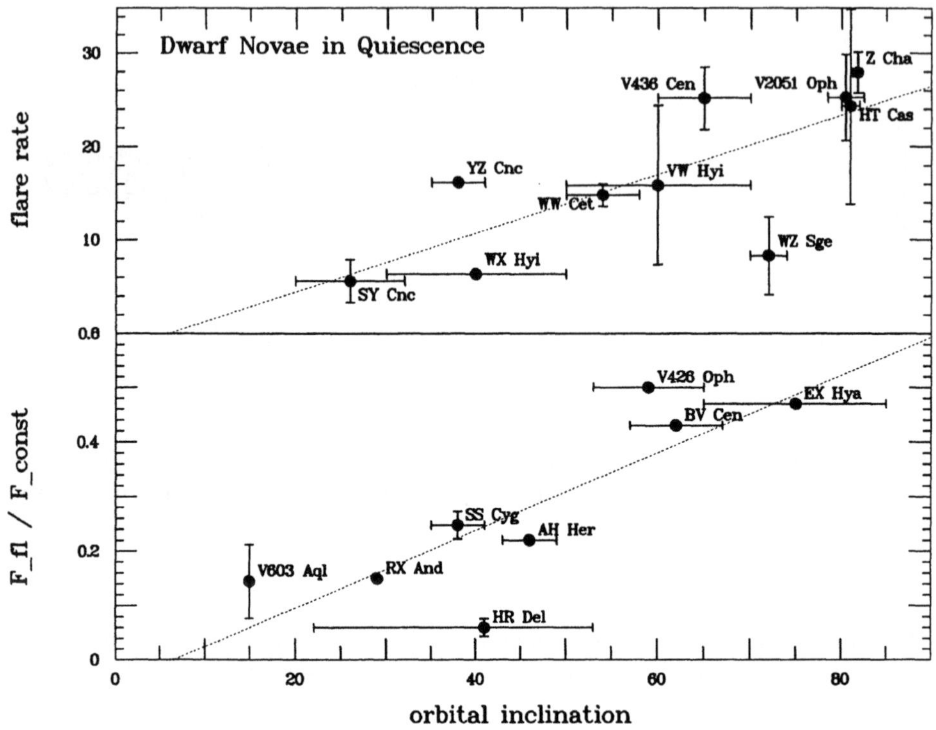

Fig. 3. The flare rate in quiescent dwarf novae (upper frame) and the ratio between the optical luminosity of the flickering light source and the constant part of the accretion disk (lower frame), both as a function of the orbital inclination.

tion between the flare rate – that is the number of individual flares per time unit – and the orbital inclination in quiescent dwarf novae (see Fig. 3; upper frame). The flare rate (measured in light curves given in magnitudes as opposed to fluxes) is large if we see the system at a high inclination. This supports a result of Bruch (1992) who found that the ratio between the optical luminosity of the flickering light source and the constant part of the accretion disk also increases with the inclination (Fig. 3; lower frame). These findings contain some important clues to the origin of flickering (see Sect. 3).

Another interesting point is the wavelength dependence of the flickering. It is normally described to be blue, but this statement is rarely quantified. Measuring the amplitudes of flares in different bands of a well-calibrated photometric system, it is possible to determine the broad band spectrum of the flickering light source. In Fig. 4 the spectrum derived from various light curves of VY Scl observed in the Walraven system is shown. The results for other CVs look similar. The solid lines indicate spectra of simple steady state accretion disk models, using the black body approximation and various mass transfer rates. All spectra are normalized to the B-band. The flickering spectrum rises steeper to the

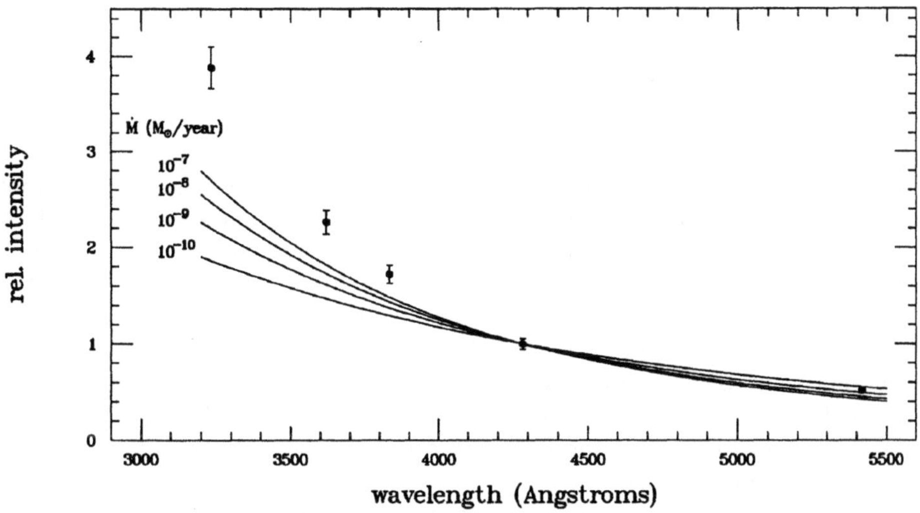

Fig. 4. Broad band spectrum of the flickering light source in the novalike variable VY Scl (dots) compared to steady state accretion disk spectra, assuming the black body approximation and different mass transfer rates (solid lines). All spectra are normalized to the mean wavelength of the B-band of the Walraven system.

UV than the simple disk model. This is also true for black bodies or power law spectra, regardless of the assumed temperatures or spectral indices, respectively. The flickering spectra suggest that the relative contribution of the corresponding light source to the total radiation is even stronger in the ultraviolet range than in the optical, enhancing the importance of the flickering light source for the total emitted radiation of a CV. UV photometry with a suitable time resolution is only available for the system AE Aqr (Eracleous et al. 1994). There, the amplitude of the flickering still increases a bit from the near UV to the middle UV, but not dramatically so. But then, AE Aqr is in many respects a very unusual system, and the behaviour of its flickering may not be representative for all CVs.

3 The site of the flickering light source

For some time, flickering has been explained as due to an unsteady transfer of matter from the secondary star, causing a variable luminosity of the hot spot. But there are many reasons why this cannot be the correct mechanism, at least not the principal one. Bruch (1992) has shown that the kinetic energy of the impact between the transferred matter and the accretion disk is not sufficient to explain the high fraction of the luminosity of the flickering light source on the total luminosity of the CVs. Moreover, the temperature of the hot spot is measured to be of the order of 10 000 ... 15 000 K, less than the temperature of the inner accretion disk which dominates the blue light of a CV. It is then hard to see how the hot spot as the flickering light source could be even bluer than the steady accretion disk.

A direct indication that the flickering arises in other parts of the system than the hot spot is seen in the light curves of some eclipsing systems which permit to distinguish the ingress and the egress of the white dwarf and the hot spot, respectively. At phases when the hot spot is eclipsed by the secondary, but the white dwarf and the inner accretion disk are not, flickering is clearly seen. In other systems, the strength of the flickering, measured as the standard deviation from the mean curve, is strongly reduced at eclipse centre, indicating that the light source of the flickering is centred on the white dwarf but has a smaller extension than the accretion disk (Horne & Stiening, 1985). Recently, Welsh & Wood (1995) have performed eclipse mapping of the flickering light source in HT Cas and found that the flickering occurs in the inner accretion disk and peaks on the white dwarf.

The ultimate energy source for the flickering must be potential energy of the gravitational field of the white dwarf. In order to release the gravitational energy, matter in the accretion disk has to sink down which has to occur unsteadily in order to explain flickering.

Of course, most gravitational energy is available close to the white dwarf. Rough energy budget calculations performed by Bruch (1992) indicate that relatively large turbulent eddies in the inner accretion disk may have just enough energy to explain the luminosity of the flickering in the optical range. Also, the time scales on which the flickering flares occur are compatible with dynamical time scales in such turbulences. However, if a correction is made for the energy which is likely emitted in the UV range, it becomes difficult to explain flickering by turbulences in the disk alone. There are no problems with the energetics, however, if an unsteady and turbulent dumping of matter onto the surface of the white dwarf is assumed. Such a scenario appears to be the only one which is not at odds with the observed properties of the flickering, although a definite physical mechanism of how this unsteady accretion occurs is not yet known. It may well be possible that the interaction of the plasma in the inner accretion disk with a magnetic field of the white dwarf (which is not necessarily strong enough to cause directly observable effects as in polars or intermediate polars) is involved.

Obviously, the structure of the boundary layer between the (Keplerian) accretion disk and the white dwarf plays an important role in this connection. The boundary layer belongs to the least understood parts of CVs. There are some theoretical calculations concerning its structure. But they all make simplifying assumptions or neglect potentially important physical effects such as the influence of a magnetic field of the white dwarf. There are good arguments for a boundary layer not being restricted to a narrow equatorial belt, but engulfing the white dwarf to a considerable extent or even entirely (Livio & Truran 1990). Unsteady accretion out of such a boundary layer could explain why the relative contribution of the flickering light source and also the flare rate increase with increasing orbital inclination: While at high inclinations the disk is seen foreshortened and thus less luminous than at low inclinations, this is not the case for the flickering light source. Thus, a larger number of flares will have an amplitude (on the magnitude scale) above a given level, and the ratio between flickering and constant light increases.

If the boundary layer would in fact be very thin, it should have a very high temperature and emit mostly in X-rays. Observations indicate that both, the temperature and the X-ray emission are significantly lower than predicted (for references, see Bruch & Duschl 1993). Duschl & Tscharnuter (1991) therefore assumed a radially extended boundary layer and investigated the consequences of this scenario. Persuing this idea further, Bruch & Duschl (1993) calculated the luminosity ratio between the Keplerian part of the accretion disk and the boundary layer as a function of the boundary layer radius. Assuming the flickering to be in fact due to a turbulent accretion of matter onto the white dwarf and thus defining a lower limit for the boundary layer luminosity, it is then possible to estimate a lower limit for the radius of the extended boundary layer by comparing the flickering luminosity with that of the quiet disk which is interpreted as the Keplerian part. In this way the radial extent of the boundary layer was found to be at least of the order of 2 or 3 white dwarf radii or even a bit more. This result compares favourably with the radius of the flickering light source determined by eclipse tomography in HT Cas (Welsh & Wood 1995).

We still cannot claim to have gained a satisfactory understanding of flickering in cataclysmic variables. But doubtlessly some progress in deriving a *systematic* picture of the empirical properties has been achieved in recent times, and certainly we will still learn quite a lot in this respect. A scenario for the flickering is beginning to emerge. Hopefully, the quantification of the observed characteristics will finally lead to a consistent physical model.

Acknowledgements: It is a pleasure to express my thanks to my collaborators S. Beckemper, T. Fritz, J. Kinscher, G. Lichtenberg, M. Niehues, C. Nohlen, M. Sand and T. Schimpke who study the flickering together with me, and to the numerous colleagues worldwide who put their light curves of cataclysmic variables at my disposal.

References

Bruch A., 1992, A&A 266,237
Bruch A., Duschl W.J., 1993, A&A 275, 219
Duschl W.J., Tscharnuter W.M., 1991, A&A 241, 153
Eracleous M., Horne K., Robinson E.L., Zhang E.-H., Marsh T.R., Wood J.H., 1994, ApJ 433, 313
Gahm G., 1995, this volume p. 203
Horne K., Stiening R.F., 1985, MNRAS 216, 933
Livio M., Truran J.W., 1990, Comm. Astrophys. 14, 221
Welsh W.F., Wood J.H., 1995, this volume p. 300

Resolved Shot Noise in the AM Her System V834 Cen

Stefan Larsson, Bengt Larsson

Stockholm Observatory, S-13336 Saltsjöbaden, Sweden

1 Introduction

Cataclysmic variables of the AM Her type are close binary systems with accretion onto a highly magnetic (a few times 10^7 G) white dwarf. In these systems the magnetic field is strong enough to prevent the formation of an accretion disc. The accretion flow is instead funneled directly onto the polar caps where the plasma is shocked and radiates intense optical cyclotron radiation as well as bremsstrahlung at X-ray energies. The cyclotron radiation component, which dominates in the optical, is highly polarized (5–35%). The cyclotron emission exhibits intensity variations on a large range of time scales. This include quasi-periodic oscillations, flickering and long term changes between brightness states. The observed source intensity is also modulated by the white dwarf rotation which is synchronized with the orbital period (101 min for V834 Cen). The flickering is believed to be due to inhomogeneities in the accretion flow. These could either be density variations in a continous flow or it could be in the form of independent blobs accreting over the polar cap region. As blobs hit the white dwarf surface and radiate their accretion energy they should be seen as shot noise. For a theoretical discussion of the fate of such blobs see Frank et al. (1988). A modeling of the optical variability in AM Herculis, in terms of a shot noise process was made already by Panek (1980). It was found that a consistent model could be constructed by randomly occurring $70^s - 90^s$ rectangular shots, with on the average a few overlapping shots. The power spectrum had a ν^{-2} shape above 0.02 Hz as expected, although it is not clear whether a break in the slope of the power spectrum was actually seen around 0.01 Hz. The shot overlap meant that individual shots could not be resolved directly in the light curve. This limits the precision and uniqueness by which shot parameters can be determined. However, in our observational data of AM Her objects we have found occasions when the character of the flickering has changed to become much more "flarelike", suggesting that individual shots might be seen in these cases. This change is also associated with a change in the shape of the power spectrum, in particular a higher power level and a steepening at the high frequency end. Most importantly, there is a distinct break in the power spectra around 0.03 Hz, indicative of a characteristic time scale in the flickering noise. In addition to a power spectrum analysis we have also made a direct modeling of individual flares in the light curve. The distribution and correlations of the fitted pulse parameters were studied and compared with simulated data. The primary aim was to investigate whether the flares are really individual shots or produced by a random superposition of more frequent smaller amplitude shots.

2 Observations and analysis

The observations of V834 Cen were made with the 3.6 m and the Danish 1.5 m telescopes at ESO at four different occasions from 1984 to 1987. The initial aim of the observations was to study the ~ 2 second quasi-periodic oscillations in this object and those results have already been published by Larsson (1992). During 5 nights of observation in January 1987, however, the object also exhibited the distinct flickering which appears to be resolved into individual pulse-shots. One orbital period was observed each night, with a time resolution of 40 ms. An example of one of these light curves can be found in Larsson (1990).

The following basic analysis was made on the time series data for V834 Cen:

- All peaks in the light curves with some height above the noise level were identified and analysed by directly fitting a few different pulse shapes. The distribution of pulse parameters, such as shape, amplitude and width were then analysed and compared to the corresponding results for simulated data.
- Power spectra were calculated from 1500 sec data segments. These were then averaged for all the five nights to produce an overall mean power spectrum.

3 Results

3.1 Fitting individual pulses

The fitting of individual pulses gave the following primary results:

- The distribution of pulse parameters indicates the presence of a distinct flickering component consisting of high amplitude narrow pulses.
- The distinct shots are best fitted by peaked pulses, such as triangular or $\exp\{-|(t-t_0)/\tau|\}$. The asymmetry of the individual pulses were calculated in terms of the rise and decay timescales as $(\tau_+ - \tau_-)/(\tau_+ + \tau_-)$. The mean asymmetry was close to zero but had a distribution with an rms deviation of ~ 0.3.
- The Full Width Half Maximum (FWHM) distribution for the distinct shots is strongly peaked around 6 – 7 sec. The distribution is shown in Fig. 1 together with an empirical distribution function fitted to the data.
- The amount of overlap for the distinct shots is small.

The result in the last point is suggested both by the observed separation in the light curve and more quantitatively by parameter distribution and correlations (e.g. between amplitude and width). Such correlations were compared with those from simulations with different amounts of overlap.

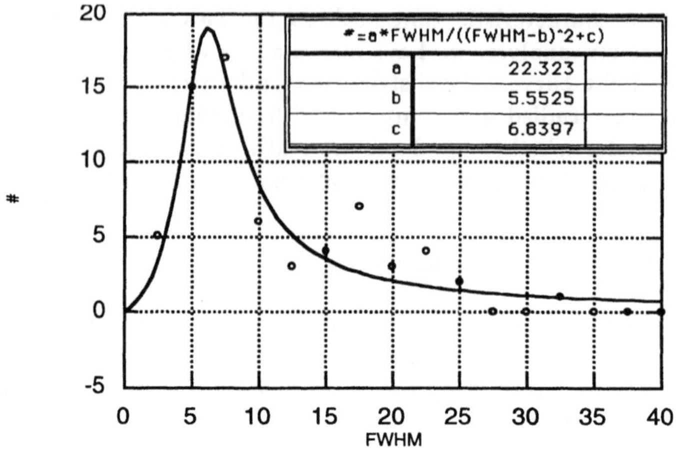

Fig. 1. The Full Width Half Maximum (FWHM) distribution for the distinct shots as determined from the fitting of pulse profiles directly to the time series data. The full drawn curve is a simple parametrization of the distribution which is used to construct a shot noise model for the overall power spectrum.

3.2 The overall power spectrum

The total average power spectrum for the 5 nights that exhibited the anomalous flickering is shown in Fig. 2. The most important feature is the break at about 0.03 Hz, which provides strong support for a shot noise model. Above this frequency the spectrum steepens to a power law of slope −2.5. This is significantly steeper than the "canonical" value of −2 that is produced by e.g. sharp edged shots, such as one-sided exponentials. For rectangular pulses the power spectrum has a sinc^2 form and therefore an envelope of ν^{-2}.

4 Discussion and model

After the extraction of pulse parameters for what appears to be a distinct shot noise component we now proceed by investigating whether the parameters are consistent with the observed time averaged power spectrum. First however, we want to point out that the pulse profiles implied by the fits in Sect. 3.1 (e.g. double sided exponentials) correspond to very steep power spectra, ν^{-4}! We have confirmed this by calculating power spectra for the individual observed 'flares'. These were then averaged in three groups with regard to pulse width. Each of these groups had power law slopes close to −4 as expected. We then calculated the power spectrum produced by double–sided exponentials with a distribution in FWHM taken from the fit in Fig. 1. The distribution in pulse widths flattens the power spectrum slope at high frequencies from −4 to −2.4 which is very close to that of the average spectrum as can be seen in Fig. 2. The turn over around $\log(\nu) = -1.6$ is also reproduced by this model. The excess

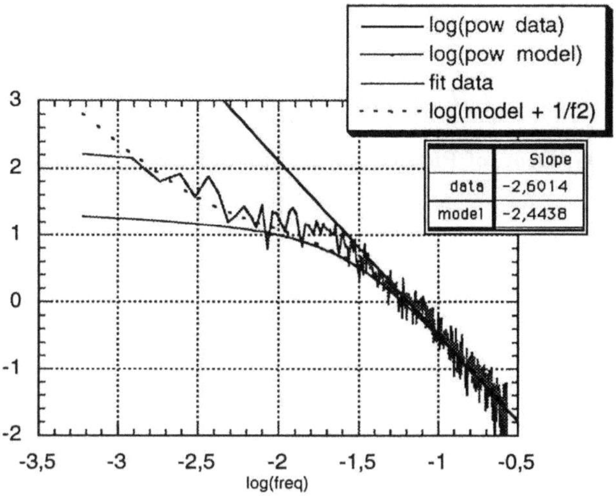

Fig. 2. The average power spectrum for the observations of V834 Cen in January 1987. The straight line is a power law fit to the steep high frequency part of the spectrum. The second solid curve, which is almost flat below $\log(\nu) = -2$, is a model consisting of double-sided exponential shots with a distribution in FWHM taken from the fit in Fig. 1. The dotted line is the same model but after adding a ν^{-2} component.

over the model at frequencies below $\log(\nu) = -2$ can be explained by adding a ν^{-2} component with the **same strength** as in our previous observations of V834 Cen in 1984 and 1985. The only parameter that was adjusted to obtain the fit in Fig. 2 was the strength of the shot noise component.

References

Frank J, King A.R., Lasota J.-P., 1988, A&A 193, 133
Larsson S., 1990, in 'Accretion-Powered Compact Binaries', C. Mauche (ed.), Cambridge University Press, Cambridge, p. 279
Larsson S., 1992, A&A 265, 133
Panek R., 1980, ApJ 241, 1077

A. Bruch: What is the reason for the different time scales of the flickering variations in nights when individual events are not resolved compared to the time scale of the resolved shots?
S. Larsson: This does probably represent a change in the amout, or scale of fragmentation in the accretion flow.

Eclipse Mapping the Flickering in HT Cas

William F. Welsh, J.H. Wood

Keele University, Dept. of Physics, Keele, Staffordshire ST5 5BG, UK

1 Introduction

Flickering is a well-known, though poorly understood, signature of accretion in cataclysmic variables. HT Cas is an excellent system in which to study flickering because its well determined characteristics allow a calibration of the experimental system (see Horne et al. 1991).

Patterson (1981) argues that the flickering in HT Cas originates from regions very near the white dwarf. In U Gem, the flickering appears to be associated with the bright spot (Warner & Nather 1971). In RW Tri, the flickering occurs throughout the disk but is centrally concentrated (Horne & Stiening 1985); the same holds true for V2051 Oph (Warner & Cropper 1983).

HT Cas is a $16^{\mathrm{m}}_{\cdot}4$ eclipsing dwarf nova with a binary period of 106 min and an inclination of $\sim 81°$. The entire eclipse last about 15 min with the white dwarf totally eclipsed for ~ 5 min. The mass ratio is 0.15 with the white dwarf having mass $\sim 0.6 M_{\odot}$ (Horne et al. 1991).

We obtained 20 eclipses of HT Cas between 1982 and 1991 in U, B and V at Mt. Palomar and also McDonald Observatory with $0.5-2$ sec integrations. These data were folded on the ephemeris given by Horne et al. (1991) and corrected for the 30 sec drift in the phasing as noted by Wood et al. (1995).

2 The flickering light curve

The mid-eclipse flux in each of the 20 eclipses is not all the same. This is most likely due to long-term secular trends. To remove this contamination we applied a small additive constant to each light curve so that the mean flux at mid-eclipse remained constant.

The twenty eclipse light curves were combined to form a mean light curve. Deviations from the mean light curve were then recorded and then the rms scatter computed at each orbital phase. Note that the rms has been debiased to remove the expected contribution from photon noise.

The rms at each orbital phase gives us an "rms light curve", which is essentially a measure of the amplitude of the flickering versus time. This rms or flickering light curve was then flattened by dividing by a spline so that the mean flux remained constant outside of eclipse.

3 Eclipse mapping

The eclipse mapping technique developed by Horne (1985) has been applied to the flickering light curves. Using the maximum entropy method, the eclipse mapping technique produces an image of the accretion disk that is an optimum solution to two conflicting constraints: (1) the predicted light curve must fit the data (χ^2 constraint), (2) the image must contain as little structure as possible (the maximum entropy constraint). In addition, the maps are further constrained to be as azimuthally smooth as possible.

4 Conclusions

For HT Cas, we have the following preliminary results:

- The color of the flickering is blue.
- The eclipse of the flickering region is total in the U band, and probably also in the B and V bands.
- The bright spot at the edge of the disk produces negligible flickering.
- Though concentrated towards the center, the flickering is not confined to the innermost region of the disk.
- The flickering brightness temperature vs. radius is more steep than the mean disk (see Wood et al. 1992), i.e., the relative amount of flickering increases toward the center of the disk.
- The brightness temperature vs. radius of the flickering is too flat for an optically thick accretion disk model.

References

Horne K., 1985, MNRAS 213, 129
Horne K., Stiening R., 1985, MNRAS 216, 933.
Horne K., Wood J.H., Stiening R., 1991, ApJ 378, 271.
Patterson J., 1981, ApJS 45, 517.
Warner B., Nather R., 1971, MNRAS 152, 219.
Warner B., Cropper M. 1983, MNRAS 203, 909.
Wood J.H., Horne K., Vennes S., 1992, ApJ 385, 294.
Wood J.H., Naylor T., Hassall B.J.M., Ramseyer T.F., 1995, MNRAS (in press)

Multiple Time Scales in Cataclysmic Variables: The Examples

Ivan L. Andronov

Department of Astronomy, Odessa State University, T.G. Shevchenko Park, 270014 Odessa, Ukraine

Abstract: Decades of patrol observations of cataclysmic variables (CVs) show variations of the accretion rate also at a time scale of a few years which may be explained by a solar-type activity of the secondary and/or by orientation changes of the magnetic axis of the white dwarf in polars. Secondary photometric periods seen in TT Ari and possibly some other CVs need theoretical explanation. Irregular variability of some objects is characterized by power-law power spectra with a power index depending on the luminosity state of the system.

1 Introduction

Periodic variations in CVs are usually attributed to: the orbital motion of the binary (Krzeminski 1965), rotation of the magnetic white dwarf (Patterson 1979), to the irradiation variations of the secondary periodically changing with a beat period between the rotation of the white dwarf and its orbital motion (Patterson & Price 1981) or to the precession of the eccentric accretion disk (Whitehurst & King 1991). Cycle-to-cycle changes like the extreme ones observed by Andronov et al. (1980, 1992b) in AM Her and MR Ser, respectively, are usually attributed to a physical variability of the accretion structure.

Here we briefly overview the results on unusual irregular and quasi-periodic variations at other time scales.

2 Year-scale variations

Transitions from "high" to "low" luminosity states in polars (Hudec & Meinunger 1976) and nova-like variables (e.g. Andronov et al. 1988) are due to accretion rate variations. They may be caused by a solar-type activity of the secondary (cf. Wenzel & Fuhrmann 1983), irradiation (King and Lasota 1984) and/or changing orientation of the magnetic pole of the white dwarf in polars (Andronov 1987, 1993). Another appearance of the magnetic activity is an UV Cet-type flare of the secondary detected by Shakhovskoy et al. (1993a).

Year and decade-scale variations seen in MV Lyr (Andronov et al. 1988) and V795 Her (Wenzel et al. 1988) may argue for more complicated character of the activity possibly corresponding to a "century"-scale solar variation. Seasonal values of the outburst cycle length in dwarf novae may vary smoothly (Bianchini 1988) or abruptly (Andronov & Shakun 1990), possibly indicating the existence of 2 sub-classes (Chinarova & Andronov 1994). Individual cycle lengths show a bimodal distribution (Szkody & Mattei 1984). Year-scale waves are also seen in low-mass X-ray binaries (Priedhorsky & Holt 1987).

An "intermediate" luminosity state may also be distinguished in some stars. MV Lyr showed "excursions" to it ($\approx 15^{\mathrm{m}}$) both from "high" ($\approx 13^{\mathrm{m}}$) and "low" ($\approx 17^{\mathrm{m}} - 18^{\mathrm{m}}$) states (Andronov et al. 1988). "Outbursts" occurring every ≈ 400 days were different from those observed in dwarf novae, and once a sequence of them with a few-day cycle and an exponentially decreasing height was observed.

An opposite behaviour shows IW And: periods of nearly constant brightness are rarely interrupted by periods of activity, when the brightness varies above and below the usual value of a day time-scale (Meinunger & Andronov 1987).

3 Week-scale beat periods

They are observed in the systems with a photometric period close to an orbital one. BY Cam is a magnetic system with a white dwarf synchronizing with the orbital motion, and the current value of the "beat period" is ≈ 15 days (Mason et al. 1994). Another example is TT Ari, where a $3\overset{\mathrm{d}}{.}76$ period was suggested (Semeniuk et al. 1987). However, in this system the photometric period is slightly shorter than the orbital one, and may be attributed neither to super-humps nor to the rotation of the white dwarf (Tremko et al. 1995), and its nature remains unclear.

4 5-7 hour waves in TT Ari

A secondary hour-scale wave with $P = 4^{\mathrm{h}}41^{\mathrm{m}}$ was suggested by Wenzel et al. (1986) based on the observations obtained during a 1986 campaign. However, the data obtained in 5 observatories in 1988 showed 4 other candidate periods, from which one "true" value is difficult to choose because the period is longer than an usual observational run (Tremko et al. 1992). To study properties of this wave, "super-long" runs starting in Japan, continued in Turkmenia and Russia, then in Europe and America are needed. Such a campaign was organized in 1994 and is planned to be continued in October during the coming years by using photometric and spectral ground-based and space observations. Particularly, in 1994, 96 hours of homogeneous observations were obtained by N.I. Dorokhov at the Dushak station of the Odessa observatory. Similar secondary waves seem to be observed in other systems (cf. Voykhanskaya 1989), and their theoretical and observational study is very important.

5 Sub-hour QPOs

Recent reviews on rapid variability are presented by Bruch (1992, 1995). Thus we briefly point out the most puzzling phenomena. Semeniuk et al. (1987) reported on the secular decrease of the cycle length in TT Ari from 27 min in 1961 to 17 min in 1985. Hollander and van Paradijs (1992) argued for a continuation of this trend in 1988 and interpreted these QPO's by interaction of the magnetosphere with an accretion disk with slowly changing parameters. Usually the cycle length of such variations undergoes cycle-to-cycle changes. Thus a best fit value significantly depends on the data set used. Tremko et al. (1995) argued for few preferred time scales, at which the peaks at the periodogram occur several (but not all) times.

6 Power-law power spectra

Aperiodic processes may be characterized by wide power spectra (cf. Box & Jenkins 1970). If QPOs are seen as wide peaks superimposed on a "featureless continuum" (cf. Larsson 1985), some types of irregular variability may be characterized by a power-law shape $S(f) \propto f^{-\gamma}$. Terebizh (1992) reviewed cases $\gamma = 0$ ("white noise"), 1 ("flicker noise") and 2 ("random walks"). Autoregressive processes studied on finite runs correspond to "noninteger" values of γ, as well as observations of the real stars AM Her, BZ Cam (Shakhovskoy et al. 1992, 1993b), HQ And (Andronov et al. 1992a), TT Ari (Tremko et al. 1995). Work on other stars is in progress showing the significance of the parameter γ for description of the fast variability.

7 "Shot-noise" models

Inhomogeneities of the accretion stream cause rapid variations in CVs. The autocorrelation function (ACF) for "shot noise" is an exponent, which may be biased due to finite length of the run and subtraction of a sample mean (Sutherland et al. 1978). Detailed analysis of the influence of the trend removal in general form onto the ACF is presented by Andronov (1994).

Middleditch (1982) and Larsson (1985) reported on second-scale QPOs in accretion columns in polars, which Langer et al. (1982) interpreted as cyclic changes of the structure . Andronov (1987) proposed a model of "oscillating spaghetti". At the column base a "boiling-type" instability occurs. Such "bombardment" solutions lead to better explanation of the "soft X-ray excess" (Thompson et al. 1986).

References

Andronov I.L., 1987, Astron. Nachr. 308, 229
Andronov I.L., 1993, Odessa Astron. Publ. 6, 21
Andronov I.L., 1994, Astron. Nachr. 315, 353
Andronov I.L., Borodina I.G., Kolesnikov S.V., Pavlenko E.P., Shakhovskoy N.M., 1992a, Commun. Spec. Astron. Obs. 69, 112
Andronov, I.L., Fuhrmann, B., Wenzel, W., 1988, Astron. Nachr. 309, 39

Andronov I.L., Pavlenko E.P., Seregina T.M., Shugarov, S.Yu., Shvechkova, N.A., 1992b, Proc. Stellar Magnetism, St.Petersburg, 160
Andronov I.L., Shakun L.I., 1990, ApSS 169, 237
Andronov I.L., Vasilieva S.V., Tsessevich V.P., 1980, Astron. Tsirk. 1142, 5
Bianchini A., 1988, IBVS 3136
Box G.E.P., Jenkins G.M., 1970, Time series Analysis. Forecasting and control, Holden Day, San Francisco
Bruch A., 1992, A&A 266, 237
Bruch A., 1995, this volume, p. 288
Chinarova L.L., Andronov I.L., 1994, Padova-Abano Conf. on Cataclysmic Variables (in press)
Hollander A., van Paradijs J., 1992, A&A 265, 77
Hudec R., Meinunger L., 1976, IBVS 1184
King A.R., Lasota J.P., 1984, A&A 140, L16
Krzeminski W., 1965, ApJ 142, 1051
Langer S.H., Chanmugam G., Shaviv G., 1982, ApJ 252, 289
Larsson S., 1985, A&A 145, L1
Mason P., Andronov I.L., Chanmugam G., Kolesnikov S. V., Pavlenko P., Shakhovskoy N.M., 1994, Padova-Abano Conf. on Cataclysmic Variables (in press)
Meinunger L., Andronov I.L., 1987, IBVS 3081
Midleditch J., 1982, ApJ 257, L71
Patterson J., 1979, ApJ 234, 978
Patterson J., Price C.M., 1981, ApJ 243, L83
Priedhorsky W., Holt S.S., 1987, SSR 45, 291
Shakhovskoy N.M., Alexeev I.Yu., Andronov I.L., Kolesnikov S.V.: 1993a, Proc. Cataclysmic Variables and Related Physics, Annals Israel Phys. Soc. 10, 237
Shakhovskoy N.M., Efimov Yu.S., Andronov I.L., Kolesnikov S.V., 1993b, IAU Symp. 155, 407
Shakhovskoy N.M., Kolesnikov S.V., Andronov I.L., 1992, Proc. Stellar Magnetism, St. Petersburg, 148
Semeniuk I., Schwarzenberg-Czerny A., Duerbeck H., Hoffman M., Smak J., Stępień K., Tremko J., 1987, ApSS 130, 167
Sutherland P.G., Weiss M.C., Kahn S.M., 1978, ApJ 219, 1029
Szkody P., Mattei J.A., 1984, PASP 96, 998
Terebizh, V.Yu., 1992, Time Series Analysis in Astrophysics (in Russ.), Moscow
Thompson A.M., Brown J.C., Kuijpers J., 1986, A&A 159, 202
Tremko J., Andronov I.L., Chinarova L.L., Kumsiashvili, M.I., Luthardt, R., Pajdosz G., Patkos L., Rößiger S., Zoła S., 1995, A&A (subm.)
Tremko J., Andronov I.L., Luthardt R., Pajdosz G., Patkos L., Roessiger S., Zola S., 1992, IBVS 3763
Voykhanskaya, N.F., 1989, Preprint Spec. Ap. Obs.43
Wenzel W., Banny M.I., Andronov I.L., 1988, Mitt. Veränderl. Sterne 11, 141
Wenzel W., Bojack W., Critescu C., Dumitrescu A., Fuhrmann B., Götz W., Grelczyk H., Hacke G, Hudec R., Huth H., Kozhevnikov V.P., Kumsiashvili M.I., Mrkos A., Oláh K., Oprescu G., Patkós L., Peřestý R., Pfau W., Reimann H.-G., Richter G., Rößiger S., Shpychka I.V., Shult R., Stecklum B., Tóth I., Tremko J., Valníček B., Verdenet M., 1986, Prepr. Astron. Inst. Czechoslovak Acad Sci. 38
Wenzel W., Fuhrmann B., 1983, Mitt. Veränderl. Sterne, 9, 175
Whitehurst R., King A., 1991, MNRAS 249, 25

HIGH-ENERGY TRANSIENTS

X-ray Bursts

Jan van Paradijs[1,2], *Walter H.G. Lewin*[3]

[1] Astronomical Institute "Anton Pannekoek", University of Amsterdam, and Center for High Energy Astrophysics, Kruislaan 403, 1098SJ Amsterdam, The Netherlands
[2] Physics Department, University of Alabama in Huntsville, Huntsville AL 35899, U.S.A.
[3] Center for Space Research and Department of Physics, Massachusetts Institute of Technology, 37-627, Cambridge MA 02139, U.S.A.

1 Introduction

An X-ray burst is a sudden increase (rise time of order seconds) of the X-ray brightness of an X-ray source, which after reaching its peak decays, generally within a minute. The sky distribution of X-ray burst sources indicates that they are galactic objects (see Fig. 1); their concentration to the direction of the galactic center tells us that they lie at typical distances of $\sim$ 8 kpc, with corresponding peak luminosities of order 10^{38} erg s^{-1}. The X-ray and optical properties of the persistent emission of X-ray burst sources show that they are low-mass X-ray binaries, in which mass is transferred from a rather normal low-mass ($< 1\ M_{\odot}$) star to a neutron star. The persistent emission is caused by the conversion of kinetic energy of the transferred matter into heat, at a rate of $\sim GM/R$ ($\sim 0.1c^2$) per gram of accreted matter. The bursts are caused by unstable thermonuclear burning of material that has accumulated on the neutron star ('thermonuclear flash').

The global properties of X-ray bursts, in particular their dependence on the mass accretion rate, are fairly well understood. Different from the case of γ-ray bursts (see the contributions by Fishman, Hartmann and Kouveliotou to this Colloquium) the relevant question about X-ray bursts is not 'What are they?', but rather 'What use are they?'. As we will argue here, X-ray bursts may provide us information on the mass and radius of a neutron star. This usefulness of X-ray bursts derives from the fact that the burst emission originates from the *surface* of the neutron star, unlike the persistent emission caused by mass accretion, of which we only know that it comes from the neutron star's near vicinity.

In the confines of this contribution we cannot discuss the properties of X-ray bursts and their dependence on important parameters, such as mass accretion rate. For an extensive discussion of many of the issues involved we refer to the recent book 'X-ray Binaries' (Lewin, Van Paradijs & Van den Heuvel 1995), in particular to the chapter on X-ray bursts. A detailed discussion of the possibilities to derive the mass and radius of a neutron star from X-ray burst observations has been given by Lewin, Van Paradijs & Taam (1993).

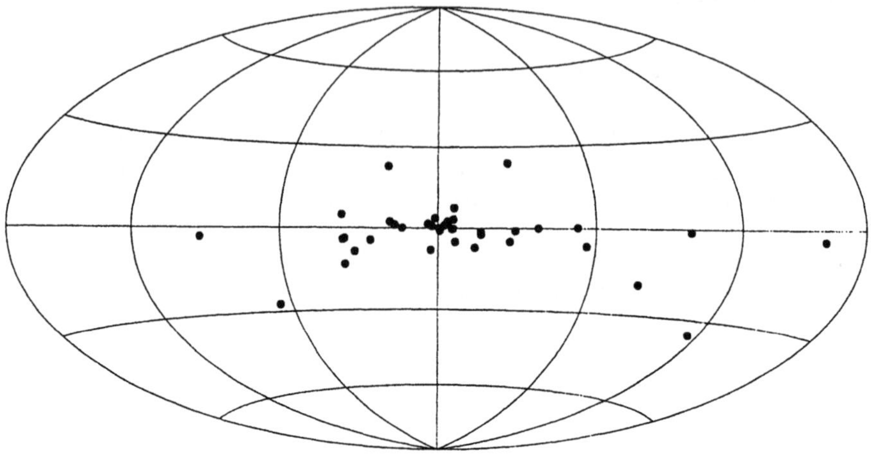

Fig. 1. Sky distribution of X-ray burst sources. The map is in galactic coordinates, with the equator representing the plane of the Milky Way; the origin of the map coincides with the direction to the galactic center.

2 Neutron star mass and radius from X-ray burst observations

A comparison of burst profiles (for the same event) as observed in different photon energy bands shows that the highter the photon energy is, the shorter is the decay part ('tail') of an X-ray burst: during the decay the X-ray burst spectrum becomes 'softer'. It was first shown by Swank et al. (1977) that burst spectra are well described by Planck functions with a temperature (kT) varying between ~ 1 and ~ 2.5 keV. The blackbody interpretation was supported by the fact that during the decay the observed burst flux F_∞ (above the persistent emission) varied approximately as the fourth power of the blackbody temperature T_∞ (the subscript ∞ indicates that the quantity is measured by a distant observer). This implies that during a burst one observes an approximately constant burst emitting area. For an assumed homogeneous spherical blackbody emitter of radius R_∞ at a distance d, we have

$$L_\infty = 4\pi R_\infty^2 \sigma T_\infty^4 = 4\pi d^2 F_\infty \,, \tag{1}$$

where L stands for luminosity. If one knew the source distance d, R_∞ would follow from observed values of F_∞ and the associated blackbody temperatures. For a distance of 10 kpc, Swank et al. (1977) and others found radii R_∞ that are roughly compatible with those expected for neutron stars. This result forms the basis for various ways to obtain information about the mass-radius relation of neutron stars.

2.1 The cooling part (decay) of X-ray bursts

Because of gravitational redshift the energy of each photon emitted from the surface of a neutron star, and also the rate at which the photons arive at a distant observer, are lower than the corresponding quantities measured at the neutron star surface, by a factor

$$1 + z_* = (1 - 2GM/R_*c^2)^{-1/2} . \tag{2}$$

Here M is the gravitational mass of the neutron star, and R_* the radius of the neutron star as observed by a local observer on its surface. As a consequence, the luminosity and blackbody temperature, as measured locally on the surface of the neutron star, are related to those measured by a distant observer as follows:

$$L_\infty = L_*(1 + z_*)^{-2} \tag{3}$$

$$T_\infty = T_*(1 + z_*)^{-1} . \tag{4}$$

Using these relations we find

$$R_\infty = R_*(1 + z_*) = R_*(1 - 2GM/R_*c^2)^{-1/2} . \tag{5}$$

These expressions are strictly valid only for non-rotating neutron stars; however, they are good approximations for neutron star spin periods larger than a few ms. The last equation shows that a measurement of R_∞ gives a relation between the mass M and radius R_* of a neutron star. It follows that R_∞ has a minimum value when $R_* = 1.5R_g = 3GM/c^2$; then

$$R_{\mathrm{min},\infty} = 1.5\sqrt{3}R_g \sim 7.7(M/M_\odot)\ \mathrm{km} . \tag{6}$$

For isotropic emission (such as blackbody emission) Eq. (5) does not hold when $R_* < 1.5\ R_g$: then a fraction of the emitted photons fall back to the neutron star surface, and the number of observed photons is reduced. When this is the case, Eq. (4) still holds, but Eq. (3) does not. The net result is that then

$$R_\infty = 1.5\sqrt{3}R_g \sim 7.7(M/M_\odot)\ \mathrm{km}\ (R_g < R_* < 1.5R_g) . \tag{7}$$

Since the radii R_* are likely larger than 1.5 R_g we do not have to be too concerned about this possibility.

In Fig. 2 we show the (M, R_*) relations for several values of R_∞. For a given value of M, R_∞ has a minimum; conversely, for a given observed value of R_∞, M cannot exceed $(R_\infty/7.7\ \mathrm{km})\ M_\odot$.

In the above it has been assumed that the source distance is known, and that the radiation is emitted isotropically. Only for burst sources in some globular clusters are the distances reasonably well known. Even if the burst emission were initially isotropic it is possible that at large distances it is not, e.g., due to the presence of the inflowing matter around the neutron star. Unfortunately, our knowledge on anisotropy of X-ray emission in X-ray binaries is very poor.

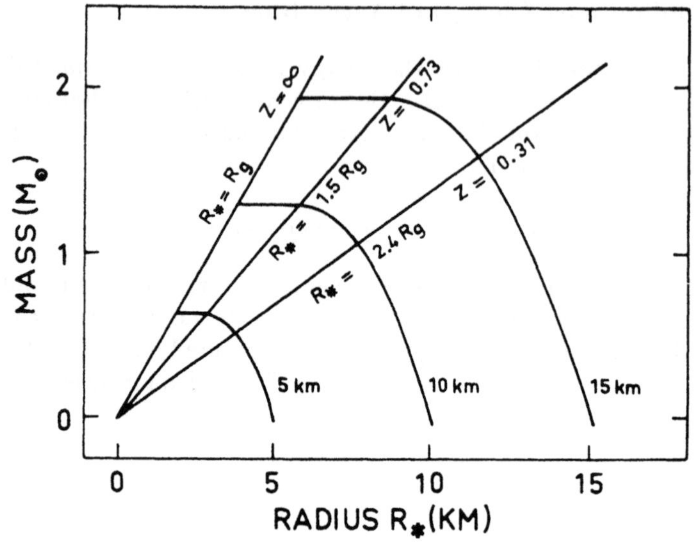

Fig. 2. Mass-radius relations for three hypothetical values of the blackbody radius R_∞ (5, 10, and 15 km). For clarity, we have not indicated error regions resulting from the uncertainties in the measurements. The straight lines indicate radii R_*, equal to the Schwarzschild radius R_g, $1.5R_g$, and $2.4R_g$, respectively. The latter could, for example, be the result of a burst with radius expansion (see text), or of the determination of the gravitational redshift of an observed spectral feature. For a given mass, the observed blackbody radius, R_∞, has a minimum value $1.5\sqrt{3}$ R_g; conversely, for a given blackbody radius R_∞ the mass cannot be larger than $R_\infty/(7.7 \text{ km})\ M_\odot$.

To account for possible anisotropy one can introduce an 'anisotropy factor' ξ according to

$$L_\infty = 4\pi R_\infty^2 \sigma T_\infty^4 = 4\pi d^2 \xi F_\infty \,. \tag{8}$$

To find R_∞ we need to know $d^2\xi$. As we discuss below, this quantity can be eliminated for bursts which show photospheric radius expansion.

2.2 Photospheric radius expansion

During very strong X-ray bursts the neutron star photosphere expands as a result of strong radiation pressure. Note that, different from the case of thermonuclear flashes on the surface of a white dwarf (which cause classical novae), the gravitational binding energy per gram of material is much larger than the amount of energy liberated by nuclear fusion; as a consequence only a small fraction of the accreted matter can be blown off the surface during an X-ray burst. Model calculations tell us that during the expansion, and subsequent contraction, of the photosphere the luminosity remains to within a fraction of a percent of the Eddington luminosity, at which the gravity force (directed inward) and the radiation force (directed outward) are in balance. As a result, during the expansion the temperature of the photosphere decreases, and the spectral energy distribution shifts toward lower photon energies. This shows up as a temporary decrease of the X-ray intensity near the peak of the burst, which is particularly prominent at the higher photon energies. When the radius expansion is very large the X-ray signal may temporarily disappear altogether (the emission is then in the EUV). During the expansion and contraction phase, at photospheric radius R, the Eddington luminosity, as observed by a distant observer is

$$L_{\mathrm{Edd},\infty} = (4\pi cGM/\kappa)(1 - 2GM/Rc^2)^{1/2} = 4\pi d^2 \xi F_{\mathrm{Edd},\infty}\,. \tag{9}$$

Here κ is the electron scattering opacity during the expansion phase (it is 0.34 $\mathrm{cm^2 g^{-1}}$ for matter with cosmic abundances, and 0.2 $\mathrm{cm^2 g^{-1}}$ for hydrogen-poor matter). Note that R is not R_*, but rather the radius of the photosphere which, at maximum expansion, can be many times the stellar radius. If the flux is measured during the part of the expansion when $R \gg R_*$, the gravitational redshift factor is unity to good approximation, and we find

$$L_{\mathrm{Edd},\infty} = (4\pi cGM/\kappa) = 4\pi d^2 \xi F_{\mathrm{Edd},\infty}\,.(R \gg R_*)\,. \tag{10}$$

Just at the end of the contraction phase, when the radius of the photosphere is again R_*, the luminosity is still Eddington limited, and we have

$$L_{\mathrm{Edd},\infty} = (4\pi cGM/\kappa)(1 - 2GM/R_* c^2)^{1/2} = 4\pi d^2 \xi F_{\mathrm{Edd},\infty}(R = R_*)\,. \tag{11}$$

Note that $F_{\mathrm{Edd},\infty}$ in Eq. (10) is the observed Eddington flux when the photospheric radius is very large, whereas $F_{\mathrm{Edd},\infty}$ in Eq. (11) is that at 'touchdown'. A measurement of these two values of $F_{\mathrm{Edd},\infty}$ leads immediately to a value for M/R_* (by dividing the two equations one eliminates $d^2\xi$; we assume here that the anisotropy remains constant throughout the burst). The basis of this method is shown in Fig. 3, which shows schematically the temperature versus flux diagram of a burst with radius expansion. A measurement of M/R_* limits the allowed values in the (M, R_*) diagram to a straight line through the origin.

2.3 Combining information from the cooling track and the radius expansion track

We can get different information about the mass and radius by combining the observations during the expansion/contraction phase (Eq. 10) with those during the cooling phase (Eq. 8); in this way we also eliminate $d^2\xi$. Again, this elimination is possible only if the anisotropy does not change during the burst. We then find with Eq. (5):

$$F_\infty / F_{\mathrm{Edd},\infty} = R_*^2 (1 - 2GM/R_* c^2)^{-1} \sigma T_\infty^4 (\kappa/cGM)\,. \tag{12}$$

This equation is a mass-radius relation for the neutron star, independent of the source distance and the anisotropy of the burst emission. A measurement of the

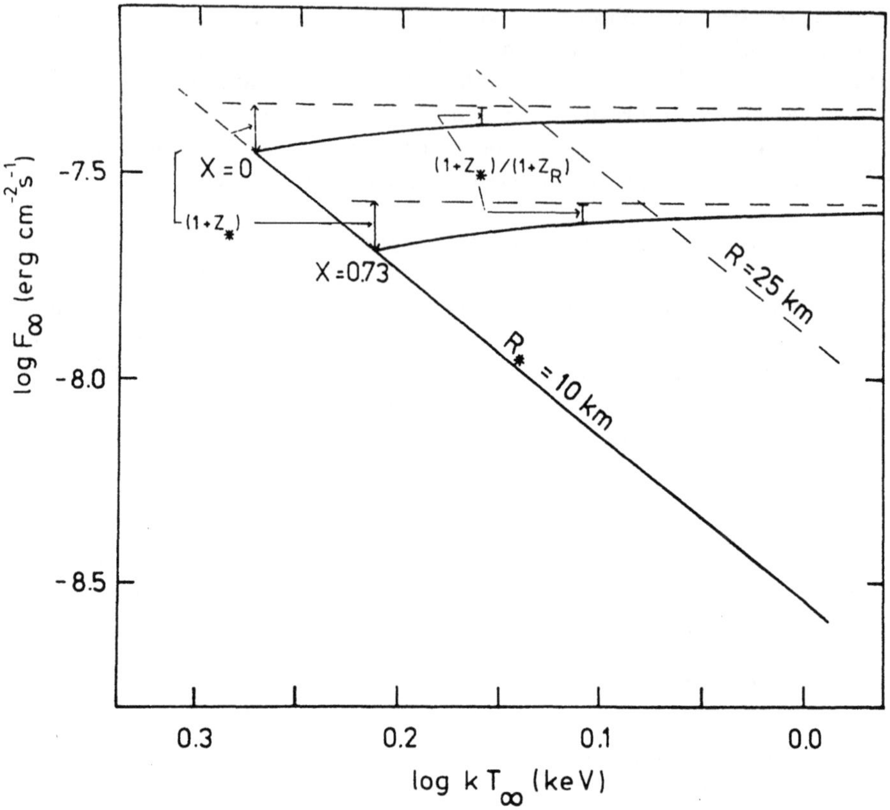

Fig. 3. Schematic diagram of the flux F_∞ versus the blackbody temperature kT_∞, showing how the gravitational redshift can be determined from the variation of the Eddington luminosity during photospheric radius expansion. The straight line (cooling track) is for a sphere with constant radius; the slope of this line is 4. The slanted dashed line holds for a sphere with a 2.5 times larger radius. Two points are indicated at which the luminosity equals the Eddington limit, for cosmic (hydrogen content by number $X = 0.73$) and for hydrogen-poor ($X = 0$) compositions; the expansion/contraction tracks are the two solid curves. The redshift factors, $1 + z_*$, are indicated (assuming no change in the composition and the anisotropy).

three observables (F_∞, $F_{\mathrm{Edd},\infty}$, and T_∞) then leads to a measurement of the quantity $A = cGF_\infty/(F_{\mathrm{Edd},\infty}\sigma T_\infty^4)$:

$$A = R_*^2(1 - 2GM/c^2R_*)^{-1}(\kappa/M)\,. \tag{13}$$

For $R < 1.5R_g$ (see Eq. 7) the quantity A becomes:

$$A = (7.7M/M_\odot \mathrm{km})^2(\kappa/M)\,. \tag{14}$$

If $R < 1.5R_g$, a value of A yields a value for M which is independent of R_*. Eqs (13, 14) (in which the distance is absent) is essentially an expression for the unchanging angular size of the burst source during the cooling phase (i.e., R_∞/d). Thus, as one moves along one of the curves in Fig. 4, the angular size of the burst source remains constant, but R_∞ and d change; this is not the case for the curves in Fig. 2.

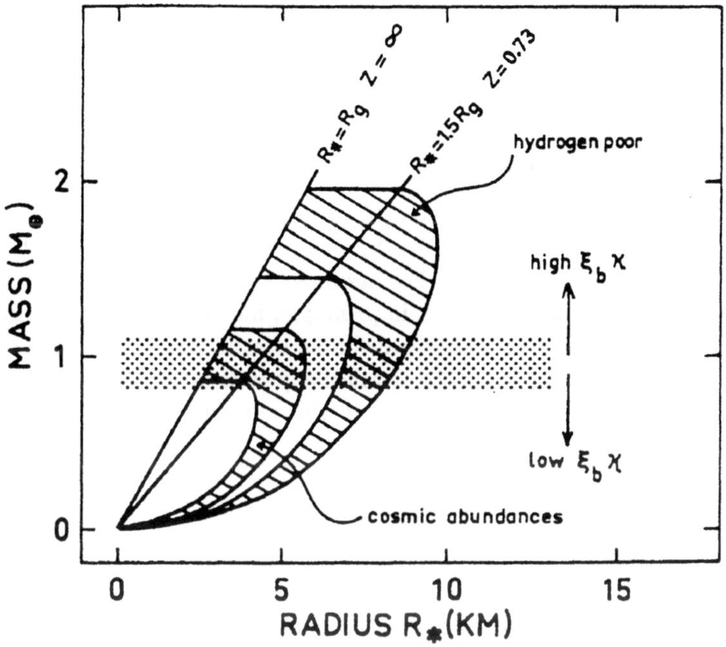

Fig. 4. Mass-radius diagram resulting from the observation of a burst with radius expansion. We have assumed, for the purpose of illustration only, that the quantity A in Eq. (13) is in the range (0.85-1.15)10^{-22} cm^4g^{-2} (see text). The allowed regions in this diagram for assumed hydrogen-poor ($\kappa = 0.2$ cm^2g^{-1}) and cosmic ($\kappa = 0.34$ cm^2g^{-1}) compositions are indicated by the hatched areas. The horizontal (shaded) band indicates schematically the constraint on the mass obtained from the measured Eddington flux ($R \gg R_*$) for a source with known distance. The width of the band reflects uncertainties in the measurements. The position of the band depends on both κ and on the anisotropy ξ, as schematically indicated.

During the decay phase of a burst with radius expansion (only one such burst per source is required), one can make several measurements of both the flux and the associated temperature. For blackbody radiation from a spherical object with a constant radius F_∞/T_∞^4 remains constant. The average value for several measurements during burst decay can then be used to construct allowed regions in the (M, R_*) diagram. In Fig. 4 we show in the (M, R_*) diagram two curves for two values of κ (0.2 and 0.34 cm^2g^{-1}), for a typical value of $A = 10^{-22}(\pm 15\%)$

2.4 Gravitational redshift from discrete spectral features

When a discrete spectral feature is present in a spectrum during times that the radiation comes from a neutron star surface, and if this feature can be identified, one has a direct measurement of the gravitational redshift at the neutron star surface, and thus of the ratio M/R_* (see Eq. 2). Lines at 4.1 ± 0.1 keV have been reported in bursts from several sources. The interpretation of these absorption features is a matter of debate. A possible origin for this line is the Lyman α transition in hydrogenic or helium-like iron, which would lead to $1 + z_* \sim 1.6$; this would seem high, but perhaps not impossible.

2.5 Combining all available information

One can combine the above methods. If, in addition, the distance to the source is known [thus, the value of $M/(\kappa\xi)$ is known (see Eq. 10)] one finds 3 equations with 4 unknowns: M, R_*, κ, and ξ. For assumed values of κ (e.g., hydrogen poor, or cosmic abundances) one can then, in principle, find values of M, R_* and ξ. This analysis has been applied to the X-ray burst source 2127+119 in the globular cluster M15, for which the distance is relatively well established on the basis of optical observations of the normal stars in this cluster (see Figs. 5 and 6).

3 Complications

Several factors complicate the above spectral analyses of X-ray bursts in terms of neutron star mass-radius relations. For instance, the opacity κ is temperature dependent; it varies by about 5% over the temperature range relevant to X-ray bursts. However, this temperature dependence can be taken into account.

A much more serious effect is that the spectra emitted by hot neutron star atmospheres are not exactly Planckian. Therefore, the observed temperature that describes the shape of the spectrum ('colour temperature' T_c) is not equal to the 'effective' temperature (T_{eff}) used in Stefan's law (Eq. 1). In principle, modelling of the radiation transfer through the neutron star atmosphere can provide a relation between T_c and T_{eff}. Available models indicate that this is a major effect, with calculated ratios of colour temperature to effective temperature around 1.5. Moreover, observational evidence indicates that the non-Planckian character of the emitted spectrum cannot be described by just a 'hardening factor' T_c/T_{eff}. For the time being the largest uncertainty in the above (M, R_*) determinations is probably due to the uncertainty in this temperature conversion.

4 Conclusions

In the absence of any knowledge of the distance of a burst source and of the anisotropy of the burst emission, a range of allowed masses and radii of the neutron star in a given burst source can be obtained from the X-ray data of a single burst which causes radius expansion of the neutron star photosphere. If,

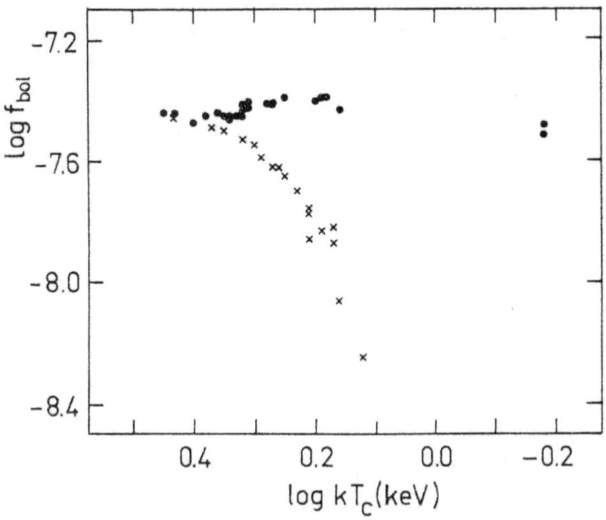

Fig. 5. Flux-temperature diagram for a burst with radius expansion observed from the source 2127+119 in the globular cluster M15. The dots indicate the expansion/contraction track, the crosses the cooling track. The two dots on the far right (very large radius expansion) are too low in flux; this is likely caused by the fact that only a minute fraction of the flux is sampled in the X-ray band, with a concomitant large uncertainty in the flux measurement. Notice that the cooling track is not a straight line with slope 4; this reflects deviations of the burst spectrum from a Planckian curve (Van Paradijs et al. 1990).

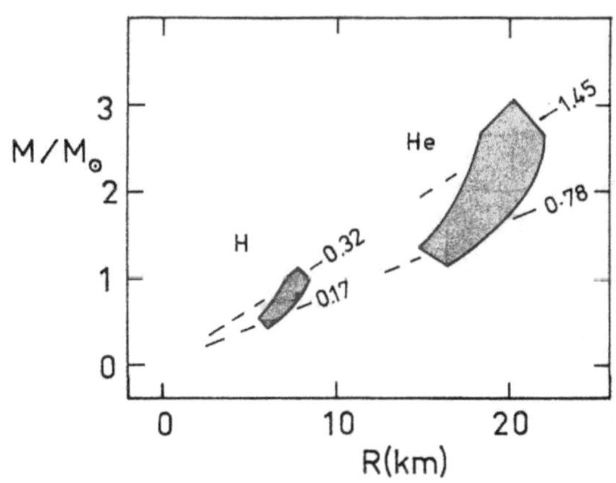

Fig. 6. Mass-radius diagram for the neutron star in 2127+119. The radius R in the figure is R_* as used in the text. The enclosed areas are allowed regions for assumed cosmic and hydrogen-poor compositions of the neutron star atmosphere, as indicated with 'He' and 'H', respectively. The numbers in the figure indicate the values derived for the anisotropy factors ξ (see text). (adapted from Van Paradijs et al. 1990).

in addition, the source distance and the composition of the neutron star photosphere are known, the mass and radius of the neutron star, and the anisotropy of the burst emission, can be determined separately. The largest uncertainty at present results from our incomplete knowledge of the conversion from the observed colour temperature to the effective temperature. Ongoing analysis of X-ray burst observations made with ASCA (Tanaka, private communication) may help improve our understanding of the spectra of hot neutron stars.

References

Lewin W.H.G., Van Paradijs J., Taam, R.E., 1993, Space Sci. Rev. 63, 223
Lewin W.H.G., Van Paradijs J., Van den Heuvel E.P.J. (eds.), 1995, X-ray Binaries, Cambridge University Press, Cambridge
Swank J.H. et al., 1977, ApJ 212, L73
Van Paradijs J., Dotani T., Tanaka Y., Tsuru T., 1990, PASJ 42, 633

D.H. Hartmann: The $M(R)$ relationship you mentioned assumes no rotation. Do the observations support this assumption?

J. van Paradijs: Yes, the theoretical $M(R)$ relation are for non-rotating neutron stars. It is generally believed that neutron stars in low-mass X-ray binaries rotate, with spin periods in the range of several milliseconds to several tens of milliseconds, due to accretion torques. However, unambiguous evidence for rotation of the low-magnetic-field neutron stars is so far lacking. The (M, R) relation would be significanty affected for spin periods below about 3 ms.

D.H. Hartmann: During atmosphere expansion one apparently has $L_{\rm Edd}(x, M) =$ const. This implies that magnetic fields in the expanding photosphere must be small. So, what do X-ray bursts tell us about magnetic field decay in neutron stars?

J. van Paradijs: According to current ideas the thermonuclear flash that gives rise to a X-ray burst, covers the whole neutron star surface when the mass accretion rate and the magnetic field strength of the neutron star are both relatively low. The limit on the magnetic field strength is not very strong (less equal $10^{10} \ldots 10^{11}$ G), but seems consistent with the mutual exclusion of X-ray bursts and X-ray pulsation. Somewhat in contrast: It is currently believed that the magnetic fields of neutron stars do not decay spontaneously, but that this decay is related to the accretion process – there is, however, no certainty about the details of the relation.

Strong X-ray Bursts with Expansion: Oscillations due to Sound Waves in Atmospheres

Iosif Lapidus[1*], *Luciano Nobili*[2], *Roberto Turolla*[2]

[1] Institute of Astronomy, University of Cambridge, Madingley Road, Cambridge CB3 0HA, UK
[2] Department of Physics, University of Padova, Via Marzolo 8, 35131 Padova, Italy

1 Oscillations in X-ray bursts with expansion

Photospheric expansion has been observed during several powerful type I X-ray bursts, and it is associated with a supersonic outflow driven by the super-Eddington energy release in the thermonuclear He–burning at the base of the envelope (Lewin et al. 1993). Such expanded atmospheres represent perfect background for the propagation of sound, and we believe there is a number of arguments in favor of the reality of these phenomena.

We have suggested an acoustic origin for the second range photospheric oscillations detected in the X-ray bursters 1608–522 and 2127+119 (Lapidus et al. 1994). On the basis of our models of radiation driven winds from neutron stars during strongest bursts (Nobili et al. 1994), we have shown that purely radial standing sound waves in expanded spherical envelopes can have periods very close to the observed ones. Both the photosphere and the sonic surface may act as a reflecting boundary; this is the reason of quite different periods in the two sources (0.65 s and $\sim$ 7 s, respectively).

Although the physics of type II bursts from the Rapid Burster is somewhat different, some of them, too, show evidence of photospheric expansion. The QPO discovered in type II bursts, are quite dissimilar to those observed in many bright LMXBs (Lewin et al. 1993). The centroid frequencies range from $\sim$ 2 to $\sim$ 7 Hz, being strongly anticorrelated with the peak flux in the burst.

The QPO in type II bursts may also have an acoustic origin. The structure of the shell is thought to be similar to that of our models of the final part of the expansion/contraction phase. The frequency of the principal mode of the "photospheric" standing wave is $\approx$ 8 Hz for the model with the lowest possible $M_{\rm env}$, while a frequency $\approx$ 2 Hz corresponds to the eigenfrequency in an atmosphere with $\dot{M} \sim 15 - 20\dot{M}_E$. The sound wave frequency is higher for weaker bursts, and this is indeed consistent with the observed anticorrelation of the QPO frequency with the average burst peak flux.

---

* The Royal Astronomical Society Sir Norman Lockyer Fellow

2 Oscillations in the Aql X–1 burst

Coherent 7.6 Hz oscillations were once detected during a type I burst of Aql X-1 (Schoelkopf & Kelley 1991). They were detected only in the 80 s portion of the light curve around the peak of the burst, and they were interpreted as reflecting the proper rotation of the underlying neutron star. In our opinion this explanation is highly implausible. The reason is that after first few seconds of the burst the burning must be completely uniform. The time periods between successive bursts on Aql X-1 are very high, so as the burst starts at some location on the star, the burning front propagates with the convective speed $\gtrsim 10^6$ cm/s (Bildsten 1995). In $\lesssim 10$ s the whole NS surface will be burning simultaneously. The asymmetry of the burst may not be due to nonuniform distribution of the nuclear fuel either. The magnetic field cannot be high enough ($\gtrsim 10^{10}$ G) to sustain such asymmetries. Should it happen, such oscillations would be observed not only in the burst but in the persistent flux as well, which is not the case. In the absence of external forces the only reason for asymmetries may be random inhomogeneities. At temperatures $\sim (1-10) \times 10^8$ K and proton thermal velocities $v_p \sim (1-3) \times 10^8$ cm/s the latter ones will be disappearing on the hydrodynamical time scale $t_{\rm hyd} \sim R_{\rm ns}/v_{\rm p} \lesssim 10^{-2}$ s, while the 7.8 Hz oscillations were claimed to persist for tens of seconds.

We propose that these oscillations also may be due to radial standing sound waves. Due to poor time resolution of *Einstein* data the photospheric expansion of the envelope can be neither excluded nor confirmed, while the peak luminosity is consistent with the Eddington limit, $\sim 7 \times 10^{-8}$ erg/(cm^2 s) at a distance $D = 4$ kpc (the most plausible value cited). Assuming that an expansion did occur, the interpretation of oscillations becomes straightforward. The main modes of sound waves have frequencies < 12 Hz, and the 7.6 Hz oscillations may be easily reproduced. Furthermore, standing waves are the eigenmodes, i.e. they are very coherent, in accordance with observations. The main mode frequency almost does not depend on parameters of the expanded atmosphere in the relevant range, i.e. not far from the end of expansion/contraction phase, so this frequency will remain constant along the burst.

References

Bildsten L., 1995, ApJ (to appear; Caltech preprint)
Lapidus I., Nobili L., Turolla R., 1994, ApJ 431, L103
Lewin W.H.G., Van Paradijs J., Taam R.E., 1993, Space Sci. Rev. 62, 223
Nobili L., Turolla R., Lapidus I., 1994, ApJ 433, 276
Schoelkopf R.J., Kelley R.L., 1991, ApJ 375, 696

Quasi-periodic Oscillations and Noise in Accreting Black Holes and Low-magnetic Field Neutron Stars

M. van der Klis

Astronomical Institute "Anton Pannekoek", University of Amsterdam, and Center for High-Energy Astrophysics, Kruislaan 403, 1098 SJ Amsterdam, The Netherlands

1 Introduction

The X-ray spectrum and the rapid X-ray variability of accreting compact objects have a common origin, and their properties can therefore be expected to be coupled. Indeed, it turns out that the X-ray spectrum and the power spectrum show correlated variations, which occur as a function of variations in the mass flux $\dot{M}$.

Stellar mass black holes and neutron stars have similar mass and size, and therefore their accretion phenomena may be expected to show similarities. Indeed, similarities exist that indicate that a unified description may be possible. If a particular phenomenon is seen in both neutron star and black-hole candidate systems this shows immediately that it cannot be due to any property that is unique to either neutron stars or black holes, such as the presence or absence of a surface, or of a strong non-aligned magnetic field. The quest for characteristics that are *unique* to black holes continues – I shall mention a few candidates below.

The power spectra of accreting compact objects can be described in terms of a small number of simple shapes (see van der Klis 1995a, c). *Power law noise* has a power distribution $\propto \nu^{-\alpha}$, *band limited noise* one that steepens towards high ν and flattens towards low ν. Band limited noise that has a maximum at $\nu > 0$ is called *peaked*; if the maximum is at $\nu = 0$ the component is called *flat-topped*. The same power spectral component can be at one time flat-topped and at another time peaked. *Quasi-periodic oscillations* (QPO) are a type of peaked noise. Usually, the term QPO is reserved for relatively narrow peaks ($\Delta\nu/\nu \lesssim 0.5$).

2 Z and atoll sources

Z and atoll sources (Hasinger and van der Klis 1989, hererafter HK89) are low magnetic-field neutron stars. Only a summary of their properties is presented here (see further van der Klis 1989, 1995c). The X-ray spectral changes are usually subtle, and *colour-colour diagrams* (CDs) and *hardness-intensity diagrams*

(HIDs), plots of X-ray hardness ratios *vs.* each other or *vs.* count rate are used to describe the X-ray spectral variations. The sources produce characteristic tracks in the CD/HIDs, and source position in the track is governed by $\dot{M}$.

The six known *Z sources* produce Z-shaped tracks. Z source power spectra show three broad noise components, *very low frequency noise* (VLFN), *low frequency noise* (LFN) and *high frequency noise* (HFN) and two QPO components, *horizontal branch oscillations* (HBO) and *normal and flaring branch oscillations* (N/FBO). HBO and LFN are a QPO and a band limited noise component that appear an disappear together, and are likely physically related. They are strongest at low $\dot{M}$ and disappear at high $\dot{M}$. HBO frequency (13–55 Hz) and LFN cut-off frequency (2–20 Hz) increase with $\dot{M}$. LFN can be flat topped or peaked, depending on the source. HBO and LFN are well explained in the *magnetospheric beat frequency model* (Alpar and Shaham 1985, Lamb et al. 1985), which requires Z sources to have a magnetosphere. Some pulsars show QPO that may be caused by a similar mechanism (Angelini et al. 1989, Finger 1995). N/FBO have a preferred frequency near 6 Hz. In Sco X-1 and GX 17+2, their frequency has been observed to increase from ~6 to ~20 Hz when $\dot{M}$ increases. *Radiation forces* at near-Eddington accretion rates probably play the key role in their formation (van der Klis et al. 1987, Hasinger 1987, Lamb 1989, Fortner et al. 1989, Miller and Lamb 1992, Alpar et al. 1992).

The 12–15 known *atoll sources* (HK89, van der Klis 1995c) show a curved track, that is often fragmented due to observational effects. increases. Their power spectra show two broad noise components called *very-low-frequency noise* (VLFN) and *high-frequency noise* (HFN). Atoll source HFN has a cut-off frequency of 0.3–20 Hz and depends strongly on $\dot{M}$: rms amplitude falls from >20% to <2% and the cut-off frequency increases when $\dot{M}$ increases (Yoshida et al. 1993, Prins et al. 1995). Atoll HFN is sometimes flat-topped and sometimes peaked.

HK89 proposed that the neutron stars in atoll sources have lower magnetic field strengths than Z sources, and are constrained to lower $\dot{M}$. The lower field explains why the (magnetospheric) HBO are not seen in atoll sources, and the lower $\dot{M}$ why the same is true for the (near-Eddington) N/FBO. The implied relation between $\dot{M}$ and magnetic field strength may have an evolutionary origin (van der Klis 1991). Predictions are that an atoll source that becomes bright will show Z source high-$\dot{M}$ properties (N/FBO and appropriate spectral branches), but never HBO, and that a Z source that becomes faint will show millisecond pulsations. The properties of Cir X-1 fit the first prediction. The source is apparently an example of an atoll source that can reach $\dot{M}_{Edd}$ (van der Klis 1991; Oosterbroek et al. 1995). As will be discussed in Section 4 Cir X-1 also shares some characteristics with black-hole candidates.

3 Black hole candidates

Three source states are distinguished in black-hole candidates (Fig. 1). In the *low state* (LS) the X-ray spectrum is a flat power law with photon spectral index 1.5–2. In the *high state* (HS) the 1–10 keV flux is much higher due to a soft component; the power law sometimes "sticks out" from under the soft component >10 kev. In the (rare) *very high state* (VHS) a similar (but brighter) soft component is seen, with perhaps an additional power law component. The VHS mainly differs from the HS by the properties of its rapid X-ray variability.

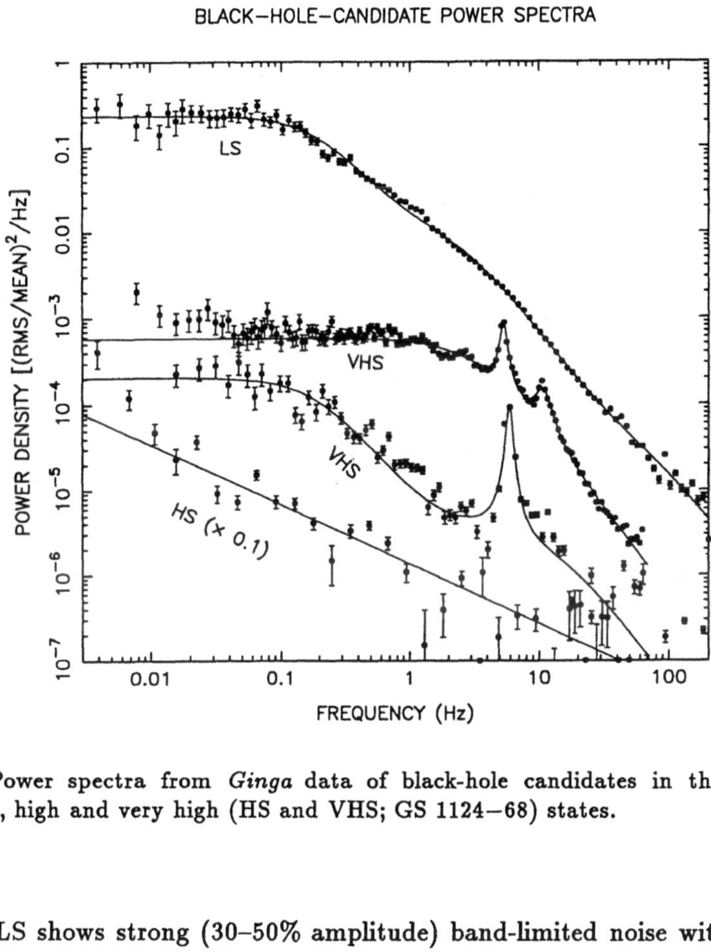

Fig. 1. Power spectra from *Ginga* data of black-hole candidates in the low (LS; Cyg X-1), high and very high (HS and VHS; GS 1124–68) states.

The LS shows strong (30–50% amplitude) band-limited noise with $\nu_{\rm cut}$ between 0.03 and 0.3 Hz which is usually flat-topped, but sometimes peaked (Vikhlinin et al. 1994). The level of the flat top and the cut-off frequency $\nu_{\rm cut}$ vary in anti-correlation, sometimes while the power spectrum above $\nu_{\rm cut}$ remains unchanged (Belloni and Hasinger 1990, Miyamoto et al. 1992a). In the HS variability is weak. Slow QPO with frequencies similar to the LS noise cut-off frequencies (~0.08–0.8 Hz; Motch et al. 1983, Ebisawa et al. 1989, Grebenev et al. 1991) and possibly related to peaked LS noise sometimes occur in LS and HS. The rare VHS shows 3–10 Hz QPO and rapidly variable broad-band noise. CD/HID branches occur in the VHS, and the power spectral parameters seem to depend on position in the branches, but these branch structures are not very similar from one epoch to the next ("messy" branches). The LS and VHS band limited noise cut-off frequency and amplitude fit one relation (van der Klis 1994b), suggesting that they form one phenomenon. The transient black-hole candidate GS 1124−68 (Nova Mus 1991 = GU Mus) in its decay went through all three states (Miyamoto et al. 1992b), strongly suggesting that the states directly follow $\dot{M}$. However, recent evidence indicates that some black hole transients, even when they are very luminous, remain in the "low" state; see Section 5.

4 Similarities between black hole candidates and low magnetic field neutron stars

There is a number of striking similarities between black-hole candidate and neutron star phenomenology (see van der Klis 1994a, 1995a). The black-hole candidate LS is very similar to the atoll source low $\dot{M}$ ("island") state. Both states occur at the lowest 1–10 keV count rates and inferred $\dot{M}$ levels. Both are dominated by strong (several 10%) band limited noise (LS noise and atoll HFN) which is sometimes flat-topped and sometimes slightly peaked. Sometimes the power spectra are nearly indistinguishable (Fig. 2, *top*; see also Yoshida et al. 1993), and the atoll source 13–80 keV X-ray spectra at low $\dot{M}$ become as hard as in black-hole candidates in the LS (Van Paradijs and van der Klis 1994). Z source LFN is also similar to black-hole candidate LS noise and atoll HFN: all 3 components are stronger at lower $\dot{M}$, disappear at higher $\dot{M}$, can be peaked and flat-topped, and have a higher $\nu_{\rm cut}$ at higher $\dot{M}$. The absence of a similar band-limited noise component in pulsars suggests that such noise arises through inhomogeneities in the inner disk, absent in pulsars. This is in accordance with the explanation of the LFN in the beat-frequency model. The black hole candidate VHS has strong similarities to the Z source high $\dot{M}$ ("normal/flaring branch") state. Both occur at the highest inferred $\dot{M}$ levels, and both show QPO, with similar frequencies (6–20 Hz in the neutron star systems, 3–10 Hz in the black-hole candidates), that depend on the position of the source in branched tracks in the HID/CDs. Clearly different is the harmonic content of the QPO (black-hole candidate VHS QPO show strong harmonics, Z source N/FBO do not) and the character of the HID/CD branches (much "messier" in BHCs). Another difference is that Z sources do not show the rapidly variable broad band noise seen in black-hole candidates.

The properties of Cir X-1 provide a further link between neutron stars and black holes. In some of its high states (Tennant 1987, Makino et al. 1992, Oosterbroek et al. 1995), this source shows a mix of Z source and black-hole candidate high $\dot{M}$ characteristics (see Fig. 2). It shows QPO with frequencies between 6 and 20 Hz and no second harmonics (both Z source characteristics) in combination with messy branches in the CD/HID and rapidly variable broad band noise (BHC characteristics). The reason, then, that Cir X-1 sometimes resembles a black hole in its rapid variability characteristics, (*e.g.* Toor 1977, Samimi 1979), while its X-ray bursts (Tennant et al. 1986a, b) show it to be a neutron

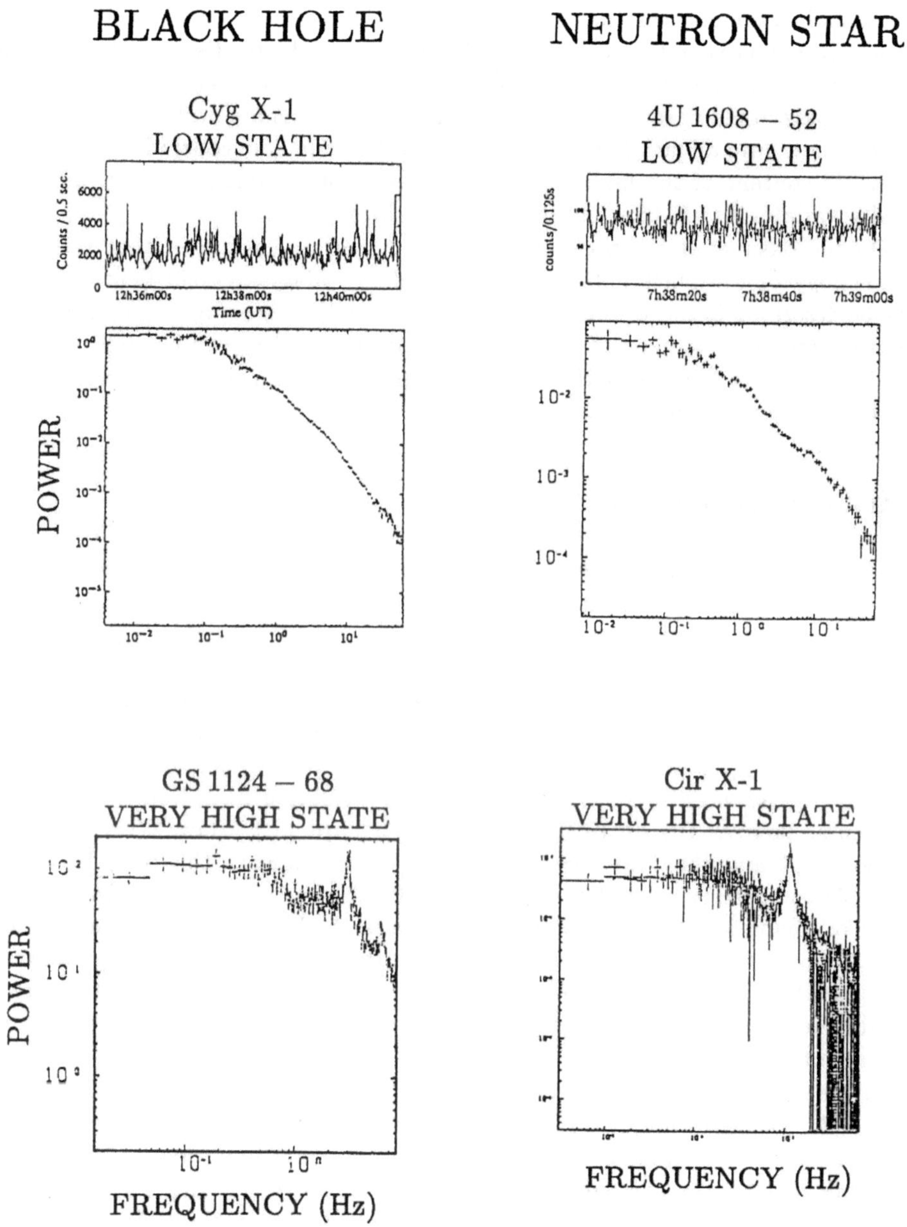

Fig. 2. Power spectra from the black-hole candidates Cyg X-1 (*top left*) and GS 1124–68 (*bottom left*) in the low state and the very high state, respectively, and from the low magnetic-field neutron stars 4U 1608–52 (*top right*) and Cir X-1 (*bottom right*) in the atoll island state and a very high X-ray brightness state, respectively, illustrate the similarity between neutron star and black-hole candidate low and very high states. Compiled from Inoue (1992), Takizawa et al. (1995) and Makino et al. (1991)

star, is that it is the only neutron star that we know that has a magnetic field as low as in atoll sources that sometimes accretes at near- or super-Eddington rates. Cir X-1 is therefore a key object, as it can help to distinguish between phenomena that are characteristic for accretion onto any compact object that has no appreciable magnetic field, and phenomena that are truly characteristic for accretion onto a black hole. From its properties one may conclude that a high harmonic content of the high $\dot{M}$ QPO may be a black hole signature, whereas rapidly variable broad band noise and messy branches are not.

On the basis of this array of similarities, it can be concluded that the phenomenology of the black-hole candidates and low magnetic-field neutron stars may be described in terms of three $\dot{M}$-driven states that are common to accreting low magnetic-field neutron stars and accreting black holes (van der Klis 1994a). Fig. 3 presents a line-up of the three common states of black-hole candidates and low magnetic-field neutron stars.

5 Inclination effects

Detailed examination of the properties of Z sources, in particular in their flaring branches, has led Kuulkers and van der Klis (1995) to propose that obscuration by a geometrically thick inner accretion disk plays a role in Z source phenomenology. The disk swells when $\dot{M}$ increases, and for higher inclination i obscuration effects already set in at lower $\dot{M}$. A similar model might apply to black-hole candidates (van der Klis 1994a, 1995b). The reason for the disappearance of the hard LS X-ray spectral component in the HS may be obscuration of a central, hot and rapidly variable region by a puffed-up accretion disk. Some black hole transients, such as GS 2023+338, show only a hard power-law X-ray spectral component, even when they are very bright. This might in this model be due to a pole-on viewing geometry. In GX 339−4 the observable energy flux in the 1–200 keV band is higher in the low state than in the high state (Grebenev et al. 1993). The increasing concentration of the hard X-rays towards the (rotation) polar axes with increasing $\dot{M}$ might explain this: at high $\dot{M}$ most of the energy would be leaving the system along the polar axis and would, in a high-inclination source, not be seen by us.

In the low magnetic-field neutron stars the X-ray flux is an unreliable indicator of $\dot{M}$; the same might turn out to be the case in the black-hole candidates. Note, that the mass flux $\dot{M}$ that by hypothesis determines source state is the mass flux *towards* the compact object. At near- and super-Eddington rates, not all of this matter may actually accrete; jets might for example be formed when $\dot{M}$ becomes high enough.

6 Conclusion

The millisecond fluctuations in black holes, and in neutron stars with various magnetic field strengths can perhaps be understood in common terms, as follows.

Phenomena Guess at B	Black hole (0 Gauss)	Atoll source ($\lesssim 10^{9-10}$ Gauss)	Z source ($\sim 10^{9-10}$ Gauss)	Guess at $\dot{M}$
High $\dot{M}$ QPO	VHS	Cir X-1 (high state)	FB	$\gtrsim 1\dot{M}_E$
Weak power-law noise		Banana	NB	$\sim 0.9\dot{M}_E$
Weak power-law noise + band-limited noise	HS		HB	$\sim 0.5\dot{M}_E$
Strong band-limited noise	LS	Island (most bursters and dippers)	(ms X-ray pulsars?)	$\sim 0.01\dot{M}_E$

VHS: very high state; HS: high state; LS: low state; FB: flaring branch; NB: normal branch; HB: horizontal branch.

Fig. 3. Proposed classification scheme for X-ray binary source states. There are three states that are common to neutron stars and black holes; in a given source the mass transfer rate $\dot{M}$ towards the compact object determines the state. The power spectral shapes that are characteristic of each state are indicated at left. The conventional names of the source states in each source type are indicated. Magnetic field strengths and mass fluxes are rough indications only; other source parameters, such as inclination, might affect the $\dot{M}/\dot{M}_E$ levels at which state transitions occur.

Two structures determine the basic physics of the accretion process, namely the magnetosphere and the inner (radiation pressure dominated) disk. Z sources are the most complex, showing HBO and N/FBO, as well as LFN and HFN, because their field is weak enough to allow the presence of an inner disk (like in black holes and atoll sources) and strong enough to form a magnetosphere (like in pulsars, but much smaller). Black holes and atoll sources have an inner disk and no appreciable magnetosphere and therefore only show an N/FBO-like component (VHS QPO) and an LFN-like component (LS noise). X-ray pulsars (not discussed here) have a magnetosphere and no inner disk and therefore show only an HBO-like and an HFN-like component. Inner disk structure causes anisotropic emission, and thereby inclination effects are introduced in the phenomenology.

Acknowledgements: This work was supported in part by the Netherlands Organization for Scientific Research (NWO) under grant PGS 78-277.

References

Alpar M.A., Shaham J., 1985, Nature 316, 239
Alpar M.A. et al., 1992, A&A 257, 627
Angelini L., Stella L., Parmar A.N., 1989, ApJ 346, 906
Belloni T., Hasinger G., 1990, A&A 227, L33
Ebisawa K., Mitsuda K., Inoue H., 1989, PASJ 41, 519
Finger M., 1995, Proc. IAU Symp. 165, The Hague, August 1994, in press
Fortner B., Lamb F.K., Miller G.S., 1989, Nature 342, 775
Grebenev S.A. et al., 1991, Sov. Astron. Lett. 17(6), 413
Grebenev S. et al., 1993, A&A Suppl. 97, 281
Hasinger G., 1987, A&A 186, 153
Hasinger G., van der Klis, M., 1989, A&A 225, 79 [HK89]
Inoue H., 1992, in Accretion Disks in Compact Stellar Systems, J.C. Wheeler (ed.), ISAS RN 518
Kuulkers E., van der Klis, M., 1995, A&A, in press
Lamb F.K., 1989, Proc. 23rd ESLAB Symposium, ESA SP-296, 215
Lamb F.K. et al., 1985, Nature 317, 681
Makino Y., Kitamoto S., Miyamoto S., 1991, poster presented at the 28th Yamada Conference, Nagoya, Japan, April 8–12 1991
Makino Y., Kitamoto S., Miyamoto S., 1992, in Frontiers of X-ray astronomy, Tanaka and Koyama (eds.), Universal Academy Press, Tokyo, p. 167
Miller G.S., Lamb F.K., 1992, ApJ 388, 541
Miyamoto S. et al., 1992a, ApJ 391, L21
Miyamoto S. et al., 1992b, Proc. Ginga Memorial Symp., Makino and Nagase (eds.), ISAS, Tokyo, p. 37
Motch C. et al., 1983, A&A 119, 171
Oosterbroek T. et al., 1995, A&A, in press
Prins S. et al., 1995, A&A, in preparation
Samimi J. et al., 1979, Nature 278, 434
Takizawa M. et al., 1995, in press
Tennant A.F., 1987, MNRAS 226, 971
Tennant A.F., Fabian A.C., Shafer R.A., 1986a, MNRAS 219, 871
Tennant A.F., Fabian A.C., Shafer R.A., 1986b, MNRAS 221, 27p
Toor A., 1977, ApJ 215, L57
Van der Klis M., 1989, ARA&A 27, 517
Van der Klis M., 1991, NATO Advanced Study Institute 344, 319
Van der Klis M., 1994a, ApJ Suppl. 92, 511
Van der Klis M., 1994b, A&A 281, L17
Van der Klis M., 1995a, NATO Advanced Study Institute 450, 301
Van der Klis M., 1995b, Proc. COSPAR Meeting on Galactic and Extragalactic Black Holes, Hamburg, July 1994, in press

Van der Klis M., 1995c, AIAP preprint 1993-009; in X-Ray Binaries, Lewin, van Paradijs and van den Heuvel (eds.), Cambridge University Press, in press
Van der Klis M. et al., 1987, ApJ 316, 411
Van Paradijs J., van der Klis M., 1994, A&A 281, L17
Vikhlinin A. et al., 1994, A&A 287, 73
Yoshida K. et al., 1993, PASJ 45, 605

D. Dravins: As you mentioned, the relevant dynamic timescale for black-hole systems may be in the range 0.1–1 milliseconds, potentially observable as variability at kilohertz frequencies. Is it possible to make any extrapolation or estimates of what phenomena one could observe there, once the observational capacity becomes available?

M. van der Klis: It is conceivable that in black holes we might be able to detect the signature of matter in its last stable orbit at 3 R_S. However, it is very hard to predict how strong the signal would be. In 1995 NASA will launch XTE, which for bright X-ray sources will detect count rates up to about 10^5 cts; this we hope will make it possible to detect such a signal, if it is there.

Detection of Nonthermal Optical Flashes with 10^{-3}–10^{-1} s Duration from Some LMXBs

G.M. Beskin[1], *S.I. Neizvestny*[1], *S.N. Mitronova*[1], *V.L. Plokhotnichenko*[1], *M.Yu. Popova*[1], *O.G. Benvenuto*[2], *C. Feinstein*[2], *M. Mendez*[2], *C. Bartolini*[3], *A. Guarnieri*[3], *R. Minarini*[1,3], *A. Piccioni*[3]

[1] Special Astrophysical Observatory, Nizhnij Arkhyz, Karachaevo-Circassia, 357147 Russia
[2] Observatorio Astronomico de la Plata, Paseo del Bosque s/n, 1900 La Plata, Argentina
[3] Dipartimento di Astronomia, Universitá di Bologna, 40126 Bologna, Italy

1 Introduction

In order to study very rapid optical variability of astrophysical objects on time scales between 10^{-7} s and 10^{2} s (Shvartsman 1977), at SAO the *MANIA* (Multichannel Analysis of Nanosecond Intensity Alterations) experiment is being used. A special photometric registration system and software has been developed (Beskin et al. 1982' Plokhotnichenko 1983, Zhuravkov et al. 1994).

One of the applications of these tools is the detection of radiation from accreted or ejected plasma near compact objects. A choice between two models of accretion onto compact objects in binary systems – hydrodynamic flow or magnetc flaring (Shakura & Sunyaev 1973, Pustil'nik & Shvartsman 1974) has not yet been made. Recent optical and X-ray observations of X-ray binaries provide information on their (fast) variability, their nonthermal radio emission, generation of very high energy particles, and nonthermal processes in rapid optical flares (Bartolini et al. 1994, Beskin et al. 1994). However, the data are not fully described by classical hydrodynamical models.

Within the context of hydrodynamical models, optical flares have a thermal origin. Thus, their brightness temperatures should not exceed 10^{7}–10^{8} K (Lipunov 1987). On the other hand, in the magnetic flare model optical radiation is generated, independent of the X-ray emission, by dissipation of magnetic fields and the interaction of electrons with these fields near "blobs" in the accretion structure (Pustil'nik 1975, Beskin & Minarini, in prep.). These processes could yield nonthermal optical flares with brightness temperatures $> 10^{8}$–10^{9} K on time scales of 10^{-6}–10^{-2} s.

Low-mass X-ray binaries (LMXBs) are the most useful systems for studies of the accretion processes in the optical range, because the normal component is faint enough to allow observations of optical emission from the accreting plasma.

2 Observational Results

We present some results of our study of very short events from some selected LMXBs: A0620–00, MXB1735–44 and Nova Per 1992 (GRO J0422+32).

A0620–00 and MXB1735-44 were observed in 1986 with the *MANIA* system on the 6 m telescope of SAO RAS and in 1991 on with the 2.15 m telescope of CASLEO (Argentina) (Shvartsman et al. 1989, Beskin et al. 1994). Nova Per 1992 was simultaneously observed in different pairs of color bands (*UB, BV, VR*) in 1992-94 with the 6 m telescope.

A0620–00. During observations of A0620–00 (without filter in one of the sets) on '986 November 13, five particularly short flashes were detected. The first two had durations of 3 and 5 ms and their rise times were 1–2 ms. The remaining events lasted $\sim$ 0.4–0.5 ms with rise times of about 0.1 ms (Fig. 1).

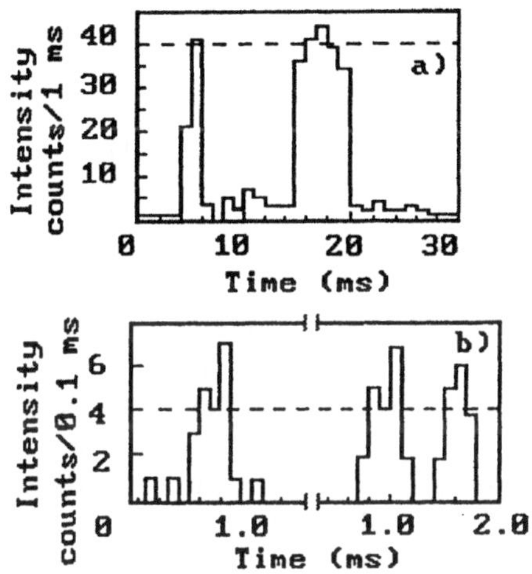

Fig. 1. Ultrashort flares of A0620–00 (Feb. 13, 1986).

MXB 1735-44. During the observations of MXB 1735–44 two flares with durations of $\sim$ 0.25 s were detected (see Fig. 2).

To study the fine structure of these events their detailed light curves were analyzed with the "splash method" (Beskin et al. 1994). These primary light curves, $I(t)$, were smoothed with a rectangular temporal window. The forward edges of both flares had very steep gradients, lasting about 0.05 – 0.06 s.

We used these two curves to construct the so called "normalized discrepancies": $\delta I(t) = \Delta I(t)/(D[\Delta I(t)]^{1/2})$, where $\Delta I(t) = I(t) - < I(t) >$ is the difference between the original and the smoothed light curve, and $D[\Delta I(t)]$ is the dispersion of this discrepancy calculated under the assumption that the data are due to a Poisson process of variable intensity. In calculating $D[\Delta I(t)]$ we took into account that $I(t)$ and $< I(t) >$ are correlated. If there is no fine time structure, $\delta I(t)$ must be close to a normal distribution with zero mean and unit dispersion. This was verified with the standard χ^2 test, and we found that the hypothesis of a normal distribution has to be rejected (at a confidence level of 95% to 99.5%). In other words, both flares have a fine structure on time scales of 5–10 ms.

GRO J0422+32. Near the maximum of the optical high state ($V < 15^m.5$)

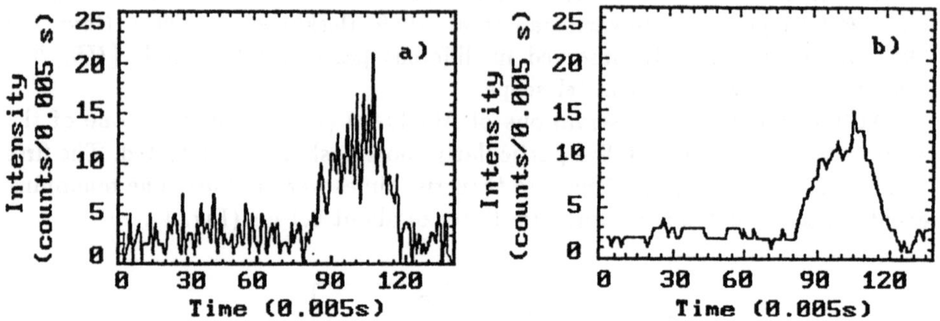

Fig. 2. One of two observed flashes of MXB1735-44: (a) original light curve, (b) smoothed light curve.

the brightness of Nova Per 1992 was irregularly variable in different colour bands on time scales from 4 ms to 200 s with amplitude factors of 0.5–4. The power spectrum of the detected variability was flat in the whole range of frequences from 0.01 to 250 Hz. On time scales from 100 ns to 1 ms variability was probably absent. During the low optical state ($V > 15^m.5$) Nova Per did not show any significant variability on time scales from 100 ns to 10 s. The shortest flares had rising times of 4–40 ms (see Fig.3), which allows us to establish an upper limit of 10^8–10^9 cm on the size of the flare region.

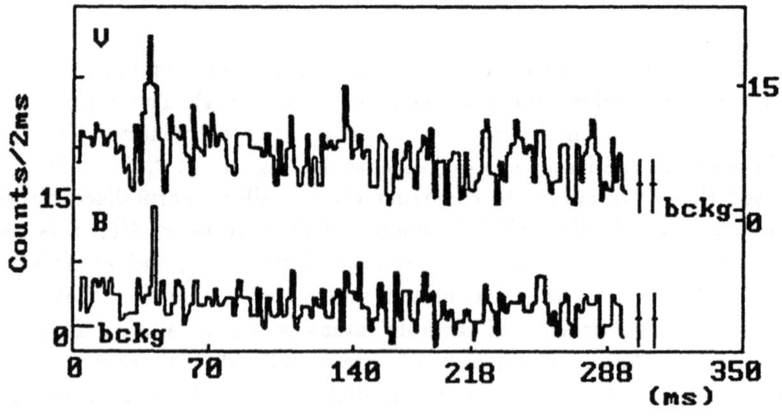

Fig. 3. Flash of Nova Per with a rise time of 4 ms in V (top) and in B (bottom) bands (January 18, 1993).

3 Interpretation

For all detected events lower limits on the brightness temperatures were estimated with the formula: $T_b = 10^8 \varrho_m t^{-2} f^{-2} D^2$, where t is the duration of the rising front in ms, f is the average frequency of observation (in units of 10^{15} Hz), D is the distance to the source in kpc, and ϱ_m is the source flux in mJy. For the calculation of T_b we assumed the following values; A0620–00: $D \sim 1$ kpc, $B \sim 19^m.3$, $A_v \sim 1^m.2$; for MXB 1735-44: $D \sim 7$ kpc, $B \sim 17\overset{m}{.}2$, $A_v \sim 0\overset{m}{.}8$; for Nova Per: $D \sim 2.4$ kpc (according to Shrader et al. (1992), $B \sim 13\overset{m}{.}9$, $A_v \sim 1\overset{m}{.}2$. We took into account the relation between the intensities of objects, the background, and the effective frequencies of the observations.

Thus, the brigthness temperatures for different flare regions of A0620–00 and MXB 1735–44 exceeded $10^8 - 10^{11}$ K and in the case of Nova Per 1992 it was between 10^8 and 10^9 K. The detected flashes clearly must have a nonthermal origin.

If we would use a thermal mechanism to explain these events, then X-ray fluxes should be very high ($> 10^{39} - 10^{40}$ erg), which is contradicted by the X-ray data obtained almost simultaneously with optical observations.

It is important to mention that for A0620–00 and MXB 1735–44 the detected events were rare, but they were common in the case of GRO J0422+32. This could perhaps be explained with these objects being observed in different states: a quiet state with low accretion rates for A0620–00 and MXB 1735–44, and an active phase with high accretion rates for GRO J0422+32.

Apparently the generation of nonthermal optical flares cannot be explained with the hydrodynamic accretion model. Various mechanisms discussed for the magnetic flaring model provide a better description of the observations.

References

Bartolini C., et al., 1994, ApJS 92, 455
Beskin G.M., et al., 1983, Sov. Astr. Lett. 14, 156
Beskin G.M., et al, 1994, A&A 289, 141
Lipunov V., 1987, Astrofyzika Nejtronnykh Zvezd, Moscow, Nauka, p. 296
Negoro H., et al, 1994, ApJ 423, L127
Plokhotnichenko V.L., 1983, Soobshcheniya SAO, 38, 29
Pustil'nik L.A., 1975, Ph.D. Thesis
Pustil'nik L.A., Shvartsman V.F., 1974, in IAU Symp. 64, Reidel, Dordrecht, p. 183
Shakura N.I., Sunyaev R.A., 1973, A&A 24, 337
Shrader C.S., et al, 1992, in Proc. of Compton Observatory Symp., St. Louis (in press)
Shvartsman V.F., 1977, Soobshcheniya SAO, 19, 5
Shvartsman V.F., et al, 1989, Sov. Astron. Lett., 15, 590
Zhuravkov A.I., et al., 1994, Bull. Spec. Astrophys. Obs. 37, 157

TTM Observations of X1755–338

H. C. Pan[1,3], *G. K. Skinner*[1], *R. A. Sunyaev*[2], *K. N. Borozdin*[2]

[1] School of Physics and Space Research, University of Birmingham, Edgbaston, Birmingham B15 2TT, UK
[2] IKI, Space Research Institute, Profsoyuznaya, 84/32, Moscow 117296, Russia
[3] Department of Physics (Theoretical Physics), University of Oxford, 1 Keble Road, Oxford OX1 3NP, UK

1 Observations

X1755–338 is an X-ray binary source which displays X-ray dips with a 4.4 hour period (White et al. 1984). It was previously noted as an unusually soft X-ray source by Jones (1977) and was suggested later as a black-hole candidate (BHC) by White & Marshall (1984), and White et al. (1984), based on the similarity of its location in an X-ray colour-colour diagram to that of a group of BHCs.

The *TTM* is a coded-mask imaging spectrometer on board the *KVANT* module of the *MIR* space station. It is capable of producing images in the $2-30$ keV band with an energy resolution of about 18% at 6 keV. The instrumental details are given in Brinkman et al. (1985).

We observed X1755–338 in 1989 March-September during the period of the *TTM* Galactic Centre Survey.

2 Results

As shown in Fig. 1, the source intensity was variable during the observations. Variation of over 30% was seen on time scales from a few hours to several days. The X-ray intensity in 1989 August-September period appeared to be nearly twice that observed in 1989 March-April. The large data gaps in the *TTM* light curve prevent us from identifying any X-ray dips.

We have studied the X-ray spectra of X1755–338 accumulated during each observational day. We find that the source spectrum consists of an ultrasoft thermal component and a hard tail above $\sim 6-10$ keV, which is similar to the high state spectrum of the well known BHCs LMC X-1, LMC X-3 and GX 339–4, also observed with the *TTM*. To our knowledge, this is the first time that a hard tail has been observed from X1755–338.

We have modelled the spectrum with a multi-temperature disc blackbody (Mitsuda et al. 1984) and a power-law. We plot in Fig. 2 the spectrum obtained on 1989 March 31. A significant hard tail is seen above 10 keV. We have found that the soft and hard X-ray components of X1755–338 vary independently, as in the spectra of LMC X-1, LMC X-3 and GX 339–4 in their high (intensity) state.

The *TTM* observations strongly suggest that X1755–338 does indeed belong to the family of BHCs.

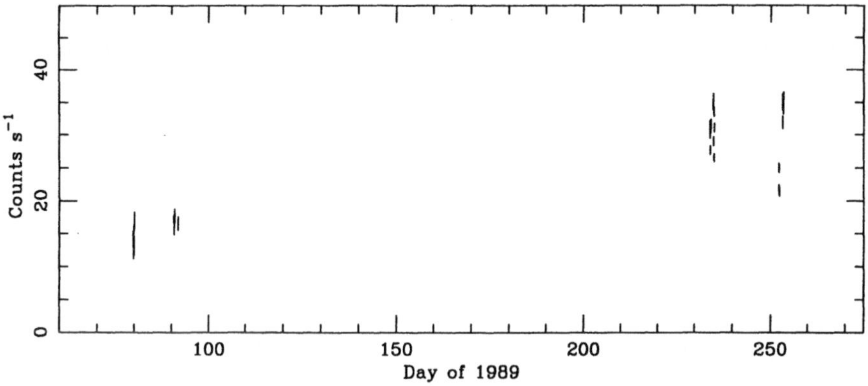

Fig. 1. X-ray (2 – 30 keV) light curve of X1755–338 in 1989 March-September.

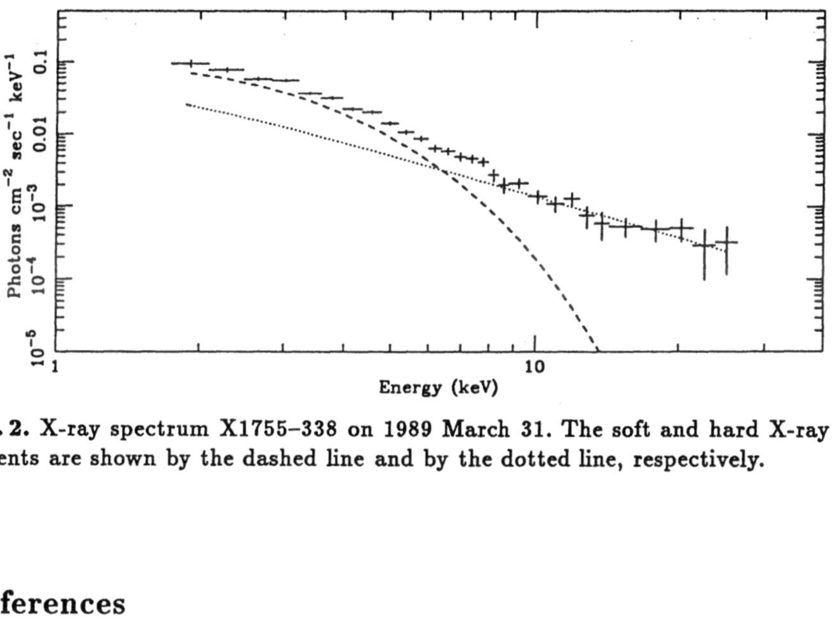

Fig. 2. X-ray spectrum X1755–338 on 1989 March 31. The soft and hard X-ray components are shown by the dashed line and by the dotted line, respectively.

References

Brinkman A.C., Dam J., Mels W.A., Skinner G.K., Willmore P.W., 1985, in Non-thermal and Very High Temperature Phenomena in X-Ray Astronomy, Perola G.C. & Salvati M. (eds), Università "La Sapienza", Roma, p. 263

Jones C., 1977, ApJ 214, 856

Mitsuda K. et al. 1984, PASJ 36, 741

White N. E., Marshall F.E., 1984, ApJ 281, 354

White N.E. et al. 1984, ApJ 283, L9

Irregular X-ray Variation of LMC X-3

H. C. Pan[1,3], *G. K. Skinner*[1], *R. A. Sunyaev*[2], *N. L. Alexandrovich*[2]

[1] School of Physics and Space Research, University of Birmingham, Edgbaston, Birmingham B15 2TT, UK
[2] IKI, Space Research Institute, Profsoyuznaya, 84/32, Moscow 117296, Russia
[3] Department of Physics (Theoretical Physics), University of Oxford, 1 Keble Road, Oxford OX1 3NP, UK

1 Introduction

LMC X-3 is an X-ray binary in the Large Magellanic Cloud. It was discovered by *UHURU* and observations with various satellites showed that the X-ray source was variable by a factor of up to 100 (e.g. Treves et al. 1988). Using the *GINGA* and *HEAO-1* observations, Cowley et al. (1991) found a strong $\sim$ 198 (or possibly $\sim$ 99) day modulation in the X-ray luminosities of LMC X-3. They suggested that this modulation may be due to an accretion disc precession, or periodic variations in the mass transfer rate, or a combination of both.

We observed LMC X-3 with the *TTM* in 1988-1990. The present paper gives some results from the analysis of the data obtained.

2 Observations and Results

The *TTM* is a coded mask telescope on board the *MIR* space station. It is capable of producing images in the energy range 2-30 keV with a spectral resolution of about 18% at 6 keV (Brinkman et al. 1985).

We monitored LMC X-3 with the *TTM* in the periods 1988 November 13 -1989 March 2, 1989 May 30-June 14, and 1990 December 22. During the observations of 1988 November 13-December 11, while the hard X-ray (7-28 keV) flux of LMC X-3 remained almost constant at about 1.5 counts s^{-1}, the 2-7 keV flux decreased gradually by 74% in 28 days. Although the X-ray spectrum on 1988 December 11 was not well determined owing to the source weakening, the analysis of the other spectra indicates that it is likely that LMC X-3 underwent a transition from a high to a low state in this period. LMC X-3 became bright again on 1989 January 3-6 and on around March 2 the 2-7 keV flux reached the highest level 14.7 ± 0.3 counts s^{-1} (55 mCrab), more than a factor of 8 larger than that observed on 1988 December 11. In 1989 May-June, the source flux in the 2-7 keV range was back to the level of 1988 November 13; however, the 7-28 keV flux was generally higher.

We have folded the 2-10 keV fluxes of LMC X-3 from the *TTM* measurements with the ephemeris given by Cowley et al. (1991). We plot the folded light curve in Fig. 1. It is clear that the *TTM* observations on 1988 November 13-15, 1989 May 30-June 6, and 1990 December 22 follow the 198 day modulation. However, the observations made from 1988 November 29 to March 2 with the *TTM*, and also with the *GINGA* LAC (on 1988 December 12 and 1989 February 10, see Fig. 1), indicate that the source showed large irregular variations with the intensity rapidly decreasing from 1988 November 29 and gradually increasing from at least 1989 January 4 until March. In this period the 198 day modulation law was largely violated by LMC X-3. If the 198 day modulation reflects the precession of the accretion disc, it is possible that the X-ray variation in the period 1988 November 29 to 1989 March 2 is caused by the change of the mass accretion rate. As the hard X-rays showed little change in this period, we assume that the variation of the soft X-rays (and the temperature of the soft X-ray emitting region) is directly related to the change in the mass accretion rate.

The *TTM* spectra of LMC X-3 can be described with a model consisting of a multi-temperature disc blackbody (Mitsuda et al. 1984) and a power-law. We find that both components are variable and there is no correlation between them.

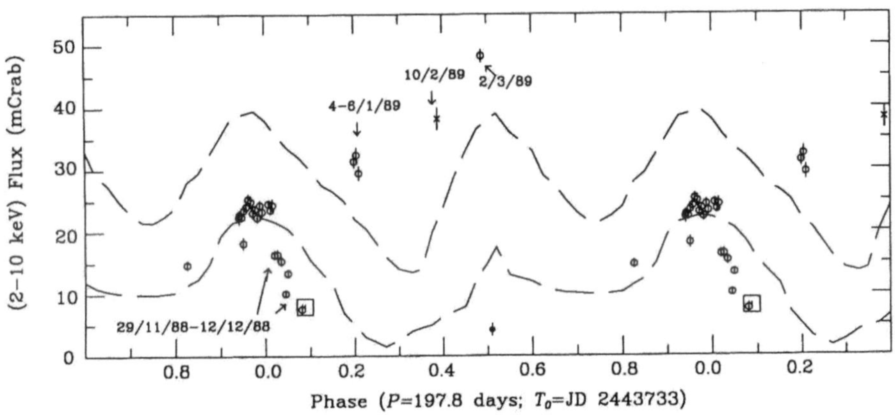

Fig. 1. X-ray (2-10 keV) intensities of LMC X-3 folded with the ephemeris of Cowley et al. (1991). The open circles represent the *TTM* measurements. The open squares and crosses are the *GINGA* fluxes. The dashed lines illustrate the 198 day modulation.

References

Brinkman A. C., Dam J., Mels W. A., Skinner G. K., Willmore P. W., 1985, in Perola G. C., Salvati M., eds, Non-thermal and Very High Temperature Phenomena in X-Ray Astronomy, Roma: Università "La Sapienza", p. 263

Cowley A.P., et al., 1991, ApJ 381, 526

Mitsuda K., et al., 1984, PASJ 36, 741

Treves A., et al., 1988, ApJ 325, 119

GAMMA-RAY BURSTS

Gamma-ray Bursts: Observational Overview, Searches for Counterparts, and BACODINE

G.J. Fishman[1], *S. Barthelmy*[2]

[1] Space Sciences Laboratory, NASA/MSFC, Huntsville, AL 35812 USA
[2] Laboratory for High Energy Astrophysics, NASA/GSFC, Greenbelt, MD 20771 USA

Abstract: The basic properties of γ-ray bursts are reviewed. Many of these properties have been known for over two decades, although new and more detailed observations have been made by Compton Observatory in the past three years. The new observation with the greatest impact has been the observed isotropic distribution of bursts along with a deficiency of weak bursts. This is not compatible with any known Galactic population of objects.

The observation of a γ-ray burst counterpart in another wavelength region has yet to be made, even though there have been substantial efforts in this pursuit from a wide variety of observations in many wavelength regions. Perhaps the greatest chance for a simultaneous or rapid observation of a γ-ray burst counterpart lies with the newly-developed BAtse COordinates DIstribution NEtwork (*BACODINE*) system. The objectives and characteristics of this system are described.

1 Introduction

It is over 25 years since the discovery of γ-ray bursts, and their origin appears as elusive as ever. The observed isotropy and inhomogeneity of these objects represent a distribution unlike any other known galactic objects. Over a hundred theories on their origin have now been cataloged (Nemiroff 1994). These models cover distance scales from the Oort cloud to cosmological distances. Never before in modern astronomy has there been so great an uncertainty in the distance to a class of objects. There is a similar level of uncertainty about the basic nature of the underlying source and the emission mechanism. Whatever the distance scale and the source of the bursts, it will most likely represent a new class of objects, processes and/or emission mechanism.

In the first section, this paper summarizes the basic observed properties of γ-ray bursts, derived from lectures presented at the Les Houches School for Theoretical Physics in January 1994. Later sections describe past and present attempts to observe counterparts of γ-ray bursts in another wavelength region and the newly-developed *BACODINE* system. The objectives and characteristics of this system offers the unprecedented opportunity to search for γ-ray burst counterparts in near-realtime.

2 Observed properties of γ-ray bursts

Considerable observational progress has been made in the past few years as more sensitive space-borne detectors have become available. Many of the observations have been made by the Burst and Transient Source Experiment (*BATSE*) on the Compton Gamma-Ray Observatory (*CGRO*). While most of these observational results are relatively straight-forward, some of the properties and interpretations of ensembles of bursts are the subject of analyses that are often debatable. Comprehensive sets of papers of the more recent observational results can be found in conference proceedings that have been published in the past three years (Paciesas & Fishman 1992, Friedlander et al. 1993, Fishman et al. 1994). Hartmann (1995) also provides a detailed list of γ-ray burst (GRB) reviews and Hurley (1994) has made available a comprehensive bibliography of the GRB literature.

Perhaps the most striking features of the time profiles of GRBs are their morphological diversity and the large range of burst durations. A cursory examination of burst profiles indicates that some are chaotic and spiky with large fluctuations on all timescales, while others show rather simple structures with few peaks. However, some bursts are seen with both characteristics present within the same burst. No periodic structures have been seen from GRBs. Coupled with this diversity is the general inability to place many GRBs into well-defined classifications. Examples of extreme differences in burst morphologies and durations are shown in a sample page from the First *BATSE* Burst Catalog (Fishman et al. 1994; Fig. 1).

The durations of γ-ray bursts range from about 10 ms to over 1000 s in the energy range in which most bursts are observed. A bi-modality is seen in the logarithmic distribution with broad, un-resolved peaks at about 0.3 s and 20 s and a minimum at around two seconds (Fig. 2). The shorter bursts are also seen to have harder spectra, as measured by a hardness ratio (Kouveliotou et al. 1994). Sub-millisecond structure has been detected in at least one burst (Bhat et al. 1992). Recent *EGRET-CGRO* observations show high energy (>100 MeV) emission over 90 minutes after the burst trigger (Hurley et al. 1994). Another general property of the γ-ray bursts time profiles is that they tend to have shorter rise-times and fall-times (sharper spikes) at higher energies. Most bursts also show an asymmetry, with shorter leading edges than trailing edges. This has been quantified (Link et al. 1993, Nemiroff et al. 1994).

An analysis of time profiles that could have extraordinary consequences is the observation by Norris et al. (1994) of a systematic widening or stretching of γ-ray burst time profiles as bursts become weaker. This analysis was performed by artificially weakening the stronger γ-ray bursts and introducing the appropriate background so that all bursts could be analyzed in a consistent manner. The observed stretching of the profiles of bursts is consistent with that expected from the effects of time-dilation from bursts at cosmological distances. However, this observation and its interpretation are not universally accepted.

Another distinguishing feature of γ-ray bursts is their high-energy emission: almost all of the power is emitted above 50 keV. Some bursts show emission as

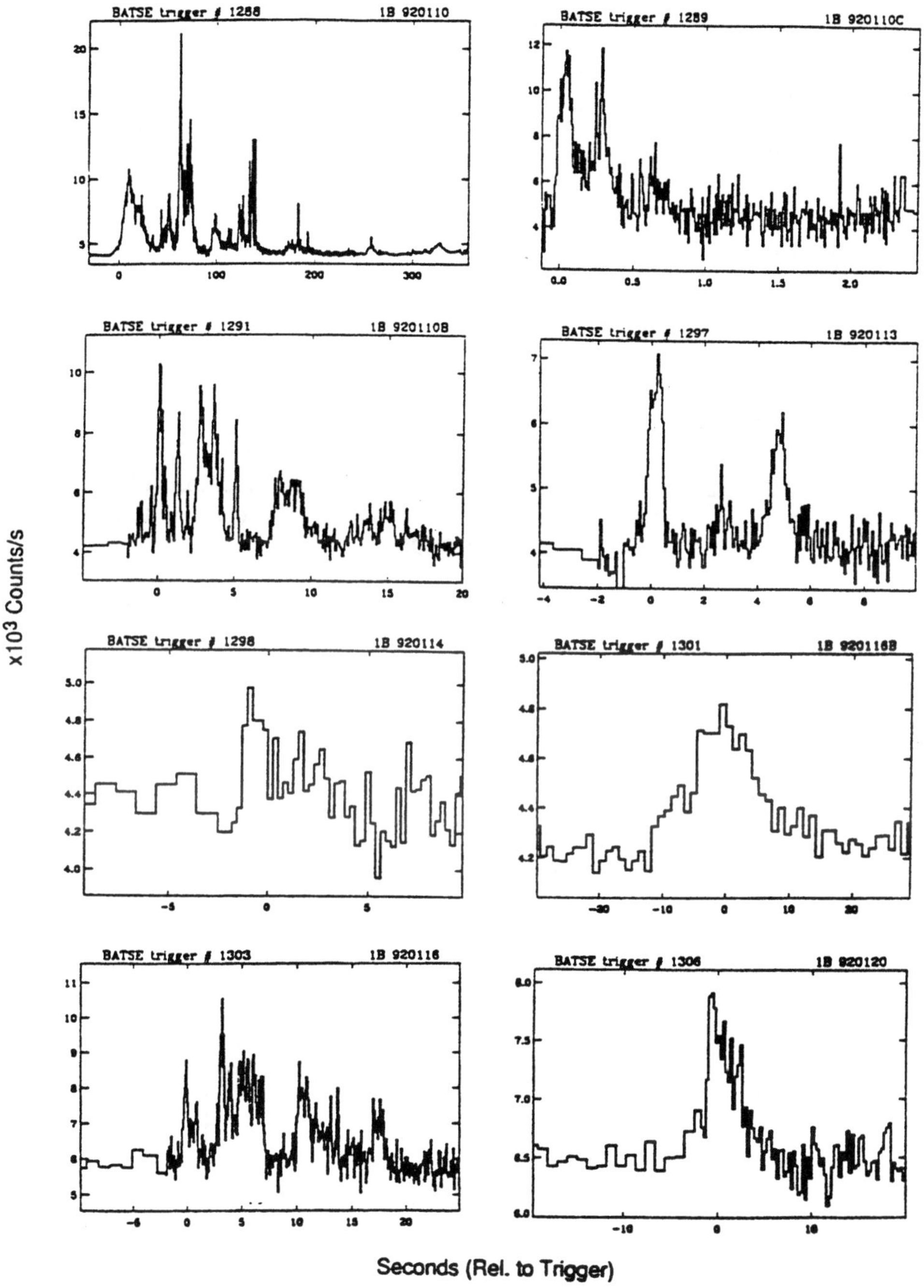

Fig. 1. A sample of eight γ-ray bursts for the First *BATSE* Catalog (Fishman et al. 1994), showing the extreme range of burst time profiles and durations.

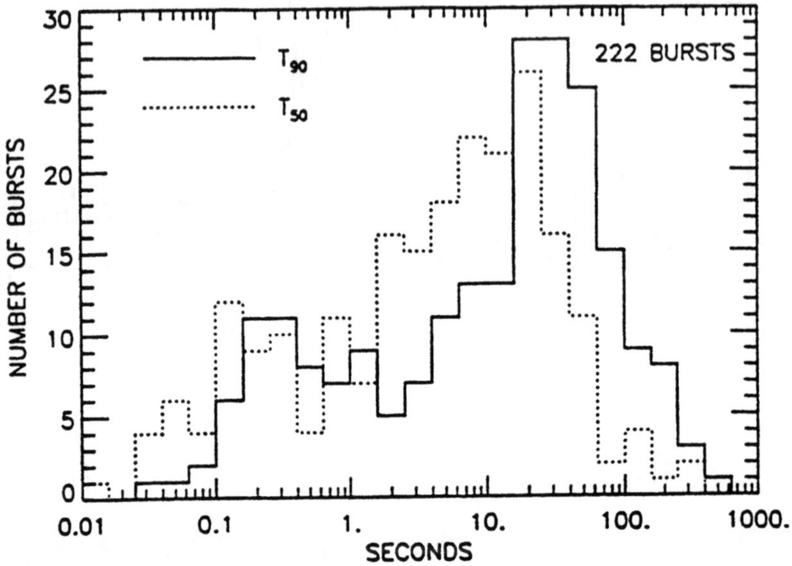

Fig. 2. The duration distribution of γ-ray bursts from the First *BATSE* Burst Catalog (Fishman et al. 1994).

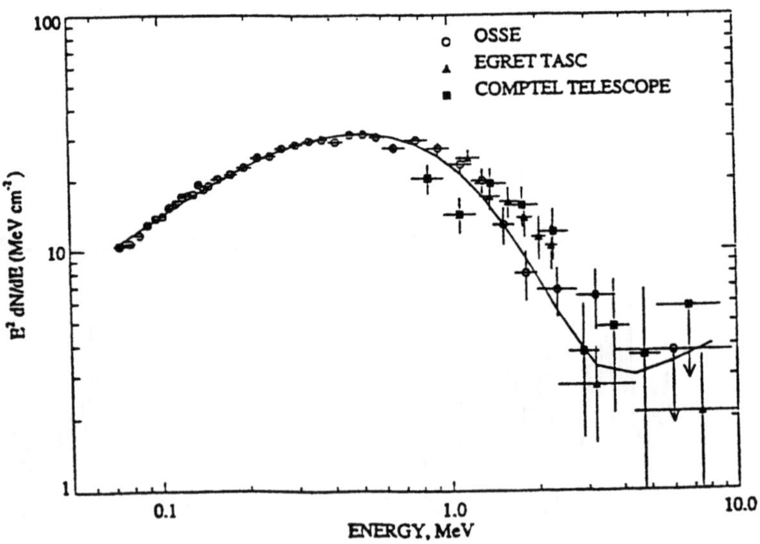

Fig. 3. The high energy spectrum from GB 910601, as measured by three of the experiments on the Compton Observatory (Share et al. 1994), integrated over a large portion of the burst. A characteristic broad spectral shape, with peak power of about 0.6 MeV is seen. (The spectral up-turn at high energies is not real.)

low as 1 keV, but the power is less than 1 or 2% of the total power (Yoshida et al. 1989). Most bursts show rather simple continua spectra which appear similar in shape when integrated over the entire burst and when sampled on various timescales within a burst. Fig. 3 shows a typical burst spectrum from 0.1–10 MeV, with the peak power at about 600 keV (Share et al. 1994). Spectral shapes which have been fit to burst spectra include broken power laws (Schaefer et al. 1992), log-normal distributions (Pendleton et al. 1994), and exponential spectra with power-law high energy tails (Band et al. 1993). Although the spectral shapes of many bursts are similar, the energy at which peak power is emitted changes greatly from burst-to-burst and rapidly within a burst. Some significant changes on time scales as short as tens of milliseconds have been observed (cf. Ford et al. 1994). In the past few years, *EGRET-GRO* has seen significant flux and power into the GeV energy range from several bursts. A very strong burst with high energy emission was detected on 17 February 1994 Hurley et al. (1994). Many of the high energy photons are delayed with respect to the bulk of the lower energy emission. There was no observed emission at lower energies from the burst region at that time, as observed by the *BATSE* experiment. A single 20 GeV photon was recorded from a burst direction which is the highest energy ever recorded from a GRB.

A search for unambiguous γ-ray line features with *BATSE/GRO* has thus far been unable to confirm the earlier reports of spectral line features from γ-ray bursts. Several papers from the 1993 Huntsville gamma-ray burst workshop proceedings (Fishman et al. 1994) and Palmer et al. (1994) discuss the preliminary *BATSE* line search analyses and their results. This work is still in progress.

The isotropy of the *BATSE* γ-ray burst distribution (Fig. 4), coupled with its inhomogeneity (as measured by the deficiency of weak γ-ray bursts) continues to be the most surprising recent observation of GRBs, and the one that has eliminated most of the expected and reasonable Galactic distribution models. The *BATSE* intensity distribution has been combined with the PVO intensity distribution to yield a combined data set of almost four decades in intensity (Fenimore et al. 1993). This composite intensity distribution matches well in the overlap region, showing a smooth transition to the $-3/2$ power law expected from a homogeneous distribution at the higher intensities and a clear deviation from this homogeneous distribution at low intensities.

The three distribution models most often discussed as being compatible with the observation of isotropy and inhomogeneity are: (1) An extended solar system model such as an Oort cloud or other sun-centered large distribution (cf. White 1993, 1994); (2) An extended Galactic halo or Galactic corona model, and (3) A cosmological model. A recent paper by Hakkila et al. (1994) shows that the parameter space available to the extended Galactic models is shrinking to unrealistic values, e.g. typical source distances of 70 kpc. Many believe that solar-system based models would likewise show an observable anisotropy. Thus, from the isotropy observations alone, one is increasingly forced to think in terms of cosmological models.

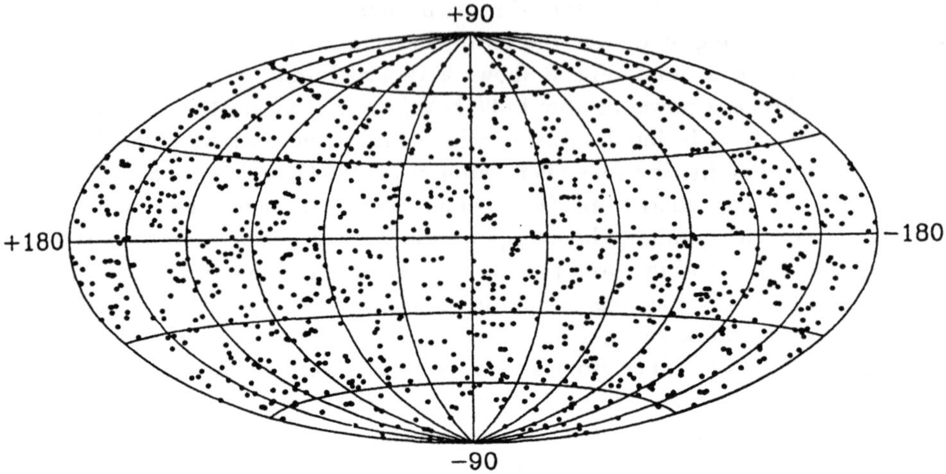

Fig. 4. The distribution of 921 GRBs observed by *BATSE* on *CGRO* in Galactic coordinates. More accurate burst locations will become available as the data are further processed for subsequent catalogs. The isotropy of the bursts is apparent.

3 Burst Counterparts

There is no doubt that a great advance in our understanding of GRBs can be attained through successful correlated observations of GRBs at other wavelengths. This fact was demonstrated recently by the combined γ-ray, X-ray, optical and radio observations of Soft γ-ray Repeaters (SGR's) (Kouveliotou et al. 1994, Murakami et al. 1994, Kulkarni et al. 1994). Within the past three years, there have been major, renewed efforts to find a counterpart to a GRB in other wavelength regions as evidenced by either simultaneous emission or afterglow emission. Comprehensive studies of archival plates also have been made. There have been several suggestions for counterparts although the results are problematic and are often debated. In view of the importance of the implied results, further observational evidence is needed before these results are accepted. Some of the world's most powerful ground-based facilities for radio and optical astronomy, high-energy air showers, atmospheric Čerenkov, and neutrino and gravitational wave astronomy are participating in these attempts for correlated burst observations. Space-borne correlated observations of well-located GRBs have also been attempted in the UV, EUV, and X-ray regions. A recent review of the present status of correlated GRB observations is given by Schaefer (1994).

Two different strategies have been employed to make follow-up observations to identify the GRB source object in the optical and other band passes. The first method is to view a fixed point in the sky in the hopes that a GRB will occur in the FoV. Because it is impossible to predict when and where a GRB will occur, very large fields-of-view (FoV) are required for the chance "observing" of a GRB. Because, in general, the large FoVs result in poor spatial resolution and crowded fields at the detector, this method has poor sensitivity.

The second method improves the sensitivity by limiting the FoV at the expense of making the observation with some time delay after the GRB. By waiting until after the GRB, rough positional information from instruments in orbit (e.g. *GRO-BATSE/COMPTEL*, *WATCH-GRANAT/EURECA*, IPN) can be obtained to direct narrower FoV instruments to the GRB error box. While the time delays between the GRBs and the observations have been steadily improving over the years – currently they are down to 5-to-36 hours range, still no positive identification of a GRB source object has been made. The general conclusion of these negative observations is that the optical emission from the source objects must have faded below $\approx$20th magnitude within several hours.

Since detector technology is unlikely to improve in the next few years to the point where the first method (stare mode) is likely to yield any new or different results, attempts are being made to improve the second method (follow-up mode). Clearly, the desired goal is to obtain accurate positional information on the GRB and to quickly get that information to instruments capable of making rapid follow-up observations.

The long-baseline interplanetary network (IPN) of burst detectors have provided the most accurate locations for these correlated observations. With the loss of both the *PVO* and *Mars Observer* spacecraft in recent years, the detectors aboard the *Ulysses* spacecraft remain the only remote, long baseline component that can be used with earth-orbiting detectors. The narrow (usually a few arcminutes) annuli derived from the *Ulysses*-Earth network observations can also be used in conjunction with coarser, single-spacecraft locations to greatly restrict the error box of some bursts. This situation will persist until the Russian *Mars 96* mission is launched. For non-IPN localizations, the *COMPTEL* and *EGRET* on the Compton Observatory can provide burst location accuracies of the order of one degree, for strong bursts with high-energy emission which happen to be in their field-of-view. *SIGMA-GRANAT* and *WATCH-GRANAT* also continue to provide accurate locations for accessible bursts. Those ground-based searches that can utilize the rather coarse *BATSE*-derived burst locations ($\approx 4^\circ$ for intense bursts) such as wide field Schmidt cameras, can also respond to these bursts quicker than ever before possible.

Since 1991, *BATSE* has had a quick alert capability that was developed to provide burst locations within several hours, under favorable conditions. A joint *BATSE-COMPTEL* capability also exists that is able to provide even more accurate ($\approx 1^\circ$) locations within several hours for those GRBs which also happen to be within the *COMPTEL* FoV. This capability was demonstrated for the intense GRB of 31 January 1993 (the Superbowl Burst), when an extraordinary effort involving over 30 instruments observed the burst region within hours and days of its occurrence (Schaefer et al. 1994). A sensitive, wide-field transient optical camera has also been operating for over three years at Kitt Peak (Vanderspek et al. 1994), responding to *BATSE*-triggered bursts.

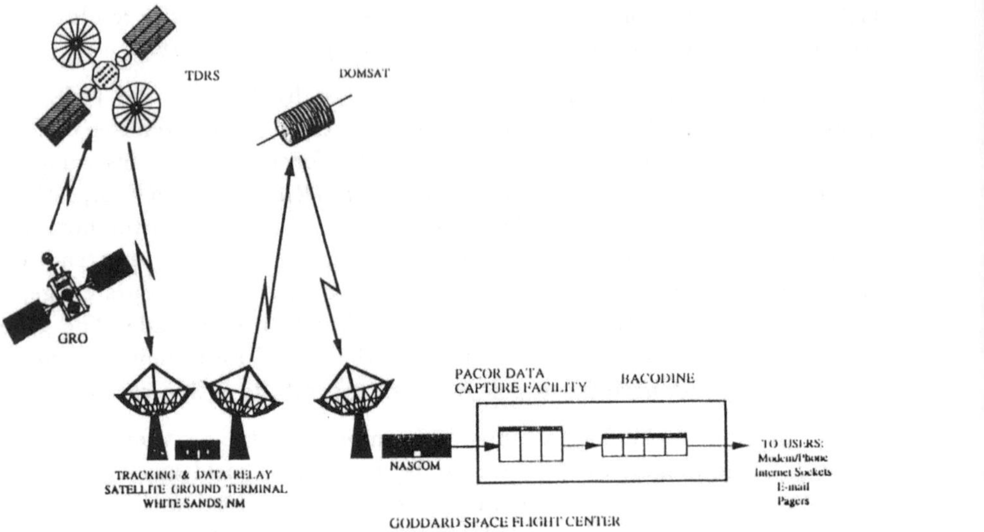

Fig. 5. Schematic diagram showing the various facilities and transmission paths and time delays in the *BACODINE* system.

4 The BACODINE System

A new, near-realtime *BATSE* burst location system called *BACODINE* (Barthelmy et al. 1994) became operational in 1993. This system, when linked to a rapid-slewing optical telescope, opens the exciting possibility of obtaining optical images of burst regions while the burst is in progress. Since the failure of the on-board tape recorders on *CGRO*, real-time transmission of the data from the spacecraft now allows for nearly continuous, real-time access to the *BATSE* data. These data can be used to make simultaneous or near-simultaneous multi-band observations of GRBs. The *BACODINE* system: (1) monitors the *CGRO* telemetry stream, (2) extracts the appropriate information from the *BATSE* portion of the data, (3) detects the occurrence of a GRB, (4) calculates the approximate coordinates for the burst, and (5) distributes those coordinates to observatories, other spacecraft and other interested parties around the world. This is done with custom hardware and software located at the Goddard Space Flight Center (GSFC) mission operations center for *CGRO*.

For those portions of the orbit where *CGRO* can get direct line-of-sight transmission of its high-gain antenna to any of the three *TDRS* satellites, the data are relayed to the NASA White Sands Ground Station in New Mexico where they are retransmitted to a domestic communications satellite (*DOMSAT*) and then transmitted back down to the GSFC Data Capture Facility (Fig. 5). There is an additional 1.0 seconds of time delay due to four hops of ground-to-geosync-orbit light-travel time and buffering within the White Sands facility. Once received at GSFC, the entire 2.048 sec of data is processed to yield GRB coordinates within 0.1 sec. The fastest method of coordinate distribution (see below) takes an additional 0.3 sec. If the GRB started during the beginning of the first 1.024-sec count-rate sample, then the total time delay between when the γ rays interacted with the *BATSE* LADs and when the coordinates are available at an instrument to make follow-up observations is 5.50 sec. If the GRB started at the end of the second 1.024-sec sample, then the time delay is 3.45 sec. More than half of the GRBs are longer than 5.5 sec, thus allowing follow-up observations to be made while the burst is still occurring.

Currently, all data from *GRO* is transmitted to ground in real-time (with essentially no time delay) with about 85% live-time. *BATSE* detects about 0.8 GRBs per day. Using the typical ground-based observational efficiencies encountered, the rate of making follow-up observations is 0.063 GRBs per day or once every 16 nights. If factors for new moon, for the brightest and for the longest GRBs are included; then the rate is 0.008 per day or once every 4.2 months.

The sequence of the processing of data in *BACODINE* can be broken down into five general steps: (1) The program monitors the telemetry stream continuously extracting the count rates for the eight *BATSE* detectors, comparing these rates to the calculated average current background rates. (It also extracts some general purpose housekeeping information in the process). (2) It monitors the "burst-in-progress" flag generated by the *BATSE* flight data processor and, when set true, it (3) takes the current count rates, subtracts the previously accumulated background rates to get the source-only rates, finds the three brightest detectors, and (4) solves the set of 3 simultaneous equations of the dot-product of the unknown burst direction and the detector normals of the three brightest detectors. (5) The burst direction is then sent to a list of instruments at various sites world-wide that are capable of making follow-up observations. Currently, the algorithm used to calculate the GRB direction assumes "ideal response" for the *BATSE* detectors. This approximation yields an uncertainty for the *BACODINE* burst position of about a 20° diameter error circle.

There are six possible methods for distributing the *BACODINE* GRB coordinates. The fastest method is the dedicated phone line. Around sunset at the instrument site (assuming it is an optical instrument), a phone/modem connection is made between the *BACODINE* computer and the computer at the instrument site. This connection is maintained throughout the night and should a burst occur during this time the coordinates (RA, Dec) are sent over the connection. At 9600 baud it takes 0.3 seconds. The second fastest method, and much less costly, is the Internet socket connection. Sockets is a technique to connect two programs running on two computers over a network. Like the dedicated phone method, the socket connection is made at some initial time and then maintained indefinitely. The time delay for the propagation of the coordinates packet varies due to the distance between the two computers, the number of routers and gateways in use, and the amount of other network traffic. However, it has been routinely shown that for a connection between Maryland and California (coast to coast US) the round-trip propagation time is 0.8 seconds average and 2.0 seconds maximum (1.2 and 5 sec, respectively, for Cambridge, England). It is difficult to make one-way travel time measurements on the Inter-

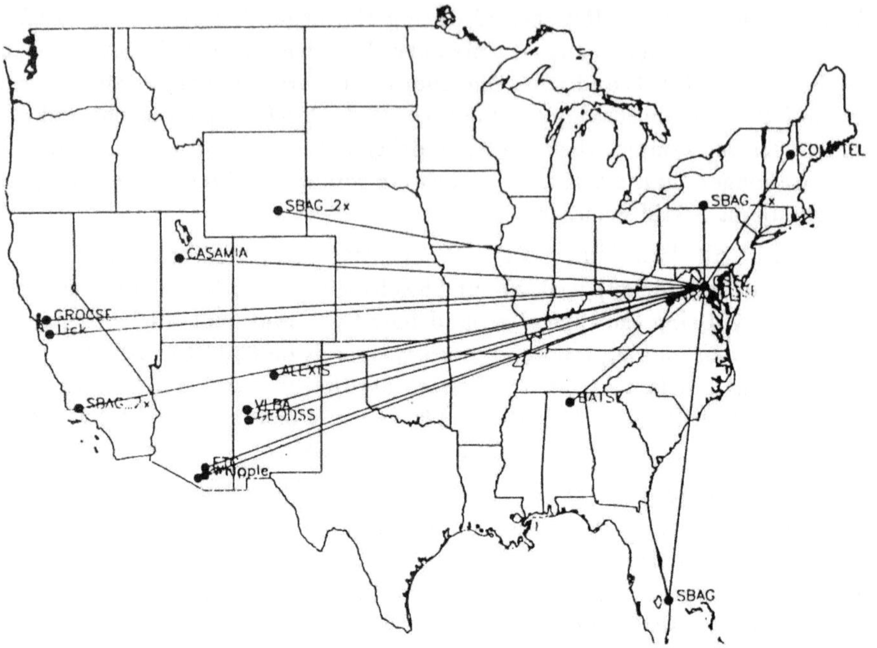

Fig. 6. A map of the US, showing the facilities that are presently receiving and responding to the *BACODINE*-distributed signals.

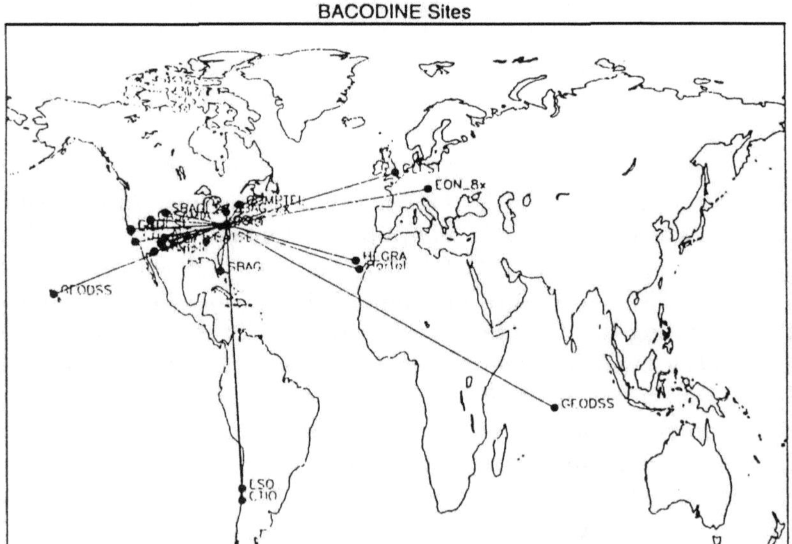

Fig. 7. World map showing the locations of *BACODINE* collaborating sites.

net, so we use round-trip travel times and assume that the one-way time is half the round-trip time. The e-mail method is relatively fast and suitable for sites with non-automated instruments (i.e. humans in the loop). The alpha-numeric pagers work well for sites without phone or Internet connections. The pager displays the RA, Dec location, the time and the initial intensity of the GRB. The pager companies accept the *BACODINE* notices via an Internet e-mail address and transmit the message to the designated pager unit. The entire process takes 1 to 2 minutes.

Currently, there are 21 groups collaborating with *BACODINE* at 34 sites around the world. Figs. 6 and 7 show US and world maps designating the locations of these ground-based instruments. Of the 21 groups, eight have optical instruments, four are radio, and three are in the very high-energy (TeV/PeV) γ-ray region. Of particular interest are several automated instruments which are fully computer-controlled and have rapid response and fast slewing times. Thus, they have the shortest delay time (5 to 30 sec range) to make an observation of a GRB error box. They are the LLNL GROCSE optical instrument in Livermore, CA; the MIT ETC optical instrument at Kitt Peak, AZ; and the CLFST radio instrument at Cambridge, England.

In addition to the normal operations with instruments that can provide rapid follow-up observations, *BACODINE* also provides rapid notification to other groups that are active in the GRB field (e.g. K. Hurley et al. for the Interplanetary Network (IPN); J. Block with the *ALEXIS* spacecraft, and the *GRO-OSSE & GRO-EGRET* teams). *BACODINE* also captures the *GRO-COMPTEL* telemetry data for those GRBs which are in the *COMPTEL* FoV. The program automatically ftp's the data to a computer at UNH where it is automatically analyzed. If the GRB is bright enough, a $1-2°$ *COMPTEL* location can be obtained in $15-30$ minutes. The *BACODINE* system also has the ability to "filter" or custom the notifications to each site according to several criteria and needs. The basic filtering is a visibility and or night-time requirement. If the site operates in the radio band-pass, then the *BACODINE* program will send that site a notice only if the calculated GRB location is visible ($> 10°$ above its local horizon). If the site is optical then it also requires that it be night at the site (the sun is more than 6° below the local horizon). An intensity threshold can also be applied. This threshold is used for instruments which have a small FoV and thus require more accurate locations to be effective.

We would like to encourage the participation of sites at many different longitudes and latitudes (in particular the Southern Hemisphere). We are currently working on improving the location accuracy by several methods. It is believed that the GRB position uncertainty will decrease to perhaps a few degrees radius for a strong burst. In the future, the *BACODINE* computers and networks will also capture and redistribute transient locations detected by the *HETE* spacecraft (to be launched in 1995) and by the IPN. Because these two GRB localization sources produce smaller error boxes, more traditional, narrow FoV telescopes with much fainter sensitivity can be used to make follow-up observations.

5 Summary

The GRB enigma appears to be as great now as it was twenty years ago (Ruderman 1975). A wealth of new data on time profiles, spectral characteristics and burst distributions has thus far failed to provided conclusive evidence on the distance scale, central object or emission mechanism for the classical GRBs. The isotropy and inhomogeneity of the bursts only shows that we are at the center of the apparent burst distribution. The recent *EGRET-CGRO* discovery of delayed GeV emission from a burst is yet another severe constraint for many of the burst models. Many feel that the identification of a burst with an object in another wavelength region may be the key to understanding these objects. The *BACODINE* system is now operational and may ultimately yield the long-awaited, unambiguous counterpart to a GRB.

References

Band D. et al., 1993, ApJ 413, 281
Barthelmy S. et al., 1994, Huntsville GRB Workshop, AIP 307, AIP, New York, p. 643
Bhat P.N., et al., 1992, Nature 359, 219
Fenimore E.E., et al., 1993, Nature 366, 40
Fishman G.J. et al., 1994, ApJS 92, 229
Fishman G.J., Brainerd J.J., Hurley K., 1994, Huntsville GRB Workshop, AIP 307, AIP, New York
Ford L., et al., 1994, ApJ 439, 305
Friedlander M. et al., 1993, Compton Gamma-Ray Observatory, AIP 280, New York
Hakkila J. et al., 1994, ApJ 422, 659
Hartmann D., 1995, The Gamma-ray Sky by SIGMA and CGRO, NATO ASI Proc.
Hurley K., 1994, Huntsville GRB Workshop, AIP 307, AIP, New York, p. 726
Hurley K., et al., 1994, Nature 372, 652
Kouveliotou C. et al., 1993, ApJ 413, L101
Kouveliotou C. et al., 1994, Nature 368, 125
Kulkarni S.R. et al., 1994, Nature 368, 129
Link B., Epstein R.I., Priedhorsky W.C., 1993, ApJ 408, L81
Murakami T. et al., 1994, Nature 368, 127
Nemiroff R.J., 1994, Huntsville GRB Workshop, AIP 307, AIP, New York, p. 730
Nemiroff R.J. et al., 1994, ApJ 423, 432
Norris J. et al., 1994, ApJ 424, 540
Paciesas W.S., Fishman G.J., 1992, Huntsville GRB Workshop, AIP 265, New York
Palmer D., et al., 1994, ApJ 433, L77
Pendleton G.N. et al., 1994, ApJ 431, 416
Ruderman M., 1975, Ann. N.Y. Acad. Sci. 262, 164
Schaefer B.E. et al., 1992, ApJ 393, L51
Schaefer B.E., 1994, Huntsville GRB Workshop, AIP 307, AIP, New York, p. 382
Schaefer B.E. et al., 1994, ApJ 422, L71
Share G. et al., 1994, Huntsville GRB Workshop, AIP 307, AIP, New York, p. 283
Vanderspek R., Krimm H., Ricker G., 1994, Huntsville GRB Workshop, AIP 307, AIP, New York, p. 438
Yoshida A. et al., 1989, PASJ 41, 509
White R.S., 1993, ApSS 208, 301
White R.S., 1994, Huntsville GRB Workshop, AIP 307 AIP, New York, p. 620

Observations of γ-ray Bursts and Solar Flares with Granat

O. Terekhov[1], *R.A. Sunyaev*[1], *D. Denisenko*[1], *A. Tkachenko*[1], *C. Barat*[2], *J.-P. Dezalay*[2], *R. Talon*[2], *N. Lund*[3], *S. Brandt*[3], *A.J. Castro-Tirado*[3]

[1] Space Research Institute (IKI), Profsoyuznaya 84/32, 117810 Moscow, Russia
[2] Centre d'Etude Spatiale des Rayonnements B.P. 4346, F-31029 Toulouse, France
[3] Danish Space Research Institute, Gl.Lundtoftevej 77, DK-2800 Lyngby, Denmark

1 Granat orbit

The *Granat* observatory was launched into a high apogee orbit on 1989 December 1 by a PROTON launcher. The initial *Granat* orbit had an apogee close to 200 000 km and a perigee of 2 000 km with an orbital period of 4 days. The satellite is entering the radiation belts for a few hours every orbit. Due to orbital evolution the perigee increased and the apogee decreased with time. After about 1.5 years of operation the perigee increased up to 2 0000 km. This has put the satellite completely outside of the proton radiation belts which makes the detector activation during the perigee passage negligible. Moreover, the satellite orbit is outside of the magnetosphere during the most parts of the mission. This makes such an orbit very attractive for high energy astrophysics missions which require low background level in order to achieve high sensitivity, and makes it essential for the investigation of the high energy transient phenomena such as cosmic γ-ray bursts and solar flares.

2 Instrumentation

The three instruments *PHEBUS* (6 BGO detectors situated at the different sites of the spacecraft), SIGMA (coded mask, position sensitive NaI detector with anticoincidence shield), and WATCH (rotation modulation collimatored all-sky monitor), are able to detect and investigate γ-ray bursts and high-energy solar flares in a broad energy range from 6 keV up to 100 MeV.

PHEBUS operates in the 100 keV – 100M eV energy range. Its sensitivity for gamma-ray burst detection is $\sim 10^{-6}$ erg cm^{-2} in almost 4π ster. The effective area of each detector is varying from 50 to 94 cm^2 depending on the GRB source position in the sky. All six detectors are recording the information about the burst simultaneously after a trigger which is provided by the *PHEBUS* electronics in the case if two detectors simultaneously detect count rate excesses at a level of 6σ in the 100 keV – 1.6 MeV energy range. Some prehistory of the background just before the γ-ray burst is also recorded.

The effective area of each Na I and Cs I scintillator detector of *WATCH* is 45 cm^2. The sensitivity of this instrument is S(6 – 120keV) $\sim 10^{-6}$ erg/cm^2. Four detector units of this instrument are mounted on the spacecraft along four tetragonal axes. The field of view of each detector is π ster.

The *SIGMA* coded mask telescope is able to detect γ-ray bursts both with Cs I anticoincidence shield detectors and with the positional sensitive detector with coded mask. The high effective area of the positional-sensitive detector gives the possibility to search (with high sensitivity of $(3-8)\times 10^{-8}$ erg/cm^2) and to investigate γ-ray bursts occurring in the 4.7×4.3 deg^2 field of view.

3 Cosmic γ-ray bursts

On the basis of the data sets published in two *PHEBUS* catalogues (Terekhov et al. 1994, 1995), the spatial distribution of the observed GRB sources was found to strongly deviate from a homogeneous one. From 118 cosmic γ-ray bursts observed by the *PHEBUS* instrument a $< V/V_{max} >= 0.366 \pm 0.012$ is derived. This clearly shows that there is deficiency of weak GRB sources.

The question about the presence of lines in γ-ray burst spectra is important for the determination of the nature of GRB sources. Using spectral information from several detectors it is possible to increase the significance of observed line features, and to discriminate against artificial lines due to scattering or absorption from nearby mechanical hardware placed on the spacecraft. Statistical analysis of large sets of γ-ray burst spectra shows that there is no evidence of statistically significant emission features in the 511 keV region and below at the level of few percent of the continuum emission intensity (Lobachev et al. 1995).

Simultaneous observations of cosmic γ-ray bursts in different energy ranges shows that strong spectral variability during GRBs takes place. The results of the analysis of GRB920723 revealed evidence for the presence of at least four differently varying spectral continuum components.

- A soft, slowly varying component is observable between a few keV – few tens of keV. This component probably is dominating the X-ray rich GRBs observed by WATCH (Castro-Tirado et al. 1994).
- Spectral analysis of the *SIGMA* (Pelaez et al. 1994) data as well as *BATSE* (Schaefer et al. 1992) results show that the energy spectra of some γ-ray bursts are more compatible with broken power law (different spectral components below and above a certain break energy) rather than with a single component law. Detailed analysis of the *PHEBUS* data for GRB920723 shows that both these two components are varying with time during the GRB event (Terekhov et al. 1995b).
- In addition to these three components an additional very hard power law component appears sporadically during the GRB event for very short (less than 1 s) time intervals (Terekhov et al. 1995b). During its presence, such a component dominates the whole energy range above 100 keV.

The *WATCH* instrument observations of γ-ray bursts show that in some cases such as GRB920723 and GRB900222, the decaying flux of soft X-ray emission in the 6 – 20 keV energy range follows the GRB event during at least 40 s after main gamma-ray pulse with a typical duration of 10 sec (Terekhov et al. 1993; Castro-Tirado et al. 1994). About 10% of the *WATCH* γ-ray bursts display X-ray tails or X-ray precursor activity. There is no preferential concentration for such X-ray rich events (for instance to the galactic plane or galactic center). The isotropic sky distribution was found for all sets of γ-ray bursts (32 events) localised by the *WATCH* instrument (Castro-Tirado et al. 1994).

WATCH is able to set upper limits for the X-ray flux from any point in the sky which is in its field of view. This means that it would detect any significant X-ray activity from γ-ray bursts which is precisely localised. The *WATCH* localisation accuracy is less than 1° error radius. No quiescent radiation is detected neither before nor after γ-ray burst events from the γ-ray burst sources. The upper limits for the quiescent emission fluxes from GRB920723 as well as from other GRB sources are $\sim$50 mCrab for time scales of hours and $\sim$20 mCrab for time scales of days.

A large fraction of the cosmic γ-ray bursts detected by the *PHEBUS* instrument are short. About 30% of the events have durations less than 2 s. Such events have energy spectra which are significantly harder than the energy spectra of the long γ-ray bursts. It is interesting to note that in spite of the short duration the time histories of such events are very complex. Time histories of short γ-ray bursts consist of individual flashes with duration 2 – 10 ms and with the separation between individual flashes of the order of 10 ms. Some individual pulses have very sharp front and decay edges with characteristic times 0.2 – 0.4 ms as in the case of GRB900320 (Terekhov et al. 1992).

The *SIGMA* telescope was pointed towards the galactic center for almost 18% of its observing time. For an object placed at the distance of the galactic center the threshold burst luminosity in the 40 – 90 keV energy range is 3×10^{38} erg s^{-1}, thus only slightly exceeding the Eddington luminosity for a neutron star. The absence of weak GRBs from the galactic center direction indicates that either the luminosity of typical GRBs is below 10^{38} erg cm^{-2} or, alternatively, the luminosity is higher than 10^{40} erg s^{-1} (Sunyaev et al. 1993). Note that recent observations of the *BATSE* instrument gave detailed constraints based on the absence of the GRB source clustering to the galactic plane and LMC reported by Meegan et al. (1994) for local disk and extended halo models (Hartmann 1995).

Observations of extragalactic sources with the *SIGMA* telescope gave constraints for the frequency of the appearance of GRB events for different extragalactic objects in a suggestion that GRBs are of extragalactic origin. Upper limits can be derived for the density of GRB sources per one solar mass of the matter for some of the extragalactic objects (Sunyaev et al. 1993).

4 High-energy solar flares

Granat carried out its observations during the solar maximum which took place in 1990-1991. During this time interval very high solar activity was observed by *Granat* instruments. The most interesting data on observations of high energy solar flares in the 100 keV – 100 MeV energy range were obtained by the *PHEBUS* instrument. Some of these bursts have also been observed up to 15 MeV by the *SIGMA* anticoincidence shield (Pelaez et al. 1992). Over 80 solar flare events with significant emission above 100 keV were detected during 1990-1993.

About 26% of these solar flares are associated with extremely intense GOES X-class events. Among these solar flares there are several extraordinal events with significant emission up to 100 MeV. Among the most interesting solar flares is SF900524 – a classical example of the high energy solar flare. During this flare a complex of high energy phenomena was observed which usually are associated with high energy solar flares: photons with energy up to 100 MeV, π^0 decayed photons, high energy neutrons, 2.2 MeV line associated with deuterium synthesis on the solar surface, fast protons appeared in a very short time after the beginning of the x-ray event (Terekhov et al. 1993,1995c; Talon et al. 1993).

In the flare SF910125 γ-rays up to 10 – 16 MeV are detected. No emission in the 65 – 124 MeV energy range is found for this flare. The flux of this flare in the energy range 1 – 4 MeV was by a factor of 30 lower than one for SF900524. It is interesting to note that according to GOES classification based on soft X-ray observations these two flares have nearly equal intensities: X9.3 for SF900524 and X10.0 for SF910125. The heliocentric angle was also not dramatically different for these flares: 81° for SF900524 and 77° for SF910125.

Some solar flares detected by the *PHEBUS* instrument have quasi-periodic pulsations of the intensity with the typical period of ~10 s. SF900524 and SF910125 also have such important feature in their light curves. The Q-factors (number of separate pulses with the typical period) of these two flares are different. While $Q = 16$ for SF910125, it is only 4-5 for SF900524. The modulation depth of the emission can be estimated as ~0.32 for SF900524 and ~0.27 for SF910125.

Deep modulation of hard X-ray and microwave emission with a period ~10 s was explained in terms of Alfvenic oscillations of a single flaring loop (Stepanov et al. 1992). In this model these pulsations are driven by Alfvenic oscillations of a flare loop excited due to upward motion of chromospheric evaporated plasma. According to this model one can estimate the density n and temperature T of the evaporated plasma as well as the flare loop magnetic field B and loop length L. It is possible to also estimate the pulsations (Stepanov et al. 1992), and that the lengths of the loops L of these two flares are nearly the same and equal to $\sim 10^9$cm ($L_9 \sim 1$). Based on this model it is possible to find that during SF900524 $T \approx 4.7 \times 10^7 L_9^2$ K which is about a factor two higher than in SF910125, $n \approx 1.5 \times 10^{12} L_9^3 G$ one order of magnitude higher and $B \approx 6.1 \times 10^2 L_9^{5/2} G$ which is 4 times higher than in the case of SF910125. These differences probably play an important role in extremely unusual properties of the SF900524 solar flare during which particles were accelerated up to very high energies.

References

Castro-Tirado A.J., Brandt S., Lund N., Lapshov I.Y., Terekhov O., Sunyaev R.A., 1994, Gamma-Ray Bursts, G. Fishman, J. Brainerd, K. Hurley (eds.), AIP Conference Proceedings 307, American Institute of Physics, New York, p. 17

Hartman D., 1995, The Lives of the Neutron Stars, NATO ASI C 450, M.A. Alpar, U. Kiziloglu, J. van Paradijs (eds.), Kluwer, Dordrecht, p. 495

Lobachev V.A., Terekhov O.V., Denisenko D.V., Sunyaev R.A., Kuznetsov A.V., Tkachenko A.Yu., Barat C., Dezalay J.-P., Talon R., 1995, Astr. Letters 21 (in press)

Meegan C., Fishman G., Wilson R., Brock M., Horack J., Paciesas W., Pendleton G., Briggs M., Koshut T., Kouveliotou C., Teegarden B., Matteson J., Hakkila J., 1994, Gamma-Ray Bursts, G. Fishman, J. Brainerd, K. Hurley (eds.), AIP Conference Proceedings 307, American Institute of Physics, New York, p. 3

Pelaez F., Mandrou P., Niel M., Mena B., Vilmer N., Trottet G., Lebrun F., Paul J., Terekhov O., Sunyaev R., Churazov E., Gilfanov M., Denisenko D., Kuznetsov A., Dyachkov A., Khavenson N., 1992, Solar Phys. 140, 121

Pelaez F., Bouchet L., Jourdain E., Niel M., Claret A., Laurent P., Lebrun F., Paul J., Terekhov O., Sunyaev R., Kuznetsov A., Denisenko D., Gilfanov M., Churazov E., Khavenson N., Diachkov A., 1994, ApJS 92, 651

Schaefer B.E., Teegarden B., Cline T., Fishman G., Meegan C., Wilson R., Paciesas W., Pendleton G., Matteson J., Band D., 1992, ApJ 393, L51

Stepanov A.V., Urpo S., Zaitsev V.V., 1992, Solar Phys. 140, 139

Sunyaev R., Churazov E., Gilfanov M., Terekhov O., D'iachkov A., Khavenson N., Kovtunenko V., Kremnev R., Claret A., Lebrun F., 1993, ApJ 402, 579

Talon R., Trottet G., Vilmer N., Barat C., Dezalay J., Sunyaev R., Terekhov O., Kuznetsov A., 1993, Solar Phys. 147, 137

Terekhov O.V., Sunyaev R.A., Kuznetsov A.V., Molchanov A.Yu., Barat C., Vedrenne G., Dezallay J.-P., Niel M., Talon R., 1992, Frontiers of X-ray Astronomy, Proceedings of the YAMADA Conference XXVIII, Frontiers Science Series - 2, Y.Tanaka, K. Koyama (eds.) Universal Academy Press, Tokyo, p. 253

Terekhov O.V., Sunyaev R.A., Kuznetsov A.V., Barat C., Talon R., Trottet G., Vilmer N., 1993, Astr. Letters 19, 163

Terekhov O.V., Lobachev V.A., Denisenko D.V., Lapshov I.Y., Sunyaev R.A., Lund N., Castro-Tirado A.J., Brandt S., 1993, Astr. Letters 19, 686

Terekhov O.V., Denisenko D.V., Lobachev V.A., Sunyaev R.A., Kovtun A.V., Kuznetsov A.V., Barat C., Dezalay J.-P., Talon R., 1994, Astr. Letters 20, 323

Terekhov O.V., Denisenko D.V., Lobachev V.A., Sunyaev R.A., Kuznetsov A.V., Tkachenko A.Yu., Barat C., Dezalay J.-P., Talon R., 1995a, Astr. Letters 21 (in press)

Terekhov O.V., Lobachev V.A., Sunyaev R.A., Kovtun A.V., Denisenko D.V., Kuznetsov A.V., Barat C., Dezalay J.-P., Talon R., 1995b, Astr. Letters 21 (in press)

Terekhov O.V. et al., 1995c, Astr. Letters 21 (in press)

Observations of High-energy Gamma-ray Bursts with EGRET

M. Merck[1], *D.L. Bertsch*[2], *B.L. Dingus*[2,3], *C.E. Fichtel*[2], *R.C. Hartman*[2], *S.D. Hunter*[2], *G. Kanbach*[1], *D.A. Kniffen*[4], *Y.C. Lin*[5], *H.A. Mayer-Hasselwander*[1], *P.F. Michelson*[5], *C. von Montigny*[2], *R. Mukherjee*[2,3], *P.L. Nolan*[5], *M. Pohl*[1], *P. Sreekumar*[2,3], *E.J. Schneid*[6], *D.J. Thompson*[2]

[1] Max-Planck-Institut für extraterrestrische Physik, D-85748 Garching, Germany
[2] NASA/Goddard Space Flight Center, Greenbelt, MD 20771, USA
[3] Universities Space Research Association, NASA/GSFC, Greenbelt, MD 20771, USA
[4] Hampden-Sydney College, P.O. Box 862, Hampden-Sydney, VA 23943, USA
[5] Hansen Experimental Physics Laboratory, Stanford University, Stanford CA 94305, USA
[6] Northrop Grumman Corporation, Mail Stop A01-26, Bethpage, NY 11714-3580, USA

1 Introduction

Gamma Ray Bursts (GRBs) have puzzled astronomers since their discovery more than 20 years ago. As no counterparts at wavelengths other than X- and γ-rays have yet been found the identification of the sources is still missing. Theoretical explanations range from colliding comets (1993) and merging neutron stars (1982) to more exotic objects, such as superconducting cosmic strings (1988). Data accumulated until now still do not discriminate between these models, although results from the *BATSE* (Burst and Transient Source Experiment) instrument aboard the Compton Gamma Ray Observatory (*CGRO*) strongly favor extragalactic models.

The Energetic Gamma Ray Experiment Telescope (*EGRET*) aboard *CGRO* has s ofar detected photons from 5 GRBs with its spark chamber. These are the highest energy γ-rays associated with GRBs to date. In this work we review previously published data and summarize the properties of these events. Elsewhere we present possible constraints from the data on the models proposed to explain GRBs.

2 EGRET as a burst detector

The *EGRET* instrument consists of a large area particle tracker, a spark chamber with interspaced tantalum foils to convert the γ-rays into electron positron pairs, a trigger telescope used to detect the presence of an event and to remove upward going particles by time of flight discrimination, and an energy calorimeter dubbed TASC (Total Absorption Shower Calorimeter) using a NaI (Tl) crystal read out with photomultipliers. An anticoincidence dome (A-dome) of plastic scintillator material encloses most of the instrument to discriminate

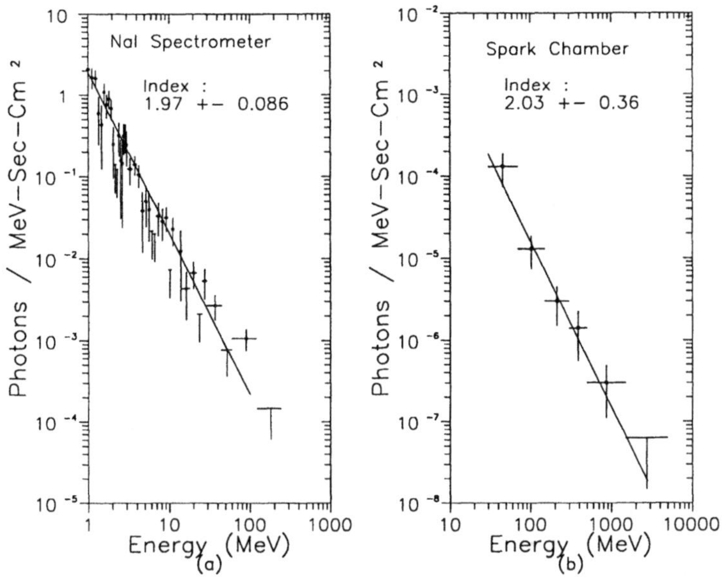

Fig. 1. Energy spectra of GRB 930131. (a) Energy spectrum from the TASC for the 1 second time interval beginning at 18:57:11.708 UT of 1993 January 31. A power law fit given by $(1.90 \pm 0.26)(E\,/\,1MeV)^{-1.97\,\pm\,0.09}$ photons/(cm^2 s MeV) is also shown. (b) Energy spectrum for the 16 γ-rays detected with the spark chamber within 25 s after the burst started. The power law fit is given by $(7.4 \pm 1.8) \cdot 10^{-6}(E\,/\,147MeV)^{-2.03\,\pm\,0.36}$ photons/(cm^2 s MeV).

against charged particles. A detailed description of *EGRET* can be found in Kanbach et al. (1988) and a discussion of the calibration and performance of the instrument was given by Thompson et al. (1993).

For the detection of GRBs, the A-dome, the TASC calorimeter, and the spark chamber can be used. The time profile of a burst can be obtained from the observed count rate in the A-dome. The plastic scintillator is sensitive to low energy γ-rays with energies higher than 50 keV. This rate is sampled with a time resolution of 0.256 seconds. The background rate in this detector is quite high and depends on the orbital position of the spacecraft. A careful treatment of the data is therefore needed to obtain the net burst time profile.

Spectra of bursts in the energy range between 1 and 100 MeV can be measured with the TASC crystal. During normal operations pulse-height spectra of 32.78 s duration are accumulated continuously and transmitted to Earth. After a BATSE trigger a special burst mode aquires four sequential spectra of 1, 2, 4 and 16 seconds duration. After background subtraction and correction with the response function of the calorimeter, derived from a detailed spacecraft mass model (for details see Schneid et al. 1993), energy spectra are derived. As an example we show in Fig. 1a the spectrum obtained during the first second after the *BATSE* trigger of GRB 930131 (Sommer et al. 1994). Although in general only a few photons are detected in the spark chamber, it is possible to derive an energy spectrum from them. In Fig. 1b we show the spectrum of GRB 930131

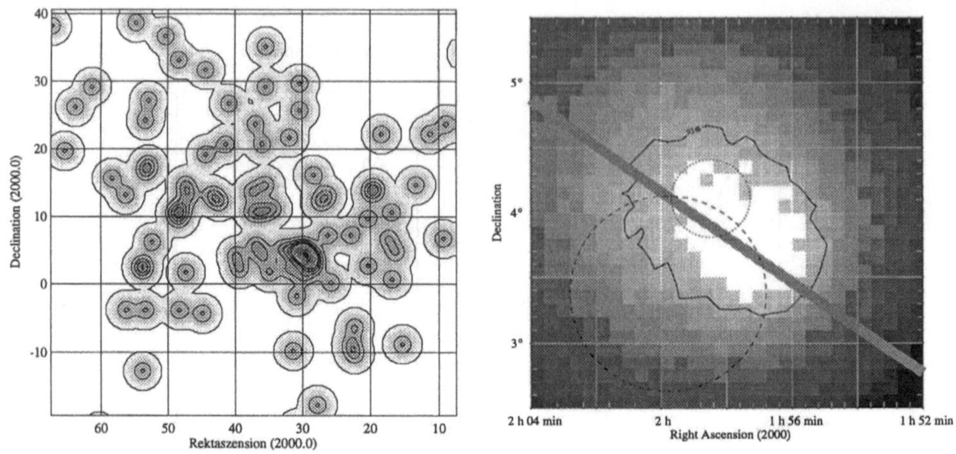

Fig. 2. Position of GRB 940217. (a) Raw image of photon arrival directions. (b) Image of likelihood analysis. The solid line corresponds to the 95% confidence contour; the small dotted circle is the error radius of the 18 GeV photon detected 4733 s after the *BATSE* trigger; the dashed circle is the 2σ *COMPTEL* burst location. The arc results from triangulation of *Ulysses* and *BATSE* arrival times. (Hurley et al. 1994)

derived from the imaged photons.

A burst location can be derived from the imaged photons detected in the spark chamber. The effective area of this detector increases from the threshold energy of 30 MeV to a maximum value of 1500 cm^2 at a few hundred MeV. As an example we show in Fig. 2a a raw image of GRB 940217 with single photon directions smoothed with a mean point spread function that was derived from calibration measurements. A few single photons are scattered over the full FoV. The GRB is clearly visible although only a few photons were detected. In Fig. 2b we show for the same burst the likelihood image after correcting for exposure. The 95% C.L. countour derived from this analysis is shown as a solid line (Hurley et al. 1994). We refer to the Figure caption for a detailed describtion of the other circles.

3 The 5 GRBs detected with EGRET

EGRET detected spark chamber photons from GRBs on May 3, 1991, June 1, 1991, January 31, 1993, Febraury 17, 1994 and March 1, 1994. Details on these detections can be found in Schneid et al. (1992) (GRB 910503), Kwok et al. (1993) (GRB 910601), Sommer et al. (1994) (GRB 930131), Hurley et al. (1994) (GRB 940217) and Schneid et al. (1995) (GRB 940301). We summarize some features of these bursts in Table 1. For the first time it was possible with *EGRET* to measure the very high energy tail of GRB spectra. Emission up to several GeV was detected. All spectra can be fit with a power law above 1 MeV. Strikingly, the spectral index derived from these fits is different for each GRB.

Table 1. Properties of the 5 GRBs detected with the *EGRET* spark chamber

GRB	Position (J2000.0) RA	Dec	Max. Energy	Specrtal Index	Duration	Delayed Emission
GRB 910503	$05^h51.2^m$	38.6°	10000 MeV	-2.2 ± 0.1	3 s	10 GeV γ 84s after burst
GRB 910601	$20^h39.6^m$	32.3°	314 MeV	-3.7 ± 0.2	33 s	10 delayed γs in 400 s
GRB 930131	$12^h15.1^m$	10.2°	1200 MeV	-2.0 ± 0.4	25 s	2 γs in next 70 s
GRB 940217	$01^h58.1^m$	4.0°	18000 MeV	-2.5 ± 0.1	180 s	18 γs up to 5400 s after burst
GRB 940301	$07^h12.0^m$	65.1°	160 MeV	-2.5 ± 0.1	21 s	no

The most surprising result was the detection of delayed emission of $\geq$30 MeV to GeV photons long after some bursts, when all lower energy activity had ceased. The most prominent example of this is GRB 940217 from which 10 photons were detected in the spark chamber during the burst duration of 180s. Another 8 γ-rays were registered from the same direction during a 600s interval following the burst, while one expects only 1.8 events from the background rate. Past this time the spacecraft entered earth occultation and data could not be taken; after emerging from the earth shadow (4700 s after the burst trigger) another 10 photons whose directions were consistent with the burst location were detected, while only 2.9 are expected from the background. During this period of delayed emission a photon with 18 GeV energy was observed. During the whole period of 5400 s of delayed high energy emission following the burst no low energy activity (25-150 keV) could be detected by the *Ulysses* γ-ray detector. More details on this extraordinary burst are given in Hurley et al. (1994). Although emission of delayed high-energy photons was most significant during this peculiar burst, there is some indication of delayed emission on several other bursts, as summarized in Table 1 (see also Dingus et al. 1994).

The typical fluence above 20 keV of the GRBs detected with *EGRET* is 10^{-5} erg cm^{-2}. These events are the most intense GRBs detected by *BATSE*, and which are also in the FoV of *EGRET*. This suggests that bursts with much lower fluences may also have spectra that extend to high energies, but perhaps at flux levels that require next generation instruments for detection.

4 Discussion

Although only 5 GRBs were detected with *EGRET* so far, some interesting physical constraints can be set. The power law spectra found up to several hundred MeV rule out all models predicting blackbody or optically thin thermal spectra. The small size of the emission region implied by the observed short temporal variations and the high photon densities derived from the observed fluxes should lead to a significant attenuation of high energy photons due to photon-photon absorption in an isotropically emitting source. The fact, that no attenuation was detected up to several GeV (e.g. in the MeV range) can only be explained if the emission comes from highly collimated beams or ultra-relativistically expanding shells. For a typical source distance of 50 kpc (e.g. locations in the Galactic halo) the observation of GeV photons during the main burst interval requires bulk Lorentz factors of $\gamma \geq 25$ whereas for cosmological sources at distances around 1 Gpc $\gamma \geq 1000$ is necessary. One should remember that the highest bulk Lorentz factors observed so far are from jets in AGN with typical values of only ~ 10. Thus the very high Lorentz factors needed in GRBs are a challenge for most of the extragalactic models which on the other hand are favored by the isotropic and homogeneous distribution of GRBs. Only models that explain GRBs with cusps on superconducting cosmic strings as discussed by Paczynski (1988), although very speculative, provide a natural way to obtain very high bulk Lorentz factors.

The observation of photons up to 18 GeV also sets an upper limit on the distance of the sources due to the lack of absorption of these photons by the microwave and infrared/optical photon fields. A maximum redshift of $z \leq 100$ is still compatible with all proposed models and only detection of TeV photons by ground based detectors could possibly constrain this distance sufficiently to rule out some cosmological interpretations.

The delayed emission observed in some GRBs has been discussed by Katz (1994) and Mészáros & Rees (1993). In their models GRBs are due to merging neutron star binaries, which leads to shock acceleration and interaction of accelerated particles with some surrounding material. These models still need some tuning to explain the widely different spectral indices found in the burst spectra and the very high bulk Lorentz factors needed. More experimental data at high photon energies would certainly advance our understanding of the GRB puzzle.

References

Dingus B.L., et al., 1994, in Gamma Ray Bursts, G. Fishman, J. Brainerd & K. Hurley (eds.), AIP Conference Procedings 307, 22
Eichler D., Livio M., Piran T., Schramm D., 1989, Nature 340, 126
Hurley K., et al., 1994, Nature 372, 652
Kanbach G., et al., 1988, SSR 49, 69
Katz J.I., 1994, ApJ 432, L27
Kwok P.W., et al., 1993, AIP Conference Procedings 280, 855
Mészáros P., Rees M.J., 1994, MNRAS 269, L41
Paczynski B., 1988, ApJ 335, 525
Schneid E.J., et al., 1992, EGRET Detection of High Energy Gamma Rays from the May 3, 1991 Gamma Ray Burst. A&A 255, L13
Schneid E.J., et al., 1993, AIP Conference Procedings 280, 850
Schneid E.J., et al., 1995, (in preparation)
Sommer M., et al., 1994, ApJ 422, L63
Thompson D.J., et al., 1993, ApJS 86, 629
White R.S., 1993, ApSS 208, 301

Population of Short γ-ray Bursters – Result of Gravitational lensing of "Quiet" γ-ray Sources in Remote Galaxies

G.M. Beskin

Special Astrophysical Observatory, Nizhnij Arkhyz, Karachaevo-Circassia, 357147 Russia

1 Introduction

We consider the hypothesis of the existence of a class of short (10^{-2}–10^{-3}s) and weak (10^{-6}–10^{-7}erg/s$\cdot$ cm^2) γ-bursts of cosmological origin. These events arise from gravitational microlensing of bright, quiet γ-ray sources (with luminosity about $3 \cdot 10^{38}$erg/s) located in distant ($z \geq 1$) galaxies by stars inside foreground galaxies. If the size of the γ-ray sources is 10^4–10^6cm, motion with respect to the observer or the intermediate galaxy may bring them close to the neigborhood of cusps along critical curves generated by individual stars (corresponding to caustics in the observer's plane). In these cases the observed fluxes increase by a factor 10^{11}–10^{13}, the temporal variation of intensity being due to the brightness distribution in the source and to the distance from the cusp (in the source plane). Using standard assumptions for the number and sizes of "quiet" γ-ray sources, the number and characteristics of galaxies, we estimate the observed burst rate to be between 1 and 10 bursts per year (see below). Their temporal and spectral properties are determined by the nature of sources (ejecting pulsar, accreting neutron star with a weak or strong magnetic field, black hole) and by their velocity relative to the cusp. Similar effects have been discussed in the context of flux variability and structure studies of quasars (e.g. Schneider & Weiss 1986, Grieger et al. 1988, Refsdal & Surdej 1994). However, in these cases the source sizes were larger than 10^{14}cm^3 and amplification did not exceed 10^2.

2 Maximum amplification and light curves

The effects of microlensing, splitting of the image of a distant object and amplification of its brightness, are determined by the combined action of a smoothed gravitational potential of the lensing galaxy and of the stars nearest to the line of sight. A system of intersecting caustics and cusps is formed in the observer's plane, while in the source plane a corresponding system of critical lines (anti-caustics) and cusps is formed (Chang & Refsdal 1984, Paczynski 1986, Schneider & Weiss 1986). We introduce the following notation: D_S – distance to a remote galaxy, D_L – distance to the intermediate galaxy, $D = (D_S - D_L)D_L/D_S$ – effective distance and R_g - Schwarzschild radius ($R_g = \frac{2GM}{c^2} = 3 \cdot 10^5 M/M_\odot$cm for a star with mass M).

The amplification, $K(x)$, for a point source near the cusp is a complex function of x(x is the distance to the cusp in the direction along the normal to the critical curve) (Blanford & Narayan 1986, Schneider & Weiss 1986), which for $x \to 0$ has the asymptotic behavior: $K_0(x) \approx (D_S/D_L) \cdot (\sqrt{2R_gD}/x)$ For the purpose of this study it is important that $K(x)$ is proportional to $A = (D_S/D_L) \cdot \sqrt{2R_gD}$, which determines the amplification factor. Let us evaluate A for $z_S \sim 2$, $z_L \sim 0.5$, employing the expression for the angular distance in a cosmological model with $\Omega = 0$ (Turner et al. 1984): $D_z = (c/2H_0) \cdot [1-(1+z)^{-2}]$, where c is the velocity of light, $H_0 = 50 km/s \cdot$ Mpc. In that case, $A \sim 5.5 \cdot 10^{16}$, and at $x \sim 10^4 - 10^5$ cm (the probable size of a "quiet" source), we have $K_0 \sim 5 \cdot 10^{12}$ - $5 \cdot 10^{11}$. The light curve of the source is determined by a convolution of its brightness distribution with the amplification function $K_0(x)$.

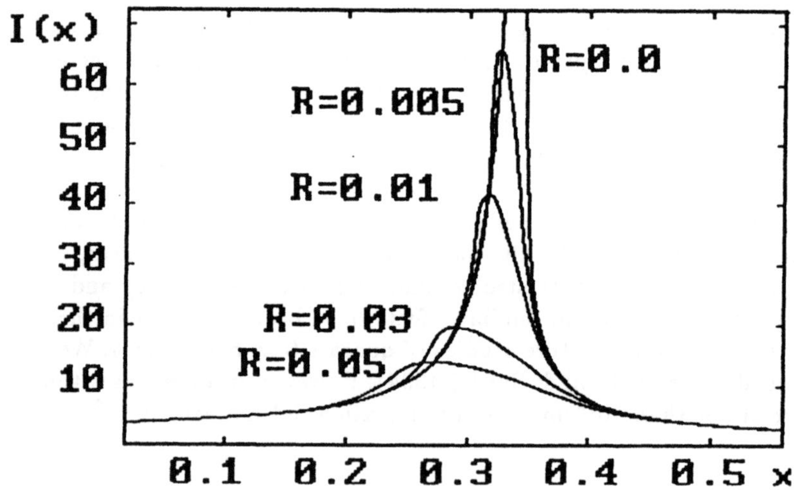

Fig. 1. Light curves of sources with uniform surface brightness (R is the size of the source region).

In the figure (taken from Schneider & Weiss 1986) curves are shown for different sizes of a source of uniform surface brightness (R and x are measured in $\sqrt{2R_gD} \sim 3 \cdot 10^{16} cm$ for the cases considered here). With decreasing R

the maximum intensity rises steeply and it is believed that at $R \sim 3 \cdot 10^{-12}$ (or 10^5 cm) amplification could reach 10^{12}. It should be emphasized that this situation may occur when the distance of a γ-ray source moving with respect to the axis "observer – lensing star" does not exceed the proper size of the source. In this case the "strong effect zone" scans the region of the observer. The dimension of this zone, d_{obs}, is simply related to the diameter of the source d_S: $d_{obs} = d_S \cdot D_L/(D_S - D_L)(1 + z_L)$, and since the velocities of source and observer along the cusp normal are related in a similar manner (Griger et al. 1988), the duration of the burst is given by the transit time of the observer across the "strong effect zone" $\tau \sim d_{obs}/v_{obs} = d_S/v_S$. Taking $v_S \sim 10^8 cm/s$, a typical velocity for intergalactic motions, we obtain $\tau \sim 10^{-3}\left(d_S/10^5\right) s$

Thus, if the real luminosity of the γ-source is close to the Eddington limit ($3 \cdot 10^{38}$erg/s), it may be observed as an amplified burst with maximum luminosity of $\sim 10^{51}$erg/s. To detect such events at cosmological distances the flux must exceed a threshold of $\sim 10^{-7}$erg/$cm^2 \cdot s$ (Mao & Paczynski 1992).

3 Number of events

The number of γ-bursts of the kind considered here registered during a time T is $N_0 = \alpha N_G N_\gamma N_C$, where N_G is the number of galaxies which contain N_γ γ-sources of Eddington luminosity each; α is the degree of overlap of remote galaxies and intermediate galaxies; N_C is the number of cusps passed by the γ-source.

It is clear that $N_C = mN_*$, where m is the mean number of cusps per star ($m \sim 3 - 6$), $N_* = n_* d_S v_S T$, where n_* is the surface density of stars in the intermediate galaxy.

At $n_* \sim 4 \cdot 10^3 pc^{-2}$ (Refsdal & Surdej 1994), $v_S \sim 10^8$ cm/s, $T \sim 1$ year $= 3 \cdot 10^7 c$, we obtain $N_c \sim 10^{-13} \cdot (d_s/10^5)$ year^{-1}. Counts of faint galaxies ($V > 25^{\mathrm{m}}$) suggest that the total number of galaxies with $z \geq 0.5$ is $10^{10} - 10^{11}$, and one can argue that the degree of their overlap is close to unity (Tyson 1988, Lilly et al. 1991, Peterson et al. 1991). Assuming that $\alpha \sim 1, N_G \sim 10^{11}$ and $N_\gamma \sim 5 \cdot 10^2$, we obtain finally $N_0 \sim 5$ year^{-1}. Taking into account the approximate character of these estimates, we suggest a possibility of detecting from 1 to 10 γ-ray bursts per year.

4 Critical tests

We point out some properties of γ-ray bursts of the type considered in this work:
(a) An inverse correlation between the duration and peak intensity should be observed.
(b) In most cases the spectrum must be harder at the peak of the intensity.

As has already been pointed out, the "strong effect zone" of gravitational lensing shifts moves to the neighbourhood of the observer, similar to the beam pattern of a pulsar, as a result of the relative motion of the γ-ray source, lensing star and observer.

If the distance "lens – γ-ray source" and "observer – lens" is of the same order, the size of the zone and the speed of its movement are close to the size of the source and the velocity of its motion relative to the lens. According to the properties of cosmological γ-ray bursts it follows that they could not be observed simultaneously by several cosmic detectors in the *KONUS* experiment. Such events have probably already been registered by single γ-ray telescopes.

To critically test this γ-burst model we suggest to measure the time delay when the same events are registered with detectors separated in space. If the increase of the counting rate is associated with the time variations of the radiation flux in the spherical wave, the time delay τ_d is determined by formula $\tau_d = l_a/c$, where l_a is the distance between detector and γ-burst and c is the velocity of light. However, when the γ-burst is caused by detectors passing through the zone of strong effects, one finds $\tau_d = l_t/v$, where l_t is the distance between detectors perpendicular to the γ-ray burst direction, v is the velocity of the strong effect zone. To carry out the test, it is necessary to use at least four detectors, located at the tips of an equilateral pyramid with sides of $\sim 10^5$ cm.

References

Blandford R., Narayan R., 1986, ApJ 310, 568
Chang K., Refsdal S., 1984, A& A 132, 168
Griger B., Kayser R., Refsdal S., 1988, A&A 194, 54
Lilly S.J., Cowie L.L., Gardner J.P., 1991, ApJ 369, 79
Mao S., Paczynski B., 1992, ApJ 388, L45
Paczynski B., 1986, ApJ, 301, 503
Peterson B.A., D'Odorico S., Tarenghi M., Wampler E.J., 1991, ESO Messenger 64, 1
Refsdal S., Surdej J., 1994, Rep. Prog. Phys. 56, 117
Schneider P., Weiss A., 1986, A&A 164, 237
Turner E.L., Ostriker J.P., Gott III J.R., 1984, ApJ 284, 1
Tyson J.A., 1988, AJ 96, 1

Multi-wavelength Flashes from GRBs

Dieter H. Hartmann

Department of Physics and Astronomy, Clemson University, Clemson, SC 29634, USA

1 Debates

The mystery of the nature and origin of gamma-ray bursts (GRBs) has persisted over two decades of intensive research, observational and theoretical. The basic problem is the absence of quiescent or transient counterparts at other wavelengths, resulting in distance scale uncertainties of more than 10 orders of magnitude. Searches for counterparts are hampered by the lack of sufficiently accurate localizations and by a lack of criteria that uniquely identify the expected appearance of counterparts (although see Meszaros & Rees 1993, Meszaros et al. 1994). The detection of archival optical transients (OT's) inside γ-ray burst error boxes suggested burst recurrence and generated hope for a breakthrough in the γ-ray burst counterpart search (e.g., Schaefer 1990). However, counterparts of these OT's have never been identified, and the reality of several archival images was questioned (Zytkow 1990, but see Schaefer 1990). The search for multi-wavelengths transients has recently been extended to neutrino- and gravitational wave emission. The status of multiwavelength surveys for quiescent or transient counterparts prior to 1988 was reviewed by Hartmann & Woosley (1988). Here we update that review and discuss "optical" emission in view of the current debate on the burster distance scale. Other recent reviews are by Schaefer (1994) and Greiner (1995).

The recent transition from the original paradigm of Galactic neutron stars to the new paradigm of cosmological sources was generated by global statistical properties of GRBs. The non-uniform brightness distribution combined with the angular isotropy argues against Galactic source distributions and favors cosmological origins. *BATSE* data improve the global statistical constraints at an approximate rate of one burst per day, but the detection and unambiguous identification of even a single GRB counterpart would carry more weight than all statistical constraints taken together. If GRBs are indeed associated with galaxies, as some models suggest, we would expect to find and identify them inside the error boxes, at least for very accurate positions. We take inventory of GRB locations and discuss implications for the models.

From their statistical properties we conclude that bursts either reside in an extended halo (100 kpc), or are extragalactic (100 Mpc) or possibly cosmological. Very local origins (Oort cloud, nearby section of Galactic spiral arms, etc.) appear to be ruled out by the data (e.g., Briggs et al. 1995). The multiwavelength signatures of GRBs obviously depend on the nature of the energy release mechanisms and the environment of the source if reprocessing of some fraction of the radiation should occur. The archival data on optical transients suggest, if one believes their reality, that a significant fraction of the burst energy is converted to low energy bands. These data also suggest that bursts do repeat on timescale of order years. Is that claim consistent with the observed angular distribution? There are only three well established cases of burst repetition: GB790305b (the "March 5 event"), GB790107 (the "soft repeater"), and GB790324. All three events have in common that they are relatively short in duration and that their mean photon energies are less than average. These characteristics led to a proposed class of "soft gamma-ray repeaters (SGRs)", that is different from the class of "classical bursts (GRBs)" (Kouveliotou 1994). Much progress has recently been made in the identification of their Galactic counterparts, but space limitations of this paper prevent a more detailed discussion of this sub-group.

There is no convincing evidence for recurrence of classical GRBs (Meegan et al. 1995), but the analysis is somewhat hampered by poor angular resolution. Model-dependent statistical estimates of recurrence from pre-*BATSE* data suggested time scales of $\sim$ 10 yrs, but analyses of early *BATSE* data led to the suggestion of much shorter recurrence times. If bursts indeed originate at large redshifts, the merger of compact stellar objects (neutron stars, black holes, white dwarfs) appears to be a natural GRB site. Since these events must occur in nature and since they are rare, we must sample a significant fraction of the universe to be consistent with the observed event rate of $\sim 10^3$ per year. Mergers are expected to occur at a rate of $\sim 10^{-6}$ yr^{-1} per host galaxy, so that recurrence over the lifetime of γ-ray detectors is not expected. Positional coincidences on the sky are thus expected to be random. Deviations from randomness can be caused by non-uniform sampling of the sky, spatial clustering of the underlying source population, large scale anisotropies, or burst repetition. Since the latter case would rule out some of the currently favored cosmological scenarios, the question of burst recurrence needs careful consideration.

Quashnock & Lamb (1993) presented arguments for clustering on small angular scales for *BATSE* burst data contained in the 1B catalog and suggested that multiple recurrences of a significant fraction of bursts were responsible for the observed excess. A possible correlation of bursts in both space and time was suggested by Wang & Lingenfelter (1993) who isolated GBS 0855−00 as a potential location of multiple recurrences. If a significant fraction of bursts indeed repeat on short timescales, the angular distribution would be distorted on scales comparable to the angular resolution of the instrument. Localization accuracy of *BATSE* is currently limited by systematic effects to $\sim 4^\circ$. Low order multipole analysis does not resolve these scales, so that different methods must be employed to test for small angle clustering. Two tools suitable for this task are the angular correlation function and the nearest neighbor statistic. Using these methods, and others, the *BATSE* 2B data were searched for recurrences (Meegan et al. 1995, Brainerd et al. 1995, Hartmann et al. 1995). Although recurrence of some bursts can not be ruled out, these studies suggest that frequent recurrence of a large fraction of all bursts does not occur.

2 Localizations

The largest and most uniform database of burst positions is now available from the *BATSE* experiment. Over 1,200 burst have been recorded since the launch of Compton Observatory (*CGRO*). *BATSE* determines directions to bursts from the relative countrates in the eight detector modules, using an essentially $\cos(\theta)$ law for the dependence of count rate on the angle of incidence. Scattering of burst photons in the spacecraft as well as reflection off the Earth atmosphere create a systematic uncertainty that amounts to about 4°. The statistical uncertainties can be less than a degree for very bright bursts, but both sources add in quadrature! Independent positions come from triangulations with the Inter Planetary Network (IPN), solar flares, and a few independently measured burst locations from *COMPTEL* and *EGRET*. In general, the *BATSE* positions alone are not accurate enough to allow deep ground-based observations. While positions with degree resolution may soon be attainable for some bursts, the majority of bursts is too faint to yield such position accuracies. However, we do not yet seek counterparts for all bursts – we are interested in finding convincing candidates for at least a few bursts.

While the positions from *CGRO* are useful in many ways, the search for counterparts requires a location accuracy that must be orders of magnitude better. There are two separate approaches to reducing the sizes of error boxes. The first method uses γ-ray data directly, and the second method attempts to obtain accurate locations through a wide field of view search for simultaneous flares in the optical band. Better positions from γ-ray data are obtained through the classical method of burst timing with a network of widely separated spacecraft (Hurley, Cline, and collaborateurs pioneered this approach). The accuracy of triangulation depends on burst brightness and structure of the lightcurves, and also on the number and sensitivity of available detectors. Despite substantial variations typical burst locations from the IPN are of order arcminutes (e.g., Hurley et al. 1993). The shapes of the error boxes also vary. Burst locations may be improved when IPN/*BATSE* positions are combined with single detector localizations from *COMPTEL*, *EGRET*, *WATCH*, or *SIGMA*. The next generation IPN could accomplish arcsecond resolution if detectors on a future Pluto flyby mission could be equipped with GRB detectors (Hurley & Cline 1994).

The other approach is based on a rapid, though coarse, localization of a burst combined with rapid ground-based searches for optical flares. The combination of *BATSE* and *COMPTEL* allows, although infrequently due to the limited FoV of *COMPTEL*, follow-up studies within hours of a burst (Kippen et al. 1994). A worldwide web of optical and radio observatories, coordinated from NMSU

(Harrison et al. 1995), is alerted by the *BATSE/COMPTEL* system and immediately surveys the *COMPTEL* field with wide FoV telescopes. The monitoring of the burst field continues for a considerable period after the burst. Detection of optical or radio transients would automatically lead to astrometric positions, which could then be studied further.

Another system with similar concepts is the *BATSE–BACODINE–GROCSE* alliance (Barthelmy et al. 1994). This system allows a rough (10°) localization of a GRB while it is bright, i.e., on a timescale of seconds. This information is used to point a small, but rapidly moving telescope with wide field of view. The prototype *GROCSE* system is located at LLNL in California. Detection of optical flares would yield the desired burst locations. If the *BACODINE/GROCSE* system (and its relatives) are not successful, hope for better positions rests on the shoulders of the *HETE* experiment (Ricker et al. 1988), which will be launched in late 1995. This experiment provides simultaneous multiwavelengths coverage in the γ-ray, X-ray, and UV bands. The detection of X-rays will yield arcminute positions, and if UV emission is detected the accuracy will be of order arcseconds! For various cosmological fireball models Meszaros et al. (1994) estimated the IR, optical, UV, X-ray, and γ-ray fluences. Their results suggest that *HETE* would detect low-energy emission from cosmological GRBs under a wide range of circumstances. In the future, new concepts such as the *ETA* (Ricker 1990), an array of detectors in solar orbit, could achieve arcsec resolution directly from γ-ray lightcurves. The *ETA* experiment would obtain precise GRB positions for $\sim 10^3$ GRBs yr^{-1}. Eventually accurate enough positions must contain uniquely identifiable counterparts, otherwise new physics is required. There could be a time in the near future when routine studies of GRB counterparts for thousands of sub-arcsec positions are used to reveal aspects of the universe that we have so far only glimpsed.

3 Quiescent counterparts

This field was reviewed at the 2nd Huntsville workshop by Schaefer (1994). The survey of GRB error boxes in the X-ray band was continued with ROSAT (Boer et al. 1993; Greiner et al. 1995). Although a few sources in or near GRB error boxes were found, no clear correlation is apparent.

Longstanding programs to survey and monitor GRB locations in the optical are being carried out at various observatories. The USNO (Flagstaff Station) survey of older IPN positions was completed (Vrba et al. 1995). While the final census of objects in this survey was consistent with the expected numbers of normal stars and galaxies several potentially interesting objects were found. In several cases moderate evidence for a GRB-QSO connection was found (Vrba et al. 1994). That such an association can not hold for all GRBs has been argued by Schaefer (1994) and Fenimore et al. (1993), based on the absence of bright host galaxies in most error boxes. The observed magnitude limits typically imply source distances in excess of ~ 1 Gpc, but a realistic luminosity function has yet to be used to derive more rigorous distance limits. Results from an optical survey program at New Mexico State's Blue Mesa Observatory were presented in recent papers (Harrison et al. 1994a,b). Again, no obvious counterparts were identified. One new development occured in the case of FY Aql, which is a Mira variable star (in the field of GRB790331) surrounded by a reflection nebula (Hartmann & Pogge 1987, Schaefer 1990). Recent observations (Irwin & Zytkow 1995) find no trace of the nebula and obtain evidence for binary nature of this star. These data may turn FY Aql again into an interesting counterpart candidate.

Searches for quiescent γ-ray burst counterparts in the extreme UV were carried out with the *ROSAT* XUV camera (Owens et al. 1993) and the *EUVE* experiment (Hurley et al. 1995). The former study surveyed the regions of 12 GRBs and 9 OTs, without evidence for emission above 10^{-11} ergs cm^{-2} s^{-1}. In the latter study the field of GRB920325 was studied 17 months after the burst, and again no steady source was found inside or near the burst error box. The upper flux limit derived from this 40ks observation was 3 10^{-14} ergs cm^{-2} s^{-1} in the 40−190 Å band. The natural source candidates we may expect to identify in the EUV are nearby compact objects with or without accretion disks around them. Detection would depend on the state of the accretion disk and the surface temperature of the star. The non-detection leads to highly model dependent constraints, but it is perhaps save to say that no normal neutron star with T $\sim$ 10^{5-6} K within a few 100 pc of the Sun would have gone undetected. Similar distance limits follow from the X-ray constraints discussed above.

4 Transient counterparts

As discussed above, accurate burst positions are now becoming available on much shorter timescales, which opens the possibility of rapid or even very rapid multiwavelengths searches for transient counterparts. Most of these programs require international collaboration and the coverage in wavelength, time, and flux relative to the burst varies dramatically from burst to burst. Within the limits of this review I select a few highlights of this very active field.

VHE: The E^2F_E spectrum of GRBs often peaks in the $\sim$ 1 MeV regime, but significant "tails" can extend to $\sim$ 100 MeV (demonstrated with *SMM*), and in some cases into the GeV range (recent *EGRET* data). This high energy emission provides severe constraints on cosmological redshifts to GRB sources because at these energies γ-rays efficiently pair-create through interaction with the CMB radiation field and the intergalactic IR background produced by galactic starlight. The *EGRET* observations of GRB940217 (Hurley et al. 1994) implies that in some cases apparently unattenuated spectra extend to $\sim$ 20 GeV. This observation emphasizes the need for ground-based searches at even higher energies. Detection in the TeV range by means of atmospheric Čerenkov detectors would severely constrain distances of cosmological scenarios. Searches have been carried out at the Whipple Observatory (Chantell et al. 1993, Connaughton et al. 1994) in the TeV band, and at the Cygnus-I air shower aray (Alexandreas et al. 1994, Schnee et al. 1995) above 100 TeV. No detection has yet been claimed. Prospects for a breakthrough in this band are very good for the *MILAGRO*

Gamma Ray Observatory currently under construction at LANL. This water-Čerenkov detector operates above ~ 250 GeV and should be the first detector in this band sensitive enough to detect all GRBs with unattenuated spectra with roughly E^{-2} extrapolation of *CGRO* data (Shoup et al. 1995).

Optical: One of the first bursts with rapid localization was GRB920501, but no unique counterpart was found among the various objects detectable in optical and X-ray observations (Hurley et al. 1994). In the case of GRB930131 an international campaign searched for optical and radio emission only 7 hours after the event (Schaefer et al. 1994). Again, no obvious counterparts were found. An even faster survey was possible for GRB940301, which was observed in the radio band one hour after the burst and in the optical band seven hours after the burst (Harrison et al. 1995). These results from the NMSU Rapid Response Network are impressive, but still did not detect transient emission that one could relate to the GRB. Clearly, simultaneous observations are the final frontier. OT searches on archival plates are a useful, but often unreliable source of information (see Hudec 1993 for a recent review), and usually the images of OTs are well removed in time from the GRB event. Greiner et al. (1994) has compiled a list of bursts for which simultaneous data are available. The absence of optical flares in these cases places an upper limit on the ratio of optical to γ-ray flux, $R_{0,\gamma} \leq 0.5$. Current efforts to further tighten these constraints are underway at LLNL, where the *GROCSE* system can respond to a *BACODINE* location in a few seconds (under favorable circumstances). Several bursts have now been observed with this system (but only with partial coverage of the error box) down to a magnitude limit of ~ 7. No optical images were found, yielding similar limits to those discussed by Greiner. Results from the *ETC* Observatory were discussed at this symposium by Vanderspek and by Greiner (1995).

Radio: If only a small fraction of the burst energy emerges in the radio band, the dispersion in the ISM of our Galaxy and the intergalactic medium (IGM) could result in arrival time delays of ~ 1 hour at radio frequencies $\nu \sim 25$ MHz. If accurate locations can be derived quickly, which is now possible with the *BACODINE* response network, there could be enough time to point radio telescopes to the burst direction. Radio detection at two frequencies would then yield the dispersion measure (DM) and thus provide an estimate of the burster distance (Palmer 1993), just as in the case of radio pulsars in our Galaxy. Paczynski & Rhoads (1993) considered the interaction of relativistically expanding fireballs with interstellar or intergalactic matter. Due to collisions between matter in the GRB wind and in the surrounding medium a fraction of the ordered kinetic energy will be randomized and some part will eventually be radiated away in the form of low energy photons. At late stages in the development of such "radio fireballs" synchrotron emission dominates over inverse Compton cooling of the electrons and one can estimate the radio spectrum assuming that the emission is incoherent. If reheating of the expanding shell generated an electron distribution function that is a power law in electron energy, $n(E) \propto E^{-p}$, synchrotron emission from optically thin regions produces a power law spectral intensity with slope $\alpha = -(p-1)/2$. Self absorption in optically thick regions changes this slope to $\alpha = 5/2$, independent of the electron distribution function and different from the thermal Rayleigh-Jeans limit $\alpha_{\rm RJ} = 2$. Paczynski & Rhoads carry out the transformation from the restframe of the emitting shell and find that GRBs could generate radio fluxes above 0.1 Jy at frequencies of a few GHz. If burst emission is strongly beamed, however, the reduced energy requirements (usually a blessing) could render the radio transients undetectable. Assuming isotropic emission, the peak of the radio emission would occur with a time delay, Δt, relative to the beginning of the fireball

$$\Delta t \sim 0.37\ E_{53}^{1/2}\ \varrho_{-24}^{1/2}\ \nu_{\rm GHz}^{-3/2} \quad {\rm yrs}\,. \tag{1}$$

While radio bursts from GRBs at cosmological distances are likely to show substantial delays, bursts in the Galactic halo require a much smaller primary energy and might thus be accompanied by simultaneous radio bursts. A radio signal at frequency ν propagates through a cold electron gas (number density n) with group velocity

$$\beta = \frac{v}{c} = \left[1 - \left(\frac{\nu_p}{\nu}\right)^2\right]^{1/2}, \tag{2}$$

where c is the speed of light in vacuum, and ν_p the plasma frequency. The time delay, relative to a wave propagating with the speed of light, is

$$\delta t(\nu) = \int_0^l dx\,(c - v)\ c^{-2} = \frac{1}{c}\int_0^l dx\ (1 - \beta)\ , \tag{3}$$

In the limit $\nu \gg \nu_p$ the time lag relates to the dispersion measure as

$$\delta t(\nu) = \frac{{\rm e}^2}{2\pi\ {\rm m_e c}\ \nu^2}\int_0^l dx\ {\rm n} \sim \frac{\rm DM}{240\ \nu_{\rm GHz}^2}\,. \tag{4}$$

where the dispersion measure (DM = total electron column density) is commonly measured in the units pc cm^{-3}. The Galactic free electron distribution gives typically DM $\sim 10^2$ pc cm^{-3} in the Galactic plane, but can reach $\sim 10^3$ pc cm^{-3} for directions close to the Galactic center. If the radio source is located at cosmological distances, the additional electrons in the intergalactic medium (IGM) add to the delay. Because the plasma frequency is density dependent, the simple δt-DM relationship does no longer hold, but it is straightforward to derive a very similar form

$$\delta t(z) = \frac{{\rm DM}_0}{240\ \nu_{\rm GHz}^2}\left(\frac{2}{\Omega}\right)$$

where ${\rm DM}_0 = 2.46\ 10^4\ {\rm h}_{75}\ \Omega_b$ is the dispersion measure scale for a Friedman universe with baryon fraction Ω_b. A radio transient emitted at $z \sim 1$ thus experiences a delay, relative to the γ-ray bursts, of over one minute. If the GRB source region itself is enshrouded by an electron cloud of significant column density, the delay would increase. Of course, radio emission could also be delayed due to intrinsic evolution of the radiating source region. A recent search for delayed radio emission was carried out for GRB940301 (Frail et al. 1994). Wide field radio images at 0.4 GHz and 1.4 GHz were obtained at the DRAO Synthesis Telescope between 3–15 and 26, 47, and 99 days after the burst. No flaring object brighter than 3.5 mJy was seen, suggesting that some model parameters were chosen too optimistically.

Gravitational waves (GW): If neutron star mergers cause GRBs, there should be a coincident burst of gravitational radiation. The brightest bursts could be detectable with *LIGO*, expected to be operational in 1999. The first phase *LIGO* may not be sensitive enough to detect such GRBs past $\sim$ 10 Mpc, but *LIGO* upgrades will increase its sampling distance with time. The knowledge of GRB arrival times reduces detector backgrounds and thus improves sensitivity. Kochanek & Piran (1993) estimate that failure to observe GRB-GW pairs with a rate of 1 yr^{-1} above a strain of $\sim 10^{-22}$ would rule out the merger hypothesis.

Neutrino bursts: The merger of two neutron stars releases $\sim 10^{53}$ ergs of energy, most of which is radiated away by thermal neutrinos of all flavors, and a small fraction of this energy may be converted to power the GRB. Neutrino energies are $\sim$ 10 MeV, and thus hard to detect. Much higher energies (GeV) could result when fireballs interact with matter, causing production of pions that decay, amongst other particles, to neutrinos (Paczynski & Xu 1994). Neutrino bursts from superconducting strings at large redshifts were discussed by Plaga (1994), but the *EGRET* observations of GeV photons argue against redshifts in excess of 10–100. Searches for simultaneous neutrino bursts were carried out by the Soudan collaboration (DeMuth et al. 1994), the *IMB* collaboration (LoSecco 1994), and at Mont Blanc (Aglietta et al. 1993). No correlation was found to a fluence level of $\sim 3\ 10^4$ ergs cm^{-2}.

References

Alexandreas D.E., 1994, ApJ 426, L1
Anglietta M. et al., 1993, 23rd ICRC, 1, 69
Barthelmy S.D., 1994, in Gamma-Ray Bursts, G.J. Fishman, K. Hurley, & J. Brainerd (eds.), AIP 307, 643
Boer M. et al., 1993, A&A 277, 503
Brainerd J.J. et al., 1995, ApJ (in press)
Briggs M. et al., 1995, ApJ (in press)
Chantell M. et al., 1993, in Compton Gamma Ray Observatory, AIP 280, 833
Connaughton V. et al., 1994, in Gamma Ray Bursts, G.J. Fishman, K. Hurley, & J. Brainerd (eds.), AIP 307, 470
DeMuth, et al., 1994, in Gamma Ray Bursts, AIP 307, 475
Fenimore E.E. et al., 1993, Nature 366, 40
Frail, D.A. et al., 1995, ApJ 437, L43
Greiner J., 1995, Proc. 17th Texas Symp. on Rel. Astrophysics (in press)
Greiner J. et al., 1994, in Gamma Ray Bursts, G.J. Fishman, J.J. Brainerd & K. Hurley (eds.), AIP 307, 408
Greiner J., Boer M., Kahabka P., Motch C., Voges W., 1995, in The Lives of Neutron Stars, A. Alpar (ed.), NATO ASI 450, Kluwer, Dordrecht, p. 519

Harrison T., McNamara B., Klemola A., 1994a, AJ 107, 254
Harrison T., McNamara B., Klemola A., 1994b, AJ 108, 600
Harrison T. et al., 1995, A&A (in press)
Hartmann, D.H., Pogge, R., 1987, ApJ 318, 363
Hartmann, D.H., Woosley, S.E., 1988, in Multiwavelengths Astrophysics, F. Córdova (ed.), Cambridge Univ. Press, Cambridge, p. 189
Hartmann, D.H. et al., 1995, ApJ (in press)
Hudec R., 1993, A&AS 97, 49
Hurley K. et al., 1993, A&AS 97, 39
Hurley K. et al., 1994, ApJS 92, 655
Hurley K. et al., 1995, ApJ (in press)
Hurley K., Cline, T., 1994, in Gamma Ray Bursts, G.J. Fishman, K. Hurley, & J. Brainerd (eds.), AIP 307, 653
Hurley K. et al., 1994, Nature 372, 652
Irwin M., Zytkow A.N., 1995, ApJ (in press)
Kippen M. et al., 1994, in Gamma Ray Bursts, eds. G.J. Fishman, K. Hurley, & J. Brainerd (eds.), AIP 307, 418
Kochanek, C.S., Piran, T., 1993, ApJ 417, L17
Kouveliotou C., 1994, Nature 370, 26
LoSecco, J.M., 1994, ApJ 425, 217
Meegan C.A. et al., 1995, ApJ (in press)
Meszaros P., Rees, M.J., 1993, ApJ 418, L59
Meszaros P., Rees, M.J., Papathanassiou, H., 1994, ApJ 432, 181
Owens A. et al., 1993, in COMPTON Observatory, M. Friedlander, N. Gehrels & D.J. Macomb (eds.), AIP 280, 798
Palmer D.M., 1993, ApJ 417, L25.
Plaga, R., 1994, ApJ 424, L9
Paczynski P., Xu G., 1994, ApJ 427, 708
Paczynski P., Rhoads J., 1993, ApJ 418, L5
Quashnock J.M., Lamb D.Q., 1993, MNRAS 265, L59
Ricker G. et al., 1988, in Nuclear Spectroscopy, N. Gehrels, & G.H. Share (eds.), AIP 170, 407
Ricker G., 1990, in High Energy Astrophysics, P. Joss (ed.), AIP 211, 365
Schaefer B.E., 1994, in Gamma Ray Bursts, G.J. Fishman, K. Hurley, & J. Brainerd (eds.), AIP 307, 382
Schaefer B., 1990, ApJ 364, 590
Schaefer B. et al., 1994, ApJ 422, L71
Schnee, R., *et al.* 1994, in Gamma Ray Bursts, G.J. Fishman, K. Hurley, & J. Brainerd (eds.), AIP 307, 481; see also BAAS 26, 1508
Shoup A. et al., 1994, BAAS 26, 1335
Vrba F.J. et al., 1994, ApJ 424, 68
Vrba F.J., Hartmann D.H., Jennings M., 1995, ApJ (in press)
Wang V., Lingenfelter R.E., 1993, ApJ 416, L13
Zytkow A., 1990, ApJ 359, 138

Optical Transients

René Hudec

Astronomical Institute, CZ–251 65 Ondřejov, Czech Republic

Abstract: We outline and discuss the recent results of optical searches for counterparts to gamma-ray burst sources as well as related problems.

1 Introduction

The searches for optical counterparts to GRB proceed in three directions, namely as searches for: **(1) Flashing counterparts** (Optical transients–OT; either real-time or archival), **(2) Fading counterparts** (OT; follow-up searches), and **(3) Quiet counterparts.** The general problem is, however, that the nature of and even the distance to GRB is unknown. Hence we actually do not know what are we looking for. This should be always taken into account during evaluation of data in all mentioned directions. Here we discuss the first two categories.

2 Searches for flashing counterparts

2.1 Archival searches

One important but not obvious assumption must be done in these analyses, namely that the GRB/OT are recurrent. Alternatively, we can estimate limits for luminosities and/or recurrencies. The recent situation in the field of archival searches (i.e. using archival astronomical plates) can be summarized as follows:

- Several very good candidates (and blue faint possible counterparts) have been detected with strong evidence for reality and light amplitudes $\gg 10^{\mathrm{m}}$ (Hudec et al. 1994a,b, Vrba et al. 1994a,b).
- The recurrence, if any, is > 1 yr (in agreement with γ-ray data).
- Archival astronomical plates are advantageous for these searches, especially for monitoring of particular GRB positions. Up to 20 000 h of exposure are available for particular positions. This would be hardly possible with other methods, e.g. modern monitoring with electronic detectors would for the same fraction of monitoring time require about 34 yr (assuming 100 nights per year and 5 hours per night), i.e. almost the full active life of one astronomer.
- The two best studied OT seems to be very colored (Hudec et al. 1994a,b)

- Coincidence of OT with QSO has been found in one case, close to the GRB position (Vrba et al. 1994a,b). A question naturally arises, namely whether or not are (at least some) QSOs sources of OT, and whether are (at least some) GRB related? The problem is in (still) large sizes of GRB error boxes as well as in the high QSO average density. At least one QSO is expected to be located in typical error box.

2.2 Real-time searches

The simultaneous optical data for GRB are extremely important because no assumption for trigger recurrency and/or delayed emission is necessary in this case. Up to now, real-time optical data are available for ~ 50 GRB but mainly with very limited sensitivity ($< 3^{m}$ for 1-sec events) and response limited to red light only (> 400 nm). The previous and recent results in this direction can be summarized as follows:

- No optical emission (> 400 nm) above mag ~ 5 (1 sec duration assumed) or $Lg/Lo > 100\ldots300$ has been detected for **a few** GRB. The faintest limit (320) exists for GRB 830313 (Hudec, 1993).
- No optical emission (> 400 nm) above magnitudes $0\ldots3$ (1 sec duration assumed) or $Lg/Lo > 0.1\ldots10$ has been detected for **many** (~ 50) GRB. These limits have been obtained mainly at the Ondřejov Observatory on meteor sky patrol plates within the framework of *GRO*-related programs (e.g. Hudec 1993, Greiner et al. 1993, 1994, 1995 and Hudec et al. 1995a)
- One can speculate whether the non-detection of simultaneous optical emission is due to (i) absence of brighter optical emission, (ii) the delayed optical emission, or (iii) the optical emission dominates in blue (this would be in agreement with results from archival searches revealing at least two reliable optical flashing candidates with extremely colored light). However, the limits recently available are still too poor even for this speculation: optical accompanying emission with $Lg/Lo \sim 100$ would remain undetectable almost in all cases. However, it should be mentioned that the time delays between optical and gamma ray emission cannot be fully excluded regarding the lack of knowledge of related physical processes. Moreover, the recently suggested possible link between OT and QSOs, and perhaps even with GRB (Vrba et al. 1994a,b), if confirmed, would result in natural expectation of delays (already known for some processes inside of QSO's, e.g. Robson et al. 1993 and von Linde et al. 1993).

3 Follow-up searches, burst alert

This category of searches for optical GRB emission is strongly related with recent achievements in satellite programs such as *GRO-BATSE*, *BACODINE*, *COMPTEL*, etc. The newly operated *BACODINE* system distributes the GRB data to ground-based observers within 0.3 to 30 sec after events (Barthelmy et al. 1994b). The optical burst alert data go usually significantly deeper than real–time data mentioned in the previous section. The present situation can be summarized as follows:

- No optical emission (> 320 nm) has been detected at times ~ hrs to ~ days after GRB down to $15\ldots20^{m}$ (e.g. Barthelmy et al. 1994a, Boer et al. 1994, Castro-Tirado et al. 1994, Kippen et al. 1994, Krimm et al. 1994, and Hudec et al. 1995b).
- One of the main problems of theses searches is the relatively high background of (mainly unknown) variable stars at a rate of ≥ 1 for a typical GRB error box area provided with current programs (*BACODINE*, *BATSE*), which are of the order of several or even several tens of degrees (Hudec & Wenzel 1995).
- The detection of these (unknown) variable stars can be used as a measure of the reliability of the method used (Hudec & Wenzel 1995).
- The variable objects found in some GRB error boxes are probably not related to GRB events but are probably newly discovered variable stars. However, we have to analyse and to classify them to avoid any kind of possible misinterpretation of real candidates.
- More than one (and better $\gg 1$) frame/plate is necessary for each trigger to avoid misinterpretation of emulsion/CCD false triggers and for comparison.

4 Strategy for the future

4.1 Archival searches

One of the main problems of past archival searches was the relatively large size of investigated GRB error boxes resulting in the not quite negligible probability that the objects found inside are just a random coincidence. A much better situation is in the case of new IPN3 error boxes with areas of order of a few arcmin^2 in the best cases (Hurley et al. 1994). Another improvement in archival searches should be the use of plates from several plate collections to increase the amount of monitoring hours available for the particular position. Then fractions of the total monitoring time available for one GRB may reach up to 2 yr and/or even more, significantly increasing the probability that the related OT will be detected if there is measurable optical emission (as already mentioned, these analyses, however, require an assumption about trigger recurrency).

4.2 Real-time searches

There are almost no simultaneous optical data except those from the Ondřejov photographic network. A more sensitive all-sky or at least very wide-field patrol service is not available at the present. The *ETC* camera (Vanderspek et al. 1994) is more sensitive (magnitude limit 8.5 for 5-sec flashes) but with a limited FoV (0.75 sr). While the Ondřejov photographic patrol has monitored about 7 000 steradian-hours during the 2.5 yr period 1991–1993, the *ETC* has monitored only about 600 steradian-hours. A promising approach is to build a double CCD monitor with large FoV and significantly better limiting magnitude than currently available photographic programs, preferably as a network (Hudec & Soldán 1995).

4.3 Burst alert searches

Here only a few papers have been published so far and the majority of obtained data is still in evaluation. It seems however to be clear that simultaneous data and/or data taken immediately (i.e. within 1 hr) after the trigger are provided only by both photographic (e.g. Hudec 1993, Greiner et al. 1993, 1994, 1995 and Hudec et al. 1995a) and CCD (e.g. Vanderspek et al. 1994 and Krimm et al. 1994) patrol experiments so far. The minimal delay in burst alert searches with larger instruments is still 5 hr or even more. There are many reasons for such delays. Initially, the GRB data were becoming available with delays of order of $\sim$ hrs but now they are available much faster. The *BACODINE* system (Barthelmy et al. 1994b) provides a very fast response. However, the optical burst alert data usually still have delays of order of hours or even more. One of the reasons is that the position of the trigger is unobservable at the time of the trigger (daytime or below the horizon) and the observers have to wait several hours. Another reason for delay is the presence of humans in the loop. Robotic telescopes with automated response to incoming information would significantly improve the delays between GRB and optical observations (see also Hudec & Soldán 1995).

In general, it is clear that more and better optical data are needed. Further, there is an obvious need for very sophisticated classification of detected triggers to exclude false events and to verify real OT.

Acknowledgement: We acknowledge supports provided by the Grant Agency of the Academy of Sciences of Czech Republic (No. 303103) and by the Grant Agency of Czech Republic (No. 202-93-0890).

References

Barthelmy S.D., Palmer D.M., Schaefer B.E., 1994a, Gamma Ray Bursts, G.J. Fishman, J.J. Brainerd, K. Hurley (eds.), AIP Conf. Proc. 307, 392–395

Barthelmy S.D. et al., 1994b, Gamma Ray Bursts, G.J. Fishman, J.J. Brainerd, K. Hurley (eds.), AIP Conf. Proc. 307, 643–647

Boer M. et al., 1994, Gamma Ray Bursts, G.J. Fishman, J.J. Brainerd, K. Hurley (eds.), AIP Conf. Proc. 307, 396–400

Castro-Tirado A.J., Brandt N., Lund N., Guziy S.S., 1994, Gamma Ray Bursts, G.J. Fishman, J.J. Brainerd, K. Hurley (eds.), AIP Conf. Proc. 307, 404–407

Greiner J. et al., 1993, Compton Gamma-Ray Observatory, M. Friedlander, N. Gehrels, D.J. Macomb (eds.), AIP Conf. Proc. 280, 828–832

Greiner J. et al., 1994, Gamma Ray Bursts, G.J. Fishman, J.J. Brainerd, K. Hurley (eds.), AIP Conf. Proc. 307, 408–412

Greiner J. et al., 1995, this volume

Hudec R., 1993, Astroph. Lett. Commun. 28, 359

Hudec R., 1993, A&A 270, 151

Hudec R., Dedoch A., Pravec P., Borovička J., 1994a, A&A 284, 839

Hudec R., Pravec P., Borovička J., 1994b, A&A 284, 499

Hudec R., Soldán J., 1994, ApJS 92, 675

Hudec R., Soldán J., 1995, this volume

Hudec R., Wenzel W., 1995, A&A (submitted)

Hudec R. et al., 1995a, in preparation
Hudec R. et al., 1995b, in preparation
Hurley K. et al., 1994, priv. communication
Kippen R.M. et al., 1994, Gamma Ray Bursts, G.J. Fishman, J.J. Brainerd, K. Hurley (eds.), AIP Conf. Proc. 307, 418
Krimm H.A., Vanderspek R.K., Ricker, G.R., 1994a, Gamma Ray Bursts, G.J. Fishman, J.J. Brainerd, K. Hurley (eds.), AIP Conf. Proc. 307, 423
von Linde J. et al., 1993, A&A 267, L23
Robson E.I. et al., 1993, MNRAS 262, 249
Vanderspek R., Krimm H.A., Ricker, G.R., 1994, Gamma Ray Bursts, G.J. Fishman, J.J. Brainerd, K. Hurley (eds.), AIP Conf. Proc. 307, 438
Vrba F.J. et al., 1994a, Gamma Ray Bursts, G.J. Fishman, J.J. Brainerd, K. Hurley (eds.), AIP Conf. Proc. 307, 448
Vrba F.J. et al., 1994b, ApJ 424, 68

Simultaneous Optical/Gamma-ray Observations of GRBs

J. Greiner[1], *W. Wenzel*[2], *R. Hudec*[3], *M. Varady*[3], *P. Štěpán*[3], *P. Spurný*[3], *J. Florián*[3], *E.I. Moskalenko*[4], *A.V. Barabanov*[4], *R. Ziener*[5], *K. Birkle*[6], *N. Bade*[7], *S.B. Tritton*[8], *T. Ichikawa*[9], *G.J. Fishman*[10], *C. Kouveliotou*[10], *C.A. Meegan*[10], *W.S. Paciesas*[10], *R.B. Wilson*[10]

[1] Max-Planck-Institut für Extraterrestrische Physik, 85740 Garching, FRG
[2] Sternwarte Sonneberg, 96515 Sonneberg, FRG
[3] Astronomical Institute Ondřejov, 25165 Ondřejov, CR
[4] Sternberg Astronomical Institute, 119899 Moscow, Russia
[5] Thüringer Landessternwarte, 07778 Tautenburg, FRG
[6] Calar Alto, German-Spanish Astronomical Centre, 04080 Almeria, Spain
[7] Hamburger Sternwarte, Gojenbergsweg 112, 21029 Hamburg, FRG
[8] Royal Observatory, Blackford Hill, Edinburgh EH9 3HJ, UK
[9] Kiso Observatory, University of Tokio, Nagano 397-01, Japan
[10] Marshall Space Flight Center, Huntsville, AL 35812, USA

Abstract: This status report presents details on the project to search for serendipitous time-correlated optical photographic observations of γ-ray bursters. The ongoing photographic observations at nine observatories are used to look for plates which have been exposed simultaneously with a γ-ray burst detected by BATSE and contain the burst position. The results for the third year of BATSE operation are presented.

1 Strategy and instruments

We are correlating two independent surveys, namely the photographic plates (optical) of the observatories involved and the BATSE (γ-ray) burst data. At present the photographic sky patrols of the observatories Sonneberg, Tautenburg and Hamburg (all Germany), Calar Alto (Spain), Ondřejov (CR), Odessa and Crimea (Ukraine), Dushanbe (Tadshikistan) and Kiso (Japan) are used for the northern hemisphere, and the UK (Siding Spring, Coonabarabran, Australia) and the ESO (La Silla, Chile) Schmidt plates for the southern hemisphere GRBs. This network is still expanding. The instruments which are used for this search are described in Greiner et al. (1994).

2 Results

With BATSE detecting and localizing $\approx$ 1 GRB per day, and a spatial and temporal optical coverage with the observatories involved of about 80%, we are presently checking 280–300 GRBs per year.

Within the first, almost three years of BATSE observations (1991 April 23 – 1994 March 31) we have identified simultaneous photographic observations for

49 GRBs. A first list of plates covering 29 GRBs was given in Greiner et al. 1994. The details for the additional 20 GRBs from the third year of BATSE operation are given in Table 1. For most of these hits, several simultaneously exposed plates from different observational stations are available. Most of these plates are taken in the framework of the Czech meteor patrol (see Hudec 1995 for details) which uses a red sensitive combination of objective and panchromatic emulsion (almost no sensitivity below 400 nm). All plates have been investigated by blink comparison. No optical flash or any brightening of an object was found which could be undoubtedly attributed to a GRB. Thus, any optical emission accompanying the high-energy burst must have been less than the minimum detectable optical flux (erg/cm^2/s) of the corresponding plate. Using the peak γ-ray flux in the 50–300 keV range reached on the 1024 ms timescale we derive a lower limit of typically $F_\gamma/F_{opt} > 0.5$–40.

Furthermore, near-simultaneous deep exposures (24 hours before or after GRB) are also registered in the course of the cross-correlation (Table 2). Since these plates have smaller fields of view, the large GRB error boxes are not always covered fully. The last column in Table 2 therefore contains the percentage of the GRB error box area which is covered by the plate. This has been calculated by adding in quadrature a systematic error of 4° to the statistical error of each burst. In addition to the partial coverage, these plates are mainly taken under programmes which are not related to variability studies; i.e. there is in most cases no plate of the same area (and same filter/emulsion combination) which could be used for blink comparison. Thus, only part of the near-simultaneous plates could be investigated for optical activity. Due to the deeper limiting magnitudes, the blink comparison of plates having suitable comparison plates revealed a number of known as well as new variable stars. However, this optical activity is presumably not related to the GRB phenomenon. We have found no optical flash event. When counting only those plates which contain 100% of the GRB error box and which have been checked by blink comparison, we can constrain the optical emission within 1–3 hours before or after the GRB to be fainter than 5–7th mag for a 1 sec duration flash or equivalently 12-14th mag for a fading counterpart.

3 Discussion

With an increasing number of simultaneous optical observations of GRB locations we can conclude with higher confidence that typical GRBs have (1) optical emission at the time of the burst at a level at least below $(F_\gamma/F_{opt})^{-1} \approx 2$ and (2) optical emission a few hours after the burst is lower by a factor of 200 than the simultaneous limits.

Acknowledgement: JG and WW are partly supported by the Deutsche Agentur für Raumfahrtangelegenheiten (DARA) GmbH under contract 50 OR 9104 3 and 50 OR 9201, and by the Deutsches Elektronen-Synchrotron (DESY-PH) under contract 05-5S0414. RH is supported by grant 303103 of the Academy of

Table 1. Simultaneous plates for BATSE GRBs in the third year

GRB	Time (UT)	Location (RA, DEC)	No. of simultaneous plates[1)]	limiting magnitude m_{pg} (for 1s flash)	Peak flux F_γ (ph/cm^2/s) 50-300 keV 1024 ms	Limit[2)] for F_γ/F_{opt}
930424	19:19:00	67.0°,40.2°	O, 4	3.0	1.64	0.7
930514	00:26:16	170.1°+71.0°	O, 8	3.0	[3)]	
930816	00:52:24	58.5°+71.4°	O, 8	3.5	[3)]	
930911	23:16:01	268.36°,69.07°	O, 5	3.0	0.82	0.3
930916	20:19:23	282.67°,65.28°	O, 3	3.0	7.90	3.5
930924	01:02:55	312.28°,12.67°	O, 7	3.0	[3)]	
931001	01:23:11	256.95°,46.09°	O, 2	0.5	0.47	0.02
931006	21:31:50	59.89°,65.89°	O, 6	2.5	[3)]	
931101	23:23:38	133.77°,6.95°	O, 2	2.0	0.14	0.03
931126	04:46:46	156.74°,67.53°	O, 1	0.0	0.81	0.02
931126	19:31:39	1.58°,-16.62°	O, 1	-1.0	[3)]	
931223	23:34:20	256.31°,41.08°	O, 2	0.0	1.84	0.05
940112	19:53:32	47.85°,64.23°	O, 4	2.5	0.26	0.07
940114	21:28:20	98.19°,-13.58°	O, 2	1.0	[3)]	
940209	20:00:01	354.87°,9.02°	O, 2	1.5	[3)]	
940214	00:38:08	121.63°,57.02°	O, 3	2.5	1.91	0.5
940217	23:02:42	29.02°,5.93°	O, 2	2.0	36.21	6.4
940227	21:04:14	356.63°,43.01°	O, 3	1.0	1.05	0.07
940310	22:00:57	84.86°,49.57°	O, 6	3.5	1.58	1.1
940327	20:29:06	24.12°,52.07°	O, 5	1.0	0.43	0.03

[1)] Letters indicate the observatory: O=Ondřejov

[2)] For the conversion of the γ-ray fluxes from ph/cm^2/s to erg/cm^2/s we have used a power law model of index −1.4 which is the most frequent value (Pendleton et al. 1994). The burst spectra with extreme deviations in slope the error in the conversion can reach up to a factor of 2. In the case of several simultaneous plates the ratio F_γ/F_{opt} is calculated for the deepest plate. Note that F_γ/F_{opt} is a true flux ratio (peak γ-ray flux in erg/cm^2/s divided by the optical flux limit in the same units) rather than the usually given "flux" ratio L_γ/L_{opt}, which is really a fluence ratio, i.e. depends on the burst duration (T_{90}) and the exposure time of the plate. F_γ/F_{opt} is always lower than L_γ/L_{opt} if the burst duration is larger than 1s.

[3)] F_γ not available due to some missing data types.

Table 2. Selected near-simultaneous (±1 day) plates for BATSE GRBs

GRB	Time (UT)	Location RA, DEC	Observatory[1]; Plate center (RA, DEC)	Time between GRB onset and exposure[2]	$m_{\rm lim}$	% of error box
910505	20:15:18	178°.6,33°.2	K 180°,30°	–7.5 h		25%
930511	3:47:38	270°.4,41°.0	Sa; 274°.96,36°.06	+19.0 h	16.0 pg	30%
930622	3:28:15	143°.4,14°.0	Od; 148°,+69°	+16.0 h	14.5 pv	100%
930724	2:21:50	287°.6,60°.0	Sp; 18h+60°	+18.9 h	13.5 pg	60%
			Sp; 18h+60°	+18.9 h	12.0 pv	60%
*930725	15:27:50	264°.9,8°.4	Sp; 18h+00°	+ 6.8 h	12.0 pg	90%
			Sp; 18h+00°	+ 6.8 h	11.0 pv	90%
			Sa; 263°.73,12°.56	–16.8 h	16.0 pg	30%
*930725	17:45:21	88°.5,65°.7	Sp; 6h+80°	+ 4.5 h	12.5 pg	50%
			Sp; 6h+80°	+ 4.5 h	11.7 pv	50%
930815	5:26:42	287°.9,–9°.3	Sp; 19h+00°	+16.6 h	12.0 pg	60%
			Sp; 19h+00°	+16.6 h	11.0 pv	60%
*930816	00:52:24	58°.5,71°.4	Sp; 2h+80°	+ 0.7 h	12.5 pg	70%
			Sp; 2h+80°	+ 0.7 h	12.0 pv	70%
931203	01:02:03	7°.63,19°.01	CA; 5°.5,+17°	– 4.7 h	19.5 pg	30%
*940310	22:00:57	84°.86,49°.57	Sp; 6h+40°	– 2.3 h	12.0 pg	100%
			Sp; 6h+60°	– 2.3 h	12.5 pg	100%
*940708	20:42:06	300°.2,22°.25	Sp; 20h+20°	+ 2.3 h	12.5 pg	100%
*940806	9:32:58	245°.2,10°.87	Sa; 246°.54,11°.41	+12.2 h	15.2 pg	80%

Notes: Plates marked with an asterisk have been checked by blink comparison.
[1] The letters indicate the observatory: CA = Calar Alto, Od = Odessa, Sa = Sonneberg astrograph, Sp=Sonneberg patrol, K = Kiso Schmidt.
[2] The shortest time is given corresponding to the start (end) of exposure for a plate taken after (before) a GRB. Typical exposure times are 30–60 min.

Sciences of the Czech Republic and by grant 202-93-0890 of the Grant Agency of the Czech Republic.

References

Greiner J., Wenzel W., Hudec R., et al. 1994, in Gamma-ray bursts, AIP 307, p. 408
Hudec R., 1995, this volume p. 376
Pendleton G.N., Paciesas W.S., Briggs M.S., et al. 1994, ApJ 431, 416

ETC Observations of the Gamma-ray Burst GRB 941014

Roland Vanderspek, George R. Ricker

Center for Space Research, Massachusetts Institute of Technology, Cambridge, USA

Abstract: The Explosive Transient Camera (*ETC*) is a dedicated wide-field sky monitor for short-duration optical transients. The *ETC* operates completely automatically, searching for transients in the night sky without the need for a human observer on site. Data products from the *ETC* include data from any transients detected, as well as CCD images of the fields-of-view observed. We report here the imaging by the *ETC* of the 2.5σ error region of GRB 941014 90 seconds after the end of the burst.

1 Introduction

The Explosive Transient Camera (*ETC*) is an automated, wide-field sky monitor instrument designed to detect short-timescale optical transients in real time. The *ETC* was developed to detect transient optical radiation associated with γ-ray bursts (GRBs); however, with its large field-of-view and high sensitivity, the *ETC* is able to detect other, perhaps previously unknown, classes of fast celestial optical transients.

The *ETC* has been operating fully automatically from the summit ridge of Kitt Peak (USA) since 1990. Since then, the *ETC* has monitored over 1000 steradian-hours of the night sky in the search for one-second optical transients as faint as V $\approx$ 10. The *ETC* has also collected and stored over 35 000 wide-field images of the night sky to limiting magnitudes ranging from V $\approx$ 11-13.

Since the launch of the Compton Gamma Ray Observatory (*CGRO*), the data collected from *ETC* observations have been compared with the locations of GRBs detected by the Burst and Transient Source Experiment (*BATSE*) on *CGRO*. If a GRB detected by *BATSE* is localized to have been in the *ETC* field-of-view during *ETC* observations and no optical radiation was detected, an upper limit on the ratio of the GRB fluence to the fluence of prompt optical radiation (S_γ/S_{opt}) can be set. To date, five such coincidences have occurred, with no correlated optical radiation (Krimm et al. 1994): the upper limits of $S_\gamma/S_{\rm opt}$ calculated for these bursts range from 2-100. However, the error regions of these GRBs are so large (typically 4°-8° radius (1σ)), that even the large *ETC* field-of-view cannot cover the 2σ error region, so only partial spatial coverage of these bursts was possible.

2 The Explosive Transient Camera

2.1 Instrumentation

The Explosive Transient Camera (*ETC*) consists of a bank of sixteen wide-field CCD cameras: each camera has a field-of-view of 20° × 15°. The *ETC* fields-of-view overlap in pairs to allow anticoincidence techniques to be used to reject local sources of optical transients: the overall field-of-view of the *ETC* system is 0.73 steradians. The characteristics of the *ETC* are given in Table 1; further details can be found in Vanderspek et al. (1992).

Table 1. The optical characteristics of the Explosive Transient Camera

Characteristic	Value
Location	Kitt Peak, USA
Number of Cameras	16
Optical System	CCD + 24 mm lens
Field-of-View per Camera	20° x 15°
Pixel Angular Size	3.1 arc-minutes
Total Field-of-View	0.73 steradian
Bandpass	4000A - 7500A
Limiting magnitude (field stars)	$V \cong 10.5$ (4σ, 5s exposure)
Limiting magnitude (1s flash)	$V \approx$ 8-10 (10σ)

2.2 Operations

The *ETC* was designed as an *automatic* instrument, to reduce the time and cost of having a human operator on site. The *ETC* control software has complete control of the *ETC* instrumentation (including the roll-off roof) and, through peripheral hardware, is aware of the local weather and sky conditions. The software has been designed to "think" as a human observer would, and so can operate with the same high efficiency a human observer would.

ETC observations consist of a continuous, contiguous series of short (5s) exposures of the night sky by all cameras simultaneously. Each image is compared, pixel-by-pixel, to the one taken immediately preceding, to find pixels which have brightened by a significant amount. If a brightening is seen at the same celestial coordinates by two *ETC* cameras in the same exposure, they are considered images of a real celestial optical transient and data from the event are stored.

At regular intervals, the*ETC* stores full images of the night sky from all CCD cameras. These images are used to check instrument health, and are later analyzed to look for long-term variability in field stars.

The *ETC* is also capable of responding to alerts from the *BACODINE* system (Barthelmy et al. 1994). When a GRB alert is received by the *ETC* system while the *ETC* is observing, the cameras are slewed to the location of the burst, and images of the burst field are taken and stored.

3 Observations of GRB 941014

On 14 October, 1994, *BATSE* detected a bright GRB: the fluence is estimated to be 10^{-5} erg cm^{-2} in the 57-318 keV energy band (C. Kouveliotou, private communication). The *BACODINE* system intercepted the *BATSE* telemetry stream, calculated the burst coordinates, and distributed them to its member sites, and electronic mail describing the burst was sent to MIT the following morning. The mail message described a burst which occurred in the field-of-view of the *ETC* cameras at a time when the *ETC* was observing: it turned out that the *ETC* was storing data to disk at the time of the *BACODINE* alert, and was, therefore, insensitive to the burst alert. However, the *ETC* restarted its observations soon thereafter and stored full images of the sky from all 16 CCD cameras. Analysis of the light curve of this burst (C. Kouveliotou, private communication) showed that these images were taken 90 seconds after the end of the burst.

Because the localization precision of *BACODINE* (8° radius (1σ)) is so large, the *ETC* images covered only a fraction of the error region calculated by *BACODINE*. When the Interplanetary Network (IPN) arc was made available (K. Hurley, private communication), the error region was reduced to the intersection of a narrow (0°.25) arc and the large *BACODINE* error circle: this arc was partially imaged by one pair of *ETC* cameras. When the more precise *BATSE* error region was made available (G. Fishman, private communication), it turned out that the entire 2.5σ error region of this bright GRB had been imaged by a pair of *ETC* cameras. This image and the 2σ *BATSE*/IPN error region are shown in Fig. 1. Because the center of the GRB error region is off-center in the *ETC* image, the coverage of the error region is asymmetric: the image contains the error region to 2.5σ on one side and to 5.5σ to the other.

4 Results and conclusions

The analysis of the *ETC* image of the 2.5σ error region of GRB 941014 revealed that the limiting magnitude of the image was $V = 10.2$ (4σ), and that four stars were within the error region. Analysis of an image taken of the same field taken two nights earlier revealed the same four stars. Thus, no new optical radiation was detected in the 2σ error region of GRB 941014. The absence of optical radiation corresponds to an upper limit on the ratio of the optical (V-band) fluence 90 seconds after the end of the burst to the γ-ray burst fluence (57-318 keV; C. Kouveliotou, private communication) of $1 \cdot 10^{-4}$.

Acknowledgements: We would like to thank the *BATSE* team for their assistance in getting the error region and light curve for GRB 941014: specifically, Gerald Fishman and Chryssa Kouveliotou were very quick in their responses to our queries. We would also like to thank Kevin Hurley for his assistance in getting the IPN coordinates. Scott Barthelmy was very quick and gracious in his relaying of *BATSE* and IPN information, in part via the *BACODINE* network. The *ETC* is supported under grants NAGW-2089 and NAG5-2165.

Fig. 1. *ETC* image of the error region of GRB 941014: the image measures 20° × 15°, with a pixel size of 3.′1. Stars as faint as V=10.5 are visible on this image. The thin arc encloses the error region of GRB 941014: the dimensions of the arc are 10° long (2σ error, as defined by *BATSE*) by 80" wide (3σ error, as defined by the Interplanetary Network). In this image, the thickness of the arc is exaggerated for clarity: the actual 3σ arc thickness is only 0.4 pixels.

References

Barthelmy S.D., et al., 1994, in Gamma-Ray Bursts, G.J. Fishman, J.J. Brainerd, K. Hurley (eds.), AIP Conf. Proc. 307, AIP, New York, p. 643

Krimm H.A., Vanderspek R., Ricker G.R., 1994 in Gamma-Ray Bursts, G.J. Fishman, J.J. Brainerd, K. Hurley (eds.), AIP Conf. Proc. 307, AIP, New York, p. 423

1 Vanderspek R., Ricker G.R., Doty J.P., 1992 in "Robotic Telescopes in the 1990s, A. V. Filippenko (ed.), Astronomical Society of the Pacific Conf. Proc. 34, A.S.P., San Francisco, p. 123

Deep Search for Optical Counterparts of Gamma-ray Bursters

V.V. Sokolov[1], *A.I. Kopylov*[1], *V.G. Kurt*[2], *S.V. Zharykov*[1]

[1] Special Astrophysical Observatory of Russian Acad. of Sci., Nizhnij Arkhyz, Karachai-Cherkessia 357147, Russia
[2] Astro Space Center of Russian Acad. of Sciences, 117810 Moscow, Russia

Abstract: Preliminary results are presented of the first (searching) stage of our observational program of the search and study of weak blue star-like objects in the two small γ-ray burst boxes GB 790418 and GB 790613. The basic search purposes are (1) the selection of star-like objects weaker than 23^{m} inside these boxes, (2) the selection of the bluest ones in $B-V$ among these objects. The first result is the fact that there are no blue (with $B-V<0$) star-like objects brighter than 24^{m} in both boxes. But in each of these GRB error boxes one blue object with smaller flux was selected for further study to elucidate their nature. In the smallest of the archival GRB error boxes (GB 790613) obtained by three satellite triangulation, the bluest star-like object with $V = 24.56(\pm 0.16)$and $B-V = -0.20(\pm 0.19)$ is near the center (at the distance $< 10''$) of the GRB box. In the GB 790418 error box, the bluest star-like object with $B = 24.40(\pm 0.20)$, $B-V = -0.40(\pm 0.40)$ is found at $\approx 50''$ from the center.

1 Introduction

If adhering to the idea of the Galactic origin of GRBs which proceeds from the concept of neutron-star type objects in the vicinity of the Sun, then such isolated, nearby objects should be found in the optical band as relatively blue, very faint ($V > 23$) ones with significant proper motions (Helfand et al. 1980). Of course, only the nearest objects can be seen at optical wavelengths. For example, in the "Local Disk models" (Hartmann 1994), these can be those up to 100 pc. A part of these nearest objects could appear as *soft* X-ray sources due to their thermal emission coming from the total surface, plus a polar cap accretion of the interstellar medium because of a strong magnetic field (B $\approx 5\times10^{12}$ G).

Since the search for GRB counterparts can be connected with the search for very faint star-like objects in visible band, deep CCD photometry with big telescopes in the smallest GRB fields is urgently needed. Up to now, such deep searches could be performed only for a few well-localized GRB events. One example is the report by Vrba et al. (1994) who studied the GB 790613 and GB 790418 error boxes at the U.S. Naval Observatory 1.0 m telescope. But the search for very faint blue star-like *candidates* in fields investigated by small telescopes can become a basic task on a large telescope.

In this respect, the observations at the 6m telescope can be considered as a part of an observational program (together with deep X-ray observing, of course) for the direct verification of one of the galactic models. Though these are called in question now because of the *CGRO* observations, they are not yet rejected completely. Even if we never find such weak and blue objects inside GRB boxes under consideration in a supposition of their relative nearness (50-100pc), then other models give even less hope to see anything in the optical band during the *quiet state* of the γ-ray bursters. Even in the "negative" case of no detections, we can say something about the upper limits of brightness of these objects for the Galactic GRB models.

Some preliminary results of the first ("searching") stage of two GRB fields can be reported. Here we only try to select the candidates for further study, or for the more precise BVR-photometry and proper motion study with the purpose of final elucidation of the nature of selected candidates.

2 Observation

CCD mosaic pointings for the two small GRB error boxes GB 790418 and GB 790613 were obtained at the 6m telescope with a 580×520 pixel CCD in the primary focus (24 m), providing 120×80″ field on the sky. The CCD chip has rectangular pixels of 24×18 μm, resulting in an image scale of $0\farcs205\times0\farcs154$/pixel. Multiple observations (or CCD mosaic) with 600 and 400 sec exposure times in the Cousins B and V filters were obtained.

Fig. 1 shows the error box for GB 790418 and the region actually covered by our CCD mosaic. The GRB error box was selected from the GRB catalogue (Atteia et al. 1987) and its configuration was taken from Golenetsky et al. (1986). The coordinates of the center of the error box are $\alpha(1950)=05^h51^m51\fs84$; $\delta(1950)=-06°57'10\farcs8$. Our telescope+CCD combination required 7 pointings to cover almost the entire error box. Inside every frame its number is shown.

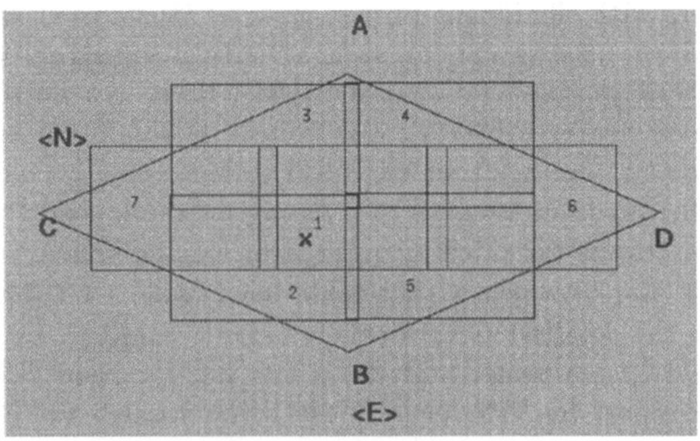

Fig. 1. CCD mosaic for the GB 790418 error box.

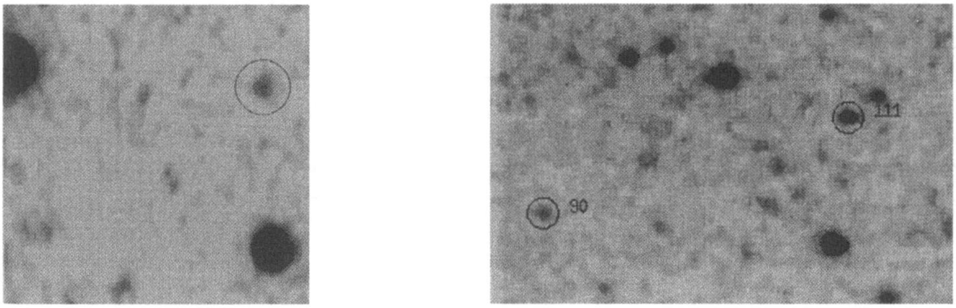

Fig. 2. Blow-ups of GB 790418 (left) and GB 790613 (right) CCD frames. Circle diameters are 6″ and 4″, respectively.

A portion of the central part of this CCD mosaic is shown in Fig. 2 (left, Gauss smoothed sum of all B and V frames). The seeing in the first night (February 6) was 2.″2–2.″3 in the central (1st and 2nd) frames, and became a little better (about 1.″9 FWHM) in the second night (Feb. 10). A relatively blue object, marked with a circle, is at a distance of approximately 50″ from the GRB error box center. It seems to be star-like, when compared with the profiles of stars. Since the object is very close to the detection threshold its stellar nature cannot be asserted. to assert its star nature now. Its brightness and colour are $B = 24.40 \pm 0.2$, $B-V = -0.40 \pm 0.40$. The object is apparently the bluest one in our CCD GRB field, in comparison with all 158 objects (including galaxies) of the field with $B < 24.4$. The cross indicates theposition: ($\alpha(1950) = 05^h51^m55^s.28$; $\delta(1950) = -06°57'04''.3$). In spite of a rather large colour error, it seems to deserve further efforts to elucidate its nature by measurements of its BVR brightness and proper motion.

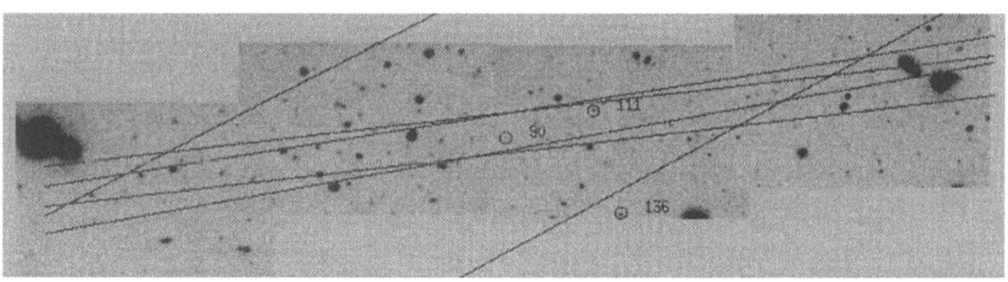

Fig. 3. The GB 790613 CCD field.

Our B and V CCD mosaic for the smallest GRB field (GB 790613) is shown in Fig. 3. The boundary of the symmetric GRB-error box is indicated according to the triangulation (see dicussion in Hack et al. 1994). Here also, each field is observed in two filters. The average seeing over the whole field did not exceed

$1''\!.4$. The observations were carried out on April 14, 1994. Three star-like and relatively blue objects are marked by circles: No. 90 with $B = 24.36 \pm 0.10$, $B - V = -0.20 \pm 0.19$, No. 111 with $B = 23.52 \pm 0.06$, $B - V = 0.18 \pm 0.08$, and the brightest, evidently star-like object, No. 136, being outside the GRB-error box ($B = 23.14 \pm 0.03$, $B - V = -0.04 \pm 0.05$). The faintest of these objects (No. 90) with a negative $B - V$ index lies within $\approx 10''$ from the center of the GRB-error box ($\alpha(1950) = 14^h 12^m 12\!\!^s\!.8$, $\delta(1950) = +78° 54' 13''\!.8$). A portion of the central CCD frame with objects No. 90 and 111 is shown in Fig. 2b.

3 Conclusion

1. Inside both GRB error boxes, no blue star-like objects brighter than $B = 24$ exist.
2. We selected the bluest candidates inside both GRB boxes for a more careful study (more precise photometry and proper motions) within our program on "The search for γ-ray bursters in a quiet state".

References

Atteia J.-L. et al., 1987, ApJS 64, 305
Golenetsky S.V. et al., 1986, preprint 1026, Ioffe PTI, Leningrad
Hartmann D.H., 1994, Proc. 2nd Huntsville GRB Workshop, AIP 307, p. 562
Hack F., et al., 1994, Proc. 2nd Huntsville GRB Workshop, AIP 307, p. 359
Helfand D.J., Chanan G.A., Novick R., 1980, Nature 289, 337
Vrba F.J., Hartmann D.H, Jennings M.C., 1994, Proc. 2nd Huntsville GRB Workshop, AIP 307, p. 443

CCD Sky Monitoring and Burst Alert

René Hudec, Jan Soldán

Astronomical Institute, CZ–251 65 Ondřejov, Czech Republic

Abstract: Two proposals for detection of optical flashes related to γ-ray bursts are briefly described. The aim of the first proposal is to develop a reliable monitoring service at optical wavelengths, and the second proposal deals with an automatically working telescope with full remote control via Internet network.

1 Optical Transient Monitor (OTM)

The idea is to get very wide-field monitoring of the sky with still reasonable quality and parameters and also to concentrate on brighter flashes, but with a high background elimination level (Hudec & Soldán 1994). Recent results (Hudec et al. 1994a,b, Hudec 1993) indicate that there are bright transients exceeding fifth magnitude (1 s duration assumed). We suggest a low-cost automatic operation experiment with good time resolution, medium sensitivity and angular resolution, and excellent background elimination. Previous experiments in this field have shown that the main problem of these investigations is the high noise (i.e.,background caused by false triggers) to signal ratio. It is rather important to be able to eliminate reasonably all background and false triggers. We suggest to develop a double CCD camera with two independent paths equipped with different filters and to develop a special software excluding false triggers and triggers caused by satellites. The OTM system consists of the following:
(1) The classical IBM PC-386, 486 computer as a host computer for the Texas Instruments Parallel DSP TMS320C40 card.
(2) The internal AT add-in card with two TMS320C40 processors with 1 MByte of static memory and 32 MBytes of dynamic memory.
(3) Two or four CCD cameras with wide–field lenses.

The basic parameters of the proposed test device are as follows: *Lens:* wide-field, $f = 6$ mm, 1 : 1.6. *Focal detector:* CCD camera ST-8 SBIG (Santa Barbara Instrument Group, Santa Barbara, CA). *Area of the focal detector:* 13.8 $\times$ 9.2 mm, 1534 $\times$ 1020 pixels (9 μm square each). *Field of view:* less than $66^\circ \times 88^\circ$ or 5808 deg^2 (28% of the hemisphere) *Angular resolution:* $0.^\circ04 - 0.^\circ1$. *Number of optical paths in one device:* 2. *Estimated limiting magnitude:* 6 (expected; for 1 s flashes). *Time resolution:* approximately 40 s (1 s exposure, rest downloading and data processing) for two cameras and $\sim$20 s for four cameras in one device (the time resolution will increase up to 10 (5) s if either a smaller area of the chip will be digitized, or, alternatively, the pixels of the whole area will be binned.

2 Robotic telescope

The goal of the project is the development of an automatically operated telescope. A new generation of control software will be developed, allowing the fully robotic operation of the system i.e. remote control of the telescope via international data network Internet without any human participation. This will minimize the time delays between the detection of high-energy phenomena on satellites and corresponding optical ground-based observations. The main scientific goal is to use the robotic telescope described here in the *BACODINE* project which calculates the positions of celestial γ-ray bursts detected by the NASA Compton Gamma Ray Observatory (*CGRO*) and distributes these positions very rapidly to participating optical ground-based observatories. The robotic telescope will receive this positional burst information by means of the Internet network and will be automatically pointed to burst positions and the wide ($\geq 10^\circ$ FoV) and narrow ($\leq 1^\circ$ FoV) CCD cameras will start observations.

The configuration of the project is based on the use of commercially available hardware components. The test configuration consists of newly developed Celestron Ultima 2000 PC-controlled mount with a 30 cm aperture telescope and a wide–field camera both with SBIG ST-8 CCD cameras, and PC-IBM 486. This enables to duplicate the system easily at other places at low cost. The corresponding software will be fully designed and developed (within $\sim$ 1 yr) and will guarantee a high level of automatization. The final goal is to establish a network of automated robotic telescopes on several sites.

An automatically operated telescope with remote control from any terminal of the Internet network can be obviously applied also in other fields of astronomy and astrophysics. The system can be used with little modifications also for other satellite astronomy projects where quasisimultaneous ground-based optical data are necessary, such as experiments *WATCH/GRANAT*, *HETE*, *INTEGRAL* etc.

Acknowledgement: We acknowledge support provided by the Grant Agency of the Czech Republic (grant No. 202-93-0890).

References

Hudec R., 1993, A&A 270, 151
Hudec R., Soldán J., 1994, ApJS 92, 675
Hudec R., Dedoch A., Pravec P., Borovička J., 1994a, A&A 284, 839
Hudec R., Pravec P., Borovička J., 1994b, A&A 284, 499

ALL–SKY MONITORING

The Plate Collection at the Harvard College Observatory

Martha L. Hazen

Harvard-Smithsonian Center for Astrophysics, Cambridge, MA 02138, U.S.A.

Abstract: This review presents an overview of the large and important photographic plate collection at the Harvard College Observatory, including its history, uniqueness, contents, storage, uses and successes.

1 History

In 1839, Louis Daguerre announced his discovery of a process by which an image made by light on silver grains could be made permanent. Very soon thereafter, in 1840, Henry Draper of New York City first successfully daguerreotyped an astronomical object, the moon; this image no longer survives. Daguerreotyping began at Harvard in 1849 with the recently installed Great Refractor (0.38 m aperture), and four daguerreotypes of the moon taken with this instrument in 1851 and 1852 remain in our collection. Four daguerreotypes of a partial eclipse of the sun in 1851 also remain.

In 1852, daguerreotyping ceased at Harvard, due to limitations on exposure time posed by an inadequate clock drive on the Great Refractor. With a new drive installed in 1857, photography began again with the then new wet-plate, or collodion, process. Many hundreds of collodion plates from 1857 to 1885 are included in Harvard's collection, including a beautiful series of an occultation of Spica by the moon in 1857, images of Saturn and Jupiter from 1857-60, images of the inner solar corona from eclipses in 1869 and 1870, several hundred detailed solar plates from the 1870s, and the majority of Benjamin Gould's plates from Córdoba, Argentina, in the 1870s and 1880s.

The research plate collection began in 1885, with the aquisition of Harvard's first photographically corrected refractor, the 0.2 m Bache telescope, with which 75 000 plates were taken up until 1955 when the telescope was closed. The arrival of this telescope allowed the Harvard astronomers to take full advantage of the new dry plate, a much easier and more sensitive photographic medium, which was developed in the early 1880s.

The late 1880s and early 1890s also saw the establishment of Harvard's southern station in Arequipa, Peru. Harvard maintained southern hemisphere telescopes until 1955; the southern station was moved from Arequipa to South

Africa in 1924. The southern patrol program was reinstated in New Zealand in the late 1970s, and ran until 1988.

2 Uniqueness

The collection at the Harvard College Observatory is unique among plate archives in at least three ways. It has the longest time span of coverage, from 1885 to 1989, with some unfortunately notable gaps. The number of plates, approximately 500 000, is the largest of any archive. It should be noted that most are direct blue plates, although some 5 to 10% are objective prism spectrum plates, and a small percentage is red or yellow plates. Finally, Harvard's collection, unlike others, covers the entire sky, beginning in 1889. Most of the other important archives of wide angle plates cover mainly the northern celestial hemisphere.

3 Contents

The collection at Harvard comprises plates taken with 50-60 different photographic telescopes; about 20 of these are responsible for the bulk of the collection. A summary of the important telescopes is shown in Table 1, where instruments are divided into groups by focal scale. In some cases, several telescopes are grouped together. For each instrument are listed aperture, focal scale, hemisphere, approximate total number of plates, and years of operation. It is of interest to mention briefly some of the telescopes and their work.

The 0.325 m telescope, the Boyden refractor, was the instrument used by Solon Bailey at the turn of the century to study the RR Lyrae stars in globular clusters like Omega Centauri. The 0.6 m Bruce refractor produced the plates used by Henrietta Leavitt to discover the period-luminosity relation for Cepheids in the Small Magellanic Cloud. And two of the three telescopes of 0.2 m aperture, the Bache and Draper refractors, took the spectral plates used for the Henry Draper Catalog of stellar spectral types. The patrol plates were taken by many different instruments that are grouped together in Table 1 by focal scale. The last entry, the 42 mm telescopes, were Harvard's most recent patrols.

As a practical indication of coverage of the sky, one can say that, for a given object of blue magnitude 15 or brighter, from 100 to several thousand images may be found in our collection. If the blue magnitude is 17.5, there will be from two or three to several hundred images.

4 Storage and care

A special fireproof building was erected in 1934 to house the collection, with extra reinforcing to support the weight of the plates. Each plate is stored in a bond paper, or more recently Tyvek, envelope. Plates stand vertically in steel cabinets. The plate stacks are temperature controlled to 20 C in the winter and

Table 1. Summary of Principal Plate Series

Aperture	Scale	N/S	No. of Plates	Years
		Scales < 80"/mm		
1.5 m	26"/mm	N	13 000	1934-1989
		S	6 000	1933-1955
0.325	42	S	19 000	1888-1951
0.6	60	S	27 000	1893-1950
		Scales 80-300"/mm		
0.2 m	170	N	76 000	1885-1976
		S	71 000	1889-1954
0.3	100	N	10 000	1905-1983
0.4	100	N	40 000	1909-1988
0.25	170	S	40 000	1915-1955
	Scales > 300"/mm = Patrol Plates			
75 mm	390	N,S	33 000	1928-1963
38	600	N,S	75 000	1898-1957
42	580	N,S	9 600	1962-1989

22 C in the summer. No attempt has been made to control the humidity, except by the air conditioning.

The current personnel comprises a quarter-time Curator and a half-time Assistant Curator. The cost of the entire care of the plates, including heating, electricity, building maintenance, etc., is roughly US$ 80K per year. About half of this comes from an endowed restricted fund of the Harvard College Observatory.

Patrol plates are filed by region, and are therefore easy to access. The remaining plates are filed by individual telescope and date (or plate number); these must be located by card catalogs, which are filed by position. We are currently working on a machine readable catalog, although the projected completion date is well into the next century.

The plate collection at Harvard is open to any astronomer by arrangement with the Curator.

5 Problems

As with any similar archival collection, Harvard's plate archives do have problems. As mentioned before, there are notable time gaps in the data, the most recent being from the early 1950s to the 1970s. However, as noted below, the collection at Sonneberg fills this in for the northern hemisphere. A second problem, as noted above, is a lack of a machine readable catalog.

Table 2. Observations of KR Aurigae

Dates	Harvard plates	Sonneberg plates
1895-99	7	—
1900-09	20	—
1910-19	32	—
1920-29	93	—
1930-39	318	54
1940-49	72	62
1950-59	13	28
1960-69	3	229
1970-79	40	$\gg$10*
1980-82	14	11

* Popova (1975) looked at 10 Sonneberg plates 1970–1971. There are no published Sonneberg data 1972–1979.

The plate stacks occupy a rather large volume of space in a choice location at Harvard. This has not escaped the notice of higher authorities, and from time to time there begins a move to relocate or store the plates somewhere else. However, the plate stacks were so beautifully and specifically built to house the collection that to convert the space to offices, etc., appears, at least up to now, to be prohibitively expensive.

6 Uses and successes

In the early days, the direct plates were used largely to help in setting up a system of photometric standards and to discover and study variable stars. The objective prism plates were used, to develop a scheme of stellar spectral classification and to apply this to stars brighter than magnitude 10; the product of this work is, of course, the Henry Draper Catalog.

More recently, the direct plates have been used to derive historical light curves of newly discovered objects and types of objects. With use of the Harvard plate collection, the rapid optical variation of the QSOs was first recognized. Light variations found for HZ Herculis from the collection first showed the turn-off and turn-on of this X-ray source (Jones et al. 1973). And the first (and only?) possible optical counterpart of a γ-ray burster was found in the plate collection (Schaefer 1981, Schaefer et al. 1984).

A study of the optical variation of the X-ray source KR Aurigae (Popova 1975, Liller 1980, Liller & Popova 1984) shows how nicely the Harvard and Sonneberg plate collections can mesh. Table 2 lists for 10 year intervals the number of plates from each observatory used in the study; clearly the major decline in numbers of plates at Harvard, beginning in the 1950s, has been matched by an increase of those at Sonneberg.

The Harvard plates are also being used currently to derive light histories of many other types of variable stars of interest. In 1994, eleven different individuals or groups each searched from several hundred to many thousands of plates to determine light curves for variables of interest.

7 Summary

The astronomical photographic plate collection at the Harvard College Observatory, because of its long time base (1885-1989), large number of plates (500 000), and equal coverage of both hemispheres of the sky, is a valuable resource for the international astronomical community. We welcome visitors interested in taking advantage of what the collection has to offer.

References

Jones C.A., Forman W., Liller W., 1973, ApJ 182, L109
Liller M.H., 1980, AJ 85, 1092
Liller M.H., Popova M.D., 1984, IBVS 2463
Popova M.D., 1975, Astrofiz. Issled. (Sofia) 1, 68
Schaefer B.E., 1981, Nature 294, 722
Schaefer B.E. et al., 1984, ApJ 286, L1

M. Tsvetkov: Do you have plans to digitize your plates in Harvard?
M. Hazen: Although I hope this may be done in the future, the sheer labor involved in such a task has discouraged us from starting this. Perhaps more important at the moment is completion of the machine readable catalog, which will take many more years at the current rate.

Optical Sky Monitoring: Past and Future

H.-J. Bräuer[1], *N. Vogt*[2,3]

[1] Sternwarte Sonneberg, 96515 Sonneberg, Germany
[2] Max-Planck-Institut für Astrophysik, 85740 Garching, Germany
[3] Universidad Catolica de Chile, Santiago, Chile

Abstract: After outlining the achievements of photographic sky monitoring at Sonneberg Observatory we describe a concept of a future CCD-based sky patrol. We suggest a world-wide network of 6 identical observing stations which cover the entire sky down to a limiting magnitude of $\sim 18^m$ and with a time resolution of between 8 and 24 hours. A preliminary technical design is proposed, and the astrophysical and general impact of such a project is briefly discussed.

1 Introduction

To be on alert for every incident in the sky is a task of mankind, an essential in any civilisation on some cultural level. Even the people of ancient times knew that: there are records left by antiquity on novae, supernovae and comets, among them the Chinese and Korean reports, which are of scientific value up to the present day.

For millenia visual observation was the only means of monitoring. Photographic techniques in astronomy became common practice just about one hundred years ago. Only since then was it possible to run what we understand by a systematic sky patrol. It was conducted first at Harvard Observatory (USA) and, a few decades later, by Cuno Hoffmeister in Sonneberg, which now is the site with the largest uninterrupted records.

2 The Sonneberg archive

The Sonneberg patrol was started in 1926 as one of the cornerstones of the Sonneberg programme of variable star research. It consists of a continuous record of the whole northern sky (Sky Patrol in the literal sense) and, in addition, of 81 fields selected along or near the Milky Way (Field Patrol). One quarter of all variables known in the Galaxy were detected by its means. Today its archive contains about 250,000 photographic plates. It is an excellent source of information on the long-term behaviour of active objects, and a valuable support of observation made from satellites at non-optical wavelengths. Its logbook data are computerised and retrievable from a database.

Table 1. Sonneberg Observatory: Photographic telescopes currently in operation. (cf. Bräuer and Fuhrmann, 1992)

	d (cm)	f (cm)	Plate Size (cm×cm)	Field Size (°×°)	(″/mm)	Plate Limit (^{m})	Number of plates	since
Schmidt	50/70	172	13×13	3.8×3.8	120	18.5 (B)	8900	1952
Field Patrol:								
Astrograph	40	200	30×30	8×8	100	17.5 (pg)	6900	1960
Astrograph	40	160	30×30	10×10	130	17.5 (pg)	12600	1938
Sky Patrol:								
7 Tessars	7.1	25	13×13	26×26	830	14.5 (pg)	96500	1956
7 Tessars	5.5	25	13×13	26×26	830	13.5 (pv)	53500	1958

For some time the Sonnebergians have been contemplating a new method of sky patrol, but have not been able to proceed because the observatory has been threatened with closure. Doubtless Sonneberg, with the involvement of the nearby Technische Universität Ilmenau, would be the place where this new patrol could materialize the fastest and in the most natural way. The new patrol can however be realized anywhere else, albeit with far more difficulty. The following is an outline of the project.

3 ASPA: The new way

An efficient sky patrol requires utilization of the most advanced technology available. These days the place of the photographic plate is being taken by digital detectors. CCDs have begun to dominate nearly all areas of observational astronomy, and now they are on the verge of conquering even the last stronghold of the traditional photographic plate: wide-field observation. Thus, since the time has come to develop new concepts of a CCD-based sky patrol, an international working group, called ASPA (*All Sky Patrol Astrophysics*) was recently founded by one of us (N.V.) together with Constanze la Dous (Madrid) and Hanns Ruder (Tübingen). The main aims of ASPA are:

1. to monitor the flux variations (stellar and non-stellar objects) in the entire sky with a time resolution of 8 - 24 hours, using latest-generation digital detectors (CCDs);
2. to continue the traditional photographic sky patrol until CCD observation is in operation;

3. to give rapid and effective on-line access to the variations observed at any sky position to the astronomical community; and
4. to transform existing photographic plate archives into digital form in order to offer the same access to data taken in past decades.

ASPA should be conducted by a world-wide network of 6 identical semi-automatic observing stations. Each station would consist of about 24 cameras ($D \approx 0.2$m, $f \approx 0.6$ m), possibly on two mountings: one for regions near the equator, the other for those near the pole. If each camera is equipped with a 2048 × 2048 pixel CCD, a limiting magnitude of 18^m with a photographic accuracy of 3 - 5 % could be reached with exposure times of the order of 6 minutes (field size $3^\circ \times 3^\circ$, scale 5"/pixel, average distance between stars: 10 pixels). The above estimates are based on Kroll and Lehmann (1993). The observing stations should be located at established observatories, preferably some 120° apart from each other in longitude. Possible locations on the northern hemisphere could be the Hawaiian Islands (Mauna Kea?), the Canary Islands and a far-east location. On the southern hemisphere they could be in Chile (Paranal area: cooperation with the Catholic University, Santiago), Namibia (Gamsberg?) and Australia. Those stations are only for observing. All data should be, possibly in compressed form, transmitted to an ASPA Centre for reduction, analysis, storage and distribution. The centre should, if possible, be an active observatory where the observing stations can be tested before deployment, and maintenance tasks as well as further technical developments can be done. The centre should also be responsible for the digitization of photographic plate archives.

4 A preliminary technical design

There are many ways of realizing a CCD-based sky patrol technically. We will only present some first ideas. (In our description we consider the northern hemisphere case.)

Each observing station could consist of two units:

Unit A: 16 cameras on a single parallactic mounting, movable only about its polar axis. With each camera adjusted to one fixed declination, they cover the declination zones $-6^\circ \ldots +39^\circ$.

Unit B: 8 additional cameras, also on a single parallactic mounting, but each one adjustable to two different declination zones (three zones for the pole camera). Each camera is equipped with a step motor for fine adjustment in right ascension. Those cameras cover the declination zones $+42^\circ \ldots +90^\circ$.

By this arrangement, each accessible sky position could be observed once per night with the following procedure: if we assume an average length of an observing night of 10 hours, and a range in right ascension covered from HA = $+4^h$ (west) at beginning of night to HA = -4^h (east) at its end, a total range of 18^h in right ascension would be monitored in a total exposure time of 600 minutes. Unit A would perform this task for the equatorial zones, resulting in 90 fields of $3^\circ \times 3^\circ$ per night and camera, with 6 minutes exposure each. Nearer

Table 2. Proposed distribution of cameras in Unit B

	Camera No	First Zone δ	N_e	Second Zone δ	N_e	Total N_e
	1	+90, +87°	9	+42°	69	78
Distribution of cameras	2	+84°	12	+45°	66	78
	3	+81°	17	+48°	63	80
(N_e = number of exposures	4	+78°	21	+51°	59	80
necessary to cover	5	+75°	26	+54°	55	81
$\Delta\alpha = 18^h$ in one night)	6	+72°	30	+57°	51	81
	7	+69°	35	+60°	47	82
	8	+66°	39	+63°	43	82

to the pole, where unit B is active, each camera would be able to cover two different declination zones. Table 2 gives a possible assignment of those zones to the cameras:

The exposures in all cameras of unit B should be synchronized. Between exposures, its mounting is pointed at the same hour angle as unit A, while the actual declination zone and the accurate field centre in right ascension are selected for each camera. During exposure the mounting just follows the daily motion of the stars.

Mechanics and optics of this design would be relatively simple, especially in unit A. This unit could work either in pointed or in tracking mode. We would obtain about 2100 frames (3° × 3°) per night and station, corresponding to about 18 GByte of information. The network of 6 stations would imply a time resolution of about 8 hours (perfect weather conditions assumed) and an overlap between northern and southern stations of 15° in declination along the equator.

5 Astrophysical impact, outlook

It is rather evident that observations of this kind open fascinating new aspects covering wide areas in present-day astrophysics. This refers not only to the 'classical' variable star work on pulsating, eruptive, or eclipsing variables. Targets of ASPA will range from objects on the fringe of the visible universe like quasars and AGNs down to tiny earth-orbit crossing asteroids nearer than the Moon. These are a few highlights: Many thousands of galaxies will be searched for supernovae and we may expect about five detections per night. One of the mysteries of modern astrophysics is the nature of gamma-ray bursts and their hitherto undetected optical counterparts. If, for instance, they would be brighter than 14^m for a few second at least, ASPA might record one coincidence of gamma-ray burst and its optical counterpart every 100 days. We also would probably observe between 50 and several hundreds of gravitational lensing events per year. Near-earth asteroids of diameters as small as 60 m would be detectable, and so on and so forth.

Of course, this project implies rather ambitious goals. Challenging features are the development and optimization of the organization of observations, especially in unit B, which requires quite sophisticated software in order to ensure efficient automatic operation. The second main challenge is the reduction and management of the large amount of data: in total some 240 million stars are to be automatically monitored, reduced, classified and, if variable, placed at the astronomical communities disposal within 24 hours. It is an immense task, but it is not impossible with today's technology: presently active gravitaional microlensing experiments (EROS, MACHO, OGLE) toward the Magellanic Clouds and the bulge of the Milky Way also produce large amounts of information on stellar and non-stellar variability. They have already developed part of the technology, and it should be utilized by ASPA. For instance, the MACHO group is able to perform the photometric reduction of a night's data (5 GByte) within 24 hours (Cook et al., 1994). This corresponds already to about 30% of the data expected from an ASPA observing station. Sophisticated reduction procedures which work automatically and assure adequate photometric accuracies are also available (Schechter et al., 1993).

In general terms, ASPA will obtain astrophysically relevant and important information at relatively low cost. It is unique information, and, if ASPA is not carried out, this information could never be recovered later, however sophisticated future technologies will be. The service function of ASPA to the astronomical community implies that a large range of scientists and institutions will benefit from the project. In particular, a very close interaction between ground-based and space astronomy is to be expected. Finally, the technological aspects of the project require close cooperation between astronomers and technology researchers at universities, as well as with industry. Interdisciplinary cooperation will be very important.

References

Bräuer H.-J., Fuhrmann B., 1992, The Messenger 68, 24
Cook K. *et al.*, 1994, IAU XXIInd General Assembly, Poster Abstracts, p. 138
Kroll P., Lehmann T., 1993, IAU WG on Wide-Field Imaging, Newsletter 3, 30
Schechter P.L., Mateo M., Saha A., 1993, PASP 105, 1342

M. Merck: 1) You will get some 60 TByte a year of data. How do you plan to store them? 2) CR events are difficult to suppress if you don't have 2 contemporaneous observations of the same field. Are you planing to implement this type of observations?
N. Vogt: 1) On optical disks or other similar, may be future, storage devices. 2) This would even double the data received. In the tracking mode these events are unimportant. In pointing mode we still have to find a solution.

The Odessa Sky Patrol Plate Collection

V.G. Karetnikov, A.K. Markina, V.P. Sotnikov

Astronomical Observatory of Odessa State University

The Odessa Astronomical Observatory has three collections of Sky Patrol negatives, which have been exposed since the beginning of the 20th century. Altogether, there are about 100,000 negatives of the stellar sky, covering a declination range from -15 to $+90°$. The limiting stellar magnitudes are 15 (photographic – pg) and 12 (photovisual – pv), respectively.

Table 1. Characteristics of the Simeiz 2-camera astrograph: Date of observations is from JD 2418405 to 2434725 (1909–1953)

Type of objective	No. of cameras	D & F in mm		Plate in cm	Field in °	Limit magn. Photom. syst.	Centre in °
UNAR	1 & 2	120	600	13x18	12×17	< 9.2 pg	0

Table 2. Characteristics of old astrographs of Odessa Observatory: Date of observations is from JD 2433812 to 2435776 (1951–1956)

Name of instrument	No. of cameras	Type of objective	D & F in mm		Field in °	Photom. system	Center in °
"Large" astrograph	1	Tessar SOI	148	1000	9×12	pg	0
"Small"	1	Industar-17	100	500	22×15	pg, pr	0
astrograph	2	Triplet Zeiss	100	500	22×15	pg, pv	0
Astrograph	1	Industar-163	67	300	20×30	pv	0
"Hedgehog"	2	Xenon	60	120	20×30	pv	-25
	3	Epostar	60	120	20×30	pv	$+25$

1. The first of the three collections is the "Simeiz" collection obtained at the 2-camera astrograph in Crimea at the beginning of the century (about 10 000 plates) to discover and study small planets and comets. The field centers are determined by the position of small planets and comets and by their expected region of presence. Coordinates are indicated on the plate envelopes.
2. The second archive of similar size consists of negatives obtained at the "old" astrographs of the Odessa Observatory. There is a small number of plates exposed in photored light (pr). For these plates we have different guide stars. These stars are indicated on the plate envelopes. The plate centers listed in Table 2 are given as offset relative to the corresponding guide star (in degrees of declination).
3. The third archive is obtained from observations of the "7-camera" astrograph in Odessa which have started in 1957 and are continued until now. It contains about 80 000 negatives. For this archive, there is an electronic version of the catalogue which contains the numbers of the negatives and exposure times. The plates were separated into three series: Old, New, III series (see Tables 3, 4). In the Old series the following 25 guide stars were used:

Table 3. Characteristics of the 7-camera astrograph: Dates of observations is from JD 2436039 (1957) until now

Name & Data	No. of cameras	Type of objective	D in mm	F in mm	Field in °	Photom. system	Center in °
The Old	1	Uran-9	100	250	30 × 40	pg,pv	+13
series.	2	Uran-9	100	250	30 × 40	pg,pv	−13
From	3	Uran-9	100	250	35 × 25	pg	+25
19.07.1957	4	Vierlinser	160	720	13 × 18	pg	+08
to	5	Vierlinser	160	720	13 × 18	pg	−08
08.04.1959	6	Uran-9	100	250	30 × 40	pg	−27
	7	Tessar SOI	148	100	9 × 12	pg	0
The New	1	Uran-9	100	250	30 × 40	pv	+10
series.	2	Uran-9	100	250	30 × 40	pv	−10
From	3	Triplet	100	500	22 × 15	pg	+30
09.04.1959	4	Vierlinser	160	720	13 × 18	pg	+18
to	5	Vierlinser	160	720	13 × 18	pg	−18
08.06.1966	6	Industar17	100	500	22 × 15	pg	−33
	7	Uran-12	200	500	18 × 24	pg,pv	0
The III	1	Uran-9	100	250	30 × 40	pv	+13
series.	2	Uran-9	100	250	30 × 40	pv	−13
From	3	UNAR	120	600	16 × 11	pg	+23
09.06.1966	4	Vierlinser	160	720	13 × 18	pg	+08
until	5	Vierlinser	160	720	13 × 18	pg	−08
now	6	UNAR	120	600	16 × 11	pg	−23
	7	Uran-12	200	500	18 × 24	pv	0

α And, δ And, β And, γ And, α Per, α Tau, α Aur, η Gem, *i* Cnc, γ Leo, δ Leo, β Leo, ε Vir, α Boo, α CrB, ζ Her, δ Her, μ Her, γ Oph, α Lyr, ζ Aql, β Cyg, ζ Cyg, β Peg. In the New series the following stars were added: 41 Ari, ζ Per, ζ Gem, λ Leo, β Vir, 70 Vir, ζ Boo, γ Her, γ Lyr, γ Cyg, ϵ Peg, but μ Her was discarded. These 34 stars have been used in the III series. The shifts of the plate centers (in degrees of declination) from the positions of guide stars are listed in Table 3. With exposure times of 30 minutes and covering a 2-hour right ascension zone at once photos can be taken of all the apparent celestial sphere each night by using guide stars located in declination between +10° to +40°.

The material of the collections is used by professionals and amateurs to study properties and brightness variations of different celestial bodies. However, precise BV-photometry is possible only with 7-camera astrograph material.

Table 4. Colour coefficient c(b) and c(v) of photometric systems of the 7-camera astrograph calculated by A.N.Rudenko in the form m(i)=V+c(i)·(B-V) and with a common error of determination dm

Cameras	1	2	3	4	5	6	7
		The Old series - JD 2436039 - 2436668					
c(b)	+0.06	-0.18	-0.21	-0.37	-0.06	-0.12	-0.23
dm	0.10	0.13	0.12	0.10	0.12	0.11	0.07
c(v)	+0.19	+0.25	from JD 2436281				
dm	0.24	0.14	from JD 2436281				
		The New series - JD 2436690 - 2439286					
c(b)'	to JD 2439038		-0.27	-0.33	-0.33	-0.14	-0.04
dm'	to JD 2439038		0.07	0.24	0.15	0.12	0.09
c(v)'	+0.23	+0.18	to JD 2439038				
dm'	0.28	0.18	to JD 2439038				
c(b)"	from JD 2439050		-0.07	-0.26	-0.31	-0.14	
dm"	from JD 2439050		0.07	0.10	0.24	0.11	
c(v)"	+0.20	+0.17	from JD 2439050 (1,2 cameras)				+0.22
dm"	0.13	0.20	from JD 2438930 (7 camera)				0.12
		The III series - JD 2439291 - up till now					
a(b)'	to JD 2441983		-0.32	-0.45	-0.31	-0.34	
dm'	to JD 2441983		0.15	0.12	0.12	0.09	
c(v)'	+0.23	+0.27	to JD 2441274				+0.14
dm'	0.16	0.17	to JD 2441274				0.11
c(b)"	to JD 2444074		-0.33	-0.03	-0.06	-0.07	
dm"	to JD 2444074		0.19	0.12	0.08	0.18	
c(v)"	+0.34	+0.34	up till now				+0.32
dm"	0.15	0.12	up till now				0.13
c(b)"'	from JD 2444075		-0.29	-0.26	-0.02	-0.07	
dm"'	from JD 2444075		0.16	0.09	0.10	0.17	

Nova Patrol at Abastumani

Givi N. Kimeridze, Omar M. Kurtanidze

Abastumani Astrophysical Observatory, 383762 Abastumani, Georgia

1 The Telescope

The Schmidt telescope (360-333-625 mm) of the Abastumani Astrophysical Observatory has been used for photographic observations for more than fifty years. It provides a field of $7^\circ\!.3$ (80 mm) and a non-vignetted field of 4° (44 mm). A limiting magnitude of $16^{\rm m}\!.5$ can be reached in 7-8 min exposure on medium sensitivity astronomical films and $18^{\rm m}$ with fine grain emulsions.

Multicolour and polarization observations of a few hundred relatively bright galaxies were carried out by M. Vashakidze during the first two decades of the telescope operation. In the mid-1950s, he also discovered the strong polarization of the Crab nebula. Since 1961, systematic photographic observations are carried out in the framework of the International Cooperation Programme of Supernova Patrol in 24 fields. The telescope is also used to study the brightness variations of eruptive and exploding variable stars.

2 The Patrol

Systematic observations of novae began in the 1920ies. Main data on the ≈ 300 novae can be found in two catalogues (Kholopov et al. 1985, Duerbeck 1987). About one fourth of novae are not classified at all, and for many more the classification is not reliable.

According to Kholopov et al., it is possible to divide novae by their brightness variations into fast, slow, very slow and recurrent novae. Some authors are also pulling out the class of nova-like stars, which includes different, nonuniform groups of variables. Novae of different classes have different maximum brightnesses, but the differences are not yet well established (Hoffmeister et al. 1990).

A most critical parameter of novae is their flare frequency. Its determination is complicated because of several reasons: (a) there is no systematic wide-field nova patrol; (b) the discovery of novae in the galactic plane and especially in the galactic center direction is limited by interstellar absorption. It is possible to estimate the flare frequency by extrapolating the value for the same type of galaxy as our own. Arp (1956) and Sharov (1993) estimated the flare frequency of novae in M31 as 26 and 31 per year, respectively, while extrapolations of nova patrols in the Galaxy lead to the value 76 per year (Liller 1987).

Fast novae are uniformly distributed along galactic longitude, while slow novae show a tendency of concentration towards the direction of the Galactic center. This fact probably indicates that there are two types of novae in galaxies like our (Duerbeck 1990), while it is not confirmed for M31 (Sharov 1993). Sharov also indicated the expediency of nova patrols in M31 and the Galaxy.

There are also much problems connected with recurrent novae. It is not yet known whether all novae are recurrent. Kaluzny and Glebowski give a list of potential recurrent novae for patrol observations. The selected objects have generally a low maximum brightness which is statistically correlated with short intervals between nova outbursts. Only one object from this list – V1017 Sgr – appears as recurrent in nova catalogues.

It should be emphasized that there are connections between novae and nova-like objects (dwarf novae, symbiotic stars, polars) and planetary nebulae. Only a few polars are known up to now, so it is desirable to carry out a nova patrol also in this respect.

We plan to initiate a nova patrol at Abastumani, using the Schmidt telescope. The 59 selected fields fully cover the following three strips: $\ell = 0° - 30°$, $|b| < 12\overset{\circ}{.}5$, $\ell = 30° - 60°$, $|b| < 7\overset{\circ}{.}5$, and $\ell = 60° - 90°$, $|b| < 5°$. We plan to use A660N photographic films with maximum sensitivity at Hα with wavelength 6000 – 6700 Å (Novae have strong Hα emission during outburst) and avoiding blue field stars and blue background. With 5 min exposures, the whole patrol cycle may be completed even during two summer nights.

This material will also be used for the identification and subsequent investigation of the variability of late type stars (M and C stars), detected in a low dispersion spectral survey (Kurtanidze & Nikolashvili 1994). Finally, we like to note that any suggestion for cooperation in Nova Patrol Programme is very welcome.

Acknowledgements: One of the autors (O.M.K.) is grateful to the SOC of the conference for financial support to attend.

References

Arp, H.C. 1956, Astr.J., 61, 15

Duerbeck, H.W. 1987, Space Science Reviews, 4

Duerbeck, H.W. 1990, in Physics of Classical Novae, IAU Coll. no.122, ed. A. Cassatella & R. Viotti, Springer, Berlin, p. 34

Hoffmeister, C. et al. 1990, Variable Stars, Nauka, Moscow

Kaluzny, J., Chlebowski, T. Preprint CfA no.2667

Kholopov, P.N. et al. 1985, General Catalogue of Variable Stars, Nauka, Moscow

Kurtanidze O.M., Nikolashvili M.G. 1994, in Astronomy from Wide-Field Imaging, IAU Symp. 161, p. 467

Liller, W. 1987, APSS 131, 449

Sharov, A.S. 1993, Pisma Astr. Zhurn. 19, 18

Wide-Field Plate Database: A Progress Report

M.K. Tsvetkov, K.Y. Stavrev, K.P. Tsvetkova, A.S. Mutafov

Institute of Astronomy, Bulgarian Academy of Sciences,
72 Tsarigradsko Shose, BG-1784 Sofia, Bulgaria

Abstract: The current status of a database containing information on wide-field observations stored in plate archives all over the world is presented. The database will permit on-line access to information on nearly $2 \cdot 10^6$ wide-field ($\gtrsim$ 1 degree) plates and films, obtained over more than a century with about 200 telescopes and cameras.

1 Introduction

The Wide-Field Plate Database (WFPDB) project started as an initiative of the IAU Commission 9 Working Group on Wide-Field Imaging in 1991. The main goals of the project are to make an inventory of all available wide-field plates and to provide on-line access to the database (Tsvetkov et al. 1994a).

2 List of Wide-Field Plate Archives

Four issues of a List of Wide-Field Plate Archives (WFPA) were distributed during the last three years (Tsvetkov 1992, Tsvetkov et al. 1994a,b). The latest version (2.0) of the WFPA contains information on 263 archives with a total of 1 764 600 wide-field plates (see Table 1), including additional information on the instrument identifier in the WFPDB and the time zone of the instrument. This and more information (such as plate size and limiting magnitude) is given in the MAIN DATA TABLE of the database (see Section 3).

Table 1. Number of wide-field plates according to archive type

Archive type	Direct plates	Spectral plates	Total
Computer-readable	542 941	19 138	562 079
Partly computer-readable	376 360	3 246	379 606
Not computer-readable	801 925	20 990	822 915
Total	1 721 226	43 374	1 764 600

3 The Wide-Field Plate Database: Structure and Content

The structure and content of the WFPDB is as follows:

Main Data Table (Index Plate Catalogue)

- plate identifier (including instrument identifier)
- coordinates of plate centre (R.A., Dec.) for equinox J2000.0
- observation date (UT)
- object or field designation
- method of observation
- emulsion
- filter
- spectral band
- plate dimensions X and Y [cm]
- observation time (UT)
- duration of exposure [min]
- pointers to tables **Quality, Notes, Observation, Availability** and **Digitization**

Table Quality
Table Notes
Table Observation
Table Availability
Table Digitization

The main data table of the WFPDB is installed on the IBM 4381 main frame computer in the Computer Centre of the Bulgarian Academy of Sciences. At present, it contains the data from 62 wide-field plate catalogues with 365 615 plates (Table 2). This constitutes 21% of the total number of wide-field plates, currently included in the WFPDB. The Index Plate Catalogue contains 40% of all existing plate catalogues in computer-readable or partly computer-readable form. As is seen from Table 2, the main contribution in the WFPDB in its present status comes from the Sonneberg plate collection (38 catalogues with information for 217 402 plates, Bräuer et al. 1994).

A special software package was developed and is used to convert the data from the original catalogues to a uniform presentation in the WFPDB, and to transform the data. Access to WFPDB is presently possible via e-mail request to WFPA@BGEARN.BITNET. In the near future, after connection to INTERNET, on-line access will be made available. As the next step in the development of WFPDB, we intend to include information on digitized plates.

Acknowledgements: We thank all astronomers who contributed to the WFPDB project providing data and advice. We especially appreciate the contribution of H.-J. Bräuer, B. Fuhrmann and P. Kroll from the Sonneberg Observatory of a large number of computer-readable catalogues from one of the largest and best kept plate collections. This project was supported by the Bulgarian National Science Foundation (grant F-311/93), the Alexander von Humboldt Foundation, the Münster Astronomical Institute (Germany), ESO, and the Computer Centre of Physics of the Bulgarian Academy of Sciences.

Table 2. List of catalogues included into the WFPDB

No.	Instrument identifier	Observatory	Aperture (m)	Tel. Type	Years of operation	Number of plates
1	AAO390	Anglo-Australian	3.90	Rfl	1974-	2424
2	ASI067	Asiago	0.67/0.92	Sch	1965-	15267
3	ASI040	Asiago	0.40/0.50	Sch	1958-	18411
4	BEI060	Beijing	0.60/0.90	Sch	1963-	1509
5	BOR033	Bordeaux	0.33	Ast	1893-	4151
6	BUC038	Bucharest	0.38	Rfr	1930-1970	7222
7	BUC016	Bucharest	0.16	Cam	1930-1961	147
8	PAL122	Cracow/Palomar	1.24	Sch	-	100
9	CRI040A	Crimea-Nauchny	2x0.40	Ast	1951-1984	508
10	CRI017A	Crimea-Nauchny	0.17	Cam	1951-1953	49
11	CRI017B	Crimea-Simeis	0.17	Cam	1948-1965	570
12	CRI040C	Crimea-Simeis	0.40	Ast	1947-1950	222
13	HEI040	Heidelberg	2x0.40	Ast	1900-1981	8900
14	KIS105	Kiso	1.05/1.50	Sch	1977-	6728
15	OND004A	Ondrejov	10x0.04	Cam	1955-1977	10060
16	OND004B	Ondrejov	13x0.04	Cam	1958-1975	16060
17	PAL122	Palomar	1.22/1.83	Sch	1947-	6904
18	ROZ200	Rozhen	2.00	RCr	1979-	1995
19	ROZ050	Rozhen	0.50/0.70	Sch	1979-	7106
20	SID124	Siding Spring-ROE	1.24/1.83	Sch	1973-	14383
21	ESO100	ESO-La Silla	1.00/1.60	Sch	1969-	9432
22	SON	Sonneberg		Cam	1950-1950	48
23	SON	Sonneberg		Cam	1950-1951	204
24	SON005	Sonneberg	0.05	Cam	1950-1956	2572
25	SON006A	Sonneberg	0.06	Cam	1941-1953	3558
26	SON006B	Sonneberg	0.06	Cam	1953-1962	3194
27	SON006C	Sonneberg	0.06	Cam	1956-1962	2781
28	SON006D	Sonneberg	0.06	Cam	1956-	12066
29	SON006E	Sonneberg	0.06	Cam	1956-	12162
30	SON006F	Sonneberg	0.06	Cam	1956-	12159
31	SON006G	Sonneberg	0.06	Cam	1957-	11056
32	SON006H	Sonneberg	0.06	Cam	1958-	10840
33	SON006I	Sonneberg	0.06	Cam	1958-	8687
34	SON006J	Sonneberg	0.06	Cam	1958-	8680
35	SON006K	Sonneberg	0.06	Cam	1958-	7914
36	SON006L	Sonneberg	0.06	Cam	1958-	8299
37	SON006M	Sonneberg	0.06	Cam	1958-	8311
38	SON006N	Sonneberg	0.06	Cam	1958-	8312
39	SON006O	Sonneberg	0.06	Cam	1958-	8431
40	SON007A	Sonneberg	0.07	Cam	1958-	8972
41	SON007B	Sonneberg	0.07	Cam	1958-	8965
42	SON007C	Sonneberg	0.07	Cam	1963-1965	925
43	SON007D	Sonneberg	0.07	Cam	1963-1965	942
44	SON007E	Sonneberg	0.07	Cam	1963-1965	906
45	SON007F	Sonneberg	0.07	Cam	1963-1965	890
46	SON008A	Sonneberg	0.08	Cam	1925-1939	300
47	SON008B	Sonneberg	0.08	Cam	1926-1928	193
48	SON009	Sonneberg	0.09	Cam	1957-1963	682
49	SON010	Sonneberg	0.10	Cam	1934-1956	7080
50	SON014A	Sonneberg	0.14	Cam	1928-1969	6248
51	SON014C	Son./Babel./Windhoek	0.14	Cam	1926-1945	5256
52	SON014D	Son./Babel./Windhoek	0.14	Cam	1926-1945	1735
53	SON014E	Sonneberg	0.14	Cam	1928-1957	11176
54	SON014F	Sonneberg	0.14	Cam	1942-1945	1325
55	SON017	Sonneberg	0.17	Cam	1923-1971	7976
56	SON030	Sonneberg	0.20	Sch	1960-1976	5322
57	SON040A	Sonneberg	0.40	Ast	1938-1945	1658
58	SON040B	Sonneberg	0.40	Ast	1960-	6858
59	SON040C	Sonneberg	0.40	Ast	1961-	10719
60	TAU134	Tautenburg	1.34/2.00	Sch	1960-	8239
61	TOR060	Torun	0.60/0.90	Sch	1962-1985	2826
62	TUR050A	Turku-Tuorla	0.50/0.50	Sch	1938-1949	5000

References

Bräuer H-J., Fuhrmann B., Kroll P., 1994, Handling and Archiving Data from Ground-based Telescopes, ESO/OAT Conf. and Workshop Proc. No. 50, M. Albrecht and F. Pasian (eds.), p. 155

Tsvetkov M.K., 1992, IAU Working group on 'Wide-field imaging', Newsletter No. 2, p. 51

Tsvetkov M.K., Stavrev K.Y., Tsvetkova K.P., Ivanov P.V., Iliev M.S., 1994a, Proc. IAU Symp. 161, Astronomy from Wide-Field Imaging, H.T. MacGillivray et al. (eds.), Kluwer, Dordrecht, p. 359

Tsvetkov M., Stavrev K., Tsvetkova K., Mutafov A., Michailov M.-E., 1994b, Proc. IAU Coll. 148, Future Utilisation of Schmidt Telescopes, ASP Conf. Ser., R. Cannon (ed.), in press

GCVS Problems with Flare Stars

Elena V. Kazarovets, Nikolai N. Samus

Institute of Astronomy (Russian Academy of Science), 48, Pyatnitskaya Str., Moscow 109017, Russia

1 Introduction

The astrometric aspect of cataloguing variable stars becomes more and more important. During the past thirty years there has been approximately an threefold increase of the number of known variables; now the rate of new discoveries becomes extremely high, and we may expect a new twofold increase in a couple of years due to discoveries made with space-borne instruments. To avoid confusion, we should continue our policy of naming new variables, and this is impossible without checking identifications with known variable stars, requiring good coordinates and/or finding charts. Operation of automatic telescopes also needs excellent coordinates just for pointing. For all these reasons, we are now working on improving considerably the accuracy standard of variable star coordinates, providing at least reasonably good coordinates for extragalactic variables (GCVS, 4th Edition, Vol. V, now in press, gives coordinates mainly to a second of arc) and variables in globular clusters, identifying variables, including those with rough published positions, with positional catalogues, etc.

Flare stars are very important variables; potentially they are one of the most abundant types of variable stars. Their identification in minimum light may be rather complicated, especially for dense regions of star clusters and associations. Most regretfully, there are many cases of papers on flare stars introducing additional confusion connected with wrong designations, wrong coordinates, missing finding charts. Thus, there appear problems preventing us from incorporating some flare stars in the GCVS. Many definite variables remain in the status of "suspects", being positionally close to known variables, and no finding chart being available either for the earlier known variable or for the new one. Sometimes we do include such a "pretender" into the GCVS, and years after a finding chart or good coordinates appear, and it turns out that we have two entries for the same star (see, for instance, numerous cases of duplicated names for flare stars in Taurus in the GCVS Vol. III).

We shall present here some typical examples of confusion. We ask the authors of papers we are going to quote to excuse us for using these particular papers as examples.

The original presentation of this paper at the Colloquium included demonstration of tables with originally published and corrected coordinates for stars mentioned in Sections 3 and 4. These tables (containing coordinates to higher accuracy than those given in the GCVS or Name-lists) may be obtained from the authors upon request. Corrected coordinates are presented in Name-lists of variable stars Nos. 71 and 72 and in the 5th Volume of the GCVS (in press).

2 An example of confused designations, coupled with absence of a finding chart

Melikian & Della Valle (1986) announced the discovery at Byurakan of two flare stars in Orion, B31 ($5^h23^m.2$, $-6°58'.5$, 1900.0) and B32 ($5^h31^m.0$, $-6°11'.0$, 1900.0). The first of them was later designated V1194 Ori. The discoverers did not publish finding charts. The catalogue of Natsvlishvili (1991) of flare stars in Orion presents photographic charts for B31 and B32; however, this is not the same B32 but the star at 5^h33^m06s, $-4°18'.4$ (1900.0), first discovered by Sugano (Kosai 1983) as an FU Ori-type object and named B32, for its flare activity discovered at Byurakan, by Mirzoyan et al. (1988). Natsvishvili (1992, private communication) now suggests to designate Sugano's star B32A; it is also known as V1143 Ori. The lack of a finding chart for the "first" B32, a star in a field rich in variables, combined with rather low accuracy of its published coordinates, does not permit us to include this star in the GCVS.

3 An example of very inaccurate coordinates and a finding chart not accessible for the majority of investigators

Kelemen (1987) announced his discovery of a number of flare stars in the Pleiades. Our discussions with several experts in the field show that people are not able to find many of these stars and express doubt on the reality of these stars at all. Upon our request, Dr. Kelemen sent to Moscow a photographic finding chart of his stars. Using it, we have been able to identify Kelemen's variables. However, the published coordinates for many of them are completely wrong. Among Kelemen's flare stars from the 71st Name-list of variable stars (Kazarovets et al. 1993), the coordinates published by Kelemen were correct only for V1036 Tau, the errors reach 50^s of time in right ascension and almost $13'$ in declination. This is probably due to misidentifications of some reference stars during positional measurements. Of 28 identifications with stars from Haro et al. (1982), suggested by Kelemen (1987), we confirm only one. Surely no one will ever find these stars without a chart.

Accurate coordinates for almost 550 flare stars in Pleiades were measured by Kazarovets (1993); this paper also discusses several cases of identification being impossible because of lack of finding charts.

4 An example of "accurate" coordinates actually being rather rough, but in the presence of a good chart

Chavira et al. (1992) discovered a number of variable stars in the Orion nebula region and published their rather rough positions for equinox 1900.0. Then, Parsamian et al. (1993) published a good field photograph for these variables. The latter paper contains coordinates of the variables for the equinox 1950.0 to $0\overset{s}{.}01$ in right ascension and to $0\overset{''}{.}01$ in declination. However, these coordinates actually are mostly quite rough. In the recently finished 72nd Name-list of variable stars (Kazarovets & Samus 1995), we present accurate coordinates, either taken from the GSC or determined by S. Antipin using reference stars from the GSC; the identifications with the published photograph met no problem. It seems that practically in all cases (notable exceptions are V1263 Ori and V1297 Ori, where it is difficult to find any explanation) rough coordinates from Chavira *et al.* 1992 were simply transformed to the new equinox, using not the best algorithm, and more digits were retained in the result than it was permitted by the original data. This could be a rather serious problem, if there were no finding chart.

5 Conclusions

These examples clearly show the importance of good finding charts for flare star identifications. Though continuously emphasized by IAU Comm. 27, the situation does not improve noticeably. Flare stars without finding charts constitute 36% of all flare stars in the Name-lists Nos. 67–71, compared to 7% of RR Lyraes without finding charts in the same lists. We expect a drastical increase of the number of known variables due to discoveries with space-borne telescopes. This will make identifications and positional work the most laborious and important part of compiling variable star catalogues. Kind cooperation of flare star discoverers is quite necessary. We ask our colleagues to take some more effort and find improved positions, publish finding charts or, at least, GSC identifications. This will spare much more effort in subsequent work.
Acknowledgments: Thanks are due to Mr. S. Antipin for GSC identifications and improved positions for new flare stars in Orion.

References

Chavira E., González G., Escamilla C., Parsamian E., 1992, IBVS 3764
Haro G., Chavira E., González G., 1982, Bull. Inst. Tonanzintla 3, No. 1
Kazarovets E.V., 1993, Peremennye Zvezdy (Variable Stars), 23, No. 3, 141
Kazarovets E.V., Samus N.N., 1995, IBVS 4140
Kazarovets E.V., Samus N.N., Goranskij V.P., 1993, IBVS 3840
Kelemen J., 1987, IBVS 3103
Kosai H., 1983, IAU Circular No. 3763
Melikian N.D., Della Valle M., 1986, IBVS 2929
Mirzoyan L.V., Melikian N.D., Natsvlishvili R.Sh., 1988, Astrofizika 28, No. 3, 540
Natsvlishvili R.Sh., 1991, Astrofizika 34, No. 1, 107
Parsamian E.S., Chavira E., González G., 1993, Rev. Mex. Astr. Astrofis. 25, No. 2, 71

Detections of Neutron Stars in the Extreme Ultraviolet

Stuart Bowyer

Center for EUV Astrophysics, University of California, Berkeley, CA 94720-5030, USA

Neutron stars were not expected to be observable in the extreme ultraviolet and hence their detection in this band is a surprise. More important is that EUV observations provide unique new information on these objects. In the accompanying Table 1, I list the neutron stars detected in the EUV with the Extreme Ultraviolet Explorer (*EUVE*). The binary object Her X-1 is the only object detected in the all sky survey. This is not surprising since it is the only neutron star system with sufficient intensity to be detectable in the 500 to 2000 second sky intergrations which are typical for the *EUVE* all sky survey. All the neutron stars were detected in the 100 Å (Lexan) bandpass. A more complete description of *EUVE* and the all sky survey are provided in Bowyer and Malina (1991), and in Bowyer et al. (1994).

Table 1. Neutron Stars Detected with *EUVE*

Object	Deep Survey cts/sec	Lexan Survey cts/sec	Age years	DM cm^{-3} pc	*d* pc	References
PSR J0437–4715	0.038		5×10^9	2.6	140	(Edelstein et al. 1995)
PSR B0656+14	0.024		1×10^5	14.0	760	(Finley et al. 1994)
Geminga	0.020		5×10^5		< 500	(Halpern 1994)
Her X-1		0.18	$< 10^7$		<5000	(Vrtilek et al. 1994)

A phased analysis of the flux from Her X-1 (Vrtilek et al. 1994) shows that the EUV flux emanates from the acretion disc and not from the neutron star itself. While these results help elucidate the character of the disc, they are not the topic of this conference and will not be discussed further.

Timing information from *EUVE* is routinely available to 8 ms accuracy, and with an easily implemented alternate mode is available to better than 1 ms. This allows the discrimination of pulse and whole-body emission from virtually all pulsars.

The pulsar B0656+14 was observed for over 100 000 seconds and this data has recently been released for general dissemination via the *EUVE* Public Archive. A preliminary examination of this data shows the data is not pulsed and hence the EUV flux must be thermal emission from the neutron star surface.

The X-ray and γ-ray (but not radio!) pulsar, Geminga, has been observed for 300 000 seconds. This data has just been released to the scientific community via the EUVE Public Archive, but as yet no analysis of this data have been published.

The millisecond pulsar J0437–4715 was observed with *EUVE* in January 1994 for about 70 000 seconds. An unrelated EUV source about 3.8 arc minutes away was well resolved with the $\approx 1'$ point spread function of the telescope. Because the data were not taken in the high time resolution mode, a pulse vs. whole-body deconvolution is not possible for this data. Nonetheless, Edelstein et al. (1995) have shown that a substantial amount of information can be derived from this observation.

The observed dispersion measure of 2.65 pc cm^{-3} implies a distance of 140 pc for this pulsar. This value is consistent with optical data on the white dwarf companion star. An estimate of the hydrogen column density to this system can be obtained from the compilation of Fruscione et al. (1994); a reasonable value is $\mathrm{N_{H\,I}} = 10^{19}\ \mathrm{cm}^2$. Combining the response of the 100 Å bandpass of the deep survey telescope with absorption due to an interstellar medium associated with this hydrogen column (Rumph et al. 1994) yields an effective area weighted wavelength of 116 Å. This is a substantially lower energy than the X-ray data obtained for this object, and leads to new insights regarding this source.

Becker & Trümper (1993) modeled the X-ray emission of J0437–4715 with a power-law model with $\alpha = -2.60$, and with a composite model. The composite model consisted of a black-body "hot-spot" ($T \sim 1.7 \times 10^6$ K, Area $= 0.05$ km^2), responsible for a possible feature at 0.8 keV, combined with a power-law source with $\alpha = -2.85$ which dominates at lower energies. The predicted EUV flux for both power laws was consistent with the observed EUV luminosity only if the absorbing column was in the narrow range of $\mathrm{N}_{HI} = (2.5 \pm 0.2) \times 10^{20}$ cm^{-2}. This column exceeds the hydrogen column out of the Galaxy (Dickey & Lockman 1990), and hence both of these models can be ruled out as the source of the EUV flux.

The EUV data were modeled as a black body to determine if this model could produce all the *ROSAT* observed X-ray flux. This could be accomplished only if the hot spot emitting area was 3 times (for $N_{\mathrm{H\,I}} = 10^{19}\ \mathrm{cm}^{-2}$) to 25 times (for $N_{\mathrm{H\,I}} = 10^{20}\ \mathrm{cm}^{-2}$) larger than the area required by the X-ray data. Edelstein et al. concluded that the observed EUV flux was not consistent with a blackbody hot-spot capable of producing the soft X-ray feature.

Edelstein et al. examined whether black-body emission could provide an alternative explanation to the 0.1 to 0.4 keV *ROSAT* flux which had been attributed to power-law emission. They concluded that both the EUV flux and the X-ray flux could be explained by a black-body with a temperature $\sim 5.7 \times 10^5$ K, an emitting area of $\sim 3\ \mathrm{km}^2$, and ISM absorption associated with a hydrogen column of $N_{\mathrm{H\,I}} = 5 \times 10^{19}\ \mathrm{cm}^{-2}$.

Alternatively, the EUV flux could be independent of the X-ray flux and originate from the entire surface of the pulsar. In this case, a minimum surface temperature of $T_{\rm min} = 1.6 \times 10^5$ K was found for a distance of 140 pc and a stellar radius of 10 km. An upper limit to the pulsar's temperature $T_{\rm max} = 4.0 \times 10^5$ K was established by using the known hydrogen column out of the Galaxy (Dickey & Lockman 1990).

Thermal EUV emission from the surface of a $\sim 5 \times 10^9$ year old neutron star is inconsistent with standard cooling models for neutron stars (Shibazaki & Lamb 1989) and would require some form of reheating. Four sources for reheating were considered: (1) frictional heating at the crust-core interface through the unpinning of internal vortex lines; (2) accretion from the ISM; (3) accretion from the white dwarf companion onto the neutron star; and (4) nucleon decay catalyzed by heavy magnetic monopoles.

Theoretical work (Shibazaki & Lamb 1989, Umeda et al. 1993) predicts that for neutron stars older than 10^9 yr, the thermal surface temperatures from frictional heating will be less than 10^5 K. Hence standard frictional heating models can be ruled out.

The gravitational accretion of ISM could heat the neutron star if the velocity of the star is less than $\sim$ 10 km s^{-1} (Paczyński 1990, Hartmann et al. 1990, Treves & Colpi 1991). A velocity of 91 ± 3 km s^{-1} has been measured for PSR 0437–4715 using timing measurements (Bell et al. 1995), and 63 ± 30 km s^{-1} using scintillation observations (Johnston & Nicastro 1995). These velocity estimates are sufficiently large that significant heating by accretion from the ISM can be ruled out.

Stellar material abated from the neutron stars' white dwarf companion by a high energy particle wind from the pulsar could accrete on the neutron star and produce heating. An upper limit to mass accretion by a high-energy particle wind from the pulsar is

$$\dot{M} \leq f L_{\rm SD} / \gamma c^2$$

where f is the fraction of the total flux from the neutron star intercepted by the companion, $L_{\rm SD}$ is the pulsar spin-down luminosity, γ is the Lorentz factor of the particles, and c is the speed of light. For parameters appropriate to this system the maximum luminosity due to this accretion is an insignificant fraction of the observed EUV luminosity.

It has been suggested that neutron stars could be heated by magnetic "monopole-catalyzed" nucleon decay (Freese et al. 1983, Kolb & Turner 1984). Magnetic monopoles which hit the surface of a neutron star will be captured by the star at a rate proportional to the monopole flux. The combination of the upper limit to the surface temperature of PSR 0437–4715 and its extreme age results in a value for the Galactic monopole flux that is three orders of magnitude lower than existing limits.

Four neutron stars (or binaries with a neutron star as one component) have been detected in the EUV to date. In three of these systems the observed emission is believed to be from the neutron star itself, but a detailed analysis has been carried out for only one of these objects. In this case substantial new information on neutron star physics has been obtained. We can expect more progress in this field when analysis of the existing data are completed, and as new neutron stars are found to be emitting in the EUV.

Acknowledgements: This work is being supported by NASA Grant NAS5-30180.

References

Becker W., Trümper J., 1993, Nature 365, 528

Bell J.F., Bailes M., Manchester R.N., Weisberg J.M., Lyne A.G., 1995 (submitted)

Bowyer S., Malina R.F., 1991, in Extreme Ultraviolet Astronomy, R.F. Malina and S. Bowyer (eds.). Pergamon Press, New York, p. 397

Bowyer S., Lieu R., Lampton M., Lewis J., Wu X., Drake J.J., Malina R.F, 1994, The First EUVE Source Catalog, ApJS 93, 569

Dickey J.M., Lockman F.J., 1990, ARA&A 28, 215

Edelstein J., Foster R.S., Bowyer S., 1995, ApJ (submitted)

Finley J.P., Ogelman H., Edelstein J., 1994, 184th AAS Bulletin

Freese K., Turner M., Schramm D.N., 1983, Phys. Rev. Lett. 51, 1625

Fruscione A., Hawkins I., Jelinsky P., Wiercigroch A., 1994, ApJS 94, 127

Halpern J., 1994, private communication

Hartmann D., Epstein R.I., Woosley S.E., 1990, ApJ 348, 625

Johnston S., Nicastro L., 1995, preprint

Kolb E., Turner M., 1984, ApJ 286, 702

Paczyński B., 1990, ApJ 348, 485

Rumph T., Bowyer S., Vennes S., 1994, AJ 107, 2108

Treves A., Colpi M., 1991, A&A 241, 107

Shibazaki N., Lamb F.K., 1989, ApJ 346, 808

Umeda H., Shibazaki N., Nomoto K., Tsuruta S., 1993, ApJ 408, 186

Vrtilek S.D., Mihara T., Primini F.A., Kahabka P., Marshal H., Agerer F., Charles P.A., Cheng F.H., Dennerl K., la Dous C., Hu E.M., Rutten R., Serlemitsos P., Soong Y., Stull J., Trümper J., Voges W., Wagner R.M., Wilson R., 1994, ApJ 436, L9

Time Variability Studies with ROSAT

Martin Kürster

Max–Planck–Institut für extraterrestrische Physik, D–85740 Garching, Germany

Abstract: An overview over the mission of the X-ray astronomy satellite *ROSAT* (Röntgensatellit) is given with an emphasis on the first soft X-ray all-sky survey carried out with an imaging telescope. Several examples of time variability studies made possible by the all-sky survey are shown for active late-type stars. Two exceptional long-duration stellar flares are presented.

1 ROSAT mission overview

At present the X-ray astronomy satellite *ROSAT* is in the fifth year of its very successful mission. *ROSAT* is an observatory in a low-earth orbit carrying an X-ray telescope (XRT; Trümper 1983, Aschenbach 1988) and an XUV telescope, the 'Wide Field Camera' (WFC; Wells et al. 1990, Kent et al. 1990). Both telescopes are assemblies of nested Wolter type I mirrors.

The focal plane instrumentation of the XRT comprises two identical Position Sensitive Proportional Counters (PSPC) built by the Max-Planck-Institut für extraterrestrische Physik (MPE; Pfeffermann et al. 1988), and the High Resolution Imager (HRI), a microchannel plate detector built by the Smithsonian Astrophysical Observatory (SAO; David et al. 1993). One of the PSPCs was lost early in the mission, however, due to the redundant second PSPC this did not constitute any impairment of the project. Recently, the supply of PSPC detector gas was used up, so that the PSPC guest observer program had to be ended in July 1994. Currently, the XRT observations are continued with the HRI.

Properties of the PSPC are a large field of view of 2° diameter, a very low intrinsic background of 2×10^{-5} cts $\sec^{-1}$ arcmin^{-2}, and a moderate spectral resolution of about 6 independent resolution elements over its sensitive energy range 0.1 – 2.4 keV. At 1 keV the PSPC yields an effective area of 220 cm^2. Contrary to the PSPC, the HRI is lacking any spectral resolution, but its spatial resolution of $\approx 3 - 4''$ surpasses that of the PSPC ($\approx 20''$ on–axis) by a factor of $5 - 7$. The field of view of the HRI has a diameter of $40'$.

After *ROSAT*'s launch on June 1, 1990, and a 2-month calibration-verification phase half a year of the mission was devoted to the first all-sky survey to be carried out with an imaging telescope in soft X-rays (see Sect. 2). Simultaneous to the XRT survey the WFC carried out the first all-sky survey in the XUV

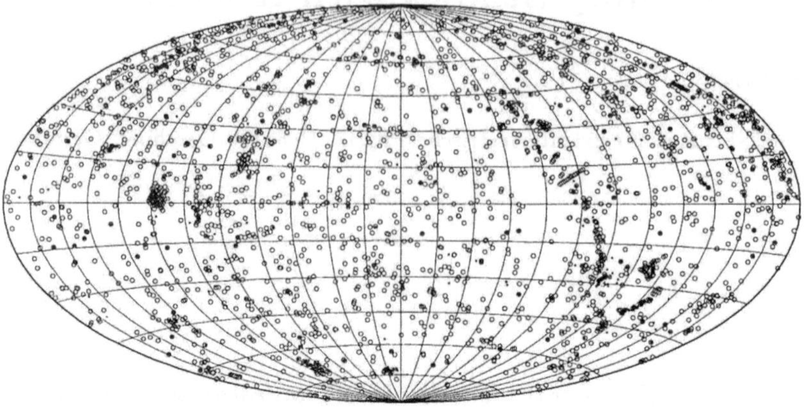

Fig. 1. Distribution of 3987 sky fields observed by ROSAT in pointed mode up to Dec 3, 1993. An Aitoff equal–area whole–sky projection is shown in equatorial coordinates; with respect to the center at RA=0, DEC=0, north is up and east is to the left.

(0.025 – 0.2 keV). However, no WFC observations will be discussed here. An overview over the XUV survey can be found in Page & Denby (1992); the WFC catalogue of bright survey sources was published by Pounds et al. (1993). Since February 1991 *ROSAT* has made pointed observations of over 5 000 fields. Fig. 1 shows their celestial distribution; it is complete up to December 3, 1993.

About 60,000 sources were detected in the XRT all-sky survey (see Sect. 2), and another 60 000 sources were discovered by pointed observations. Proposals for pointed *ROSAT* observations were accepted from about 540 PIs from 24 countries. After one year of PI proprietary time all pointed data enter public archives at MPE, GSFC, and in Leicester. Currently about 600 archival requests per month demonstrate the high interest of the scientific community in *ROSAT* data. More than 500 publications in refereed journals have reported *ROSAT* results.

With respect to the topic of this conference it should be pointed out that *ROSAT* was not primarily devised for time variability studies. There are two complications for any kind of time dependent observation. First, the fixed mount of the solar panels constrains telescope orientation to a 30° wide strip on the sky perpendicular to the solar direction; and second, the low earth orbit causes frequent earth blocks of targets typically once per orbit. For this reason only 5 % of the total pointed observing time is reserved for time critical observations.

Fortunately, observations that are more or less randomly distributed in time have proved to allow very successful studies of time variable phenomena; one observation of this type will be shown in Sect. 3. Additionally, the scanning approach adopted in the all-sky survey provided another unique opportunity for time variability studies; three examples taken from the survey will be given in Sects. 2 and 3. All examples concern coronally active late-type stars.

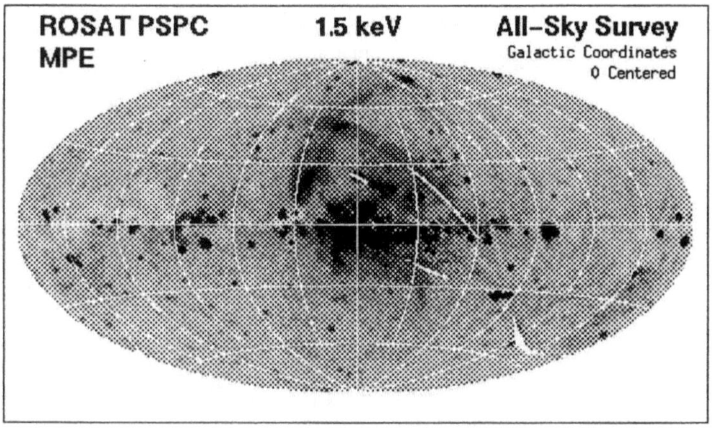

Fig. 2. The *ROSAT* X-ray sky in the hard energy band (1.0 – 2.0 keV). This image from the all-sky survey is displayed in galactic coordinates centered on latitude 0° and longitude 0°. The hard band was selected because of its low background.

2 The ROSAT XRT all-sky survey

For the *ROSAT* X-ray all-sky survey the sky was scanned in great circles (one circle per 96 min satellite orbit) whose normal pointed at the Sun. Following the Sun over half a year the whole sky was covered. Each scan collected photons from a sky strip of width 2° equal to the size of the PSPC's field of view.

Due to this scanning geometry the poles of the ecliptic were contained in every scan and observed in 96 min steps over half a year whereas sky regions near the ecliptic equator were covered only during a 2 day interval. Exposure times for the ecliptic poles were $\approx$ 50 000 sec vs. $\approx$ 400 sec for the ecliptic equator. Satellite passages through the Earth's radiation belts whose high particle background requires to switch off the detector, decreased the exposure locally making the overall distribution inhomogeneous. Due to the South Atlantic Anomaly (SAA) southern hemisphere exposure times were shorter than northern hemisphere ones. More details on the survey are given by Voges (1992, 1993).

Fig. 2 shows the exposure corrected all-sky image obtained from the survey in the hard energy band (1.0 – 2.0 keV) displayed in galactic coordinates (Snowden et al. 1995). The bright point sources (mostly accreting binaries) can be seen to accumulate in the galactic plane and in the Large Magellanic Cloud (lower right). Other prominent sources are supernova remnants and clusters of galaxies, e.g. the Virgo cluster at high galactic latitudes (top right). Galactic diffuse emission is most pronounced near the center region, in the Cygnus Loop (ringlike feature at center left) and in the North Polar Spur (arclike region in upper half of image).

The celestial distribution of survey point sources as detected by the *ROSAT* Standard Analysis System (SASS) in the total energy band is shown in Fig. 3.

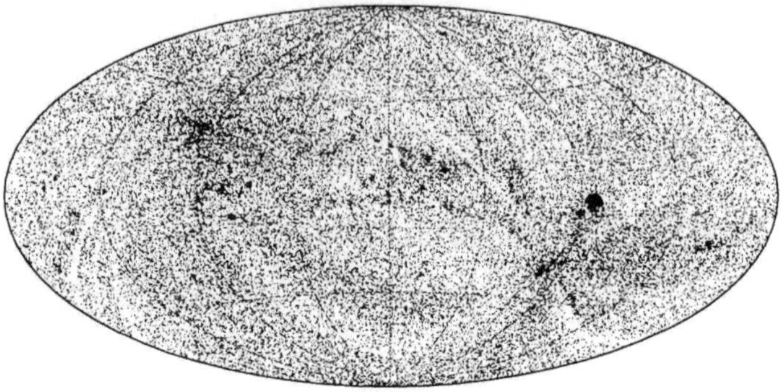

Fig. 3. Point sources detected in the *ROSAT* X-ray all-sky survey in the total energy band (0.1 – 2.4 keV) and displayed in galactic coordinates (as Fig. 2). The exposure time varies strongly over the sky with a mean value of 400 sec corresponding to a flux limit of 5×10^{-13} erg $cm^{-2}sec^{-1}$. This preliminary image containing 58 359 sources will be improved after a reprocessing of the survey data.

Enhanced sensitivity at the ecliptic poles shows as particularly dense source regions near the centers of the 2nd and 4th quadrants (upper left and lower right).

Figs. 4 and 5 demonstrate the typical temporal coverage of the sky showing survey light curves of two X-ray sources which are located near the ecliptic equator and near the southern ecliptic pole, respectively. Both objects are rapidly rotating, coronally active stars and, among these, prime candidates for time variability to occur in the form of flares and of rotational modulation caused by an inhomogeneous distribution of coronal emission regions.

Fig. 4 shows the light curve of HR 1099 (=V711 Tau), an evolved RS CVn-type binary (spectral types G5IV+K1IV) in a 2.8 day synchronized orbit. Data points correspond to single survey scans in an observing window 2 days wide. A short-duration flare is seen in a single scan in an otherwise constant light curve. This low level of variability in a very active binary came as a surprise.

Fig. 5 shows the light curve of the single pre-main sequence K0V dwarf AB Dor (=HD 36705) which was observed over 35 days, however, with frequent data gaps mostly due to the SAA. The source is very variable exhibiting flaring activity as well as variability on time scales of a few hours. A period analysis failed to find the 13 hour rotation period implying that most of the variability is 'irregular,' i.e. attributable to low-level flaring and/or restructuring of coronal emission regions on time scales shorter than the rotation period (see Kürster et al. 1992). Pagano et al. (1993) compare this data set with simultaneous observations made in other wavelength bands. Rotational modulation in AB Dor is revealed in a pointed observation presented by Kürster & Dennerl (1993).

More on survey observations of active late-type stars can be found in papers by Dempsey et al. (1993a,b), Schmitt (1994), and Haisch & Schmitt (1994).

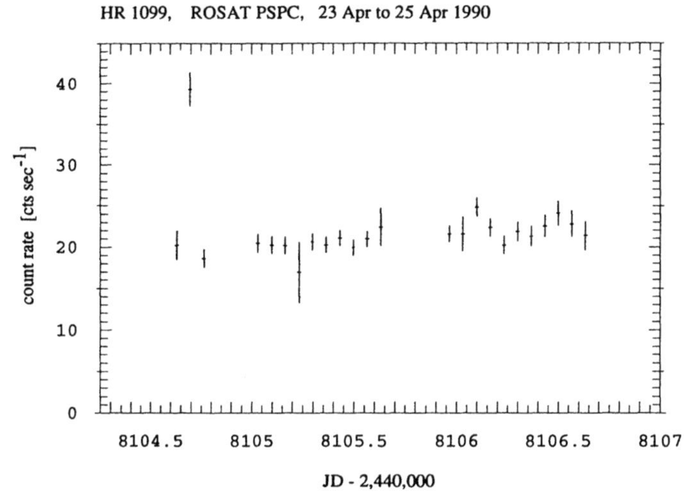

Fig. 4. *ROSAT* all-sky survey light curve of the RS CVn-type binary HR 1099 (=V711 Tau). Data points are spaced at 96 min intervals and represent individual scans. Being located at a low ecliptic latitude ($-18.3°$) HR 1099 was visible only for 2 days. A short-duration flare occured in an otherwise non-variable episode of the star.

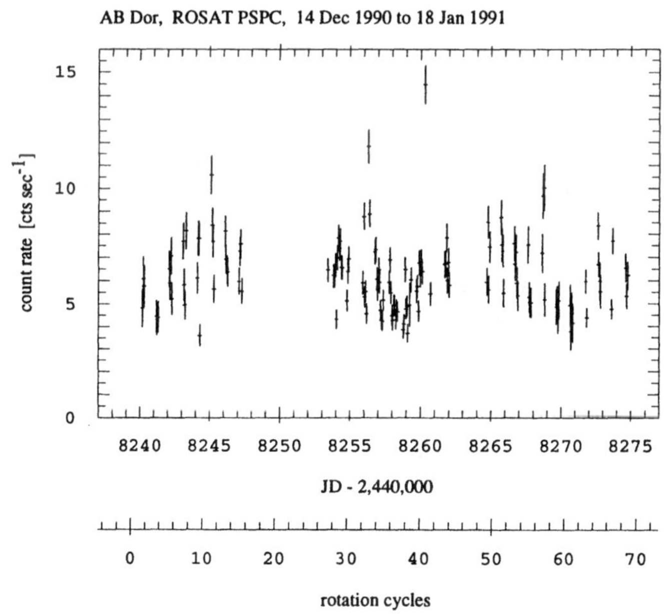

Fig. 5. *ROSAT* all-sky survey light curve of the young K0V star AB Dor (=HD 36705) which is located within 3.4° of the south ecliptic pole and was visible during the survey for about 35 days or almost 70 rotational cycles ($P = 0.51479$ day; Innis et al. 1988). Most of the many data gaps are due to detector switch–offs in the SAA.

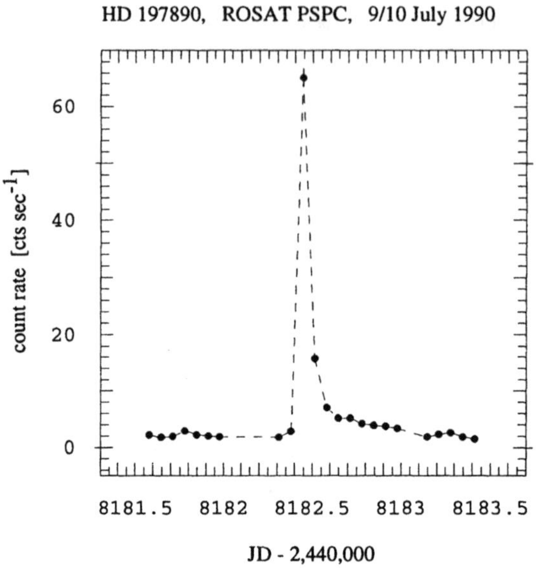

Fig. 6. *ROSAT* all-sky survey observation of the AB Dor-type K0V star HD 197890 (='Speedy Mic') showing the brightest stellar X-ray flare seen in the survey. The total flare duration is about twice the stellar rotation period.

3 Two exceptional long-duration stellar flares

This section presents two highlights among time variability studies carried out with *ROSAT*. Both are long-duration flares on active stars. To begin with the conclusion, in both cases the flares lasted several stellar rotations with no strong modulation of the flare light curve visible; deep depressions in the light curve are expected if the rotating star takes the flaring region to its invisible hemisphere. In both events the flaring regions must either have been very extended vertically, so that they were not totally occulted, or located in the small circumpolar regions of these high inclination stars. Both scenarios are quite different from solar flares which occur in rather compact regions and never near the poles.

The first observation revealed the strongest (in terms of the peak count rate) stellar X-ray flare found in the all-sky survey. It occured on the young AB Dor–type K0V star HD 197890 in Microscopium, nicknamed 'Speedy Mic' due to its very high rotational velocity. This event was first discovered by Matthews et al. (1994) in the WFC all-sky survey in which the source was detectable only during the flare phase. The XRT survey data taken over a 2 day interval revealed flaring emission as well as quiescent emission before and after the outburst (Fig. 6; see also Kürster 1994a). The flare released an energy of $\approx 2.2 \times 10^{36}$ erg sec^{-1} in the PSPC band. Its duration of 0.7 day is about twice the stellar rotation period of 0.3 day (Anders et al. 1993). No occultation-type modulation is seen during the flare implying that it either occured in a

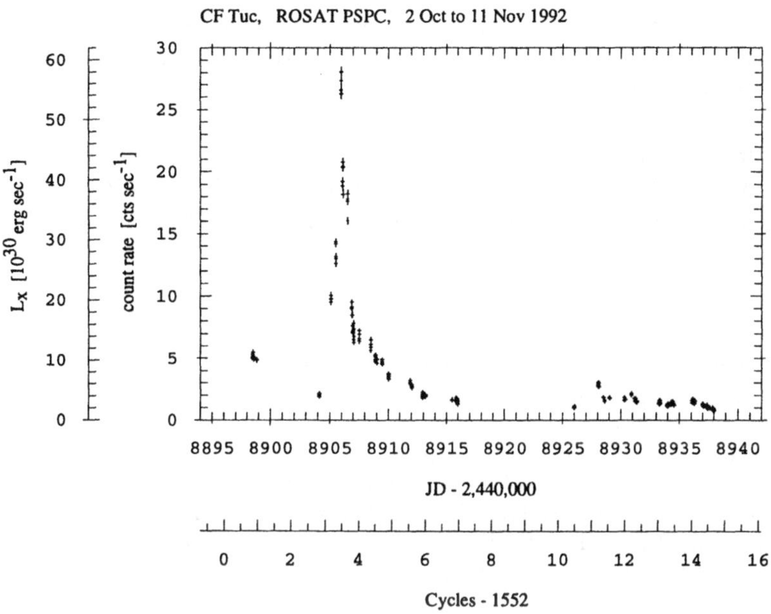

Fig. 7. A pointed observation of the RS CVn-type binary CF Tuc (=HD 5303) showing the longest known stellar X-ray flare which lasted more than 3 rotational/orbital binary cycles (ephemeris HJD 2,444,555.009+2.797672 day; period from Budding 1985; time of primary eclipse refined). The X-ray luminosity assuming a constant flux conversion factor of 6.0×10^{-12} erg cm^{-2} cts^{-1} is indicated.

very extended region or on the pole. However, the quiescent emission (before JD 2,448,182 and after JD 2,448,183) shows a slight modulation compatible with the 0.3 day period indicating an inhomogeneous corona.

In a pointed observation extending over 40 days the second flare (Fig. 7) was discovered on the partially eclipsing RS CVn-type binary CF Tuc (=HD 5303; spectral types G0IV+K4IV). It is the longest X-ray flare ever observed with a total duration of 9 days or more than three times the 2.8 day rotational/orbital period of this synchronized binary. With an energy release of $\approx 7.8 \times 10^{36}$ erg sec^{-1} in the PSPC band this outburst is also among the strongest known X-ray flares. As for 'Speedy Mic', there are no indications for rotational occultations of the flaring region nor for eclipses by the companion star (see Kürster 1994a,b).

These temporally extended observations made by *ROSAT* in the all-sky survey or in widespread pointed observations provide unique opportunities for detailed studies of long-duration stellar flares. I have shown two examples that yield important information on the structure of active-star coronae from their mere duration. Other stellar flares observed with ROSAT in pointed mode can be found in papers by Ottmann & Schmitt (1994), and Preibisch et al. (1993).

Acknowledgements: The *ROSAT* project has been a collaborative effort of many people at MPE, AIT, GSFC, SAO, and the UK WFC consortium. The work described in this paper was most directly supported by J. Trümper, W. Voges, and J. Schmitt. K. Dennerl and S. Snowden provided some figures. Special thanks to the MPE ROSAT data center, in particular R. Gruber, W. Pietsch, C. Rosso, F. Haberl, and U. Zimmermann. The ROSAT project is supported by the German Bundesministerium für Forschung und Technologie (BMFT/DARA) and the Max Planck Society.

References

Anders G.J., Jeffries R.D., Kellett B.J., Coates D.W., 1993, MNRAS 265, 941

Aschenbach B., 1988, Appl. Optics 27, 1404

Budding E., 1985, IBVS 2779

David L.P., Harnden F.R., Kearns K.E., Zombeck M.V., 1993, The ROSAT High Resolution Imager (HRI), U.S. ROSAT Science Data Center/SAO document

Dempsey R.C., Linsky J.L., Fleming T.A., Schmitt J.H.M.M., 1993a, ApJSS 86, 599

Dempsey R.C., Linsky J.L., Schmitt J.H.M.M., Fleming T.A., 1993b, ApJ 413, 333

Haisch B., Schmitt J.H.M.M., 1994, ApJ 426, 716

Innis J.L., Thompson K., Coates D.W., Lloyd Evans T., 1988b, MNRAS, 235, 1411

Kent B.J., Reading D.H., Swinyard, B.M., Graper E.B., Spurett P.H., 1990, Proc. SPIE, Vol. 1344, p. 255

Kürster M., 1994a, AG Abstract Series 10, 22

Kürster M., 1994b, in Proc. 8th Cambridge Workshop on Cool Stars, Stellar Systems, and the Sun, J. Linsky (ed.), A.S.P. Conf. Ser. 64, p. 104

Kürster M., Dennerl K., 1993, in *Physics of Solar and Stellar Coronae,* ed. J. Linsky and S. Serio (Kluwer, Dordrecht), p. 443

Kürster M., Schmitt J.H.M.M., Fleming T.A., 1992, in Proc. 7th Cambridge Workshop on Cool Stars, Stellar systems and the Sun, M. Giampapa and J. Bookbinder (eds.), A.S.P. Conf. Ser. 26, p. 109

Matthews L., et al., 1994, MNRAS 266, 757

Ottmann R., Schmitt J.H.M.M., 1994, A&A 283, 871

Pagano I., Rodonó M., Cutispoto G., Collier Cameron A., Kürster M., Kellet B.J., Bromage G.E., Jeffries R., Foing B., Ehrenfreund P., 1993, in Physics of Solar and Stellar Coronae, J. Linsky & S. Serio (eds.), Kluwer, Dordrecht, p. 457

Page C.G. Denby M., 1992, in Proc. 1st Ann. Conf. on Astonomical Data Analysis Software and Systems, (in press)

Pfeffermann, E., Briel U.G., Hippmann H., Kettenring G., Metzner G., Predehl P., Reger G., Stephan K.H., Zombeck M.V., Chappell J., & Murray S.S., 1988, Proc. SPIE 733, 519

Pounds K.A, et al., 1993, MNRAS 260, 77

Preibisch Th., Zinnecker H., Schmitt J.H.M.M., 1993, A&A 279, L33

Schmitt J.H.M.M., 1994, ApJS 90, 735

Snowden S.L., Freyberg M.J., Schmitt J.H.M.M., Voges W., Trümper J., Edgar R.J., McCammon D., Plucinsky P.P., Sanders W.T., 1995, ApJ (in press)

Trümper J., 1983, Adv. Space Res. Vol. 2, No. 4, p. 241

Voges W., 1992, in Proc. ISY Conf. "Space Science", ISY-3, ESA publications, p. 9

Voges W., 1993, Adv. Space Res., Vol 13, No. 12, p. 391

Wells A. et al., 1990, Proc. SPIE, 1344, p. 230

Wide-field Cameras for GRB Observations in X-rays and EUV

Eugene Moskalenko

Sternberg State Astronomical Institute, Moscow State University,
119899 Moscow, Russia

1 Introduction

Recent observations of the *ASCA* satellite resulted in the first identification of a GB source (Murakami et al. 1994). This success confirmed the importance of simultaneous observations in different wavelength bands for GB studies. Besides the *ASCA* results, there were several observations of GBs in X-ray band with the *Ginga* (Yoshida et al,.1989), V 78/1 (Laros et al. 1984) and other satellites. It became clear that GBs emit 4 – 8% of their energy in the 2 – 10 keV range. The main task now is to have an equipment which will be able to monitor the sky in X-rays in a mode similar to that of GRO observations, i.e. the telescope should have an all-sky field-of-view (FoV) and should work continuously.

A telescope with these features but operating at soft X-ray energies may directly determine the GB distance scale, due to interstellar absorption of the photons with energies less than 2 keV, as was pointed out first by Schaefer (1993). Flaring sources similar to GBs in time scale may be found also in the EUV (hundreds of angstroms) with the help of very wide-field cameras. Of course each such device – in X-ray, soft X-ray and EUV bands – will discover many transient objects, flaring events, will study time variability of bright "stationary" sources etc. In this paper we describe several instrumental approaches in these fields.

2 The "BARS" all-sky X-ray monitor

We have designed an all-sky X-ray monitor for continuous observations in the 2 – 25 keV energy band. The detectors are 6 standard proportional counters. Each detector has a 100 μm Be window of 170 cm^2 area. The counters are filled with 90% Xe + 10% methane gas mixture under 400 mm Hg pressure so that the resulting quantum efficiency is greater than 15% in the total energy range. Such devices have been used onboard the space station "Astron" for several years. Detectors used in the "Bars" experiment have no collimation. The counters are to be placed at the spacecraft at such a mode that their fields-of-view will look along 6 orthogonal axes, so that they observe the whole sky. The effective area of each counter varies as cosine of an incidence angle of photons, so it is possible o estimate roughly the direction of each event. For the brightest sources the error radii should be of the order of $5-6^{\circ}$. Several electronic boxes supply all the voltages and analyze spectral and timing information. There are two operational modes:

(a) all the pulses are divided in four energy channels: 2 – 3.5, 3.5 – 6, 6 – 10 and 10 – 25 keV, and the accumulated number of pulses for each channel from each detector with time resolution of 64 ms goes to the telemetry system,
(b) the time of arrival and amplitude are determined for each pulse from each counter and are telemetered to Earth.

The maximal count rate from each counter is 40 000 c/s. This telescope may operate only in the Earth's shadow. The total mass of the telescope is 60 kilogram.

At a low-Earth orbit and at latitudes below 40^{circ} the main background component is diffuse X-rays, if a counter is oriented looking to the galactic poles. In this case the background countrate is about 1000 c/s in the 2–10 keV band. While looking to the galactic center region the countrate rises up to 5000 c/s and consists mainly of the sum of many discrete source fluxes.

For sensitivity estimates we use a composite spectral form of a typical GB:

$$\begin{aligned} dN/dE &= A \cdot E^{-0.7} \text{ photons/cm}^2 \text{ s keV} \quad \text{for E} < 10keV, \\ dN/dE &= 2A \cdot E^{-1} \text{ photons/cm}^2 \text{ s keV} \quad \text{for E} = 10 - 100keV, \\ dN/dE &= 200A \cdot E^{-2} \text{ photons/cm}^2 \text{ s keV} \quad \text{for E} > 100keV. \end{aligned}$$

Estimated background levels together with such a spectrum means that, e.g., if the duration of a GB is 1 s, our sensitivity at the level of 5σ is about 10^{-8} ergs/cm^2 s for the 2 – 10 keV energy range; it corresponds roughly to bursts with $7 \cdot 10^{-7}$ ergs/cm^2 fluence detected by the *BATSE* experiment onboard the Compton Observatory in the 50 – 300 keV band.

This spectrometer if produced and fully tested, but economical difficulties prevent its actual launch to space.

3 Soft X-ray counter with a thin beryllium window

For soft X-ray observations of discrete sources, proportional counters usually have thin organic windows of several μm thickness. But if the problem is to design a very wide-field telescope for sky monitoring and detection of short-lived events in the 0.1 – 1 keV range, this approach can hardly be applied, for the following reasons:

(a) if the counter has no collimation, the soft diffuse X-ray background becomes the main background component. But while in the region of $E > 1$ keV its photon spectrum has a power law with an exponent of -1.4, in the range $E < 1$ keV, this exponent becomes -2 (McCammon & Sanders 1990), which substantially reduces the signal to noise ratio. Contrary to this, the GB spectra

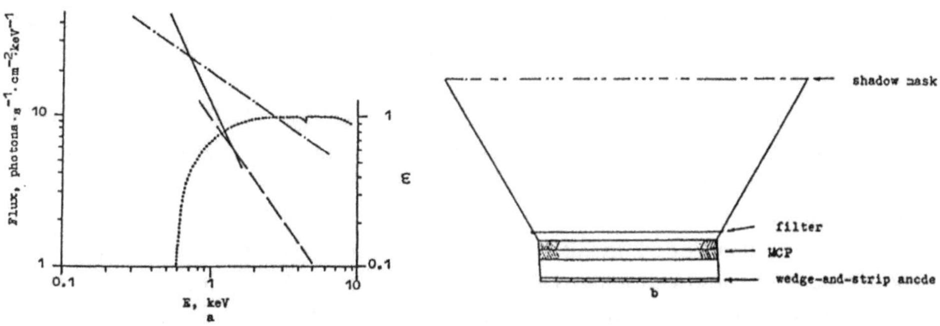

Fig. 1. a – solid line: soft X-ray diffuse flux for a detector with 1 steradian FoV, dashed line: the same for a diffuse background in the 2 – 10 keV range, dash-dotted line: flux from the GB with fluence of 10^{-5} ergs/cm^2 and 1 sec duration. Dotted curve: quantum efficiency of a counter with a 4 μm Be window. b – schematic view of the EUV shadow camera.

become flatter with decreasing photon energy down to 1 keV and less, which worsens the situation.
(b) The energy resolution of a counter is proportional to $E^{0.5}$ and becomes greater than 50% with energy decreasing below 1 keV. This means that the softest photons from very intense background intensities pollute the background at higher energies.
(c) The typical efficiency of a thin film counter has non-zero values lower than carbon K-edge of 0.28 keV and is close to zero between 0.28 and 0.7 keV, i.e. exactly where it is very important to analyse GB spectra for the determination of a probable Galactic absorption.

We propose to use a counter with a 4 μm Beryllium window as a detector for observations of soft X-ray GB components. Its efficiency is shown on Fig. 1a. The advantage of using a thin Beryllium window is clearly seen – it has a significant efficiency exactly in the $E > 0.6$ keV range and has no sensitivity at lower energies, where the background is high. An additional and very important technical advantage of such a counter is that there is no need for a heavy and complicated gas-flow system.

The *ASCA* team has successfully designed an imaging gas scintillation counter with a 10 μm Be window (Makishima 1993), and it is expected that present-day technology permits the construction of a detector with a 4 μm Be foil.

Based on the composite spectrum, we estimat that such a counter with an area of 100 cm^2 will detect the soft components of GBs with 1 s duration and $5 \cdot 10^{-6}$ ergs/cm^2 fluence at the 5σ level, and will determine $N_{\rm H}$ quantities with a precision of about 30% at the 1σ level for $N_{\rm H} > 10^{21}$ atoms/cm^2.

4 Wide-field shadow camera for EUV observations

Microchannel plates (MCP) are widely used as detectors in EUV astronomy. Until now, there are no telescopes with FoV of more than several degrees working in the EUV band.

The main idea of a proposed device is to use a large MCP together with a two-dimensional shadow mask (the latter is used commonly in X-ray telescopes). MCPs are now produced with dimensions up to about 6×6 cm^2 and achieve a linear resolution of better than 100 mum (Lampton 1991). We propose to use such a MCP and a coded mask with 12×12 cm^2 dimensions consisting of a 0.1×0.1 mm^2 random cell array to obtain a 60° FoV with 6′ angular resolution. To achieve this, the mask should be placed 5.3 cm above the MCP, and the read-out of pulses produced by incident photons will be performed by an appropriate anode. The overall efficiency of such a camera for 100 – 300 Å photons is expected to be greater than 0.2.

The main background component in the EUV band will again be the diffuse photon background. There are powerful geocoronal emission lines at 304 Å, 584 Å etc. (Labov & Bowyer 1991). To get rid of this emission and to achieve the lowest background level of about 2000 photons/cm^2 s sr the preferrable range is 100 – 200 AA, which may be selected by a wide-band filter. The schematic view of the wide-field camera is shown on Fig. 1b. Sensitivity estimations show that it will be about 10^{-8} ergs/cm^2 s for a 1 s flash, which corresponds to $4 \cdot 10^{-5}$ ergs/cm^2 fluence in the 50-300 Å band for a GB. On the other hand this sensitivity corresponds to 100 photons/cm^2 s, i.e. the brightest sources like HZ 43 will be discovered after only 10 seconds of observations. This indicates a very attractive future for such cameras.

Acknowledgements: The author is very grateful to Dr. G. Fishman and Prof. S. Bowyer for helpful discussions, and to E. Gunaze for help in the calculations.

References

Labov S., Bowyer S., 1991, Adv. Space Res. 11, No. 11, 149

Lampton M.L., 1991, EUV Astronomy, D.F. Malina and S. Bowyer (eds.), Pergamon, N.Y., p. 353

Laros J.G., Evans W.D., Fenimore E.E. et al., 1984, ApJ 286, 681

Makishima K., 1993, ASCA News, No. 1, 6

McCammon D., Sanders W., 1990, ARA&A 28, 657

Murakami T., Tanaka Y., Kulkarni S.R., et al., 1994, Nature 368, 127

Schaefer B.E., 1993, Compton Gamma-Ray Observatory, M. Friedlander, N. Gehrels, D.J. Macomb (eds.), AIP 280, AIP, New York, p. 803

Yoshida A., Murakami T., Itoh M. et al., 1989, PASJ 41, 509

Wide-field X-ray Optics

René Hudec[1], *Adolf Inneman*[2], *Ladislav Pina*[3], *Petr Řehák*[4], *Paul Gorenstein*[5]

[1] Astronomical Institute, Academy of Sciences of Czech Republic, 251 65 Ondřejov, Czech Republic
[2] KOMA, Composite Materials, K lesu 965, CZ-142 00 Praha 4, Czech Republic
[3] Faculty of Nuclear Engineering, Czech Technical University, Břehová 7, CZ-115 19 Praha 1, Czech Republic
[4] Development-Optical Workshop of the Academy of Sciences of Czech Republic, Skálova 89, CZ-511 01 Turnov, Czech Republic
[5] Harvard-Smithsonian Center for Astrophysics, 60 Garden Street, Cambridge, MA 02138, USA

Abstract: Classical X-ray mirror optics usually has a limited Field of View (FoV) of $\sim 1^\circ$. Wide-field imaging can be achieved by lobster-eye type reflecting X-ray optics. We summarise several different approaches and suggest an innovative technology for the production of X-ray reflecting flats and cells necessary to develop one- or two-dimensional wide-field X-ray optics. The technology is based on double-sided, replicated reflecting foils produced by electroforming and CF/composite technologies.

1 Introduction

The use of double reflecting mirrors for focusing X-rays was suggested by Hans Wolter (1952). The grazing incidence mirrors of the Wolter 1 type appear to be most advantageous for space astronomy applications. However, they have a limited FoV of order 1 °. Lobster-eye X-ray optics, suggested more than 20 years later by Schmidt and Angel (Schmidt 1975, Angel 1979), represents an alternative to achieve large FoV. However, in this case the production poses difficulties. No lobster-eye x-ray objectives have been constructed to date. We outline possible geometries as well as possible production methods. A very promising new method is based on replica technology. The idea is to create a perfect copy of a negatively shaped master. We have carried out numerous tests and shown that this can even be done several times for particular masters, and that the degradation of shape and/or surface quality is within acceptable limits (Hudec et al. 1988).

The replication process begins with the production (with optical grinding and polishing) of high quality glass or glass ceramic masters (of negative shape). Surface smoothness of the masters can be enhanced using either superpolishing or lacquer coating. The final mirrors are produced by using electrodeposition (galvanoplastics) of a nickel layer on a polished glass master which is then removed. In some cases, the electroformed nickel deposit will be reinforced by composite material (external epoxy-carbon fibre layer). The material of the reflecting surfaces is electroformed nickel with good reflectivity (e.g. 60% at 0.83

nm and 1° incidence angle). An application of supplementary reflective film (e.g. gold coating) is also possible. For flat mirrors, float glass can be used instead of polished masters. Recently, we considered different high-quality float glasses with surface micro-roughness scales below 1 nm.

2 The lobster-eye X-ray optics

The X-ray optics of the lobster-eye geometry offers an excellent opportunity to achieve very large fields of view for energies between 0.1 and 10 keV. Two somewhat different systems have been suggested.

2.1 One dimensional systems

A one dimensional lobster-eye geometry was originally suggested by Schmidt (1975), based upon flat reflectors. The device consists of a set of flat reflecting surfaces. The plane reflectors are arranged in an uniform radial pattern around the perimeter of a cylinder of radius R. X-rays from a given direction are focussed to a line on the surface of a cylinder of radius $R/2$. The azimuthal angle is determined directly from the centroid of the focused image. At a glancing incidence angle for X-rays of wavelength 1 nm (and longer), this device can be used to focus a sizable fraction of an intercepted beam of X-rays incident in parallel. Focussing is not perfect and the image size is finite. However, a one dimensional focusing device offers a wide FoV, up to maximum of 2π with a coded aperture. It appears possible to achieve an angular resolution of $0\overset{\circ}{.}1$ or better.

Two such systems in sequence, to form a double–focusing device, should offer a FoV of up to 1000 square degrees at moderate angular resolution. We propose the use of such devices in X-ray astronomy to survey or monitor the sky.

Innovative very wide–field X-ray telescopes have been suggested based on these optical elements but have sofar not been flown in space. One of the proposals is the All Sky Supernova and Transient Explorer (ASTRE) (Gorenstein, 1987). This proposal also includes a cylindrical coded aperture outside of the reflectors which provide angular resolution along the cylinder axis. The coded aperture contains circumferential open slits (1 mm wide) in a pseudo–random pattern. The line image is modulated along its length by the coded aperture. The image is cross-correlated with the coded aperture to determine the polar angle of one or more sources. The field of view of this system can, in principle, be as large as 360° in the azimuthal direction and nearly 90% of the solid angle in the polar direction.

To create this mirror system, a development of double–sided flats is necessary, while recent X-ray foils are usually one–sided. There is potential for extending the wide field imaging system to higher energy with multilayer coatings analog to those described by Joensen et al. (1994) for flat reflectors in the Kirkpatrick-Baez geometry. These coatings excert great stress on the substrate. The system must meet severe weight limitations and so the new development of double-sided flats reinforced by composite material (to keep weight minimal while preserving mechanical stability) must be initiated.

This is the goal of the new development in which innovative technologies for double–sided flats are tested. The basis of the sandwich-type construction of the X-ray flats is an electroformed nickel layer which is deposited on plates of float glass. The nickel-coated plates of float glass are connected by means of carbon/fibre composite material and after hardening the set of connected plates will be cut/ground off. Subsequently, the plates of glass and the produced composite sandwiches with double-sided nickel mirror foils and inner composite reinforcement will be separated. These foils are lacquered on both sides so that the surface microroughness can reach values under 1 nm. The foils will be covered by a thin gold layer at the final stage.

2.2 Two dimensional systems

The idea of two dimensional lobster-eye type wide-field X-ray optics was first mentioned by Angel (1979). The lobster-eye optical grazing incidence X-ray objective consists of numerous tiny square cells located on the sphere and is similar to the reflective eyes of macruran crustaceans, such as lobsters. The FoV can be made as large as desired, and it is possible to achieve high efficiency for photon energies up to 10 keV. Spatial resolution of a few seconds of arc over the full field of view is possible, in principle, if very small reflecting cells can be fabricated.

This idea has not been developed further because of difficulties in the production of numerous polished square cells of very small size (about 1×1 mm or less at lengths of order of tens of mm). On the other hand, the very wide field imaging of the sky in X-rays would have very important consequences for a number of applications in X-ray and γ-ray astronomy.

This demand can be satisfied by electroformed replication and first test cells have already been successfully fabricated. This innovative technology is now under intensified study and a proposal was submitted to the Grant Agency of the Czech Republic. The recent approach is based on electroforming and composite material technology to produce identical triangular segments with square cells to be aligned in quadrants on a sphere. First test cells of 2×2 mm were successfully produced, indicating that electroforming can be a valuable tool in this very complicated and not yet solved technical problem.

3 Discussion

Without doubt, the use of very wide-field X-ray imaging systems would have a significant impact in many areas of X-ray and γ-ray astrophysics. The production of corresponding optical elements can be achieved by methods of electroforming and composite replication as an alternative to other methods. Results obtained with the developed new technologies for production of large area and high quality one-sided X-ray foils are very promising and together with composite material technologies represent important input for the development of double-sided flats needed for lobster-eye geometries of X-ray optics.

Acknowledgements: The development of double-sided X-ray reflecting foils is supported by a grant within the US-Czech Science and Technology Program, No. 930 37.

References

Angel J.R.P., 1979, ApJ 233, 364

Gorenstein P., 1987, in *Variability of Galactic and Extragalactic X-ray Sources*, A. Treves (ed.), Associazione per L'Avanzamento dell'Astronomia, Milano-Bologna

Hudec R., Valniček B., 1984, Adv. Space Res. 3, No. 10-12, 545

Hudec R., Valniček B., 1986, SPIE Proc. 597, 111

Hudec R., Valniček B., Aschenbach B., Braeuninger H., Burkert W., 1988, Appl. Optics 27, 1453

Hudec R., Valniček B., Svátek L., Landa V., 1989, Proc. of the Conference X-Ray Instrumentation in Medicine and Biology, Plasma Physics, Astrophysics and Synchrotron Radiation, Paris, 24-28 April 1989

Joensen K. D., Gorenstein P., Wood J., Christensen F. E., Høghøj P., 1994, SPIE 2279, (in press)

Schmidt W.K.H., 1975, Nucl. Instr. and Methods 127, 285

Wolter H., 1952, Ann. Phys. 10, 94

FLARES AND FLASHES:

PAST AND FUTURE

ROUNDTABLE DISCUSSION

Flares And Flashes: The Future

Round Table Discussion
Moderator: R. Vanderspek
Contributing authors: H.W. Duerbeck, G.J. Fishman, J. Greiner, R. Hudec, M. Merck, A. Poveda, N.N. Samus, W.C. Seitter, R. Vanderspek, J. van Paradijs, H. Zinnecker

Vanderspek: Introduction

The goal of this round-table discussion is to bring up issues which may be of mutual interest to those working on specific flare and variable stars and those performing wide-field searches for, among other things, counterparts to gamma-ray bursts (GRBs). Admittedly, however, trying to unify the interests of the flare people and the GRB people will be an interesting task. Many of the presentations this week have been devoted to future instrumentation and future modes of observations, but there have been also interesting discussions of specific questions regarding flare stars, variable stars and GRBs. Certainly, wide-field observations of the night sky will continue for the foreseeable future: there are a lot of instruments running now, and there are many instruments being planned. Most of these observations cover many square degrees of sky for long periods with a reasonable magnitude limit. How could these observations could be modified or improved to also allow useful data on on variable stars or flare stars to be taken? What sort of information, statistics, periodicities could be gleaned from wide-field searches that will be happening in the future and searches that are going on today? Are there methods in common use by one group of observers that may be useful to the other?

The speakers in this session have volunteered to give a short presentation on issues which may benefit from discussion within this group.

Merck: Ground-based TeV observations of GRBs

I want to tell you about the efforts that have been made in TeV astronomy to follow up GRBs. Before I came to *EGRET*, I was working on the *HEGRA* experiment, the High-Energy Gamma-Ray Astronomy observatory on the island of La Palma. It is one of the big arrays, and it is now complemented with a system of Čerenkov telescopes and all the inventory of techniques of ground-based gamma-ray astronomy. All groups working in ground based gamma ray astronomy got very excited with the results of *BATSE*, especially because of the delayed high energy emission seen by *EGRET*. The high rate of cosmic ray events normally is a big background for ground-based gamma ray astronomy. On the other hand, bursts have very short duration and one may easily detect them with ground-based instruments, as they are essentially background-free on these small timescales.

To search for bursts *HEGRA* has been connected to the *BACODINE* system. Of the system of five imaging Čerenkov telescopes, two are operational. The remaining three will come on-line in 1995. The telescopes are similar to the Whipple telescope but with less mirror area, so that we have a higher energy threshold (around 1 TeV). The burst observation mode is the following: after a *BATSE* trigger, we determine if the position is visible from our location. If this is true we will point the telescopes to the given best position. All Čerenkov observations are only possible during nights without moonlight so only bursts during these time intervals can be followed up. The telescopes will be pointed to the region of the sky where the burst happened for a second night after the burst to cover all possible delayed emission. After this, the telescopes will return to their target assigned by the normal observational program. The scintillator array has been operational for five years: very interesting limits on GRBs were presented at the Huntsville meeting.

The third instrument in operation at La Palma is a very interesting Čerenkov array. Its performance is much better than that of the scintillator array: it has a much better angular resolution (5′) and a lower energy threshold (25 TeV). I think we can lower this threshold to 10 TeV with the present design by improving the electronics and we think that with an improved second generation experiment an energy threshold of 1 TeV is feasible. This Čerenkov array consists of nearly normal photomultipliers with a Winston cone assembly to collect the Čerenkov photons. Each photomultiplier views one steradian of the sky the whole time. A normal photomultiplier would be destroyed by this high background light level in about 10 minutes, thus specially designed multiplier with a reduced amplification, having only 6 dynode stages, is used. The rest of the amplification is done electronically. The noise is further reduced by a linear gate which picks out a very short time around the photomultiplier signal, thus reducing the contribution of the night sky background. Using a coincidence trigger of 12 such detector stations, we can reconstruct the Čerenkov light front of the air shower. This front is very well defined, with a thickness of a few nanoseconds. With special filters it could become possible to use this technique also during nights with moonlight.

To push the energy threshold down to 1 TeV we are testing the Quasar type photomultipliers from Russia, the ones used in the Lake Baikal experiment, which are much bigger (25 inch diameter). The number of Čerenkov photons from a 1 TeV air shower collected with such a multiplier should be sufficient to dicriminate against the night sky background noise. We are considering an experiment consisting of a 1 km^2 array of this type. Technically the event rate of a few hundred million events per night is challenging, but solvable. We don't have to move them to a *BACODINE* location, because they will have one steradian, *i.e.* a section of the sky with $\approx 40^\circ$ opening angle, in the field-of-view of this detector the whole time.

I think that if there is emission in the TeV range at the level you may expect by extrapolating the GRB spectra, we should be sensitive with this type of instrument to all of the *BATSE* bursts. Those events with the highest fluence, those which were also detected by *EGRET*, should be detectable with our present system if the spectra can be extrapolated to 10 TeV.

Moving the telescopes is a problem, as we were thinking about using our Čerenkov telescopes as an optical monitor for bursts. If an optical flash happens, and it is very short in duration, say 1 millisecond (the typical timescale for the collapse of the core of a merging neutron star into a black hole), the photomultipliers are the best way to detect this. You will be much more sensitive than with photographic plates, which integrate over a long time. But the problem is that you can't move your telescope very fast and the probability is very low for getting a burst in a $3^\circ \times 3^\circ$ field-of-view by chance: my calculation gave 1 coincidence for 15 years of *BATSE* triggers.

With the wide field-of-view Čerenkov array we did search for emission from AE Aquarii, and we didn't find anything. We had an observing campaign this summer with the Čerenkov telescopes, which was supposed to be in collaboration with the Potchefstroom group. They didn't get their telescopes working so no correlations are possible. A very preliminary analysis of our data showed no obvious signal. I think we should have seen a signal of the size the Potchefstroom people claimed as a 100σ excess. It should be quite obvious in a first analysis and no such signal was found.

Hudec: Future strategies for optical searches for GRBs.
Monitoring individual γ-ray burst error boxes for transient optical radiation is difficult in that there are so many of them and the repetition rates are low. An alternative approach is to monitor many gamma ray positions simultaneously. We have developed the idea of the *OTM* (optical transient monitor) experiment to get a very large field-of-view of $\approx 50^\circ$ diameter, in which case ≈ 30 or more gamma ray positions can be monitored simultaneously. The *OTM* is based on wide field optics and a CCD camera, which we expect to have something like 1000^2 pixel, and a sensitivity for stars of between $10^{\rm m}$ and $12^{\rm m}$.

Simultaneous optical observations of GRBs to date almost all have a limiting magnitude for a one-second flash of between $0^{\rm m}$ and $3^{\rm m}$. With the *OTM*, we expect to have a one-second limiting magnitude of between $5^{\rm m}$ and $6^{\rm m}$. Our idea is to look for a final confirmation that optical transients are indeed real. We would like to operate two such stations, separated by about 100 km to get very good parallax, to be able to eliminate all triggers which could be coming from the Earth's atmosphere (such as satellite flashes). The real flash must be located on the frames from both stations, and must be at the same time and the same position: then we can be finally sure it is the optical counterpart.

In addition to this system, which is optimized in the red part of the spectrum (as are most wide-field surveys using CCDs and lenses), we are trying to develop a wide-field optical system which is more sensitive in the blue. Because our analysis of the Schaefer 1928 OT shows that the image is real, not a plate defect, we conclude that the differences it shows from neighboring stars is due to the fact that the flash is very blue. The system we are developing will use a CCD with blue-enhancing coating and wide-field optics that are optimized for better blue transmission.

Vanderspek: The Explosive Transient Camera

I would like to briefly review the Explosive Transient Camera (*ETC*) and its possible use to look for flare stars and long-term variables. Each night, the *ETC* stores several sets of full $20° \times 15°$ images of the night sky from all sixteen cameras for later analysis. Among the images stored in this manner during the course of a year, one can find each part of the sky between $-15°$ and $+60°$ imaged roughly sixty times. To date, we have collected over 36 000 images over the last four years. We are in the process of analysing these images. We process each image by looking for stars in the field, and perform astrometry and photometry on each star detected in the field. We have right ascension, declination, magnitude and its error for 1000-2000 objects on each of these fields down to a limiting magnitude between 9^m and 12^m, depending on the night and the optical system used. When the analysis of these images is complete, we will have 30 000 000 brightness measurements of roughly 100 000 objects, and we are sure to see transient phenomena among the data.

In the beginning of 1995, we will install some new CCDs which will increase the field-of-view by a factor three. We will have about the same limiting magnitude, and the pixel size is slightly more than $3'$, but we will be able to collect data from significantly more objects over a significantly larger declination range.

I would like to mention that when the High Energy Transient Experiment (*HETE*) launches and all work wells, we will be taking full images of the night sky in a near UV band between 230 and 270 nm with a FoV of $40° \times 40°$ roughly $10 - 40$ times per night. We plan to collect a large database of UV images, and the potential for detecting variability from flare stars and other sources is enormous.

Vanderspek: GRB error circles and wide-field observations

The situation for looking for optical counterparts for GRBs so far is not very encouraging: many workers have spent many years looking for them, and none have been seen. However, when Rene Hudec and Jochen Greiner described their work, one result really stood out: they have 55 full sky plates of GRB error regions taken during outburst and they have not seen any transient optical radiation. This tells me that we aren't going to see anything if we continue to look with such wide fields. The future is smaller fields, where you can go to deeper limiting magnitude. One difficulty in going to smaller fields-of-view is that the error circles derived by *BATSE* alone are rather large ($4°$-$8°$ 1σ radius). The *ETC* has imaged the region of the sky where *BATSE* detected a GRB at the time of the GRB on five separate occasions, but we have not been able to make a firm statement about the presence of transient optical radiation because of the size of the error circles. Gerry Fishman has told me that there is some promise in reducing the size of the *BATSE* error circles.

Fishman: BATSE error circles

Before saying something about the error circles, I would like to tell you about the prospects for *BATSE*, because there is so much activity with coordinated ground-based observations. You might wonder if this all will be for nought. Will *BATSE* continue to operate? I don't have a crystal ball. I don't know what the future political or economical situation will be. The ongoing astrophysical missions will continue to be reviewed every 2 or 3 years. As my personal feeling the prospects are very good that *BATSE* and the *CGRO* will operate for at least another 5 years. There is a fair chance that it may operate 10 years.

As far as the error boxes from *BATSE* are concerned, I don't need to apologize. The best we can do right now is of the order of $3-4^\circ$. There is some prospect for improving that of the order of 2°. The basic limitation is that *BATSE* is uncollimated: there is no focussing or lensing, there are just flat plates. There are systematic errors that are being improved: the largest of those are atmospheric scattering, non-linear energy calibration, and spacecraft scattering. As we are getting more bursts and as we are getting more calibrational location from the interplanetary network and solar flares, we will be able to reduce this. I doubt that we ever get better than 2°.

Concerning burst location determination in real time, we can't do that well because we don't have access to all of the data that is required to make all this corrections. The best we will be ever do with the real time *BACODINE* locations probably will be of the order of $6-8^\circ$. Maybe there is a chance of improving that to maybe $4-5^\circ$ at 1σ radius.

Seitter: The Muenster Redshift Project (MSRP)

I came here to listen and to learn, but some of the contributions made me aware of the fact that at Muenster we are actually doing things which might be of interest to this audience, especially in view of some pilot projects which we have already carried out.

The Muenster Redshift Project (MRSP), aimed at answering questions of observational cosmology, as a by-product assembled catalogs of stars from the ESO/SERC Sky Atlas R-plates with red magnitudes $r_F \leq 20^m$ for more than 20 million stars, and from the J-plates with blue magnitudes $b_J \leq 21\overset{m}{.}5$ for about half a million. The blue survey continues.

The corresponding color indices permit us to detect the very red dwarf and giant populations. The means of separating the dwarfs are proper motion measurements, which, so far, have been obtained for some 40 000 stars. The Hertzsprung-Luyten-Diagram (HLD) is shown in Fig. 1. In this diagram, intrinsic magnitudes are represented by proper motion absolute magnitudes M_μ. They are related to the absolute magnitudes M by $M_\mu = m + 5\log\mu + 5 = M + 5\log T - 3.379$, where T is the tangential velocity in km/s.

In addition to color measurements, very low-dispersion (245 nm/mm) objective prism plates are available. So far, 1 300 M stars have been classified, while spectral data for more than another 30 000 are partially reduced. From our objective prism work we found that the space density of M dwarfs as a function of distance from the galactic plane up to z=150 pc is on average larger by a factor 1.8 than suggested in the Bahcall-Soneira model. Also apparent is the somewhat steeper decline of space densities with distance from the galactic plane.

The fact that less than 2000 known flare stars have been registered so far (Tsvetkova et al. 1995, this volume p. 121) suggests that one might use the above methods to search our catalogs for potential flare candidates.

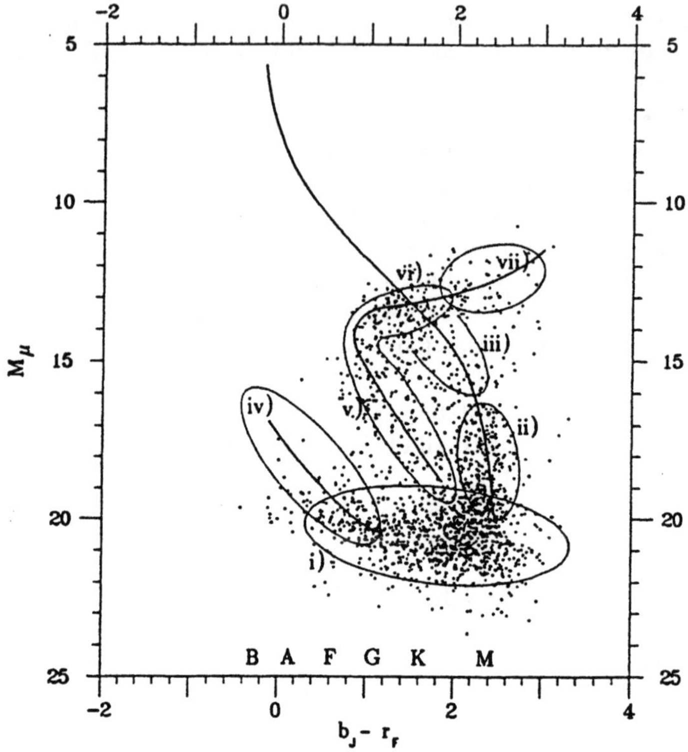

Fig. 1. The location of major spectral types and luminosity classes in the HLD: i = artefacts, ii = M dwarfs, iii = G and K dwarfs, iv = white dwarfs, v = G and K metal-poor stars, vi = halo G and K giants, vii = halo M giants.

Samus: The future of the General Catalog of Variable Stars (GCVS)

I would like to give a brief contribution regarding future work on the GCVS. Financial support through IAU commission 27 is still available, and this support was unanimously expressed at the IAU this year. The fifth and final volume of the GCVS (4th edition), devoted to extragalactic variables and supernovae, is now at the printers, will appear this winter, and will be distributed in the normal way to the users. The next step is a supplement to the Catalogue of Suspected Variables: it is practically ready, and we shall look through all the material once again this spring. I don't know exactly what will follow. I am not sure we are going to print a booklet. It is rather expensive to distribute the book, but at least we will make this catalog available to the users in electronic form. In the present situation with new variables and with the approaching *HIPPARCOS* wave of information one of the most important fields of our activity will be the namelist of variable stars. All the variables should be named, so that you will not confuse them with already known variables. This work will be continued. The *HIPPARCOS* team expressed their wish and the IAU expressed their wish that variables should be designated before their material is published. If they really give us an idea which their variables are, we will try to do that. But it is a tremendous job, but we are going to publish the 72nd namelist in about one month (IBVS 4140, January 2, 1995).

The magnetic tape version of what I would call version 4.1 of the GCVS is now available. There are some mistakes which are to be corrected, and then we are going to start with version 4.2, where we are going to introduce more serious corrections. The classification system is now rather obsolete, I must admit. We must introduce new ephemerides for periodic variables, which is also necessary for users. We have all that stuff not available in datacenters (remarks, reference list), which has to be added and it will be also available to the users. Another direction of our work is work on coordinates. We undertake the serious effort to identify variable stars with positional catalogs and guide star catalogs. Our first attempt to do it automatically showed that all automatic identifications are 80% wrong, and so it has to be done by checking everything on the screen. We also have the necessary visualisation software, and an enthusiastic young man is now trying to work through it. He is very active and good, and he has done several thousand stars in two months.

The next possibility of our work: we have a database in form of a card catalog, and we have started work on a computerized database of the GCVS. Unfortunately, because the volume of information in our card catalog is large, and we do not have enough staff and enough time to do this. We fear that the software we have for this work is not quite adequate, and we lose time when we work with it. Although we have the first version, and can use it inside the group, it is insufficient now in its completeness and scope to be presented to users. I don't know what the future of this work will be: it strongly depends on the future of our institute, which is also not quite clear.

The last point is the funding. The group working on the GCVS is presently funded by Russian astronomical institutions. Compared to other astronomical groups that means "very poorly". We are very grateful to several international organisations, who gave us grants in the recent year. We had the grant of the ESO. We had a small but very necessary grant from the AAVSO which was used for the least paid members of our group. And then we have now applied for a grant of the Russian foundation for fundamental research and I hope we will get it. It will give us a possibility to continue all this work at normal scale. I am not pessimistic, though I am not quite optimistic.

Zinnecker: Flare and flash terminology

I would like to bring up the definition of terminologies. During this conference, I heard about flares, flashes, flickering, transients, bursts, events, and there are even subspecifications like surface flares (*e.g.* on WTTS), accretion flares (on CTTS), outbursts (on novae), eruptions (of fuors), explosions (of supernovae), explosive transients, and even echos (of jets). I think that it might be useful to define what we mean.

Shouldn't we try to adopt a more coherent classification which could be physical, according to timescales, or according to the amplitudes of these "events". I wanted to at least propose to define some nomenclature. Concerning the flares and flashes, which is the title of our conference, I was surprised that none of the speakers of the first day had attempted to even define for us what flares and flashes were. I start from an intuitive point of view: a flare for me is something that rises very quickly, but I do not care how it declines, whereas a flash is something that rises quickly and comes down quickly. I do not know whether it has to be repetitive or periodic. Dr. Poveda told me the difference between flares and flashes: a flare is on a dMe star and a flash – introduced by Haro – is a flare on a star in an aggregate, or cluster. So flares and flashes are the essentially same: it is just the location that determines the terminology. So that is the right definition.

We should talk about this, and whether we accept it. In any case, it also raises the question of whether the above mentioned terminology should be according to the timescales of rise and decline, or in terms of the amplitude, or in terms of physics.

Duerbeck: Flare and flash terminology

Hans and I have briefly discussed this problem. I think the work of a good astronomer is to convert terminology of variable star activity into the physics of the phenomenon, and of course, there are a lot of phenomena known.

In the variable star catalog, we also see some problems of terminology: We have the cataclysmic variables (C) and the eruptive variables (E), and the difference is at times very difficult to define. Some eruptive variables indeed show eruptions (CTTS), others show flares (WTTS), others show outbursts (FUORS). Consider the cataclysmic variables: 'dwarf nova eruptions'. The instability which is the physical reason of variability in dwarf novae is very similar, maybe related, to the instability in accreting pre main-sequence stars. Then you can connect cataclysmic and eruptive variables in the variable star catalogs. Either one has to say 'dwarf nova outbursts' are not real outbursts like those of novae, they are 'dwarf nova eruptions' instead. But I do not know if one can reshape our terminology if we have a (hopefully) better theory to explain the phenomena.

I have written down here for example the known physical reasons for these varabilities:

GCVS class	Terminology	GCVS Var. type or other	physics of phenomenon
C	Explosion	SN	gravitational collapse / bounce
C	Burst	neutron star	explosive TN burning
C	Outburst	N	TNR in outer layer of star
C	Outburst	UG	Disk instability
E	Flare	late type star	magnetic activity in outer layer
C	Flickering	catacl. var.	activity in inner accretion disk (magnetic?)
E	Eruption	pre-MS-star	disc changes, $\dot{M}$
?	Flash		

If the physical processes are unknown, some neutral word, for example transient, should be used because the source is at some time there and at another it isn't (or it is weak), *e.g.* a gamma-optical transient. I think a fast transient can be termed flash, but of course we have to properly define the timescale: what is "fast"? I leave this question open.

Poveda: The original definitions of 'flare' and 'flash'

When I came to this meeting "Flares and flashes", I thought the definition was very clear. For quite a number of years flash stars have been identified with those young stars in clusters, that have some sort of flare activity. When Haro first discovered this kind of outburst in stars in Orion, he introduced the term flash star to distinguish from the flare stars of UV Ceti type in the solar vicinity, to avoid any possible confusion between the two types of activities. With the passage of time, evidence has been growing that this activity in young stars in Orion, in the Pleiades, in the Hyades is similar to the activity of UV Ceti stars in the solar vicinity. Nevertheless a few years ago, I remembered that George Herbig expressed the concern that one should pay more attention to the differences between UV Ceti activity and flash activity in young clusters. In the paper I presented I presented some evidence that the UV Ceti stars in the solar vicinity have the same kinematics as some galactic clusters. It appears that some of the UV Ceti stars are members of superclusters: the Hyades or Pleiades supercluster. It appears, although it needs more evidence, that the UV Ceti stars in solar vicinity are just the oldest members of the flare population of galactic clusters that have been broken down long ago. We cannot recognize most of them, some of them we recognize, and this may be the evolutionary, the cosmogonic relation between one group and the other. However, for the time being it would be a good idea to keep separate names for those activities in galactic clusters from the activity in the solar vicinity.

van Paradijs: Precise terminology is not necessary

I would like to make a general point, perhaps a little bit philosophical. The meaning of words of course is only given by their users in general. As long as you do not know what the phenomenon that you are observing means in physical terms you have to base yourself on phenomenology, and it is a social phenomenon whether or not there is confusion about terms. I think your suggestion to try to make these phenomenological terms more precise by introducing timescales is extremely dangerous, because talking about a rapid rise and a slow decay clearly is a very relative thing. You see supernovae come up in a few days, and you see it go down in a couple of months: rapid rise, slow decay. And clearly a rise of a day must be considered extremly slow when you are dealing with M stars which optical magnitude increases. So either you have an understanding what is going on in a physical sense, and you do not need phenomenology anymore to worry about your semantics, or you don't know what you are talking about, for instance GRB's, but there the phenomenology is clear enough to avoid, I think, any confusion. So I think to some extent it is a non-issue.

Vanderspek: Summary

I would like to thank the participants in this round-table discussion for their contributions and insights. Clearly, there are regions of overlap between the variable star observers and those looking for counterparts to GRBs. The experiments that search for counterparts to GRBs, independent of energy, have been designed to look for variability in celestial sources, and thus are "tuned" to the characteristics of "flare" and "flash" sources: these experiments may be able to contribute to the understanding of variable stars of many classes. In particular, the wide-field experiments described here should be able to provide information about the rate of rare, high-amplitude flares.

There are other areas of overlap between flare stars and GRBs which have only been touched upon briefly in this discussion by J. Greiner. When addressing the flare star topic from the GRB point of view, one would like to know answers on the following questions and unsolved problems: (1) Is it possible that flare stars are responsible for a fraction of the detected GRBs? (2) Flare star theory predicts hard X-rays as the primary emission, but no positive detections have been made to date despite first correlations of monitoring X-ray data and strong, known flare events. (3) Can a flare star make MeV photons? (4) Is there a maximum energy which a flare star can release, either optical or bolometric? Perhaps this is a situation where two seemingly unrelated phenomena truly do have a common source, and perhaps, in time, we will be able to determine the link between the two.

After Dinner Address

G.A. Richter

It was exactly 22 days before Cuno Hoffmeister's 65th birthday when I began my work at Sonneberg Observatory, following my studies at the University of Jena. In those days, all Sonneberg astronomers were busy preparing scientific papers which they were going to dedicate to their leader, with the exception of myself. I was not able to write a scientific paper in such a short time.

And I was quite ashamed when on the 2nd of February 1957 I had to offer my birthday congratulations empty-handed. But I distinctly remember how extremely happy he looked. Hoffmeister was near the zenith of his scientific career. New buildings and domes were planned or under construction, and the Zeiss company of Jena had promised to loan a 40 cm astrograph in compensation for the loss of a telescope to the Soviet Union as part of the reparations agreement of 1945.

Today, too, I am a little embarrassed. However, this time by your noticing the coincidence of this colloquium and the completion of our, Wolfgang Wenzel's and mine, 65th year of our lives. Also, by turning this mere coincidence into an occasion to bestow honors on us, while to Hoffmeister, whom we take as a shining example still today, an international conference in Sonneberg was denied at his time.

We did not find 10,000 variables, as Hoffmeister did, but we made an effort to continue the time honored tradition of the systematic sky patrol and the field patrol. To some the word "tradition" has become discredited; following a tradition is often regarded synonymous with desperately clinging to antiquated and out-of-date things. But one easily overlooks the fact that there are traditions which should be continued because they are essential, as is the case for continuous monitoring of the sky. For example, in the weather service nobody would conceive it as a "bright idea" to close all weather stations because 150 years of world weather record is sufficient! During guided tours through Sonneberg Observatory I could make every child realize the necessity of a permanent sky patrol. But to convince a politician ···

Sometimes I can't help thinking that there are also some astronomers who believe it possible to predict the next Tunguska event or the appearance of the next Galactic supernova by means of a computer.

Exactly 69 years ago, in December 1925, Cuno Hoffmeister, a son of the town of Sonneberg, founded the observatory on the mountain Erbisbühl, 640 m above sea level and 260 m above the city of Sonneberg. There was a close co-operation with Ernst Hartwig and Ernst Zinner at Bamberg and with Paul Guthnick at Babelsberg. Thus, from 1930 to 1945, Sonneberg Observatory was a branch of the Berlin-Babelsberg University Observatory, 350 km away from Sonneberg. Nowadays, in the era of modern traffic and electronic communication, it is sometimes claimed that Sonneberg is too far from the nearest university. But there are quite a number within a distance of 100 km!

The principal work at Sonneberg Observatory has been the study of variable stars in connection with the sky patrol (observation of the whole sky) and the field patrol (observations of selected fields with a deeper limiting magnitude in order to get a larger sample of different classes of variable stars for statistical investigations).

Furthermore, since the early fifties photoelectric measurements have been carried out in order to observe objects of special interest. For example, multiple-periodic pulsating stars like AC And, extraordinary X-ray binaries like HZ Her and TT Ari, or the quite exclusive He-flash star FG Sge. Before the fall of the wall, these measurements were adapted to the technical possibilities of that time, now one of our 60 cm telescopes works with new CCD camera technology.

In addition to variable stars, the research program at Sonneberg also covered such diverse areas as interplanetary matter and upper atmospheric phenomena, including studies of asteroids, comets, meteors and meteor streams, zodiacal light and gegenschein, airglow, noctilucent clouds, and luminous bands.

It should be remembered that Hoffmeister investigated, in the early sixties, the angle between the tail of a comet and the radius-vector from the sun. He concluded that there must be a particle flow from the sun. Thus, independent from Ludwig Biermann, Hoffmeister discovered the solar wind.

For some time, extragalactic research was performed at Sonneberg as well.

When, after the Second World War, Sonneberg Observatory was separated from Berlin-Babelsberg, the only way for the institution to survive was an affiliation with the Deutsche Akademie der Wissenschaften zu Berlin. Owing to this affiliation it became possible that, in the early sixties, Sonneberg could start a modern sky patrol, which stood the test of time.

Although, in later years modern equipment was lacking, good results were achieved. Hoffmeister's wisdom "in the long run it is not the size of the means that is decisive for success but the impetus behind it" proved to be correct.

In 1961 our work was further complicated by the incorporation of Sonneberg into a restricted area along the Iron Curtain and, as a consequence, by strict prohibition of any contact with scientists in western countries, including the Federal Republic of Germany. Only now and then, astronomers from the eastern countries received permission to visit Sonneberg Observatory. Worst of all were the reprisals taken against Cuno Hoffmeister by the communist firebrands of the Academy, which finally were the cause of his death in 1968.

In appreciation of the work at Sonneberg Observatory it is indispensable to remember the late coworkers who essentially shaped the character of the institute. I refer to Paul Ahnert, Nikolaus Richter, Johannes Hoppe, Sergej Gaposchkin, Heribert Schneller, Rudolf Brandt, Otto Morgenroth, Artur Teichgraeber, Hans Huth, and many others. Also to be mentioned are the coworkers who had to leave the Observatory after the reunification of Germany.

Now, some words have to be said about the observers at the sky patrol and field patrol. Year in, year out, and night in, night out they left their more or less sympathetic wives alone at home at night and, rather than winning gratitude, were criticised whenever they were not on deck on time, while frequently it were others who reaped the fruits of their labor. – I think, it is the late Sonneberg astronomer Hans Huth who holds the world record in the number of sky plates taken by one person.

When Hoffmeister, who was an enthusiastic hobby-seaman, became aware that one of the observers was not on deck, i.e., not at his telescope, the delinquent never had to fear violent words. But the ironical comments of the captain could be very painful. Occasionally, it was claimed that Hoffmeister was a misogynist. I cannot confirm this, but certainly Hoffmeister was not a man who appreciated flirtation. But he did get along very well with his female collaborators, especially if they spoke English. But whenever one of his male collaborators were visited by ladies during working hours or even during observation periods, Hoffmeister stated very angrily: Nymph aboard! In this respect Hoffmeister kept a close watch on Sergej Gaposchkin, who liked ladies. But finally Gaposchkin emigrated from Germany to America in order to marry Cecilia Payne, and both are known to have become famous scientists. Hoffmeister attributed great importance to his collaborators' taking good care of fixtures and fittings. When one morning he was told that one of his people had dropped from the observer's ladder, Hoffmeister was horrified and said: I do hope nothing happened to the instrument!

Concerning present and future work at Sonneberg Observatory I think I need not say much because you are learning a lot during this conference. I only want to remind you of the working group ASPA. Projects in progress are the search for optical flashes of gamma-ray bursts and the optical identification and classification of ROSAT X-ray sources.

In conclusion, I would like to tell you that near the end of my official career as an astronomer I am delighted to see a long-cherished wish of mine coming true: one day to enjoy a great international colloquium in Sonneberg. My wish is fulfilled. Thank you very much.

Resolution

An Appeal to Save Sonneberg Observatory

The participants of the IAU Colloquium No. 151 on "Flares and Flashes", held in Sonneberg December 5–9, 1994, express their utmost concern about the threat of closure of the Sonneberg Observatory as a research institute by the end of this year. Sonneberg is the only place in the world where routine sky patrol observations are being performed and such a step should be a dramatic loss for astronomical research which can never be recovered in the future. The participants are aware that a continuation of a sky patrol in the traditional form, on the basis of photographic plates, is neither feasible (for reasons of the imminent discontinuation of the production of such plates), nor practical, as technologically much more advanced detectors (CCDs) are readily available. They consider the ASPA project — "All Sky Patrol Astrophysics" — as the best means of efficient continuation, and in fact extension, of sky patrols with new technologies. In particular they consider the Sonneberg Observatory a most adequate site for the logistical center needed for ASPA, in view of the long tradition and expertise in this field, the large unique plate archive, and the deep involvement in high-technology research at the Technical University of Ilmenau. In view of this, the participants of IAU Colloquium No. 151 urge all involved local authorities to provide for the continuation of the scientific work at Sonneberg Observatory.

The participants of
IAU Colloquium 151 "Flares and Flashes"

APPENDIX

Addresses of IAU Colloquium Attendees

Dr. Meil Abada-Simon
Sterrekundig Instituut
Universiteit Utrecht
Postbus 80.000
3508 TA Utrecht
The Netherlands

Mr. Tobias Aigner
Max-Planck-Institut für Extraterrestrische Physik
85740 Garching
Germany

Dr. Ivan Andronov
Department of Astronomy
Odessa State University
T.G. Shevchenko Park
Odessa 270014
Ukraine

Mr. Bill Ball
University of Central Lancashire
Department Physics and Astronomy
Preston, Lancashire
United Kingdom

Prof. Arnold O. Benz
Institut für Astronomie
Eidgenössische Techn. Hochschule
CH-8092 Zürich
Switzerland

Dr. Grigory M. Beskin
Special Astrophysical Observatory
Russian Academy of Sciences
Nizhnij Arkhyz
Karachajevo-Cherkessija 357147
Russia

Dr. Nina Beskrovnaya
Central Astronomical Observatory
Pulkovo, St.Petersburg, 196140
Russia

Prof. Stuart Bowyer
University of California
Center for EUV Astrophysics
2150 Kittredge Street
Berkeley CA 94720-5030
U.S.A.

Mr. Wolfgang Brandner
Astronomisches Institut der Universität Würzburg
Am Hubland
97074 Würzburg
Germany

Dr. Hans-Jürgen Bräuer
Sonneberg Observatory
96515 Sonneberg
Germany

Dr. Albert Bruch
Astronomisches Institut
Universität Münster
Wilhelm-Klemm-Str. 10
48149 Münster
Germany

Dr. P. Brendan Byrne
Armagh Observatory
College Hill
Armagh BY61 9DG
Northern Ireland

Dr. Don Casperson
Los Alamos National Laboratory
Astrophysics and Radiation
Measurements
Group NIS-2, Mail Stop D436
Los Alamos, NM 87545
U.S.A.

Dr. Alberto J. Castro-Tirado
Lab. de Astrofisica Espacial Madrid
LAEFF
Apdo 50727
E-28080 Madrid
Spain

Dr. György Csepura
Heliophysical Observatory
Hungarian Academy of Sciences
P.O. Box 30
H-4010 Debrecen
Hungary

Prof. Dainis Dravins
Lund Observatory
Box 43
S-22100 Lund
Sweden

Dr. Hilmar Duerbeck
Astronomisches Institut
Universität Münster
Wilhelm-Klemm-Str. 10
48149 Münster
Germany

Dr. Christian Fendt
Lund Observatory
Box 43
S-22100 Lund
Sweden

Dr. Gerald Fishman
University Space Research Association
NASA/Marshall Space Flight Center,
ES-81
Huntsville, AL 35812
U.S.A.

Dr. Elena Franciosini
Department of Astronomy and Space
Science
University of Florence
Largo e. Fermi 5
I-50125 Florence
Italy

Dr. Gösta F. Gahm
Stockholm Observatory
S-13336 Saltsjöbaden
Sweden

Dr. Roald E. Gershberg
Crimean Astrophysical Observatory
Crimea Nauchny 334413
Ukraine

Dr. Alexander Glukhov
Rostov State University
Department of Astrophysics
Physics Faculty
5 Zorge Str.
344104 Rostov-on-Don
Russia

Dr. Woldemar Götz
Sternwartestr. 23a
96515 Sonneberg
Germany

Dr. Jochen Greiner
Max-Planck-Institut für Extra-
terrestrische Physik
85740 Garching
Germany

Mr. Erik Gullbring
Stockholm Observatory
S-13336 Saltsjöbaden
Sweden

Dr. Dieter H. Hartmann
Department of Physics and Astrophysics
Kinard Laboratory of Physics
Clemson University
Clemson, SC 29634-1911
U.S.A.

Dr. Martha L. Hazen
Harvard College Observatory
60 Garden Street
Cambridge, MA 02138
U.S.A.

Dr. Joachim Hildebrandt
Astrophysikalisches Institut Potsdam
An der Sternwarte 16
14482 Potsdam
Germany

Prof. Alisher Hojaev
Ulugbek Astronomical Institute
Astronomicheskaya Str. 33
Tashkent 700052
Uzbekistan

Dr. Rene Hudec
Astronomical Institute Ondrejov
CZ-251 65 Ondrejov
Czech Republik

Dr. Nasar Ikhsanov
Central Astronomical Observatory
Pulkovo
196140 St. Petersburg
Russia

Prof. Dr. Valentin Karetnikov
Astronomical Observatory
Odessa State University
T.G. Shevchenko Park
270014 Odessa
Ukraine

Dr. Mariko Kato
Department of Astronomy
Keio University
Hiyoshi Kouhoku-ku
Yokohama 223
Japan

Dr. Maria Katsova
Sternberg State Astronomical Institute
Moscow State University
13 Universitetsky Prospect
119899 Moscow V-234
Russia

Prof. Dr. H.V. Klapdor-Kleingrothaus
Max-Planck-Institut für Kernphysik
PF 103980
69029 Heidelberg
Germany

Dr. R. Konstantinova-Antova
Institute of Astronomy
Bulgarian Academy of Sciences
Tsarigradsko Shose 72
BG-1784 Sofia
Bulgaria

Dr. Chryssa Kouveliotou
University Space Research Association
NASA/MSFC, ES-84
Huntsville, AL 35812
U.S.A.

Dr. Irina Krivosheina
Radiophysical Research Institute
603600 GSP 51,
st. B.-Percherskaya 25
Nizhnij Novgorod
Russia

Mr. Peter Kroll
Sonneberg Observatory
96515 Sonneberg
Germany

Dr. Albrecht Krüger
Astrophysikalisches Institut Potsdam
An der Sternwarte 16
14482 Potsdam
Germany

Dr. Martin Kürster
Max-Planck-Institut für Extraterrestrische Physik
85740 Garching
Germany

Mr. Omar M. Kurtanidse
Astrophysical Observatory Abastumani
383762 Abastumani
Georgia

Dr. Constanze La Dous
ESA/IUE Observatory
Villafranca del Castillo
Apartado 50727
E-28080 Madrid
Spain

Dr. Iosif Lapidus
Institute of Astronomy
University of Cambridge
Madingley Road
Cambridge CB3 OHA
United Kingdom

Dr. Stefan Larsson
Stockholm Observatory
S-13336 Saltsjöbaden
Sweden

Thomas Lehmann
ESO
La Silla
Chile

Prof. Dr. Moissei A. Livshits
Institut of Terrestrial Magnetism, Ionosphere and Radio Wave Propagation
Russian Academy of Sciences
142092 Moscow, Region Troitsk
Izmiran
Russia

Mr. S. Lührs
Astronomisches Institut
Universität Münster
Wilhelm-Klemm-Strasse 10
48149 Münster
Germany

Dr. Elena Luthardt
Department of Astronomy
Odessa State University
T.G. Shevchenko Park
Odessa 270014
Ukraine

Dr. Rainer Luthardt
Sonneberg Observatory
96515 Sonneberg
Germany

Dr. Farouk Mahmoud
Department of Astronomy
National Research Institute
of Astronomy and Geophysics
Helwan, Cairo
Egypt

Dr. Ala Markina
Astronomical Observatory
Odessa State University
T.G. Shevchenko Park
270014 Odessa
Ukraine

Dr. Isolde Meinunger
Sonneberg Observatory
96515 Sonneberg
Germany

Dr. Martin Merck
Max-Planck-Institut für Extraterrestrische Physik
85740 Garching
Germany

Dr. Roberto Minarini
Special Astrophysical Observatory
of Russian Academy of Sciences
Nizhnij Arkhyz
Zelenchukskij Rajon
Karachajevo-Cherkessija 357147
Russia

Dr. I. Felix Mirabel
Service d'Astrophysique
Centre d'Etudes de Saclay
F-91191 Gif-sur-Yvelle Cedex
France

Prof. Ludwig Mirzoyan
Byurakan Astrophysical Observatory
Armenia

Dr. E.I. Moskalenko
Sternberg State Astronomical Institute
Moscow State University
13 Universitetsky Prospect
119899 Moscow V-234
Russia

Dr. Sergei Neizvestny
Special Astrophysical Observatory
Zelenchukskij Rajon
Nizhnij Arkhyz
Karachajevo-Cherkessija 357147
Russia

Dr. Peter Notni
Universität Potsdam
Astrophysikalisches Institut Potsdam
An der Sternwarte 16
14482 Potsdam
Germany

Prof. Hans Oleak
Universität Potsdam
Astrophysikalisches Institut Potsdam
An der Sternwarte 16
14482 Potsdam
Germany

Dr. Renate Ottmann
Max-Planck-Institut für Extraterrestrische Physik
85740 Garching
Germany

Dr. M. Türker Özkan
Istanbul University Observatory
34452 University Istanbul
Turkey

Dr. William S. Paciesas
Marshall Space Flight Center
NASA ES62
Huntsville, AL 35812
U.S.A.

Dr. Isabella Pagano
Osservatorio Astrofisico di Catania
Viale Andrea Doria 6
I-95125 Catania
Italy

Dr. Roberto Pallavicini
Osservatorio Astrofisico di Arcetri
Largo e. Fermi 5
I-50125 Firenze
Italy

Dr. Hongchao Pan
University of Oxford
Theoretical Physics
Department of Physics
Keble Road
Oxford OX1 3NP
United Kingdom

Dr. Alexander Panferov
Special Astrophysical Observatory
Nizhnij Arkhyz
Stavropol Territory 357147
Russia

Prof. Arcadio R. Poveda
Instituto de Astronomia - UNAM
National University of Mexico
Allende 156 (Tlalpan) - Mexico, D.F.
14000 Mexico D.F.
Mexico

Mr. Thomas Preibisch
Astronomisches Institut
Universität Würzburg
Am Hubland
97074 Würzburg
Germany

Prof. Eric R. Priest
University of St. Andrews
Department of Mathematical and
Computational Sciences
St. Andrews KY16 9SS
United Kingdom

Dr. Lev Pustil'nik
Jordan Valley Regional College
Ha-Sheisaf, 11/6
P.O. Box 3024
Qatsrin 12900
Israel

Dr. Izold Pustyl'nik
Institute for Astrophysics &
Atmospheric Physics Tartu
Tartu Astrophysical Observatory
Toravere EE 2444
Estonia

Mr. Olaf Reimer
Universität Siegen
FB 7 Physik
Reichweinstraße
57068 Siegen
Germany

Dr. Gerold A. Richter
Sonneberg Observatory
96515 Sonneberg
Germany

Dr. George Ricker
Massachusetts Institut of Technology
77 Mass. Ave.
Cambridge, MA 02139-4307
U.S.A

Dr. Gennady Roizman
Jordan Valley Regional College
Observatory
Ravakia 6/22
14200 Tiberias
Israel

Dr. Siegfried Rössiger
Sonneberg Observatory
96515 Sonneberg
Germany

Dr. Nikolai Samus
Institute for Astronomy
Ul. Pyatnitskaya 48
109017 Moscow
Russia

Dr. B.B. Sanwal
Uttar Pradesh State Observatory
Manora Peak
Naini Tal 263 129
India

Mr. Robert Schwarz
Astrophysikalisches Institut Potsdam
An der Sternwarte 16
14482 Potsdam
Germany

Prof. Waltraut Seitter
Astronomisches Institut
Universität Münster
Wilhelm-Klemm-Str. 10
48149 Münster
Germany

Dr. Evgeni H. Semkov
Institute of Astronomy
Bulgarian Academy of Sciences
72 Tsarigradsko Shose Blvd.
BG 1784 Sofia
Bulgaria

Dr. Nadezda Shakhovskaya
Crimean Astrophysical Observatory
Crimea Nauchny 334413
Ukraine

Prof. Maurice M. Shapiro
University of Maryland
205 Yoakum Pkwy 2-1514
Alexandria, VA 22304
U.S.A.

Dr. Vladimir Sokolov
Special Astrophysical Observatory
of Russian Academy of Sciences
Nizhnij Arkhyz
Karachajevo-Cherkessija 357147
Russia

Dr. Jörg S.W. Stegert
Sternwarte der Universität Bonn
Auf dem Hügel 71
53121 Bonn
Germany

Dr. Alexander V. Stepanov
Crimean Astrophysical Observatory
RT-22 Katzevily Crimea 334247
Ukraine

Prof. Rashid Sunyaev
Space Research Institute (IKI)
Profsoyusnaya 84/32
117810 Moscow
Russia

Dr. Oleg Terekhov
Space Research Institute (IKI)
Profsoyusnaya 84/32
11 7810 Moscow
Russia

Dr. Guinetta Torricelli
Osservatorio di Arcetri
Largo e. Fermi 5
I-50125 Firenze
Italy

Dr. Corrado Trigilio
Instituto di Radioastronomia
del C.N.R.
VLBI-Station Noto, CP 169
96017 Noto
Italy

Dr. Milcho Tsvetkov
Institut of Astronomy
Bulgarian Academy of Sciences
Tsarigradsko Shose 72
BG-1784 Sofia
Bulgaria

Dr. Katya Tsvetkova
Institut of Astronomy
Bulgarian Academy of Sciences
Tsarigradsko Shose 72
BG-1784 Sofia
Bulgaria

Dr. Grazia Umana
Instituto di Radioastronomia
del C.N.R.
VLBI-Station Noto, CP 169
96017 Noto
Italy

Dr. Michiel van der Klis
Astronomical Institute 'Anton Pannekoek'
University of Amsterdam
Kruislaan 403
NL-1098 SJ Amsterdam
the Netherlands

Dr. Roland Vanderspek
Massachusetts Institut of Technology
77 Mass. Ave., Rm 37-527
Cambridge, MA 02139-4307
U.S.A

Prof. Jan van Paradijs
Astronomical Institute 'Anton Pannekoek'
University of Amsterdam
Kruislaan 403
NL-1098 SJ Amsterdam
the Netherlands

Dr. Rita Ventura
Osservatorio Astrofisico di Catania
Citta Universitaria
Viale Andrea Doria 6
I-95125 Catania
Italy

Dr. Irina Verlyuk
Main Astronomical Observatory
National Academy of Sciences
Goloseevo
Kiev 252650
Ukraine

Dr. Nikolaus Vogt
Universidad Catolica
Santiago
Chile

Dr. Jun-jie Wang
Beijing Astronomical Observatory
Chinese Academy of Sciences
Beijing 100080
P.R. China

Dr. William Welsh
Keele University
Department of Physics
Keele, Staffordshire ST5 5BG
United Kingdom

Jürgen Winterberg
Astronomisches Institut
Universität Münster
Wilhelm-Klemm-Str. 10
48149 Münster
Germany

Dr. Ruslan Yudin
Central Astronomical Observatory
Russian Academy of Sciences
194140 Pulkovo, St. Petersburg
Russia

Dr. Boris E. Zhilyaev
Main Astronomical Observatory
Academy of Science of Ukraine
Department of Experimental Astrophysics
Goloseevo
Kiev 252650
Ukraine

Dr. Hans F. Zinnecker
Astronomisches Institut der Universität Würzburg
Lehrstuhl Astronomie
Am Hubland
97074 Würzburg
Germany

E-mail Addresses of IAU Colloquium Attendees

Abada-Simon, Meil	abada@fys.ruu.nl
Aigner, Tobias	tba@mpe-garching.mpg.de
Andronov, Ivan L.	root@astro.odessa.ua
Ball, Bill	wnb@sa1.star.uclan.ac.uk
Benz, Arnold O.	benz@astro.phys.ethz.ch
Beskin, Grigory M.	beskin@sao.stavropol.su
Beskrovnaya, Nina	beskr@gaoran.spb.su
Bowyer, Stuart	bowyer@cea.berkeley.edu
Brandner, Wolfgang	wolfgang@astro.uni-wuerzburg.de
Bruch, Albert	albert@cygnus.uni-muenster.de
Byrne, P. Brendan	pbb@star.arm.ac.uk
Casperson, Donald E.	dcasperson@lanl.gov
Castro-Tirado, Alberto	ajct@laeff.esa.es
Csepura, György	csepura@tigris.klte.hu
Dravins, Dainis	dainis@astro.lu.se
Duerbeck, Hilmar	hilmar@cygnus.uni-muenster.de
Fendt, Christian	
Fishman, Gerald J.	fishman@ssl.msfc.nasa.gov
Franciosini, Elena	francio@arcetri.astro.it
Gahm, Gösta	gahm@emil.astro.su.se
Gershberg, Roald	gershberg@crao.crimea.ua
Glukhov, Alexander	cosmos@physfac.rostov-na-donu.su
Götz, Woldemar	
Greiner, Jochen	jcg@mpe-garching.mpg.de
Gullbring, Erik	erik@jarlabanke.astro.su.se
Hartmann, Dieter	hartmann@biophy.phys.clemson.edu
Hazen, Martha L.	martha@cfa.harvard.edu
Hildebrandt, Joachim	jhildebrandt@aip.de
Hojaev, Alisher	root@ipmce.tashkent.su
Hudec, Rene	rhudec@asu.cas.cz
Ikhsanov, Nazar	nri@pulkovo.spb.su
Karetnikov, Valentin G.	root@astro.odessa.ua
Kato, Mariko	mariko@educ.cc.keio.ac.jp

Katsova, Maria M.	maria@sai.msk.su
Klapdor-Kleingrothaus, H.V.	
Klis, Michiel van der	michiel@astro.uva.nl
Konstantinova-Antova, Renada	belogr%bgearn.bitnet@vm.gmd.de
Kouveliotou, Chryssa	kouveliotou@batse.msfc.nasa.gov
Krivosheina, Irina	
Kroll, Peter	kroll@nova.tat.physik.uni-tuebingen.de
Krüger, Albrecht	akrueger@aip.de
Kürster, Martin	mak@mpe-garching.mpg.de
Kurtanidze, Omar M.	gmelik@dtapha.kheta.georgia.su
laDous, Constanze	cld@vilspa.esa.es
Lapidus, Iosif	lapidus@mail.ast.cam.ac.uk
Larsson, Stefan	stefan@astro.su.se
Lehmann, Thomas	tlehmann@eso.org
Livshits, Moissei	
Lührs, S.	
Luthardt, Elena	root@astro.odessa.ua
Luthardt, Rainer	
Mahmoud, Farouk M.	
Markina, Ala	root@astro.odessa.ua
Meinunger, Isolde	
Merck, Martin	mnm@mpe-garching.mpg.de
Minarini, Roberto	minarini@sao.stavropol.su
Mirabel, Felix	33590::MIRABEL
Mirzoyan, Ludwig V.	tarkh@acsci.armenia.su
Moskalenko, Evgeni	moskal@sai.msk.su
Neizvestny, Sergei	
Notni, Peter	pnotni@aip.de
Oleak, Hans	holeak@aip.de
Ottmann, Renate	reo@mpe.mpe-garching.mpg.de
Özkan, Türker	ik001@triuvm11.bitnet
Paciesas, William S.	paciesas@fender.msfc.nasa.gov
Pagano, Isabella	isabella@peroni.ct.astro.it
Pallavicini, Roberto	pallavic@arcetri.astro.it
Pan, Hongchao	hpan@thphys.ox.ac.uk
Panferov, Alexander	
Paradijs, Jan van	jvp@astro.uva.nl
Poveda, Arcadio Ricalde	poveda@redvax1.dgsca.unam.mx
Preibisch, Thomas	preib@astro.uni-wuerzburg.de
Priest, Eric R.	eric@dcs.st-andrews.ac.uk
Pustil'nik, Lev	lev@yarden.yarden.ac.il
Pustyl'nik, Izold	izold@jupiter.aai.ee
Reimer, Olaf	reimer@sicip1.physik.uni-siegen.de
Richter, Gerold A.	
Ricker, George	grr@space.mit.edu

Roizman, Genady	lev@yarden.yarden.ac.il
Rößiger, Siegfried	
Samus, Nikolai	samus@sai.msu.su
Sanwal, B.B.	
Schwarz, Robert	rschwarz@aip.de
Seitter, Waltraut	waltraut@cygnus.uni-muenster.de
Shakhovskaya, Nadezhda I.	nish@crao.crimea.ua
Shapiro, Maurice	
Sokolov, Vladimir V.	sokolov@sao.stavropol.su
Stegert, Jörg	jstegert@astro.uni-bonn.de
Stepanov, Alexander V.	stepanov@rt22.crimea.ua
Sunyaev, Rashid	sunyaev@[]
Terekhov, Oleg	oterekho@esoc1.bitnet
Torricelli, Guinetta	gtorricelli@arcetri.astro.it
Trigilio, Corrado	ctrigilio@astrct.ct.astro.it
Tsvetkov, Milcho	tsvetkov@bgearn.bitnet
Tsvetkova, Katya	tsvetkova@bgearn.bitnet
Umana, Grazia	gumana@ct.astro.it
Vanderspek, Roland	roland@blitz.mit.edu
Ventura, Rita	rventura@astrct.ct.astro.it
Verlyuk, Irina A.	maouas@gluk.apc.org
Vogt, Nikolaus	vogt@tat.physik.uni-tuebingen.de
Wang, Jun-jie	yac@bao01.bao.ac.cn
Welsh, William F.	wfw@astro.keele.ac.uk
Winterberg, Juergen	juergen@cygnus.uni-muenster.de
Yudin, Ruslan	ruslan@gaoran.spb.su
Zhilyaev, Boris E.	maouas@gluk.apc.org
Zinnecker, Hans	hans@astro.uni-wuerzburg.de

Subject and Object Index

Author Index

MIX
Papier aus verantwortungsvollen Quellen
Paper from responsible sources
FSC® C105338

If you have any concerns about our products,
you can contact us on
ProductSafety@springernature.com

In case Publisher is established outside the EU,
the EU authorized representative is:
Springer Nature Customer Service Center GmbH
Europaplatz 3, 69115 Heidelberg, Germany

Printed by Libri Plureos GmbH
in Hamburg, Germany